From optical to THz control of materials

Burlington House, London, UK and online

23–25 May 2022

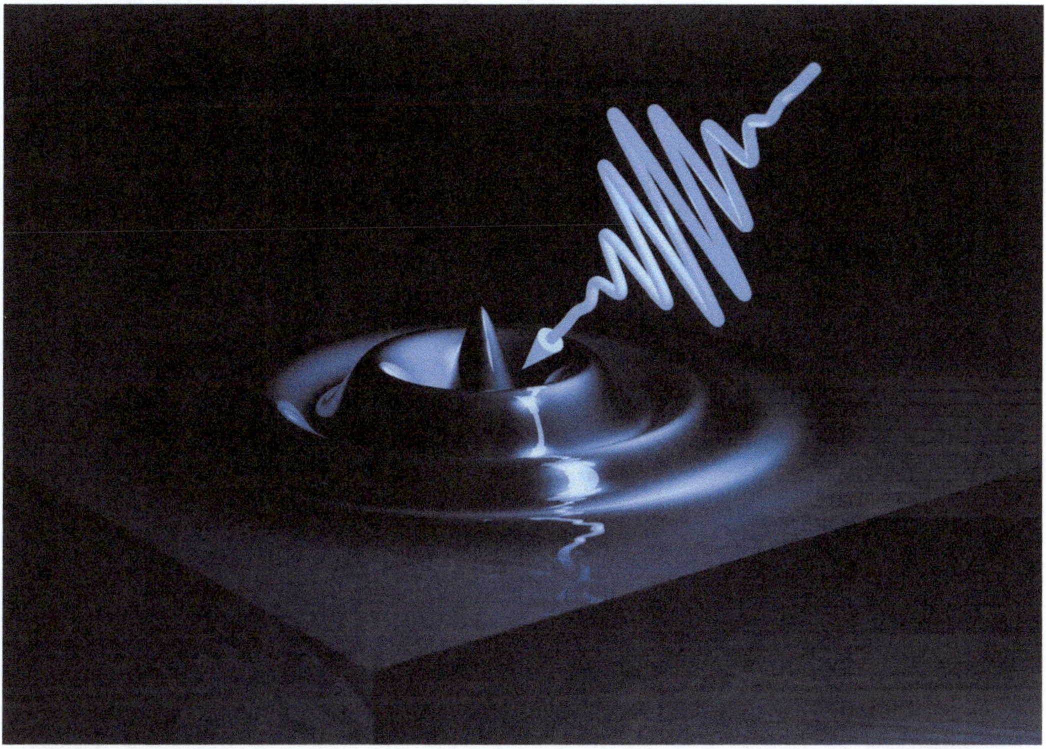

FARADAY DISCUSSIONS

Volume 237, 2022

ROYAL SOCIETY OF CHEMISTRY

The **Faraday Division** of the Royal Society of Chemistry, previously the Faraday Society, was founded in 1903 to promote the study of sciences lying between chemistry, physics and biology.

Faraday Discussions (Print ISSN 1359-6640, Electronic ISSN 1364-5498) is published 8 times a year by the Royal Society of Chemistry, Thomas Graham House, Science Park, Milton Road, Cambridge, UK CB4 0WF.

Volume 237 ISBN 978-1-83916-694-5

2022 annual subscription price: print+electronic £1133 US $1994; electronic only £1079, US $1899.
Customers in Canada will be subject to a surcharge to cover GST. Customers in the EU subscribing to the electronic version only will be charged VAT.

All orders, with cheques made payable to the Royal Society of Chemistry, should be sent to the Royal Society of Chemistry Order Department, Royal Society of Chemistry, Thomas Graham House, Science Park, Milton Road, Cambridge, CB4 0WF, UK
Tel +44 (0)1223 432398; E-mail **orders@rsc.org**

If you take an institutional subscription to any Royal Society of Chemistry journal you are entitled to free, site-wide web access to that journal. You can arrange access via Internet Protocol (IP) address at **www.rsc.org/ip**

Customers should make payments by cheque in sterling payable on a UK clearing bank or in US dollars payable on a US clearing bank.

Printed in the UK

MIX
Paper from responsible sources
FSC® C013604

ROYAL SOCIETY OF CHEMISTRY

Faraday Discussions

Faraday Discussions are unique international discussion meetings that focus on rapidly developing areas of chemistry and its interfaces with other scientific disciplines.

Information for Authors

From optical to THz control of materials

Faraday Discussions

www.rsc.org/faraday_d

A General Discussion on optical to THz control of materials was held in London, UK and online on the 23rd, 24th and 25th May 2022.

The Royal Society of Chemistry is the world's leading chemistry community. Through our high impact journals and publications we connect the world with the chemical sciences and invest the profits back into the chemistry community.

CONTENTS

ISSN 1359-6640; ISBN 978-1-83916-694-5

Cover
See Gierster *et al.*, *Faraday Discuss.*, 2022, **237**, 58–79.

Ultrashort laser pulse metallizing a semiconductor surface.

Image reproduced by permission of Samuel Palato from *Faraday Discuss.*, 2022, **237**, 58–79.

INTRODUCTORY LECTURE

PAPERS AND DISCUSSIONS

EXHIBITORS

AMPLITUDE

Amplitude

POSTER SPONSORSHIP

PCCP

Faraday Discussions

PAPER

Spiers Memorial Lecture: From optical to THz control of materials

Steven L. Johnson[ab]

Received 10th May 2022, Accepted 6th June 2022
DOI: 10.1039/d2fd00098a

Advances over the past decade have presented new avenues to achieve control over material properties using intense pulses of electromagnetic radiation, with frequencies ranging from optical (approximately 1 PHz, or 10^{15} Hz) down to below 1 THz (10^{12} Hz). Some of these new developments have arisen from new experimental methods to drive and observe transient material properties, while others have emerged from new computational techniques that have made nonequilibrium dynamics more tractable to our understanding. One common issue with most attempts to realize control using electromagnetic pulses is the dissipation of energy, which in many cases poses a limit due to uncontrolled heating and has led to strong interest in using lower frequency and/or highly specific excitations to minimize this effect. Emergent developments in experimental tools using shaped X-ray pulses may in the future offer new possibilities for material control, provided that the issue of heat dissipation can be resolved for higher frequency light.

1 Introduction

The idea of using intense pulses of light to control the outcome of chemical reactions has been a subject of active interest for some time, and in fact has been a prominent subject of several past *Faraday Discussions* (FD 113 and FD 153). Here, the core idea is that for at least some kinds of chemical processes, it is possible to use appropriately tailored electromagnetic radiation to influence the outcome in some way, which usually relies on the persistent coherence of excited electronic states in the molecules involved. While this work on chemical processes has yielded important new insights and potential applications,[1–3] the rapid energy redistribution and decoherence of electronic states observed in most molecular systems has presented challenges.[4]

In view of this, it might seem problematic to extend the idea of light-driven control to solid-state materials, where the size of the system becomes much larger than the small molecules that have been most amenable to coherent

[a]*Institute for Quantum Electronics, ETH Zürich, Auguste-Piccard-Hof 1, 8093 Zürich, Switzerland. E-mail: johnson@phys.ethz.ch; Fax: +41 44 633 10 54; Tel: +41 44 633 76 31*
[b]*SwissFEL, Paul Scherrer Institute, Forschungsstrasse 111, 5232 Villigen PSI, Switzerland*

control in chemistry. While this might be true in many cases, in materials where there is already in equilibrium strong competition between incompatible ordering mechanisms, a tiny nudge toward one direction can produce outsized effects. One example of such a competition is between superconductivity and charge or spin order in many unconventional superconductors, which in equilibrium results in complex phase diagrams.[5,6] Another is the related phenomenon of charge density waves, which results also in solid–solid phase transitions from a competition between vibrational and electronic energies.[7] In other cases, emergent many-body effects can lead to surprisingly long coherence times that can in principle be exploited for control.[8] Electronic coherence may not even be necessary for some control concepts, as long-lived vibrational coherences can also be manipulated by appropriately shaped light.[9]

Recent methodological advances over the last few decades have now made possible the first explorations of these possibilities. One important development is the advent of intense sources of short-pulse radiation in the THz and mid-infrared ranges.[10,11] There are several reasons why this is useful for the control of materials. For insulators and semiconductors, excitation at low frequencies can be used to drive specific dipole-active excitations without the delivery of a significant amount of energy density to delocalized electronic states. If the driven excitations couple in some way to a change in order, this may be used to change this order with minimal uncontrolled heating of the material. In materials where several different excitations couple to the desired material control parameter, it may be more efficient and less damaging to drive the excitation with the lowest energy. In yet other materials, such as Mott insulators, sufficiently high electric fields can drive a phase transition; high intensity quasi-single-cycle THz-frequency pulses may offer a way to achieve this on sub-picosecond timescales.[12] In superconductors, nonlinear interactions in the THz frequency range make it in principle possible to excite the Higgs mode, an oscillation of the amplitude of the superconducting order parameter.[13]

Another important enabling technology is the development of experimental methods to study transient material properties on femtosecond and picosecond time scales. Both electron and X-ray diffraction have now been successfully extended to achieve femtosecond time resolution in pump–probe experiments, offering the possibility to observe directly transient structural changes on atomic length scales.[14–19] In parallel to this, angle-resolved photoemission (ARPES) is also now an established tool for the direct measurement of electronic structure.[20–22] Taken together, these methods are a unique tool for observing and understanding the interactions of the lattice, charge and spin degrees of freedom in materials, which is essential for the development and validation of light-driven control mechanisms.

In parallel to this experimental work, there have also been significant advances in theoretical concepts and computational tools for understanding the potential for material control with light. For molecular solids, new methods to approximate the rates of intersystem crossing have been developed and applied.[23,24] The phenomenon of nucleation and growth of a new phase in a molecular crystal was also successfully modeled using coupling between localized electronic excitations and vibrational modes, both showing the importance of vibrational coupling in molecular crystals and providing an indication of how to treat such phenomena more generally.[25] In strongly correlated systems, an important new tool is the extension of dynamical mean field theory to nonequilibrium cases, which has allowed for fundamental study of nonequilibrium phenomena in quantum materials.[26] On a more applied level, there have also been several developments in applying density functional theory to the study of several phenomena that have been observed experimentally, including ionic Raman scattering (also called nonlinear phononics)[27] and interactions between spins and time-dependent polarizations that arise from particular coherent vibrational motions.[28]

Here, I give a brief overview of selected aspects of light-driven material control research, with an emphasis on topics discussed in papers in this *Faraday Discussion*. As an organizing principle, I separate potential control schemes by the initial mechanism of light–matter coupling. First, I discuss control *via* the driving of electronic transitions, which has to date been the most experimentally studied mechanism for material control. Then, I discuss possibilities for control using the direct excitation of vibrations with light, made possible with intense radiation at lower frequencies in the mid-infrared and THz ranges. Next are analogous studies where magnetic or hybrid magnetic-vibrational excitations are driven. Lastly, I talk about some other control schemes where the electric field of an intense laser pulse initiates some change to a material that is not included in the previously discussed categories. In all cases, I indicate some of the key factors relevant to current and future work in the area.

2 Control using electronic transitions

One of the most actively investigated mechanisms for material control is the direct excitation of electric-dipole active electronic transitions. These transitions can have very high cross sections and are often accessible at near-optical wavelengths, and so are a relatively easy way to deliver a considerable amount of energy density into a material. Depending on the interactions of either the initial or final electronic state with other excitations in the material, this initial electronic excitation can then lead to other changes to the material. An understanding of the interactions of the excited electronic transition with the full array of other electronic and vibrational excitations are often needed to predict how the material can respond.

A very general phenomenon and a prominent example of this is the generation of coherent vibrational excitations *via* electron–phonon coupling, often called "coherent phonons" in the literature when discussing crystalline materials.[29] Coherent phonon generation is often described as having two "mechanisms": impulsive stimulated Raman scattering[30] or displacive excitation.[31] This is not strictly accurate, since in both cases the microscopic mechanisms are very similar, but the resulting dynamics manifest in different forms.[32] In general, a particular electronic state transition can drive coherent oscillations of a vibrational mode if the state change causes a shift of the local minimum of the potential energy along the vibrational mode coordinate.

Fig. 1 presents a simplified view of how coherent phonons are generated in both the impulsive and displacive limits. For a light pulse with a duration much shorter than the vibrational period, the lifetime of the electronic states populated by the light interaction relative to the period of the vibration changes the form of the dynamics. If the lifetime is very short compared to the light pulse duration, the potential energy shift occurs only during the light pulse interaction. This

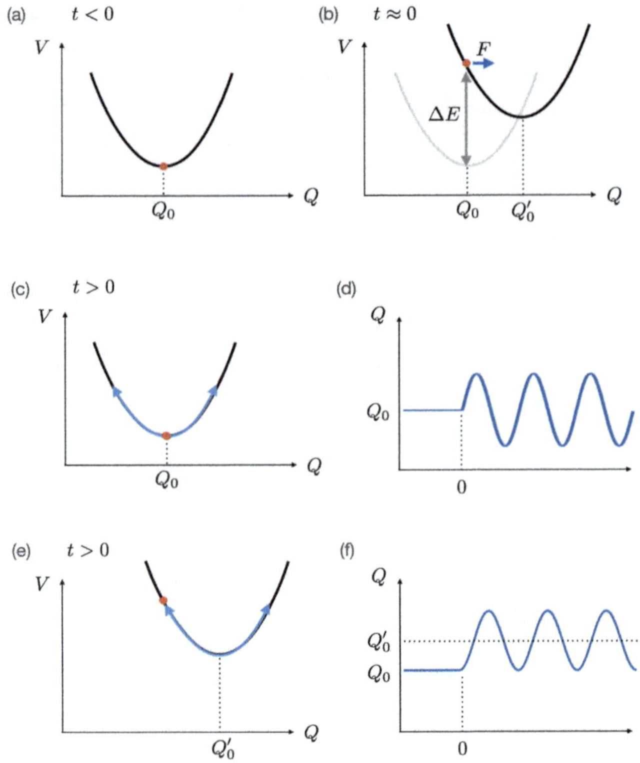

Fig. 1 Schematic depiction of coherent phonon generation in solids. (a) The effective interatomic potential energy as a function of a normal mode with coordinate Q. In equilibrium (before excitation), the expectation value of the coordinate is at Q_0, where the potential energy has a minimum. (b) During interaction with an intense light pulse, the interatomic potential energy changes due to electron–phonon interaction mediated by the electric field. In this case, the new potential energy surface has a different potential energy minimum with respect to the Q coordinate, located at Q'_0. In a classical treatment of the mode dynamics, this results in a force that drives the coordinate toward the new minimum. (c) In the "impulsive" limit, the electronic state relaxes quickly back to the ground state. The transient force on the mode during the light interaction nonetheless results in a coherent oscillation about the position Q_0 with a sine-like phase, as shown in (d). (e) In the "displacive" limit, the material is brought into an excited electronic state that persists for a time much longer than the inverse frequency of the mode. This causes a coherent oscillation of the expectation value of the coordinate about Q'_0 with a cosine-like phase, as shown in (f).

results in an effective force along the vibrational mode coordinate that is impulsive in nature, leading to coherent oscillations with a sine-like phase around the initial equilibrium structure. This is the limit that corresponds to impulsive stimulated Raman scattering. If instead the electronic state (or in some cases, collections of excited states) that causes the potential energy shift has a long lifetime compared to the period of vibration, the resulting effective force is step-like. This leads to coherent oscillations about the shifted potential energy minimum with a cosine-like phase. This limit is called displacive excitation. In a more general case (where the state lifetime is neither very short nor very long compared to the vibrational period), the phase of the oscillation is somewhere in between. It is also possible to drive similar dynamics in the variance of the vibrational coordinate by exciting electronic states that change the frequency of vibrational modes, a phenomenon sometimes called phonon squeezing due to similarities with light squeezing in quantum optics.[33–35] Tracking such dynamics using diffuse X-ray scattering is in fact a newly developed method to determine electron–phonon coupling and phonon dispersion in strongly photoexcited materials, a method called Fourier transform inelastic X-ray scattering.[36]

A class of systems where these dynamics play a key role is molecular spin-crossover materials,[37] which serves as one key example of light-driven material control. The spin-crossover phenomenon refers to a change in the spin state of a 3d-transition metal ion component by changing the temperature, pressure or photoexcitation. The typical photoexcitation pathway, depicted in Fig. 2, involves an initial excitation into a manifold of so-called MLCT (metal-to-ligand change transfer) states which, as the name suggests, involves a shift of charge density from the ion to a nearby ligand. This MLCT manifold then relaxes to a high-spin state. Due to the rapidity of these decay processes, the local structure of the metal–ligand complex does not coincide with a local minimum of the interatomic potential energy surface in the excited high-spin electronic state. This results in an elongation of the metal–ligand bond distance, which ends up effectively trapping the ion in a metastable configuration.[38] In crystalline materials, the structural component of the transition results in cooperative structural transformations over multiple length and time scales, which are possible to resolve

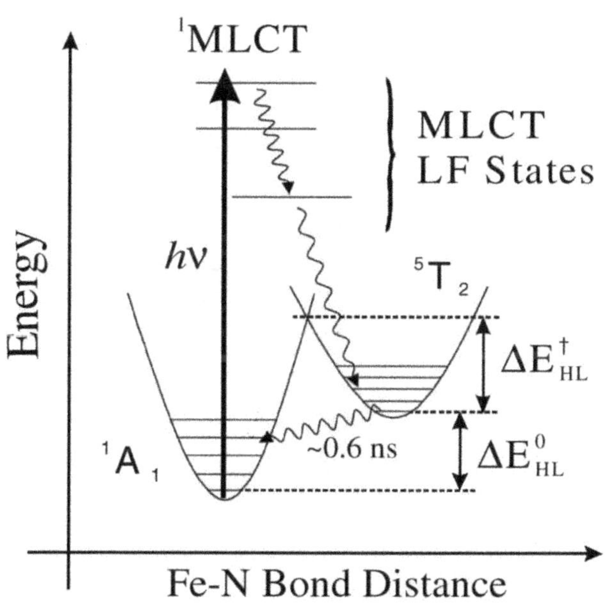

Fig. 2 Configuration-coordinate diagram for photoexcited $Fe(bpy)_3$, one example of a spin-crossover molecule. The transition to a high-spin electronic state of the Fe ion occurs *via* a higher energy MLCT state that is directly excited by the absorption of a visible or UV photon. Reprinted with permission from ref. 40 [Gawelda *et al.*, *Phys. Rev. Lett.*, 2007, **98**, 054401]. Copyright 2007 by the American Physical Society.

using a variety of methods, including time-resolved microscopy, diffraction and spectroscopy.[39] Note also that typically a large fraction of the energy initially deposited by the excitation is dissipated during the relaxation process, resulting in an increase in the effective temperature of the metastable state relative to the initial low-spin state, which also has an influence on the structure.

Control has also been explored in materials with more delocalized electronic states. In semiconductors, excitation can result in the formation, destruction or modulation of excitons (bound pairs of electrons and holes) that can be tracked using ultrafast spectroscopy.[41–43] As with the molecular solids, electron–phonon coupling often strongly affects the dynamics.[44] In some low-dimensional metals, another manifestation of electron–phonon coupling is the formation of charge density waves (CDWs), a periodic modulation of the charge and an associated lattice distortion that results in the formation of a gap in the electronic density of states at the Fermi energy. Photon-driven electronic excitation across the CDW band gap shifts this balance, leading to displacive excitation of the amplitude mode of the distortion or, at higher excitation levels, a transient phase change to an unmodulated "melted" state observable by optical spectroscopy and X-ray diffraction.[45–49] Electronic excitation has also been used to drive other kinds of structural symmetry changes in various materials.[50,51]

Another new development in this arena is the interest in Higgs mode spectroscopy of superconductors, recently stimulated due to the ability to generate intense THz pulses suitable for nonlinear interactions at low frequencies.[13] The Higgs mode in superconductors refers to an oscillation of the amplitude of the complex-valued superconducting order parameter. The name of the mode is derived from a mathematical analogy to the Higgs mechanism in particle physics. The Higgs mode carries no charge moments, and so usually does not couple to light in any linear-response regime. It is also a very low energy excitation that exists in the THz frequency regime. It can be shown that the Higgs mode can in some circumstances contribute to an enhancement of third-harmonic generation for THz frequency pulses when the THz frequency is half of the superconducting pair energy gap, and experimental evidence for such an enhancement has been observed in both BCS and cuprate superconductors.[89–91] An alternative method that has been proposed to excite and observe the Higgs mode is to use pump–probe THz methods to very suddenly disrupt ("quench") the pairing interactions by using a THz pulse to inject quasiparticles near the superconducting gap, leading to a change of the free energy as a function of the amplitude of the condensate. This can then launch coherent oscillations of the Higgs mode that can be observed as a modulation of the reflectance of a probe (see Fig. 3). Such an experiment was performed by Matsunaga *et al.*, which indeed observed coherent oscillations at the Higgs mode frequency in a BCS superconductor.[88]

It is also possible to strongly control magnetic order in materials using electronic excitation. One of the earliest and still much discussed phenomena is ultrafast demagnetization, reported first by Beaurepaire *et al.*[52] Intense electronic excitation of several different ferromagnetic transition metals and alloys results in a rapid drop of the magnetization density on a time scale of tens of femtoseconds.[53,54] The precise mechanism in all observed cases is still under debate but appears to be a combination of rapid spin transport[55] and spin-flip scattering with phonons[56] that occurs during the extremely high transient electron temperatures caused by laser heating of the conduction electrons. Ultrafast X-ray scattering measurements have demonstrated that a significant fraction of the angular momentum of the spins in laser excited iron is transferred to the lattice in the form of transverse strain launched from the surface.[57]

A related phenomenon observed in some metallic ferrimagnetic systems with a magnetic compensation point is all-optical switching of the net magnetization.[58,59] In these materials, it is possible to understand the switching dynamics as primarily driven by changes in the transient temperature of the metal that occur in response to short pulse excitation. This was first observed in the FeGdCo alloy,[60] an amorphous alloy where the Gd and FeCo spins are antiferromagnetically coupled. At room temperature, the magnetic moment contributed by the Gd is smaller and the alloy has a net magnetization along the direction of the FeCo. The sudden heating of the electrons by light-pulse excitation causes a faster and larger demagnetization of the Fe component, resulting in a transient net magnetization aligned to the Gd component. During the slow cooling of the system, angular momentum conservation within the spin system then leads to a reversal of the total ferrimagnetic order relative to the initial configuration.

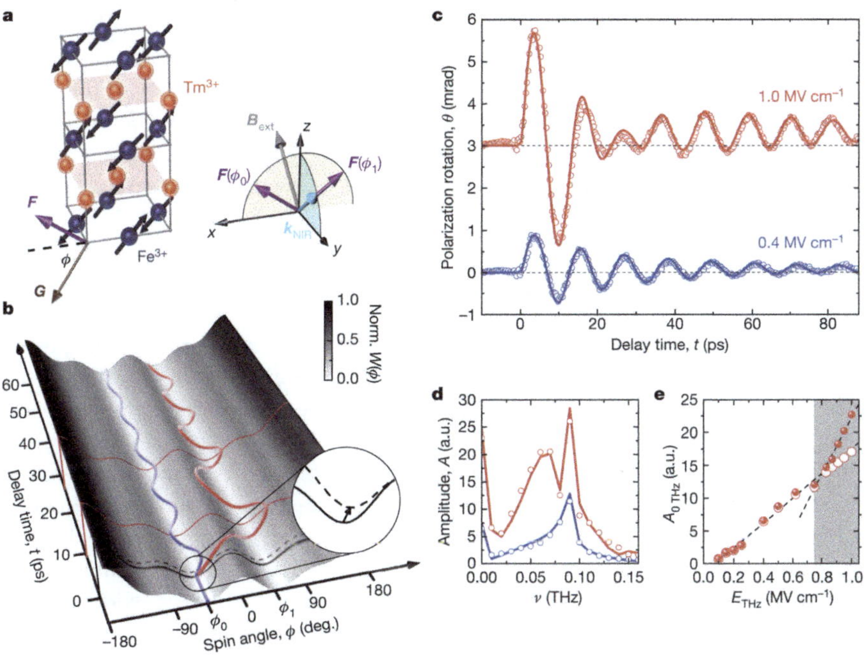

Fig. 3 THz-driven switching in the canted antiferromagnet $TmFeO_3$. (a) The structure of $TmFeO_3$, showing the sublattice spin orientations and the direction of the net magnetic moment. (b) A sketch of the magnetic potential energy and spin dynamics for low (blue) and high (red) field excitations. (c) Measured dynamics of the Faraday rotation angle, an experimental measure of the spin angle orientation. (d) Amplitude of the Fourier transform of the data in panel c. (e) Simulated behavior of the long-lived offset in the Faraday rotation as a function of the excitation level for two alignments of the near-infrared probe (red circles and red spheres). The difference at high fields is a consequence of the metastable transfer of spins over the potential energy barrier. Reprinted with permission from *Nature*[65] [Schlauderer *et al.*, Temporal and spectral fingerprints of ultrafast all-coherent spin switching]. Copyright 2019 Springer Nature.

It is also possible to excite coherent spin dynamics in magnetically ordered systems in a manner analogous to the excitation of coherent vibrational dynamics. This has been of particular interest recently in antiferromagnetic systems, which have attracted interest for their potential for fast dynamics not limited by angular momentum conservation.[61,62] One example is offered by the orthoferrites, where electronic excitation can launch spin dynamics by inducing a sudden transient change to the magnetic anisotropy that determines the orientation of the sublattice magnetisations.[63] In a semi-classical treatment, the Lagrangian for antiferromagnetic dynamics contains a term mathematically analogous to the kinetic energy of a massive particle in a simple harmonic oscillator, and thus shows inertial motion that can be exploited to achieve actual switching of the magnetic order in response to the impulsive excitation of coherent magnetic dynamics.[64]

The manipulation of magnetic anisotropy is another means to control materials with a non-zero net magnetic moment, as has been demonstrated by exciting d–d transitions in insulating ferrimagnetic garnets, where the dissipation of energy into general heating is very low compared to similar switching in metallic systems.[66] This can also be extended to very low frequencies. The orthoferrite $TmFeO_3$ is an antiferromagnet with a small ferromagnetic spin canting.[67] By strong near-field enhancement of an electromagnetic pulse in the THz frequency range, Schlauderer *et al.* were able to demonstrate coherent switching of the ferromagnetic orientation *via* excitation of the local electronic states of the Tm^{3+} ions (see Fig. 3).[65] The idea that such low energy excitations can be effectively exploited to achieve ferromagnetic switching is highly promising, since this appears to approach a minimal dissipation limit for such control schemes.

3 Control by vibrational excitations

For material control where a structural coordinate is either an order parameter or strongly coupled to an order parameter, the direct excitation of vibrations with electromagnetic fields can offer an attractive alternative. If the order parameter itself happens to couple directly to the electromagnetic radiation, it may be possible to very directly drive some metastable change. An example of this would be soft-mode structural ferroelectrics, with typical soft mode frequencies in the THz frequency range. Time-resolved X-ray diffraction experiments have been applied to directly measure such dynamics.[68,69] Appropriately shaped and timed THz pulses have been predicted as a means to drive domain reversal.[70] In $SrTiO_3$, there is evidence that very high transient THz fields can drive the structure to a metastable ferroelectric state, possibly *via* a soft mode excitation, which has been inferred from time-resolved measurements of second harmonic generation.[71]

In other materials, it may not be possible to directly affect the order parameter with electromagnetic radiation, but it is nevertheless coupled nonlinearly to particular vibrational modes that it is possible to drive directly, usually at frequencies raging from 1–40 THz. This is broadly known as nonlinear phononics, which has been extensively reviewed elsewhere.[73] Analogously to electron–phonon driven mechanisms for coherent phonon generation, it is also possible to drive coherent vibrational modes by exciting high frequency vibrational modes with appropriate symmetry.[72,74] To illustrate the basic idea, assume that the

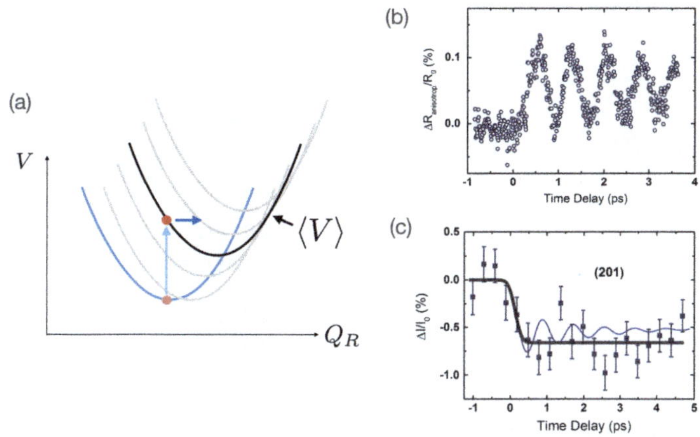

Fig. 4 Displacive excitation of coherent phonons using vibrational excitation and nonlinear vibrational coupling. (a) Potential energy as a function of a Raman-active phonon mode with coordinate Q_R for different fixed values of a driven infrared coordinate Q_{IR}. The blue curve shows the $Q_{IR} = 0$ potential energy relation, while the grey curves are fixed non-zero values of Q_{IR} with shifted minima due to $Q_R Q_{IR}^2$ coupling. (b) Measured anisotropic reflectivity changes in $La_{0.7}Sr_{0.3}MnO_3$ due to coherent excitation of an E_g mode *via* driving of a high-frequency E_u mode. (c) Corresponding changes in X-ray diffraction from the (201) lattice planes, demonstrating unambiguously the displacement of the E_g mode coordinate. Panels (b) and (c) are reprinted from ref. 72 [Först *et al.*, Displacive lattice excitation through nonlinear phononics viewed by femtosecond X-ray diffraction, *Solid State Commun.*, 2013, **169**, 24]. Copyright 2013, with permission from Elsevier.

interatomic potential energy can be written as a function of two coupled normal mode coordinates as

$$V(Q_R, Q_{IR}) = \frac{1}{2}\Omega_R{}^2 Q_R{}^2 + \frac{1}{2}\Omega_{IR}{}^2 Q_{IR}{}^2 + aQ_R Q_{IR}{}^2, \tag{1}$$

where Q_R is the coordinate of a Raman-active mode with frequency Ω_R, Q_{IR} is the coordinate of an infrared-active mode with frequency Ω_{IR}, and a is a coupling constant. If we rewrite this as

$$V(Q_R, Q_{IR}) = \frac{1}{2}\Omega_R{}^2\left(Q_R + \frac{a}{\Omega_R{}^2}Q_{IR}{}^2\right)^2 + \frac{1}{2}\Omega_{IR}{}^2 Q_{IR}{}^2 - \frac{a^2 Q_{IR}{}^4}{2\Omega_R{}^2}, \tag{2}$$

it is evident that the $aQ_R Q_{IR}{}^2$ nonlinear coupling term appearing in eqn (1) can be considered as causing a shift in the minimum of the potential energy with respect to Q_R by an amount $-aQ_{IR}{}^2/\Omega_R{}^2$. In the case where $\Omega_{IR} \gg \Omega_R$, the dynamics of the Q_R coordinate can be described by an effective potential energy that is displacively shifted from equilibrium by $\Delta Q_0 = -a\langle Q_{IR}{}^2\rangle/\Omega_R{}^2$ (see Fig. 4). A more generalized version of this concept has been applied to explain experiments observing ferroelectric polarisation dynamics in $LiNbO_3$, where optical second harmonic generation was used to infer a transient reversal of the polarisation.[75]

A variety of material phase transitions using nonlinear phononics have been reported. One early example is the driving of an insulator–metal transition in a manganite by vibrational excitation, which was interpreted as a nonlinear phononics phenomenon.[76] A later example of a similar phase transition was

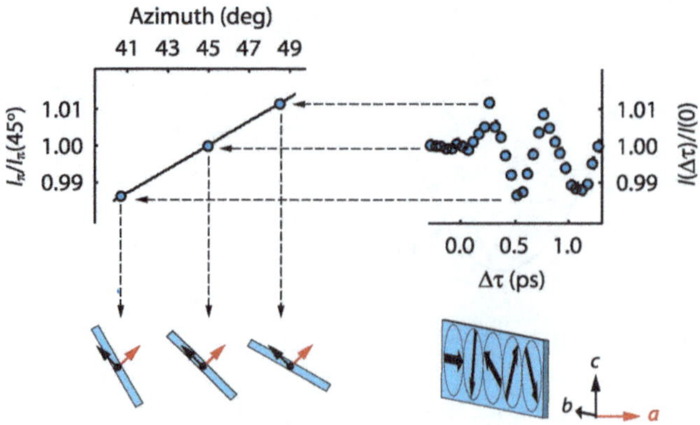

Fig. 5 Direct measurement of coherent rotations of the cycloidal spin order in $TbMnO_3$, driven by THz excitation of an electromagnon excitation. The left side shows the dependence of the diffracted X-ray intensity from a magnetic order peak as a function of the angle of the spin cycloid plane. On the right is shown example data of the time-dependence of the diffracted magnetic intensity, showing the coherent oscillations of the magnetic order. This figure is reproduced from Kubacka *et al.*[87] with permission from the American Association for the Advancement of Science.

explained instead by a more direct nonlinear coupling between an excited mode and the electronic band gap.[77] It is also possible to destroy electronic and magnetic ordering using vibrational excitation, in a manner qualitatively similar to electronic excitation.[78–80] Nonlinear phononics is also a promising mechanism to explain experiments on cuprates where the resonant excitation of selected vibrational modes has been shown to enhance spectroscopic signatures of superconductivity.[81] Similar excitations have been shown to enhance superconducting transport in other materials as well, although it is not at present clear whether nonlinear phononics or some other mechanism leads to the enhancement of the superconducting order in general.[82,83]

Much of the interest in using the excitation of vibrational modes to control materials is connected to the idea that such a mechanism can lead to more selective deposition of energy into the system, minimizing possibly undesirable heating that inevitably leads to a state of higher entropy. Typical experiments nevertheless require very high intensities for the stimulating pump pulse, which can also drive nonlinear optical processes that compete with the excitation of the vibrational mode and can lead to undesirable nonspecific excitations.[84] It is probably still necessary in most cases to carefully select materials that are amenable to such control by seeking situations where strongly competing orders already exist in equilibrium and can be rebalanced *via* specific structural distortions.

4 Driving magnetic and hybrid vibrational–magnetic excitations

It is also possible to use the magnetic field component of pulsed radiation to drive magnetic excitations *via* magnetic dipole interactions, although due to the relative weakness of the magnetic Zeeman interaction, this usually leads to fairly

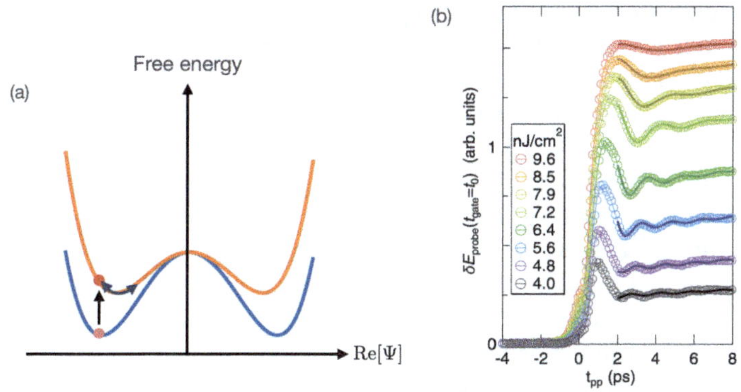

Fig. 6 Quench-driving of the Higgs mode in a superconductor. (a) The Landau–Ginzburg free energy in a superconductor as a function of the real part of the condensate wavefunction Ψ. The blue curve shows the initial free energy, which is suddenly changed to the orange curve by quasiparticle injection from a THz-pulse. This displacively excites coherent oscillations of the Higgs mode. (b) Experimental data showing changes in the electric field of a THz probe pulse at the Higgs mode frequency in superconducting NbN. Panel (b) is reprinted with permission from ref. 88 [Matsunaga *et al.*, *Phys. Rev. Lett.*, 2013, **111**, 057002]. Copyright 2013 by the American Physical Society.

weak perturbations compared to other control methods. The use of broadband single-cycle THz pulses to drive and control the amplitude of coherent antiferromagnetic spin excitations has been demonstrated in NiO.[85] In this work, Kampfrath *et al.* estimate that peak magnetic fields of more than 10 T are needed to drive dynamics out of the linear regime, which would imply peak electric fields on the order of 3 GV m^{-1}, where many gapped materials experience strong nonlinear electronic interactions from effects like Zener tunneling that lead to massive Joule heating from field-generated carriers.[86] It might therefore be quite challenging to exploit Zeeman interactions for magnetization control into new metastable configurations for general magnetic systems.

An alternative is to drive hybrid excitations of spin and vibration, where the vibrational component confers an electric dipole moment that can more strongly couple to the light. These are called electromagnon modes and appear in several materials, including some multiferroics. One example is $TbMnO_3$, where Mochizuki *et al.* predicted that intense THz pulses with peak field strengths on the order of 1 GV m^{-1} can lead to persistent switching of the antiferromagnetic order. Ultrafast magnetic X-ray scattering measurements of similar THz-induced motion validated the basic idea that such pulses can drive large amplitude coherent electromagnons in this material (see Fig. 5), and suggested that even more moderate peak fields of 200 MV m^{-1} could lead to spin reorientations to new metastable configurations.[87] While fields this high might also lead to transient temperature increases *via* Joule heating, the fields are considerably lower than those that would be needed for similar effects from Zeeman coupling.

5 Electric-field-driven control

In recent years there has been discussion of several mechanisms for transient electric-field-driven control of materials that do not fit neatly into any of the previous categories. One is the concept of using periodic electric fields to create Floquet states in materials, steady-state solutions to a periodically time-dependent Hamiltonian. Much recent discussion, mostly on the theory side, has focused on the question of whether it is possible to "engineer" new kinds of quantum order in periodically driven materials.[92] Experimental verification of this idea in real materials is fairly sparse, although there are claims of evidence for some phenomena related to Floquet states, such as dynamical localization[93] and the formation of Floquet–Bloch bands seen in ARPES.[94] One challenge here is again the fact that to see sizable changes, very large electric field amplitudes may be required. While in gapped systems with minimal impurities it may be possible to operate at a frequency where linear interactions of the electric field with the material are negligible, at high fields, nonlinear field-driven effects other than the periodic modulation needed for Floquet engineering may appear and lead to unwanted effects. Even in idealized systems used in simulations, the dissipation and thermalization of energy is a serious issue, although in some cases may be manageable.[95–97]

The advent of extremely short, sub-femtosecond pulses generated by high-harmonics of ultrafast lasers[98] or free-electron lasers[99] has also offered new opportunities for studying field driven phenomena in materials at near-optical frequencies.[100,101] Experiments in this regime are highly challenging, but offer a potential opportunity to study and even guide pure charge dynamics on nanometer length scales.

On the other extreme end of the frequency spectrum, there are some methods of material control where quasi-static electric fields have been shown to drive material transformations. There has recently been interest in exploring the time scale of these transformations, which is now possible using quasi-half-cycle THz radiation.[102] In several materials, including Mott insulators, fast changes in conductivity in response to intense THz pulses have been observed.[12,103,104] The mechanisms for these changes are not in all cases well understood, but one plausible idea is that the intense electric field causes the excitation of electrons across the Mott gap by a phenomenon analogous to Zener tunneling in semiconductor junctions,[105] followed by avalanche ionization and Joule heating from the subsequent acceleration of carriers in the THz electric field.[12] This causes a highly inhomogeneous heating along thin fibrous regions of the material aligned to the electric field of the THz pulse. The expansion of the heated fiber and the compression of the surrounding material may lead to the metastable metallization of portions of the material.

6 Conclusions

In this introductory overview for this *Faraday Discussion*, I have tried to give a unified framework for the diverse methods currently being pursued for material control. One common theme in many aspects of this research is the role of the thermalization of the energy deposited from light–matter interaction. Sometimes, this thermalization process is desirable, for example when the goal is to drive the material into a higher entropy state. This is the case for the melting of charge density waves or ultrafast demagnetization. This photoinduced state is not necessarily identical to a state obtained by a conventional increase of the temperature, since some thermally driven changes, such as lattice expansion, can take much longer than the thermalization of other degrees of freedom. The exact path to thermalization might also result in the trapping of the system in a metastable configuration.

In many cases, however, the thermal redistribution of energy can lead to undesired effects. Thus, a recent trend is to try to achieve material control *via* highly specific excitations that are maximally coupled to the desired dynamics, and minimally coupled to dissipative excitations that lead to thermalization. This has led to strong interest in using THz and mid-infrared light sources, since these are the frequencies where in gapped materials it is possible to drive infrared active collective excitations of vibrations, spins or local electronic transitions without exciting delocalized electronic states that rapidly dissipate. This has generated some spectacular successes, but the ultimate potential of this concept is yet to be determined. In particular, it is not yet clear how generally such control can be applied without requiring such high intensities that strong nonlinear field driven effects compete with the targeted linear or low-order nonlinear interactions.

The applications of new seeded femtosecond and even sub-femtosecond X-rays from free electron lasers are still mostly only conceptual, but may offer unique new opportunities in material understanding and control. One such opportunity is the ability to drive and observe purely electronic dynamics on attosecond time scales, which has already begun to be explored using existing sources. Another potential application to materials would exploit the short wavelength of such radiation, which is comparable to interatomic distances. The short wavelength, combined with high intensities and transform limited pulses, makes possible the manipulation of short-wavelength excitations in materials, spanning the Brillouin zone of crystals. Of course, with intense X-rays, parasitic heating is an even bigger issue than for optical pulses, since the inevitable ionization of core levels leads inexorably to a cascade of electronic and other excitations. It may therefore be necessary for such control concepts to focus predominantly on very fast time scales, before thermalization from those cascading excitations takes hold. It is nevertheless an interesting new frontier to explore.

Conflicts of interest

There are no conflicts to declare.

Acknowledgements

The author acknowledges financial support from project grant 200020_192337 of the Swiss National Science Foundation.

Notes and references

1 D. Goswami, *Phys. Rep.*, 2003, **374**, 385–481.
2 M. Shapiro and P. Brumer, *Rep. Prog. Phys.*, 2003, **66**, 859.
3 Y. Silberberg, *Annu. Rev. Phys. Chem.*, 2009, **60**, 277–292.
4 W. S. Warren, H. Rabitz and M. Dahleh, *Science*, 1993, **259**, 1581–1589.

5 J. Chang, E. Blackburn, A. T. Holmes, N. B. Christensen, J. Larsen, J. Mesot, R. Liang, D. A. Bonn, W. N. Hardy, A. Watenphul, M. v. Zimmermann, E. M. Forgan and S. M. Hayden, *Nat. Phys.*, 2012, **8**, 871–876.
6 P. Dai, *Rev. Mod. Phys.*, 2015, **87**, 855–896.
7 G. Grüner, *Density Waves in Solids*, Westview Press, 2009.
8 A. Streltsov, G. Adesso and M. B. Plenio, *Rev. Mod. Phys.*, 2017, **89**, 041003.
9 S. L. Johnson, in *X-Ray Free Electron Lasers: Applications in Materials, Chemistry and Biology*, RSC Energy and Environment Series, 2017, pp. 284–300.
10 M. Knorr, J. Raab, M. Tauer, P. Merkl, D. Peller, E. Wittmann, E. Riedle, C. Lange and R. Huber, *Opt. Lett.*, 2017, **42**, 4367.
11 N. Stojanovic and M. Drescher, *J. Phys. B: At., Mol. Opt. Phys.*, 2013, **46**, 192001.
12 F. Giorgianni, J. Sakai and S. Lupi, *Nat. Commun.*, 2019, **10**, 1159.
13 R. Shimano and N. Tsuji, *Annu. Rev. Condens. Matter Phys.*, 2019, **11**, 1–22.
14 G. Sciaini and R. J. D. Miller, *Rep. Prog. Phys.*, 2011, **74**, 096101.
15 S. Lahme, C. Kealhofer, F. Krausz and P. Baum, *Struct. Dyn.*, 2014, **1**, 034303.
16 P. Zhu, Y. Zhu, Y. Hidaka, L. Wu, J. Cao, H. Berger, J. Geck, R. Kraus, S. Pjerov, Y. Shen, R. I. Tobey, J. P. Hill and X. J. Wang, *New J. Phys.*, 2015, **17**, 063004.
17 R. J. D. Miller, *Science*, 2014, **343**, 1108–1116.
18 R. Schoenlein, T. Elsaesser, K. Holldack, Z. Huang, H. Kapteyn, M. Murnane and M. Woerner, *Philos. Trans. R. Soc. London, Ser. A*, 2019, **377**, 20180384.
19 A. M. Lindenberg, S. L. Johnson and D. A. Reis, in *Annual Review of Materials Research*, Annual Review of Materials Research, 2017, vol. 47, pp. 1–25.
20 E. J. Sie, T. Rohwer, C. Lee and N. Gedik, *Nat. Commun.*, 2019, **10**, 3535.
21 Y. Ishida, T. Togashi, K. Yamamoto, M. Tanaka, T. Kiss, T. Otsu, Y. Kobayashi and S. Shin, *Rev. Sci. Instrum.*, 2014, **85**, 123904.
22 E. Carpene, E. Mancini, C. Dallera, G. Ghiringhelli, C. Manzoni, G. Cerullo and S. D. Silvestri, *Rev. Sci. Instrum.*, 2009, **80**, 055101.
23 C. M. Marian, *Wiley Interdiscip. Rev.: Comput. Mol. Sci.*, 2012, **2**, 187–203.
24 C. M. Marian, *Annu. Rev.Phys. Chem.*, 2021, **72**, 617–640.
25 K. Ishida and K. Nasu, *Phys. Rev. Lett.*, 2008, **100**, 116403.
26 H. Aoki, N. Tsuji, M. Eckstein, M. Kollar, T. Oka and P. Werner, *Rev. Mod. Phys.*, 2014, **86**, 779–837.
27 A. Subedi, A. Cavalleri and A. Georges, *Phys. Rev. B*, 2014, **89**, 220301.
28 D. M. Juraschek, M. Fechner, A. V. Balatsky and N. A. Spaldin, *Phys. Rev. Mater.*, 2017, **1**, 014401.
29 R. Merlin, *Solid State Commun.*, 1997, **102**, 207–220.
30 Y.-X. Yan, E. B. Gamble and K. A. Nelson, *J. Chem. Phys.*, 1985, **83**, 5391.
31 H. J. Zeiger, J. Vidal, T. K. Cheng, E. P. Ippen, G. Dresselhaus and M. S. Dresselhaus, *Phys. Rev. B: Condens. Matter Mater. Phys.*, 1992, **45**, 768–778.
32 T. Stevens, J. Kuhl and R. Merlin, *Phys. Rev. B: Condens. Matter Mater. Phys.*, 2002, **65**, 144304.
33 G. A. Garrett, A. G. Rojo, A. K. Sood, J. F. Whitaker and R. Merlin, *Science*, 1997, **275**, 1638–1640.
34 X. Hu and F. Nori, *Phys. Rev. Lett.*, 1996, **76**, 2294.
35 S. Johnson, P. Beaud, E. Vorobeva, C. Milne, E. Murray, S. Fahy and G. Ingold, *Phys. Rev. Lett.*, 2009, **102**, 175503.
36 M. Trigo, M. Fuchs, J. Chen, M. P. Jiang, M. Cammarata, S. Fahy, D. M. Fritz, K. Gaffney, S. Ghimire, A. Higginbotham, S. L. Johnson, M. E. Kozina,

J. Larsson, H. Lemke, A. M. Lindenberg, G. Ndabashimiye, F. Quirin, K. Sokolowski-Tinten, C. Uher, G. Wang, J. S. Wark, D. Zhu and D. A. Reis, *Nat. Phys.*, 2013, **9**, 790–794.

37 *Spin-crossover Materials: Properties and Applications*, ed. M. A. Halcrow, John Wiley & Sons, Ltd, Chichester, UK, 2013.

38 S. Decurtins, P. Gutlich, K. M. Hasselbach, A. Hauser and H. Spiering, *Inorg. Chem.*, 1985, **24**, 2174–2178.

39 E. Collet and P. Guionneau, *C. R. Chim.*, 2018, **21**, 1133–1151.

40 W. Gawelda, V.-T. Pham, M. Benfatto, Y. Zaushitsyn, M. Kaiser, D. Grolimund, S. L. Johnson, R. Abela, A. Hauser, C. Bressler and M. Chergui, *Phys. Rev. Lett.*, 2007, **98**, 057401.

41 C. Trovatello, F. Katsch, N. J. Borys, M. Selig, K. Yao, R. Borrego-Varillas, F. Scotognella, I. Kriegel, A. Yan, A. Zettl, P. J. Schuck, A. Knorr, G. Cerullo and S. D. Conte, *Nat. Commun.*, 2020, **11**, 5277.

42 P. Merkl, F. Mooshammer, P. Steinleitner, A. Girnghuber, K.-Q. Lin, P. Nagler, J. Holler, C. Schüller, J. M. Lupton, T. Korn, S. Ovesen, S. Brem, E. Malic and R. Huber, *Nat. Mater.*, 2019, **18**, 691–696.

43 H. M. Bretscher, P. Andrich, P. Telang, A. Singh, L. Harnagea, A. K. Sood and A. Rao, *Nat. Commun.*, 2021, **12**, 1699.

44 G. Antonius and S. G. Louie, *Phys. Rev. B*, 2022, **105**, 085111.

45 A. Tomeljak, H. SCHAFER, D. Städter, M. Beyer, K. Biljaković and J. Demsar, *Phys. Rev. Lett.*, 2009, **102**, 066404.

46 E. Möhr-Vorobeva, S. L. Johnson, P. Beaud, U. Staub, R. D. Souza, C. Milne, G. Ingold, J. Demsar, H. Schaefer and A. Titov, *Phys. Rev. Lett.*, 2011, **107**, 036403.

47 T. Huber, S. O. Mariager, A. Ferrer, H. Schaefer, J. A. Johnson, S. Gruebel, A. Luebcke, L. Huber, T. Kubacka, C. Dornes, C. Laulhe, S. Ravy, G. Ingold, P. Beaud, J. Demsar and S. L. Johnson, *Phys. Rev. Lett.*, 2014, **113**, 026401.

48 R. Yusupov, T. Mertelj, V. V. Kabanov, S. Brazovskii, P. Kusar, J.-H. Chu, I. R. Fisher and D. Mihailovic, *Nat. Phys.*, 2010, **6**, 681–684.

49 M. Trigo, P. Giraldo-Gallo, M. E. Kozina, T. Henighan, M. P. Jiang, H. Liu, J. N. Clark, M. Chollet, J. M. Glownia, D. Zhu, T. Katayama, D. Leuenberger, P. S. Kirchmann, I. R. Fisher, Z. X. Shen and D. A. Reis, *Phys. Rev. B*, 2019, **99**, 104111.

50 E. J. Sie, C. M. Nyby, C. D. Pemmaraju, S. J. Park, X. Shen, J. Yang, M. C. Hoffmann, B. K. Ofori-Okai, R. Li, A. H. Reid, S. Weathersby, E. Mannebach, N. Finney, D. Rhodes, D. Chenet, A. Antony, L. Balicas, J. Hone, T. P. Devereaux, T. F. Heinz, X. Wang and A. M. Lindenberg, *Nature*, 2019, **565**, 61–66.

51 P. Zalden, F. Quirin, M. Schumacher, J. Siegel, S. Wei, A. Koc, M. Nicoul, M. Trigo, P. Andreasson, H. Enquist, M. J. Shu, T. Pardini, M. Chollet, D. Zhu, H. Lemke, I. Ronneberger, J. Larsson, A. M. Lindenberg, H. E. Fischer, S. Hau-Riege, D. A. Reis, R. Mazzarello, M. Wuttig and K. Sokolowski-Tinten, *Science*, 2019, **364**, 1062–1067.

52 E. Beaurepaire, J.-C. Merle, A. Daunois and J.-Y. Bigot, *Phys. Rev. Lett.*, 1996, **76**, 4250–4253.

53 E. Carpene, E. Mancini, C. Dallera, M. Brenna, E. Puppin and S. D. Silvestri, *Phys. Rev. B*, 2008, **78**, 174422.

54 M. Fähnle, *Am. J. Mod. Phys.*, 2018, **7**, 68.

55 M. Battiato, K. Carva and P. M. Oppeneer, *Phys. Rev. Lett.*, 2010, **105**, 027203.
56 G. P. Zhang and W. Hübner, *Phys. Rev. Lett.*, 2000, **85**, 3025–3028.
57 C. Dornes, Y. Acremann, M. Savoini, M. Kubli, M. J. Neugebauer, E. Abreu, L. Huber, G. Lantz, C. A. F. Vaz, H. Lemke, E. M. Bothschafter, M. Porer, V. Esposito, L. Rettig, M. Buzzi, A. Alberca, Y. W. Windsor, P. Beaud, U. Staub, D. Zhu, S. Song, J. M. Glownia and S. L. Johnson, *Nature*, 2019, **565**, 209–212.
58 A. V. Kimel and M. Li, *Nat. Rev. Mater.*, 2019, **4**, 189–200.
59 C. Banerjee, N. Teichert, K. E. Siewierska, Z. Gercsi, G. Y. P. Atcheson, P. Stamenov, K. Rode, J. M. D. Coey and J. Besbas, *Nat. Commun.*, 2020, **11**, 4444.
60 C. D. Stanciu, F. Hansteen, A. V. Kimel, A. Kirilyuk, A. Tsukamoto, A. Itoh and T. Rasing, *Phys. Rev. Lett.*, 2007, **99**, 047601.
61 B. A. Ivanov, *J. Exp. Theor. Phys.*, 2020, **131**, 95–112.
62 D. Bossini, M. Pancaldi, L. Soumah, M. Basini, F. Mertens, M. Cinchetti, T. Satoh, O. Gomonay and S. Bonetti, *Phys. Rev. Lett.*, 2021, **127**, 077202.
63 A. V. Kimel, A. Kirilyuk, A. Tsvetkov, R. V. Pisarev and T. Rasing, *Nature*, 1967, **429**, 850–853.
64 A. V. Kimel, B. A. Ivanov, R. V. Pisarev, P. A. Usachev, A. Kirilyuk and T. Rasing, *Nat. Phys.*, 2009, **5**, 727–731.
65 S. Schlauderer, C. Lange, S. Baierl, T. Ebnet, C. P. Schmid, D. C. Valovcin, A. K. Zvezdin, A. V. Kimel, R. V. Mikhaylovskiy and R. Huber, *Nature*, 2019, **569**, 383–387.
66 A. Stupakiewicz, K. Szerenos, M. D. Davydova, K. A. Zvezdin, A. K. Zvezdin, A. Kirilyuk and A. V. Kimel, *Nat. Commun.*, 2019, **10**, 612.
67 J. A. Leake, G. Shirane and J. P. Remeika, *Solid State Commun.*, 1968, **6**, 15–17.
68 M. Kozina, M. Fechner, P. Marsik, T. v. Driel, J. M. Glownia, C. Bernhard, M. Radovic, D. Zhu, S. Bonetti, U. Staub and M. C. Hoffmann, *Nat. Phys.*, 2019, **15**, 387–392.
69 S. Grübel, PhD thesis, ETH Zurich, 2016.
70 T. Qi, Y.-H. Shin, K.-L. Yeh, K. A. Nelson and A. M. Rappe, *Phys. Rev. Lett.*, 2009, **102**, 247603.
71 X. Li, T. Qiu, J. Zhang, E. Baldini, J. Lu, A. M. Rappe and K. A. Nelson, *Science*, 2019, **364**, 1079–1082.
72 M. Först, R. Mankowsky, H. Bromberger, D. M. Fritz, H. Lemke, D. Zhu, M. Chollet, Y. Tomioka, Y. Tokura, R. Merlin, J. P. Hill, S. L. Johnson and A. Cavalleri, *Solid State Commun.*, 2013, **169**, 24–27.
73 R. Mankowsky, M. Först and A. Cavalleri, *Rep. Prog. Phys.*, 2016, **79**, 064503.
74 M. Foerst, C. Manzoni, S. Kaiser, Y. Tomioka, Y. Tokura, R. Merlin and A. Cavalleri, *Nat. Phys.*, 2011, **7**, 854–856.
75 R. Mankowsky, A. v. Hoegen, M. Först and A. Cavalleri, *Phys. Rev. Lett.*, 2017, **118**, 197601.
76 M. Rini, R. Tobey, N. Dean, J. Itatani, Y. Tomioka, Y. Tokura, R. W. Schoenlein and A. Cavalleri, *Nature*, 2007, **449**, 72–74.
77 V. Esposito, L. Rettig, E. M. Bothschafter, Y. Deng, C. Dornes, L. Huber, T. Huber, G. Ingold, Y. Inubushi, T. Katayama, T. Kawaguchi, H. Lemke, K. Ogawa, S. Owada, M. Radovic, M. Ramakrishnan, Z. Ristic, V. Scagnoli, Y. Tanaka, T. Togashi, K. Tono, I. Usov, Y. W. Windsor, M. Yabashi, S. L. Johnson, P. Beaud and U. Staub, *Struct. Dyn.*, 2018, **5**, 064501.

78 R. I. Tobey, D. Prabhakaran, A. T. Boothroyd and A. Cavalleri, *Phys. Rev. Lett.*, 2008, **101**, 197404.
79 T. Ogasawara, T. Kimura, T. Ishikawa, M. Kuwata-Gonokami and Y. Tokura, *Phys. Rev. B: Condens. Matter Mater. Phys.*, 2001, **63**, 4.
80 M. Först, R. I. Tobey, S. Wall, H. Bromberger, V. Khanna, A. L. Cavalieri, Y.-D. Chuang, W. S. Lee, R. Moore, W. F. Schlotter, J. J. Turner, O. Krupin, M. Trigo, H. Zheng, J. F. Mitchell, S. S. Dhesi, J. P. Hill and A. Cavalleri, *Phys. Rev. B*, 2011, **84**, 241104.
81 R. Mankowsky, A. Subedi, M. Först, S. O. Mariager, M. Chollet, H. T. Lemke, J. S. Robinson, J. M. Glownia, M. P. Minitti, A. Frano, M. Fechner, N. A. Spaldin, T. Loew, B. Keimer, A. Georges and A. Cavalleri, *Nature*, 2014, **516**, 71–73.
82 M. Mitrano, A. Cantaluppi, D. Nicoletti, S. Kaiser, A. Perucchi, S. Lupi, P. D. Pietro, D. Pontiroli, M. Riccò, S. R. Clark, D. Jaksch and A. Cavalleri, *Nature*, 2016, **530**, 461–464.
83 D. Fausti, R. I. Tobey, N. Dean, S. Kaiser, A. Dienst, M. C. Hoffmann, S. Pyon, T. Takayama, H. Takagi and A. Cavalleri, *Science*, 2011, **331**, 189–191.
84 M. J. Neugebauer, D. M. Juraschek, M. Savoini, P. Engeler, L. Boie, E. Abreu, N. A. Spaldin and S. L. Johnson, *Phys. Rev. Res.*, 2021, **3**, 013126.
85 T. Kampfrath, A. Sell, G. Klatt, A. Pashkin, S. Mährlein, T. Dekorsy, M. Wolf, M. Fiebig, A. Leitenstorfer and R. Huber, *Nat. Photonics*, 2011, **5**, 31–34.
86 Y. Sanari, T. Tachizaki, Y. Saito, K. Makino, P. Fons, A. V. Kolobov, J. Tominaga, K. Tanaka, Y. Kanemitsu, M. Hase and H. Hirori, *Phys. Rev. Lett.*, 2018, **121**, 165702.
87 T. Kubacka, J. A. Johnson, M. C. Hoffmann, C. Vicario, S. d. Jong, P. Beaud, S. Grübel, S. W. Huang, L. Huber, L. Patthey, Y. D. Chuang, J. J. Turner, G. L. Dakovski, W. S. Lee, M. P. Minitti, W. Schlotter, R. G. Moore, C. P. Hauri, S. M. Koohpayeh, V. Scagnoli, G. Ingold, S. L. Johnson and U. Staub, *Science*, 2014, **343**, 1333–1336.
88 R. Matsunaga, Y. I. Hamada, K. Makise, Y. Uzawa, H. Terai, Z. Wang and R. Shimano, *Phys. Rev. Lett.*, 2013, **111**, 057002.
89 R. Matsunaga, N. Tsuji, H. Fujita, A. Sugioka, K. Makise, Y. Uzawa, H. Terai, Z. Wang, H. Aoki and R. Shimano, *Science*, 2014, **345**, 1145–1149.
90 K. Katsumi, N. Tsuji, Y. I. Hamada, R. Matsunaga, J. Schneeloch, R. D. Zhong, G. D. Gu, H. Aoki, Y. Gallais and R. Shimano, *Phys. Rev. Lett.*, 2018, **120**, 117001.
91 H. Chu, M.-J. Kim, K. Katsumi, S. Kovalev, R. D. Dawson, L. Schwarz, N. Yoshikawa, G. Kim, D. Putzky, Z. Z. Li, H. Raffy, S. Germanskiy, J.-C. Deinert, N. Awari, I. Ilyakov, B. Green, M. Chen, M. Bawatna, G. Cristiani, G. Logvenov, Y. Gallais, A. V. Boris, B. Keimer, A. P. Schnyder, D. Manske, M. Gensch, Z. Wang, R. Shimano and S. Kaiser, *Nat. Commun.*, 2020, **11**, 1793.
92 T. Oka and S. Kitamura, *Annu. Rev. Condens. Matter Phys.*, 2019, **10**, 387–408.
93 T. Ishikawa, Y. Sagae, Y. Naitoh, Y. Kawakami, H. Itoh, K. Yamamoto, K. Yakushi, H. Kishida, T. Sasaki, S. Ishihara, Y. Tanaka, K. Yonemitsu and S. Iwai, *Nat. Commun.*, 2014, **5**, 5528.
94 Y. H. Wang, H. Steinberg, P. Jarillo-Herrero and N. Gedik, *Science*, 2013, **342**, 453–457.
95 R. Moessner and S. L. Sondhi, *Nat. Phys.*, 2017, **13**, 424–428.

96 K. Viebahn, J. Minguzzi, K. Sandholzer, A.-S. Walter, M. Sajnani, F. Görg and T. Esslinger, *Phys. Rev. X*, 2021, **11**, 011057.
97 O. R. Diermann and M. Holthaus, *Sci. Rep.*, 2019, **9**, 17614.
98 A. Baltuška, T. Udem, M. Uiberacker, M. Hentschel, E. Goulielmakis, C. Gohle, R. Holzwarth, V. S. Yakovlev, A. Scrinzi, T. W. Hänsch and F. Krausz, *Nature*, 2003, **421**, 611–615.
99 J. Duris, S. Li, T. Driver, E. G. Champenois, J. P. MacArthur, A. A. Lutman, Z. Zhang, P. Rosenberger, J. W. Aldrich, R. Coffee, G. Coslovich, F.-J. Decker, J. M. Glownia, G. Hartmann, W. Helml, A. Kamalov, J. Knurr, J. Krzywinski, M.-F. Lin, J. P. Marangos, M. Nantel, A. Natan, J. T. O'Neal, N. Shivaram, P. Walter, A. L. Wang, J. J. Welch, T. J. A. Wolf, J. Z. Xu, M. F. Kling, P. H. Bucksbaum, A. Zholents, Z. Huang, J. P. Cryan and A. Marinelli, *Nat. Photonics*, 2020, **14**, 30–36.
100 S. Ghimire, *Nat. Photonics*, 2022, **16**, 7–9.
101 M. F. Ciappina, J. A. Pérez-Hernández, A. S. Landsman, W. Okell, S. Zherebtsov, B. Förg, J. Schötz, J. L. Seiffert, T. Fennel, T. Shaaran, T. Zimmermann, A. Chacón, R. Guichard, A. Zaïr, J. W. G. Tisch, J. P. Marangos, T. Witting, A. Braun, S. A. Maier, L. Roso, M. Krüger, P. Hommelhoff, M. F. Kling, F. Krausz and M. Lewenstein, *arXiv*, 2016, **80**, 054401.
102 M. Zhang, R. A. McKinnon, G. Viola, B. Yang, D. Zhang, M. J. Reece, I. Abrahams and H. Yan, *ACS Photonics*, 2021, **8**, 147–151.
103 P. Zalden, M. J. Shu, F. Chen, X. Wu, Y. Zhu, H. Wen, S. Johnston, Z.-X. Shen, P. Landreman, M. Brongersma, S. W. Fong, H.-S. P. Wong, M.-J. Sher, P. Jost, M. Kaes, M. Salinga, A. v. Hoegen, M. Wuttig and A. M. Lindenberg, *Phys. Rev. Lett.*, 2016, **117**, 067601.
104 E. Abreu, D. Babich, E. Janod, S. Houver, B. Corraze, L. Cario and S. Johnson, *2019 44th International Conference on Infrared, Millimeter, and Terahertz Waves (IRMMW-THz)*, 2019, pp. 1–2.
105 T. Oka, R. Arita and H. Aoki, *Phys. Rev. Lett.*, 2003, **91**, 066406.

Faraday Discussions

ROYAL SOCIETY OF CHEMISTRY

PAPER

Visualizing optically-induced strains by five-dimensional ultrafast electron microscopy

A. Nakamura, *[a] T. Shimojima [a] and K. Ishizaka [ab]

Received 14th March 2022, Accepted 28th March 2022

DOI: 10.1039/d2fd00062h

Ultrafast optical control of strain is crucial for the future development of nanometric acoustic devices. Although ultrafast electron microscopy has played an important role in the visualization of strain dynamics in the GHz frequency region, quantitative strain evaluation with nm × ps spatio-temporal resolution is still challenging. Five-dimensional scanning transmission electron microscopy (5D-STEM) is a powerful technique that measures time-dependent diffraction or deflection of the electron beam at the respective two-dimensional sample positions in real space. In this paper, we demonstrate that convergent beam electron diffraction (CBED) measurements using 5D-STEM are capable of quantitative time-dependent strain mapping in the nm × ps scale. We observe the generation and propagation of acoustic waves in a nanofabricated silicon thin plate of 100 nm thickness. The polarization and amplitude of the acoustic waves propagating in the silicon plate are quantitatively determined from the CBED analysis. Further Fourier-transformation analysis reveals the strain distribution in the momentum-frequency space, which gives the dispersion relation in arbitrary directions along the plate. Versatility of 5D-STEM-CBED analysis enables quantitative strain mapping even in complex nanofabricated samples, as demonstrated in this study.

1 Introduction

Transmission electron microscopy (TEM) and scanning transmission electron microscopy (STEM) are fundamental tools for modern material and biological science due to their potential to visualize nanoscale structures. Although time resolution of conventional TEM and STEM is limited by the shutter speed of the camera and scan speed of the electron beam (∼ms), recent advances in ultrafast electron microscopy that uses ultrashort electron pulse, enables the observation of

[a]*RIKEN Center for Emergent Matter Science, Wako, Saitama 351-0198, Japan. E-mail: asuka.nakamura@riken.jp*

[b]*Quantum-Phase Electronics Center and Department of Applied Physics, The University of Tokyo, Bunkyo, Tokyo 113-8656, Japan*

ultrafast dynamics after photoexcitation on the ultrafast (fs–ns) temporal scale.[1,2] In the last couple of decades, ultrafast electron microscopy and diffraction have been utilized to reveal various structural, electronic and magnetic phenomena such as structural phase transition,[3,4] acoustic wave propagation,[5] dielectric response,[6] and dynamics in magnets.[7,8] In particular, dynamics of strain, *i.e.*, acoustic waves, is one of the fundamental excitations in a solid system and therefore should exist in all samples: indeed, confined and propagating acoustic waves have been observed in many solid systems.[5,9–11] Ultrafast electron microscopy and diffraction is a unique method that can observe nanometric acoustic waves, and therefore play a significant role in novel acoustic device development in nano-fabricated systems such as phononic crystals. In previous ultrafast TEM experiments, bright-field imaging and electron diffraction have been mainly used to elucidate acoustic responses.[5,9] However, quantitative evaluation of amplitude and polarization of acoustic waves in these samples has been difficult using such methods, due to the rather indirect relation between the strain and the diffraction intensities,[12] although extensive effort in this field has been made recently.[13]

In conventional (static) TEM, quantitative evaluation of local strain has been achieved by analyzing electron diffraction patterns such as convergent beam electron diffraction (CBED). In this method, the electron beam is focused on the nanometer–micrometer scale with ~10 mrad convergent angle, and diffracted electrons appearing as disks and lines are recorded by a camera. By calculating the scattering angle of electrons from the two-dimensional diffraction pattern in momentum space (k_x,k_y), local lattice strain can be quantitatively evaluated. Recently, development of two-dimensional pixelated detectors with high sensitivity and fast (~ms) readout time has enabled us to obtain two-dimensional mapping of CBED information by scanning the sample in real space (x,y), which is called four-dimensional scanning transmission electron microscopy (4D-STEM). By analyzing four-dimensional data by high-performance computers, various structural properties in solid samples, such as crystalline orientation and lattice parameters are mapped in two-dimensional real space.[14] Therefore, time-resolved 4D-STEM using ultrafast TEM, *i.e.* five-dimensional scanning transmission electron microscopy (5D-STEM),[15] is a promising method to quantitatively obtain two-dimensional strain movies in nm × ps spatiotemporal resolution. By combining the STEM and pump–probe method using the femtosecond laser, the time resolution of STEM, which has been determined by the scan speed, can be significantly enhanced from the seconds to picoseconds. Realization of 5D-STEM is crucial to quantitatively analyze phononic phenomena especially in nanometric acoustic devices, because its time scale becomes the ps range as is determined by sound velocity (~nm ps^{-1}).

In this study, we investigate the time (t) dependent strain mapping at two-dimensional sample positions (x,y) by analyzing the 5D-STEM data for single-crystalline silicon. Although a previous study already shows one-dimensional scan of strain in graphite,[16] two-dimensional mapping requires qualitative change in software as well as hardware. As a demonstration, we perform 5D-STEM on a nanofabricated silicon single-crystalline plate. We also develop software that can efficiently handle the huge amount of 5D-STEM data – more than hundreds of gigabytes. The displacement gradient tensor $\boldsymbol{D}(x,y,t) = \nabla \boldsymbol{u}(x,y,t)$ is evaluated by detecting the scattering angle of electrons in CBED images, where $\boldsymbol{u}$ is the atomic displacement vector. The movie of $\Delta \boldsymbol{D} = \boldsymbol{D}(x,y,t) - \boldsymbol{D}(x,y,0)$ clearly visualizes the photoinduced generation and propagation of acoustic waves. We further apply Fourier-transformation analysis on $\Delta \boldsymbol{D}(x,y,t)$ and obtain the spectrum $\Delta\tilde{\boldsymbol{D}}(\kappa_x,\kappa_y,\omega)$ in momentum (κ_x,κ_y) and frequency (ω) space. Typical application of 5D-STEM on nano-fabricated samples indicates that the 5D-STEM-CBED is a powerful tool for the quantitative evaluation of optically-induced strain on the nm × ps scale.

2 Materials and method

2.1 Setup of 5D-STEM

Fig. 1(a) shows the setup of our 5D-STEM measurement system, based on the pump–probe method utilizing a combination of laser pulses shorter than 1 ps and TEM (Technai Femto, Thermo Fisher Scientific). The trail of optical pulses with $\omega = 1.2$ eV photon energy is generated by a Yb:KGd$(WO_4)_2$ solid-state laser (PHAROS, Light Conversion). They are separated into pump and probe beams. The pump beam passes through the delay stage and excites the sample mounted in the electron microscope. Typical spot size of the pump beam at the sample position is 100 μm, which is much larger than the typical field of view of STEM (∼10 μm at maximum). The photon energy of the probe beam is quadrupled by two β-$Ba_2B_2O_4$ (BBO) crystals. The generated $4\omega = 4.8$ eV optical pulse is irradiated onto the carbon-coated LaB_6 photocathode (Applied Physics Technologies) with a diameter of 50 μm, which is mounted at the electron source of TEM. The optical pulse is converted into an electron pulse by the photoelectric effect and then accelerated to 200 keV. In the present CBED measurement setup, the electron pulse is focused on the sample to less than 15 nm, with a convergence angle of 30 mrad. Two-dimensional CBED patterns are recorded by a pixelated detector that can be synchronized with the scan clock signal of the scan coil. For such a pixelated detector, we employed a MerlinEM direct electron detector (Quantum Detectors) whose readout speed of 256 × 256 pixels is less than 2 ms.

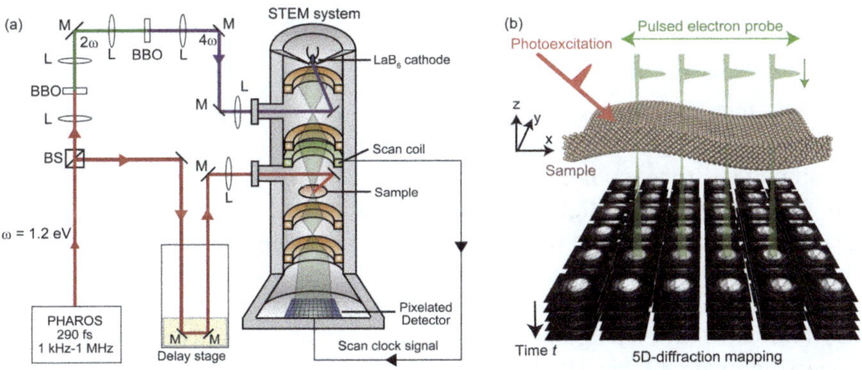

Fig. 1 (a) Schematic of the 5D-STEM system. L, M, and BS denote lens, mirror and beam splitter, respectively. Fundamental 1.2 eV light pulse is converted to 4.8 eV by two β-$Ba_2B_2O_4$ (BBO) crystals and then focused on the LaB_6 cathode in the STEM system. The CBED pattern of the photoexcited sample is recorded by the pixelated detector, which is electrically synchronized with the scan coil of STEM. (b) Schematic of the 5D-STEM. The photoexcited sample is probed by scanning the convergent electron pulse in two-dimensional real space (x,y). The time (t)-dependence of two-dimensional (k_x,k_y) CBED images are obtained at the respective positions.

Fig. 1(b) shows the schematic of strain imaging using the CBED setup in our 5D-STEM system. The sample is excited by the optical pump pulse with 290 fs duration, and strains are induced *via* a variety of mechanisms, *e.g.*, thermoelasticity.[17,18] The pulsed convergent electron beam propagating along the z direction is focused onto a certain position (x,y) of the sample. In this paper, we define the Cartesian coordinates x, y, and z axes based on the propagation direction of the probe electron beam, which is fixed to the laboratory system. There, the two-dimensional (k_x,k_y) CBED pattern is recorded by the pixelated detector. By scanning the position of the electron beam in the x and y direction, two-dimensional (x, y) mapping of the two-dimensional (k_x,k_y) CBED patterns can be obtained. Such 4D-STEM data are repeatedly obtained by changing the relative arrival time (t) of the optical pulse (pump) and the electron pulse (probe). This enables us to record the time-evolution of the 4D-STEM data with the time resolution determined by the duration of electron and optical pulses. By analyzing the obtained five-dimensional (x,y,k_x,k_y,t) dataset, the ultrafast movies of displacement gradient tensor $\boldsymbol{D}(x,y,t)$ can be obtained, as described in Sections 2.3 and 2.4.

We should note that the temporal resolution of the ultrafast electron microscope (and consequently the 5D-STEM) is mainly dominated by the duration of electron pulses, which tend to elongate by the repulsion of electrons within a single pulse, *i.e.* the space charge effect.[19] Therefore, the number of electrons in a single electron pulse must be eliminated to achieve a high enough time resolution. We confirmed that the best temporal resolution of our 5D-STEM system is below 2 ps when we set the number of electrons per pulse to be less than ~1000.[9] In our system, the spatial resolution (15 nm) is restricted by the drift motion of the sample during data acquisition, in contrast to conventional static STEM where it is determined by the spot size of the probe electron beam. This is because the typical 5D-STEM-CBED measurement in our setup requires more than a day for the data acquisition due to the reduced intensity of the photoelectron beam.

2.2 Sample preparation

We use a [001] oriented single-crystalline silicon plate 100 μm × 100 μm × 100 nm as a sample. We deposited a metallic tungsten patterned onto the silicon plate, with ~100 nm thickness, by the focused-ion beam (FIB) method using NB5000 (Hitachi High-Tech). This metallic tungsten absorbs the pump optical pulse (1.2 eV) while the silicon part shows a very low absorption rate.[20] Thus, the patterned tungsten part immediately gets photo-thermalized by the pump and acts as the acoustic phonon generator at the silicon interface. We also fabricated the triangular patterns of voids in the silicon plate by FIB (appearing in Fig. 5). For all of these nanofabrications, we set the acceleration voltage to 40 kV and used the small (15 μm) aperture to avoid damage to the sample.

2.3 Calculation of displacement gradient tensor from the CBED pattern

Here, we express the lattice vectors $\boldsymbol{a},\boldsymbol{b},\boldsymbol{c}$ and reciprocal lattice vectors $\boldsymbol{a}^*,\boldsymbol{b}^*,\boldsymbol{c}^*$ of the original non-deformed crystal, using matrices $\boldsymbol{A}$ and $\boldsymbol{A}^*$ as

$$A = \begin{pmatrix} \boldsymbol{a} \\ \boldsymbol{b} \\ \boldsymbol{c} \end{pmatrix} = \begin{pmatrix} a_1 & a_2 & a_3 \\ b_1 & b_2 & b_3 \\ c_1 & c_2 & c_3 \end{pmatrix}, \ A^* = \begin{pmatrix} \boldsymbol{a}^* \\ \boldsymbol{b}^* \\ \boldsymbol{c}^* \end{pmatrix} = \begin{pmatrix} a_1^* & a_2^* & a_3^* \\ b_1^* & b_2^* & b_3^* \\ c_1^* & c_2^* & c_3^* \end{pmatrix}.$$

When we assume that $\boldsymbol{R}$ is a rotation matrix that represents the rotation of the crystal with respect to the probe electron beam, and $\boldsymbol{D}$ is the displacement gradient tensor that describes the deformation of the crystal, the lattice parameter of the deformed crystal $\boldsymbol{A}_\mathrm{d}$ can be newly expressed as

$$\boldsymbol{A}_\mathrm{d} = \boldsymbol{R}(\boldsymbol{E} + \boldsymbol{D})\boldsymbol{A},$$

where $\boldsymbol{E}$ is the identity matrix. When we ignore the second- and higher-order terms of $\boldsymbol{D}$, the new reciprocal lattice vectors $\boldsymbol{A}_\mathrm{d}^* = (\boldsymbol{a}_\mathrm{d}^*, \boldsymbol{b}_\mathrm{d}^*, \boldsymbol{c}_\mathrm{d}^*)^T$ can be approximately expressed by

$$\boldsymbol{A}_\mathrm{d}^* = 2\pi\left[\boldsymbol{A}_\mathrm{d}^{-1}\right]^T \simeq \boldsymbol{R}\left(\boldsymbol{E} - \boldsymbol{D}^T\right)\boldsymbol{A}^*. \tag{1}$$

We use the relations $\boldsymbol{A}^* = 2\pi[\boldsymbol{A}^{-1}]^\mathrm{T}$, $\boldsymbol{R}^{-1} = \boldsymbol{R}^\mathrm{T}$, and $(\boldsymbol{E} + \boldsymbol{D})^{-1} \simeq \boldsymbol{E} - \boldsymbol{D}$ to derive eqn (1). The superscript T indicates the transpose matrix.

In conventional CBED analysis, the displacement gradient tensor $\boldsymbol{D}$ is calculated from the excess and deficit line positions appearing in the CBED patterns. When the propagation direction of the convergent electron beam is parallel to the z axis, the relations $\boldsymbol{k}' - \boldsymbol{k} = \boldsymbol{g}$ and $\boldsymbol{k}^2 = \boldsymbol{k}'^2$ are satisfied, where $k = (k_x, k_y, -\sqrt{{k_0}^2 - {k_x}^2 - {k_y}^2})$ and $\boldsymbol{k}'$ are the wave vectors of the incident and scattered electron beams, and k_0 is the wave number of the incident electron beam. Note that $\sqrt{{k_x}^2 + {k_y}^2}/k_0$ should be smaller than the convergence angle of the probe electron beam. $\boldsymbol{g}$ is the scattering vector that satisfies

$$\boldsymbol{g} = (g_x, g_y, g_z) = h\boldsymbol{a}_\mathrm{d}^* + k\boldsymbol{b}_\mathrm{d}^* + l\boldsymbol{c}_\mathrm{d}^*, \tag{2}$$

where h, k, l are the Miller indices. The deficit and excess line satisfies the conditions

$$\text{Deficit line: } \boldsymbol{k}\cdot\boldsymbol{g} = -|\boldsymbol{g}|^2/2,$$

$$\text{Excess line: } \boldsymbol{k}'\cdot\boldsymbol{g} = |\boldsymbol{g}|^2/2,$$

which give the set of diffracted lines $[k_x, k_y]$ and $[k'_x, k'_y]$ in reciprocal space. In addition, the relations $k_x, k_y \ll k_0$ and ${g_z}^2 \ll {g_x}^2 + {g_y}^2$ are satisfied considering that the electrons are accelerated to 200 keV. Using these relations, the positions of the deficit and excess lines with respect to the origin of reciprocal space, r_d and r_e, are expressed by

$$\text{Deficit line}: r_\mathrm{d} \simeq -\frac{k_0 g_z}{\sqrt{{g_x}^2 + {g_y}^2}} + \frac{\sqrt{{g_x}^2 + {g_y}^2}}{2}, \tag{3}$$

$$\text{Excess line}: r_\mathrm{e} \simeq -\frac{k_0 g_z}{\sqrt{{g_x}^2 + {g_y}^2}} - \frac{\sqrt{{g_x}^2 + {g_y}^2}}{2}. \tag{4}$$

Using eqn (1)–(4), the displacement gradient tensor $\boldsymbol{D}$ can be calculated from the excess and deficit line positions in the CBED patterns.

In this study, we approximate $\boldsymbol{R} \simeq \boldsymbol{E}$ since the tilting angle of the sample with respect to the electron beam is less than 2° (see Section 3.1). For single-crystalline silicon, $\boldsymbol{A}^* = 2\pi\boldsymbol{E}/a$ holds, where a is the lattice constant of the cubic crystal. When we focus on h00 diffraction in a single-crystalline silicon and ignore the second- and higher-order terms of $\boldsymbol{D}$, the positions of the deficit and excess lines can be calculated as

$$\text{Deficit line}: r_\mathrm{d} = k_0 D_{zx} + \frac{h\pi}{a}(1 - D_{xx}), \tag{5}$$

$$\text{Excess line}: r_\mathrm{e} = k_0 D_{zx} - \frac{h\pi}{a}(1 - D_{xx}). \tag{6}$$

Therefore, D_{xx} and D_{zx} can be calculated form the positions of excess and deficit lines experimentally. Note that all 9 components of $\boldsymbol{D}$ can be determined by measuring the multiple excess and deficit lines at different tilting angles, in principle.[21] We determined D_{xx}, D_{yy}, D_{zx}, D_{zy} as we scanned the beam position and the delay time (x,y,t), by recording the positions of the excess (deficit) lines obtained from the local maxima (minima) of $\bar{8}00$ and 080 diffraction intensities.

2.4 Derivation of time-dependent strain from 5D-STEM-CBED

Calculation of the photoinduced time-dependent displacement gradient tensor $\Delta\boldsymbol{D}(x,y,t) = \boldsymbol{D}(x,y,t) - \boldsymbol{D}(x,y,0)$ from the 5D-STEM data $I(x,y,k_x,k_y,t)$ can be achieved by simply repeating the CBED analysis for 4D-STEM data, in principle. However, in practice, an additional pre/postprocess is required due to the very low intensity of the probe electron beam. As the number of electrons in a single electron pulse for 5D-STEM is limited to less than <1000 for high time resolution measurements (see Section 2.1), the data acquisition time of more than several weeks is typically required if we simply apply the above analysis for $I(x,y,k_x,k_y,t)$ at 100 × 100 × 256 × 256 × 100 points. To avoid this problem, we applied an additional process to the 5D-STEM data: (1) 1 × 1 × 3 × 3 × 1 median filter and (2) 5 × 5 × 1 × 1 × 3 average filter are applied to the 5D $I(x,y,k_x,k_y,t)$ data. Then $\Delta\boldsymbol{D}$ is derived by the CBED analysis as described in Section 2.3. Afterwards, (3) 3 × 3 × 1 median filter is applied to the $\Delta\boldsymbol{D}(x,y,t)$ data. The two median filters [(1), (3)] are used to remove the impulse noise caused during the measurement and the analysis, respectively. Owing to the average filter (2), the signal-to-noise ratio at a respective point (x,y,t) is improved by ~100 times. Although the total space- and time-resolutions become slightly worse through these processes, the statistics of the analyzed data greatly improves. These processes are indispensable in practice, since a long acquisition time suppresses the effective space resolution due to the sample drift. It should be noted that owing to these pre/postprocesses, the experimental results appearing in this paper require only about 1 day of data acquisition time. Yet this limits the total space resolution to ~15 nm, as mentioned above.

We parallelized these data processes on a high-performance computer with sufficient memory and calculation cores. This is because the data size of the typical 5D-STEM data (100 × 100 × 256 × 256 × 100 of 32-bit float) is about 300 gigabytes and applying the CBED analysis takes more than a few hours when a single-core system is used for calculation. We employed the shared-memory high-performance computer with 3 terabytes memory and 112 cores for the analysis. For parallelization, we employed Python-based library *Dask*,[22,23] which enables the parallel computation of the above-mentioned processes. Using *Dask*, the calculation can be easily extended to the much larger data size by utilizing the distributed-memory system, if required.

3 Results and discussion

3.1 Static CBED measurement on 100 nm thick Si plate

To begin with, we show the result of static CBED measurements on a single-crystalline silicon plate. As shown in Fig. 2(a), we set the [001] oriented silicon plate with [001] parallel to the probe electron beam, and then rotate it by $\alpha = 1.9°$ and $\beta = 0.8°$ around the x and y axes, respectively. As a result, a, b, and c axes of the single-crystalline silicon are slightly deviated from the x, y, and z axes. The typical CBED pattern under this setup is shown in Fig. 2(b). In the transmitted electron disk, we observe several dark deficit lines, where the electrons are diffracted while travelling through the silicon plate. Correspondingly, the diffracted electrons form the bright excess lines. To determine the Miller indices of these excess and deficit lines, we simulate the CBED pattern of the single crystalline silicon as shown in Fig. 2(c), based on the method described in ref. 16. Most of the observed CBED patterns are reproduced by the simulation, and thus we can assign the Miller indices of the respective lines. Hereafter, we focus on the time dependence of $\bar{8}00$ and 080 deficit/excess lines to determine the time-dependent displacement gradient tensor $\Delta\boldsymbol{D}(x,y,t) = \boldsymbol{D}(x,y,t) - \boldsymbol{D}(x,y,0)$.

3.2 Strain dynamics in Si plate obtained by 5D-STEM-CBED

In this section, we demonstrate the quantitative evaluation of the displacement gradient tensor $\Delta\boldsymbol{D}(x,y,t)$, derived from the single 5D-STEM data. Here we use the

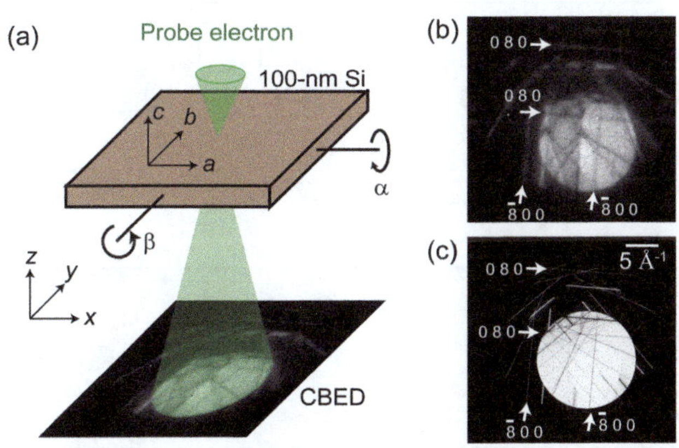

Fig. 2 (a) Schematic of the CBED experimental setup. x, y, and z axes are fixed to the laboratory system, and $-z$ corresponds to the propagation direction of the convergent electron beam. The a, b, and c axes of the single-crystalline silicon sample are aligned parallel to x, y, and z axes, and then the sample is rotated by $\alpha = 1.9°$ and $\beta = 0.8°$ around x and y axes, respectively. (b) Typical CBED pattern for the single-crystalline silicon with a thickness of 100 nm. $\bar{8}00$ and 080 excess and deficit lines are indicated by white arrows. (c) Simulated CBED pattern for the single-crystalline silicon sample.

100 nm-thick silicon plate with a small tungsten nanodisk deposited at the center, which acts as a single point-like source of the photoinduced strain. Fig. 3(a) shows the bright-field image of the sample reconstructed from the 5D-STEM data ($t < 0$). The metallic tungsten nanodisk (diameter: 700 nm, thickness: 100 nm) is deposited at the center using FIB. The x and y directions are depicted in Fig. 3(a). When the sample is photoexcited by the 1.2 eV pump optical pulse, only the tungsten nanodisk is immediately photo-thermalized because silicon hardly absorbs the 1.2 eV light.[20] As a result, the strain is first locally generated around the tungsten nanodisk and starts to propagate in the silicon plate as the radial acoustic waves.[18] We quantitatively estimated the amplitude of such acoustic waves by analyzing the t-dependent CBED patterns. The CBED pattern at the position indicated by the green cross is displayed in Fig. 3(b). We analyzed the time dependence of $\bar{8}00$ excess line [Fig. 3(c)] and estimated $\Delta D(x,y,t)$ from the peak position of $\bar{8}00$ excess line, following the equations in Section 2.3. We confirmed that the shift of the excess line is identical with the deficit line, which indicates that the observed $\bar{8}00$ line shift is mainly caused by ΔD_{zx}, and the contribution from ΔD_{xx} is invisibly small in the present setup. Resulting time dependent imaging of $\Delta D_{zx}(x,y,t)$ is shown in Fig. 3(d). The generation and radial propagation of the strain is clearly visualized quantitatively. In the same way, we

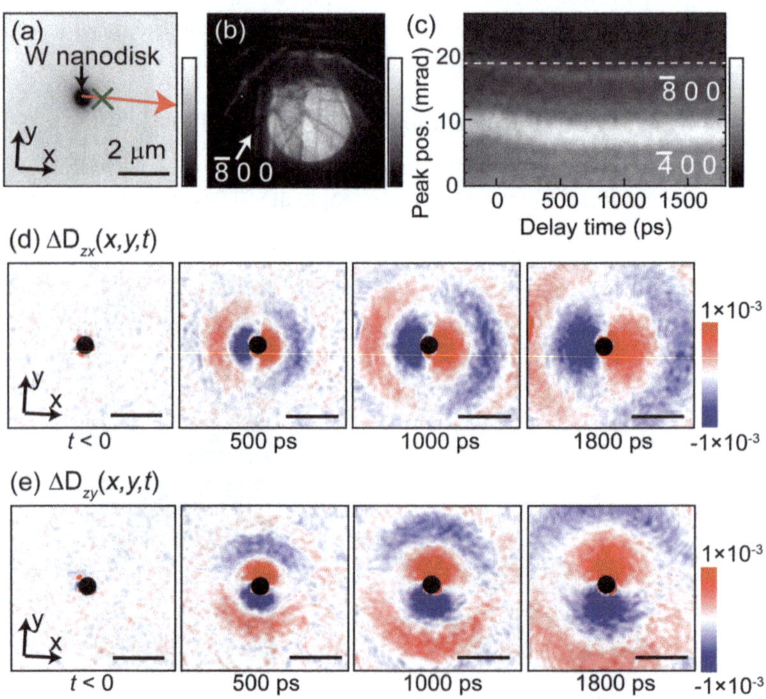

Fig. 3 (a) The bright-field image of silicon single crystal on which a tungsten nanodisk (700 nm diameter) is deposited. (b) CBED pattern at the position depicted by the green cross in (a). (c) Time dependence of $\bar{8}00$ excess line. The position of $\bar{8}00$ excess line is shifted after photoexcitation. (d, and e) Time dependence of displacement gradient tensor component $\Delta D_{zx}(x,y,t)$ and $\Delta D_{zy}(x,y,t)$ calculated from the 5D-STEM data. The scale bars indicate 2 μm. The black circle at the center indicates the tungsten nanodisk.

estimated $\Delta D_{zy}(x,y,t)$ from the 080 excess line position as shown in Fig. 3(e). As the (100) and (010) planes are crystallographically equivalent, the amplitude of $\Delta D_{zx}(x,y,t)$ is almost identical with $\Delta D_{zy}(x,y,t)$ when the image is rotated by 90°. These measurements can determine the amplitude and polarization of the propagating acoustic waves experimentally, which had been technically difficult in the conventional time-dependent bright-field image and diffractions.[5,9]

Fourier-transformation analysis on $\Delta D_{zx}(x,y,t)$ and $\Delta D_{zy}(x,y,t)$ reveals the distribution of strains in the momentum–frequency (κ–ω) space. Fig. 4(a) shows the time dependence of $\Delta D_{zx}(x,y,t)$ recorded along the red arrow in Fig. 3(a). The wavelength of the strain wave seemingly changes as a function of position and t, indicating that the oscillatory behavior is composed of multiple modes with different sound velocities. We applied the x, y, t Fourier transformation to $\Delta D_{zx}(x,y,t)$ and obtained $\Delta\tilde{D}_{zx}(\kappa_x,\kappa_y,\omega)$, which shows the strain distribution in κ–ω space. In Fig. 4(b), we show $\Delta\tilde{D}_{zx}$ corresponding to the Fourier transformation of Fig. 4(a). The red curves show the theoretical acoustic eigen modes in a 100 nm thick silicon plate, calculated using the partial wave technique.[24] In this frequency region, there are two dispersion curves corresponding to the so-called symmetric (S_0) and asymmetric (A_0) modes, which describe the plate-wave mode with symmetric and asymmetric strain distributions along the thickness direction, respectively. The obtained strain distribution in the κ–ω space suggests that the asymmetric A_0 mode is mainly excited in the frequency region of $\omega < 0.5$ GHz. Since the tungsten is deposited only on the top side of the plate sample, the asymmetric A_0 mode is more likely to be excited by the photo-thermalization of the tungsten disk. In addition, we further obtained the constant frequency cut of $\tilde{S}(\kappa_x,\kappa_y,\omega) = [\Delta\tilde{D}_{zx}{}^2(\kappa_x,\kappa_y,\omega) + \Delta\tilde{D}_{zy}{}^2(\kappa_x,\kappa_y,\omega)]$, which successfully visualizes the in-plane anisotropy of the sound (A_0 mode) velocity. To remove the contribution from the static structure at $\omega = 0$, we show $\tilde{S}(\kappa_x,\kappa_y,\omega)/\tilde{S}((\kappa_x,\kappa_y,0)$ in Fig. 4(c–e). Moreover, we applied four-fold symmetrization based on the crystallographic symmetry of the silicon single crystal. Although the constant frequency cut shows the almost isotropic circular feature of the A_0 mode dispersion relation, slight 4-fold anisotropy appears at 1.43 GHz. This may be caused by the crystallographic anisotropy of the cubic silicon single crystal.

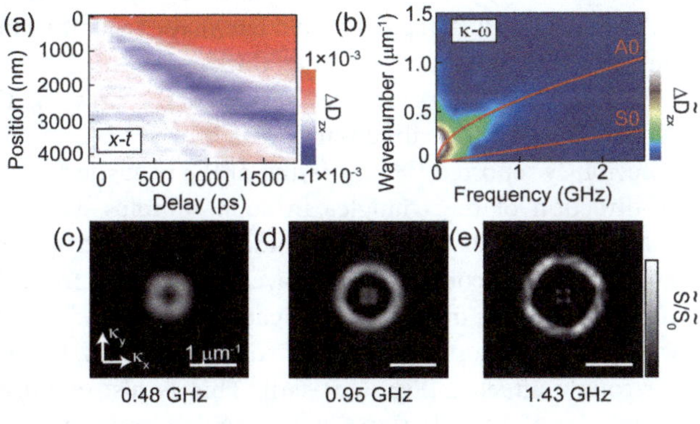

Fig. 4 (a) Space-time dependence of ΔD_{zx} along the red arrow in Fig. 3(a). (b) Fourier transformation of the ΔD_{zx} in (a). Red curves indicate the theoretical acoustic dispersion relation of zeroth-order asymmetric (A_0) and symmetric (S_0) modes in the 100 nm silicon thin plate. (c–e) $\tilde{S}/\tilde{S}_0 = \tilde{S}(\kappa_x,\kappa_y,\omega)/\tilde{S}((\kappa_x,\kappa_y,0)$ at 0.48, 0.95, and 1.43 GHz, respectively.

3.3 Pros and cons of the 5D-STEM-CBED method

The data in Fig. 3 and 4 demonstrate that the quantitative estimation of strains with nm × ps resolution can be achieved by analyzing 5D-STEM-CBED. We can also extract the strain distribution in κ–ω space, from which the in-plane anisotropy of the sound velocities and acoustic dispersion curves can be obtained. We emphasize that all images in Fig. 3 and 4 are obtained by analyzing a single 5D-STEM piece of data with 5 axes (x,y,k_x,k_y,t). Once the single 5D-STEM data is obtained, the displacement gradient tensor $\boldsymbol{D}(x,y,t)$ at arbitrary position and time can be evaluated by postprocessing. As compared to ref. 16 where ultrafast CBED-STEM analysis has been done along a single axis in real space, our 5D-STEM system is capable of taking a full two-dimensional ultrafast movie of the strain. As a result, propagation of strain along the arbitrary direction can be evaluated at the same time as demonstrated in Fig. 3(d, e) and 4(c–e). We also note that the quantitative estimation of the displacement gradient tensor can be applied to all types of single crystals. Such a method should play an important role in exploring the unique ultrafast strain dynamics in future acoustic nanodevices.

On the other hand, 5D-STEM-CBED suffers from long data acquisition time as compared to the usual ultrafast bright-field imaging method, where the spatial resolution also tends to get worse due to the sample drift. Although increasing the repetition rate of the laser should partly solve this problem, there is a clear limit because of sample damage becoming serious at higher repetition rates. Therefore, practically, compensating such low S/N ratio by data processing is one of the key techniques in 5D-STEM. As described in Section 2.4, we applied several median and average filters along the x, y, and t axes which significantly improve the data quality. Increasing the number of electrons per pulse without spreading the pulse duration may be necessary to achieve ~1 nm space resolution in future. Such improvement of a future 5D-STEM system may become possible by utilizing pulse compression techniques.[25,26]

3.4 Practical application of 5D-STEM-CBED on nano-patterned Si plate

Finally, we show a practical demonstration of 5D-STEM-CBED for nanofabricated samples. We prepared two types of samples as shown in Fig. 5(a and b). Similar to Fig. 3(a), the metallic tungsten is deposited in a bar shape (400 nm width, 100 nm thickness) at the left side of each sample. They work as the line-like source of the photoinduced strain waves. Moreover, we fabricated the array of the 700 nm triangular voids at the center, the direction of which in Sample #1 is opposite to Sample #2. As previously reported,[27,28] scattering of the acoustic waves is strongly affected by the direction of the triangles in such samples. We quantitatively evaluated the t-dependent displacement gradient tensor $\Delta D(x,y,t)$ in these samples. In Sample #1, the acoustic strain waves are emitted from the metallic tungsten bar at the left edge and start to propagate horizontally [Fig. 5(c1–c3)]. As the acoustic wave passes through the array of triangular voids, the amplitude is seemingly not strongly affected [Fig. 5(c4–c6)]. This is also confirmed by the space–time dependence of $\Delta D_{zx}{}^{av}(x,t)$ in Fig. 5(e), which is obtained by averaging $\Delta D_{xz}(x,y,t)$ along the vertical (y) direction. In Sample #2, we also observed similar emission of the acoustic wave from the tungsten bar in the t = 0–2 ns region [Fig. 5(d1–d3)]. However, as the strain wave propagate through the triangular

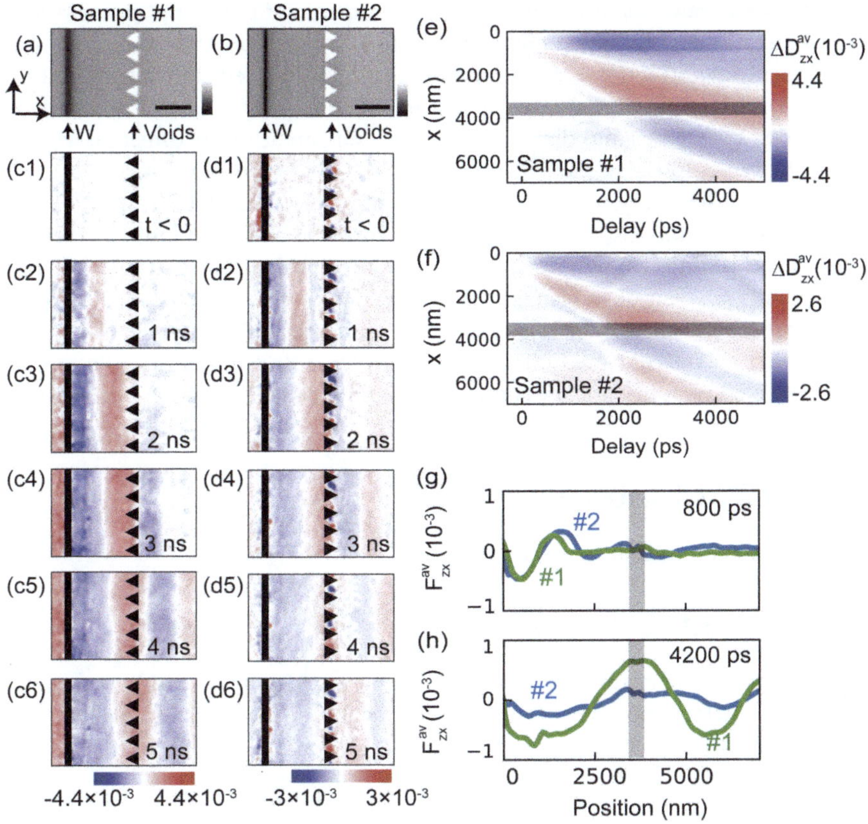

Fig. 5 (a and b) Bright-field images of Samples #1 and #2. The scale bar denotes 2 μm. The black region corresponds to the deposited tungsten bar. The white triangles are voids fabricated by FIB. (c and d) Time dependence of $\Delta D_{zx}(x,y,t)$ calculated by the 5D-STEM-CBED analysis. The black line and triangles indicate the deposited tungsten bars and the triangular voids, respectively. (e and f) Position and time dependence of $\Delta D_{zx}{}^{av}(x,t)$ for Samples #1 and #2, obtained by averaging $\Delta D_{zx}(x,y,t)$ along the y direction. (g and h) $\Delta D_{zx}{}^{av}(x,t)$ at 800 and 4200 ps, respectively. The gray shaded regions in (e–j) denote the position where the triangular void patterns are fabricated.

voids, the amplitude apparently becomes suppressed [Fig. 5(d4–d6)]. To quantitatively compare the amplitudes of strain waves in Samples #1 and #2, we show $\Delta D_{zx}{}^{av}$ at $t = 800$ ps in Fig. 5(g). ΔD_{zx} in Samples #1 and #2 are comparable in this t-range, indicating that the photogeneration of strains similarly occurs for both samples. In contrast, the amplitude of the acoustic wave in Sample #2 at 4200 ps [Fig. 5(h)] is significantly suppressed as compared to that of Sample #1. These results suggest that the scattering process of the acoustic strain is quantitatively different in such directionally nano-patterned artificial systems.

4 Conclusions

We have demonstrated 5D-STEM-CBED analysis for the quantitative time-dependent mapping of optically-induced strains. 5D-STEM measurement on 100 nm silicon single crystal with the metallic tungsten patterns deposited as the acoustic wave source demonstrated that time-dependent strain and polarization can be quantitatively mapped in two-dimensional space from the CBED analysis. By Fourier-transformation analysis, we further characterized the distribution of strains in the momentum-frequency space, which can be used to reveal anisotropic sound velocities and dispersion curves along an arbitrary direction within the plate. We emphasize that the analysis of 5D-STEM requires additional pre/postprocesses to compensate the low S/N ratio. The data size of 5D-STEM is more than a few hundreds of gigabytes, and thus the analysis requires efficient parallel computing. Nonetheless, the quantitative evaluation of strains by 5D-STEM-CBED enabled us to visualize the directional properties of acoustic waves realized in the Si plate that was artificially patterned with arrays of triangular voids. Such a 5D-STEM method will play a versatile and important role in the future development of optically controlled nanometer-scale acoustic phenomena and functions.

Conflicts of interest

There are no conflicts of interest to declare.

Acknowledgements

This work was partially supported by grant-in-aid for Scientific Research (KAKENHI) (Grant No. 21K14488), MEXT Quantum Leap Flagship Program (MEXT Q-LEAP).

References

1 A. H. Zewail, *Annu. Rev. Phys. Chem.*, 2006, **57**, 65.
2 A. Adhikari, J. K. Eliason, J. Sun, R. Bose, D. J. Flannigan and O. F. Mohammed, *ACS Appl. Mater. Interfaces*, 2017, **9**, 3.
3 P. Baum, D.-S. Yang and A. H. Zewail, *Science*, 2007, **318**, 788.
4 M. Eichberger, H. Schäfer, M. Krumova, M. Beyer, J. Demsar, H. Berger, G. Moriena, G. Sciaini and R. J. D. Miller, *Nature*, 2010, **468**, 799.
5 D. R. Cremons, D. A. Plemmons and D. J. Flannigan, *Nat. Commun.*, 2016, **7**, 11230.
6 X. Fu, F. Barantani, S. Gargiulo, I. Madan, G. Berruto, T. LaGrange, L. Jin, J. Wu, G. M. Vanacore, F. Carbone and Y. Zhu, *Nat. Commun.*, 2020, **11**, 5770.
7 G. Berruto, I. Madan, Y. Murooka, G. M. Vanacore, E. Pomarico, J. Rajeswari, R. Lamb, P. Huang, A. J. Kruchkov, Y. Togawa, T. Lagrange, D. Mcgrouther, H. M. Rønnow and F. Carbone, *Phys. Rev. Lett.*, 2018, **120**, 117201.
8 T. Shimojima, A. Nakamura, X. Yu, K. Karube, Y. Taguchi, Y. Tokura and K. Ishizaka, *Sci. Adv.*, 2021, **7**, 1322.
9 A. Nakamura, T. Shimojima, Y. Chiashi, M. Kamitani, H. Sakai, S. Ishiwata, H. Li and K. Ishizaka, *Nano Lett.*, 2020, **20**, 4932.
10 M. Harb, W. Peng, G. Sciaini, C. T. Hebeisen, R. Ernstorfer, M. A. Eriksson, M. G. Lagally, S. G. Kruglik and R. J. D. Miller, *Phys. Rev. B: Condens. Matter Mater. Phys.*, 2009, **79**, 94301.
11 D. R. Cremons, D. A. Plemmons and D. J. Flannigan, *Struct. Dyn.*, 2017, **4**, 44019.

12 K. Maaz, *The Transmission Electron Microscope - Theory and Applications*, InTech, 2015.
13 D. X. Du and D. J. Flannigan, *Struct. Dyn.*, 2020, **7**, 24103.
14 C. Ophus, *Microsc. Microanal.*, 2019, **25**, 563.
15 T. Shimojima, A. Nakamura, and K. Ishizaka, in preparation.
16 A. Feist, N. Rubiano da Silva, W. Liang, C. Ropers and S. Schäfer, *Struct. Dyn.*, 2018, **5**, 14302.
17 C. Thomsen, H. T. Grahn, H. J. Maris and J. Tauc, *Phys. Rev. B: Condens. Matter Mater. Phys.*, 1986, **34**, 4129.
18 P. Ruello and V. E. Gusev, *Ultrasonics*, 2015, **56**, 21.
19 B. J. Siwick, J. R. Dwyer, R. E. Jordan and R. J. D. Miller, *Chem. Phys.*, 2004, **299**, 285.
20 C. Schinke, P. Christian Peest, J. Schmidt, R. Brendel, K. Bothe, M. R. Vogt, I. Kröger, S. Winter, A. Schirmacher, S. Lim, H. T. Nguyen and D. Macdonald, *AIP Adv.*, 2015, **5**, 67168.
21 Y. Martin, J. L. Rouviere, J. M. Zuo and V. Favre-Nicolin, *Ultramicroscopy*, 2016, **160**, 64.
22 M. Rocklin, PROC. 14th PYTHON Sci. CONF, 2015.
23 C. R. Harris, K. Jarrod Millman, S. J. van der Walt, R. Gommers, P. Virtanen, D. Cournapeau, E. Wieser, J. Taylor, S. Berg, N. J. Smith, R. Kern, M. Picus, S. Hoyer, M. H. van Kerkwijk, M. Brett, A. Haldane, J. Fernández del Río, M. Wiebe, P. Peterson, P. Gérard-Marchant, K. Sheppard, T. Reddy, W. Weckesser, H. Abbasi, C. Gohlke and T. E. Oliphant, *Nature*, 2020, **585**, 357.
24 J. Rose, *Ultrasonic Waves in Solid Media*, Cambridge University Press, Cambridge, 1999.
25 T. Van Oudheusden, E. F. De Jong, S. B. Van Der Geer, W. P. E. M. Op T Root, O. J. Luiten and B. J. Siwick, *J. Appl. Phys.*, 2007, **102**, 93501.
26 R. P. Chatelain, V. R. Morrison, C. Godbout and B. J. Siwick, *Appl. Phys. Lett.*, 2012, **101**, 81901.
27 S. Shirota, R. Krishnan, Y. Tanaka and N. Nishiguchi, *Jpn. J. Appl. Phys.*, 2007, **46**, L1025.
28 S. Danworaphong, T. A. Kelf, O. Matsuda, M. Tomoda, Y. Tanaka, N. Nishiguchi, O. B. Wright, Y. Nishijima, K. Ueno, S. Juodkazis and H. Misawa, *Appl. Phys. Lett.*, 2011, **99**, 201910.

Faraday Discussions

ROYAL SOCIETY OF CHEMISTRY

PAPER

Ultrafast dynamics in $(TaSe_4)_2I$ triggered by valence and core-level excitation

Wibke Bronsch,*[a] Manuel Tuniz,[b] Giuseppe Crupi,[b] Michela De Col,[b] Denny Puntel,[b] Davide Soranzio,[b] Alessandro Giammarino,[b] Michele Perlangeli,[b] Helmuth Berger,[c] Dario De Angelis,[a] Danny Fainozzi,[a] Ettore Paltanin,[ab] Jacopo Stefano Pelli Cresi,[a] Gabor Kurdi,[a] Laura Foglia,[a] Riccardo Mincigrucci,[a] Fulvio Parmigiani,[a] Filippo Bencivenga[a] and Federico Cilento[a]

Received 27th January 2022, Accepted 2nd March 2022

DOI: 10.1039/d2fd00019a

Dimensionality plays a key role in the emergence of ordered phases, such as charge density-waves (CDW), which can couple to, and modulate, the topological properties of matter. In this work, we study the out-of-equilibrium dynamics of the paradigmatic quasi-one-dimensional material $(TaSe_4)_2I$, which exhibits a transition into an incommensurate CDW phase when cooled to just below room temperature, namely at $T_{CDW} = 263$ K. We make use of both optical laser and free-electron laser (FEL) based time-resolved spectroscopies in order to study the effect of a selective excitation on the normal-state and on the CDW phases by probing the near-infrared/visible optical properties both along and perpendicularly to the direction of the CDW, where the system is metallic and insulating, respectively. Excitation of the core-levels by ultrashort X-ray FEL pulses at 47 eV and 119 eV induces reflectivity transients resembling those recorded when only exciting the valence band of the compound – by near-infrared pulses at 1.55 eV – in the case of the insulating sub-system. Conversely, the metallic sub-system displays relaxation dynamics which depend on the energy of photo-excitation. Moreover, excitation of the CDW amplitude mode is recorded only for excitation at a low-photon-energy. This fact suggests that the coupling of light to ordered states of matter can predominantly be achieved when directly injecting delocalized carriers in the valence band, rather than localized excitations in the core levels. Complementing this, table-top experiments allow us to prove the quasi-unidirectional nature of the CDW phase in $(TaSe_4)_2I$, whose fingerprints are detected along its c-axis only. Our results provide new insights into the symmetry of the ordered phase of $(TaSe_4)_2I$ perturbed by a selective excitation, and suggest a novel approach based on complementary table-top and FEL spectroscopies for the study of complex materials.

[a]*Elettra - Sincrotrone Trieste S.C.p.A., Strada Statale 14, km 163.5, 34149 Basovizza, Trieste, Italy. E-mail: wibke.bronsch@elettra.eu; federico.cilento@elettra.eu*

[b]*Dipartimento di Fisica, Università degli Studi di Trieste, 34127 Trieste, Italy*

[c]*Institute of Physics, Ecole Polytechnique Fédérale de Lausanne (EPFL), CH-1015 Lausanne, Switzerland*

1 Introduction

Non-equilibrium spectroscopy techniques have developed in a short period of time from avant-garde experiments to established tools for studying the properties of complex materials, delivering comprehensive datasets which combine both spectroscopic and temporal information. This allows the revealing and disentangling of electron, lattice and spin dynamics, facilitating the comprehension of cause–effect relationships and the interplay among the various degrees of freedom.[1] A common approach to gain such information is perturbing an ordered state by an intense, ultrashort light pulse, below the sample damage threshold, and tracing the material response during its relaxation back to equilibrium conditions. Especially in connection to materials displaying phase transitions, non-equilibrium spectroscopies have received great interest in order to gain insights about the driving forces responsible for the transitions. Examples include the study of correlated materials and superconductors,[2] charge density-wave (CDW) systems[3,4] and exciton-related dynamics.[5,6] In several systems, photo-induced phase-transitions[7] can switch the physical and functional properties of the material on ultrafast timescales. Optical pump–probe spectroscopy, which allows for studying the effects on the valence band structure, delivers comprehensive information about various effects, *e.g.* band gap renormalization,[8] the excitation of coherent phonons[9] or the excitation of the amplitude mode in CDW materials.[10] Depending on the ground state of the compound under scrutiny, and the intensity (and photon energy) of the photoexcitation, the system can evolve in a new state by following different pathways, which may or may not have a counterpart at equilibrium.[11,12]

X-ray pulses, in contrast to light pulses in the visible range, open up the possibility to study many-body effects triggered by the creation of a core-hole. Ultrafast changes of the optical reflectivity induced by X-ray pulses have aroused great interest as a tool for femtosecond X-ray/optical cross-correlation, which serves as a beam characterization tool for more widely spread IR-pump/X-ray probe experiments.[13,14] Less explored than the use of X-ray probes, in order to make use of the element selectivity of synchrotron and free-electron laser radiation, is the ability of intense X-ray pulses to alter the refractive index of solids on an ultrafast timescale. Because of the absorption of X-rays, in addition to the excitation of the valence band, photoelectrons and core holes are generated. The presence of core level holes induces Auger or fluorescence processes, which ultimately alter the carrier density in the conduction and valence bands and, thus, perturb the optical reflectivity.[15] Even though optical reflectivity changes induced by intense X-ray pulses offer the possibility to study ultrafast many body responses, so far the variety of materials studied by applying this technique is rather small and the interpretation and analysis of the experimental data is a recently discussed topic.[14–17]

In the present work, we report a comprehensive study on the effect of core-level and valence band excitations on the paradigmatic quasi-one-dimensional material[18] $(TaSe_4)_2I$, tracing the changes in its optical reflectivity induced by photoexcitation with ultrashort light pulses. In particular, we compare the influence of core-level and valence band excitations on the optical properties of $(TaSe_4)_2I$ single crystals along both their *a*- and *c*-axes. By doing so, we can study the effect of photoexcitation at considerably different photon energies simultaneously on a metallic and an insulating band structure.

$(TaSe_4)_2I$ has recently received significant attention due to its rich topology.[19–21] For our aim, $(TaSe_4)_2I$ constitutes an ideal model system because it allows us to compare the influence of valence band and core-level excitations on different electronic structures. Indeed, as anticipated above, in the normal state, the material shows a metallic band structure along its *c*-axis and insulating properties along the *a*-axes.[19] Fig. 1a and b show a top and a side view of the unit cell of the crystal structure, respectively. Below a critical temperature of $T_c = 263$ K, the material undergoes a phase transition into a CDW phase.[22–24] Due to the quasi one-dimensional character of the phase-transition, the periodic lattice distortion related to the CDW phase sets-in along the *c*-axis of the crystal and leaves the *a*-axes non-impacted. Schaefer *et al.*[25] first detected fingerprints of the phase transition from the metallic to the CDW phase of $(TaSe_4)_2I$ in the time-resolved optical properties when the material is excited in the

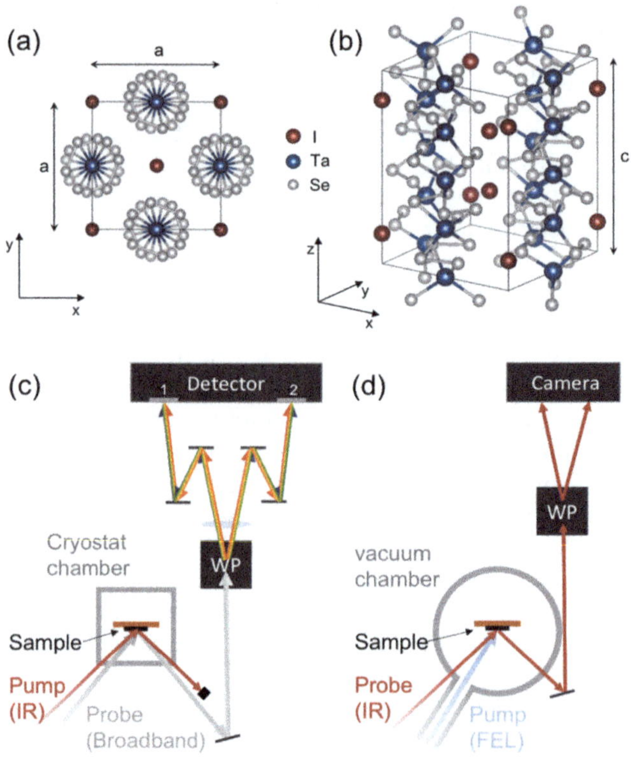

Fig. 1 Crystal structure and experimental geometries. (a) Top view of the unit cell of $(TaSe_4)_2I$. (b) Side view of the unit cell of $(TaSe_4)_2I$; the in-plane axes, *a*, and the *c* axis (direction of the quasi-1D CDW) are indicated. The crystal structure displayed in (a and b) has been produced by using VESTA. All the lattice parameters used to obtain these plots are freely accessible at https://materialsproject.org/materials/mp-30531/. (c) Setup for table-top optical pump–probe spectroscopy with the supercontinuum probe. (d) Setup for pumping with FEL light and probing at $\lambda = 800$ nm. With both setups, the *a*-axis and *c*-axis time-resolved optical properties are measured simultaneously. Note that in panels (c and d), the beams are depicted at arbitrary incident angles for simplifying the sketches. In both setups, the probe beam is impinging on the sample in near-normal incidence geometry.

near-infrared range (at 800 nm or at 1300 nm). On the contrary, the effect of core-hole excitations on the valence and conduction band structure of $(TaSe_4)_2I$ has not been investigated yet. Here, we show for the first time the direct comparison of the non-equilibrium dynamics along the different crystal axis and for different excitation schemes, fluences and temperatures.

2 Experimental setups

We studied the optical response of $(TaSe_4)_2I$ to valence band and core-level excitations by using two different setups, which are schematically shown in Figs. 1c and d, respectively. Fig. 1c illustrates the table-top optical pump–probe spectroscopy setup available at the T-ReX facility at Elettra – Sincrotrone Trieste. We worked with a pulsed Ti–Sapphire source at a repetition rate of 250 kHz. As a pump wavelength, we used the fundamental of the source centered at $\approx$800 nm, whereas for the probe beam, we used a supercontinuum beam to be able to probe changes in the reflectivity of the sample in the spectral range $\approx$ 450–900 nm. The broadband probe beam hit the sample at near-normal incidence, while the pump beam impinged on the sample with an angle of $\approx 20°$ to the surface normal. We measured the spot-size to be around 40 μm FWHM (full width at half maximum) for the probe beam and $\approx$140 μm FWHM for the pump beam. The reflected beam from the sample was lead through a Wollaston prism, which allowed for simultaneous detection of the sample response along and perpendicular to the direction of the CDW. Details about the setup and the supercontinuum source are published in ref. 26. During the measurements, we kept the sample mounted in a high vacuum chamber. The samples were glued to a copper stick attached to a cryostat, which allowed for measurements at variable temperatures down to 15 K in the case of liquid helium cooling. Fig. 1d schematically shows the setup used for core-level excitations. Free electron laser (FEL) pump and 800 nm probe experiments were performed at the FEL FERMI in Trieste.[27] Due to the opacity of the investigated samples to IR light, all experiments were performed in reflection geometry. The IR-probe beam impinged on the sample in a near-normal incidence geometry, whereas the FEL pump beam arrived with an angle of $\approx 40°$ with respect to the surface normal in the case of a photon energy of 47 eV and of 13.5° in the case of 119 eV photon energy. The reflected IR light from the sample was split by a Wollaston prism into two components, which relate to the reflectance along and perpendicular to the CDW chains. In this way, we were able to simultaneously record the optical response of the sample to the FEL excitation along the two high-symmetry axes of $(TaSe_4)_2I$. The FEL beam had a spot-size of 150 μm × 200 μm FWHM, whereas the 800 nm probe had a FWHM of 150 μm × 150 μm. The repetition rate of the FEL amounted to 50 Hz. The needle-like $(TaSe_4)_2I$ crystalline samples were synthesized by the CVD (chemical vapor deposition) technique, starting from a stoichiometric mixture of high-purity elements. Their size was in the order of $6.0 \times 0.3 \times 0.3$ mm^3.

3 Results and discussion

3.1 Optical excitation

We performed time-resolved optical spectroscopy experiments on high-quality $(TaSe_4)_2I$ crystals, in order to study the effect of photoexcitation at $h\nu_P$ = 1.55 eV on the electronic band structure of the compound, for different temperature (above and below the critical temperature $T_c \sim 263$ K) and fluence conditions. By following the sketches presented in Fig. 1, and carefully aligning the sample, the setup allowed us to simultaneously acquire the transient reflectivity variation, $\Delta R/R$, for both the *c*-axis component (*z* direction, termed in the following as the "parallel" direction, that is, the direction parallel to the quasi-1D CDW) and the *a*-axis component (*x* or *y* direction, termed the "perpendicular" direction). Spectroscopic measurements are shown in Fig. 2(a–d) as bi-dimensional maps where the $\Delta R/R$, indicated as a colorscale, is plotted *versus* the pump–probe delay, t, and probe photon energy, $h\nu$, that is: $\Delta R/R(t,h\nu)$. Panels (a and b) of Fig. 2 show the data acquired above T_c, at $T = 300$ K, for the parallel and perpendicular components, respectively. Each panel comprises two measurements acquired at different (absorbed) fluences, 800 μJ cm^{-2} and 200 μJ cm^{-2}. The extrema of the colorscale for the low-fluence measurements are set to 1/4 of the ones of the high-fluence measurements, in order to facilitate comparisons. Similarly, panels (c and d) of Fig. 2 display the data acquired well below T_c, at $T = 15$ K, for the parallel and perpendicular components, respectively. The parallel, metallic-like component and the perpendicular, insulating-like component show markedly different behaviors that will be addressed by a dedicated analysis described in the following. Briefly, the parallel component displays a positive signal which peaks in the near-infrared side of the measured spectral range and decreases smoothly towards the visible range. Conversely, the perpendicular component shows a spectrally-structured $\Delta R/R$ signal with sign changes, which is smaller in amplitude than the parallel one. Upon cooling, the spectral features become sharper, and coherent oscillations linked to the excitation of the amplitude mode of the CDW are detected along the parallel direction at low fluence. We want to point out here that the $\Delta R/R(t,h\nu)$ maps displayed in panels (a–d) of Fig. 2 have been corrected for the supercontinuum probe pulse chirp. The correction consists of the alignment of the arrival time of each spectral component of the pulse. However, in the case of the parallel component, it is clear from the data of panels (a–d) that a proper chirp correction in the vicinity of $h\nu =$ 1.55 eV ($\lambda = 800$ nm) cannot be achieved. For this reason, for the parallel component data only, the energy region 1.5–1.65 eV is excluded from the analysis. For a qualitative analysis of the response of $(TaSe_4)_2I$ to optical photoexcitation, time profiles (at constant $h\nu$) and spectral profiles (for a given delay t) are extracted from the $\Delta R/R(t,h\nu)$ maps. In panels (e and f) of Fig. 2, we present example time profiles taken at $h\nu = 1.7$ eV ($\sim$730 nm). The traces are scaled by the fluence's ratio. We observe that at $T = 300$ K, both components show a sub-linear dependence on the fluence. Conversely, at $T = 15$ K, the trend is opposite, and the signal is super-linear. For the metallic, parallel component, we observe a double-exponential decay (faster at low temperatures), while for the insulating perpendicular component, a positive plateau is reached after a fast, negative transient. The most important fluence-dependent effect is observed for the parallel component at $T = 15$ K, where the coherent oscillations are almost completely washed out after photoexcitation at 800 μJ cm^{-2}, signaling that the CDW phase can be melted by impulsive photoexcitation. The spectral profiles extracted at $t =$ 200 fs are reported in panels (g and h) of Fig. 2, and confirm that the spectral features in the perpendicular component become sharper at lower temperature. The results of the analysis of the amplitude mode, as obtained from the dataset reported in Fig. 2c at 200 μJ cm^{-2} fluence, are reported in Fig. 3. In particular, the map $I(\omega,h\nu)$ shows the spectral evolution of the FFT (fast Fourier transform) of the

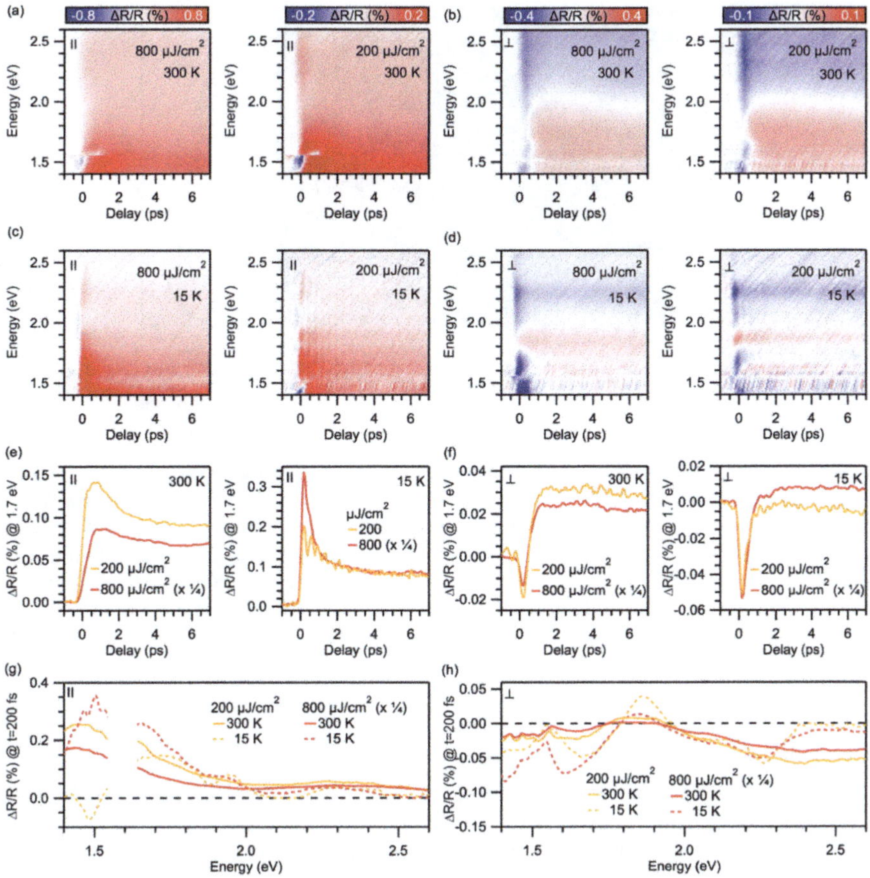

Fig. 2 Time-resolved optical spectroscopy of $(TaSe_4)_2I$. Time-and-frequency resolved normalized reflectivity variations at $T = 300$ K, (a) for the parallel direction (c-axis) and (b) for the perpendicular direction (a-axis). Panels (c and d) refer to the case $T = 15$ K. All panels comprise a measurement acquired at an absorbed fluence of 800 μJ cm^{-2} and at an absorbed fluence of 200 μJ cm^{-2}. The extrema of the colorscale of the measurements acquired at low fluence are set to 1/4 of the ones of the measurements at high fluence for ease of comparison. All the measurements acquired at the same fluence and along the same direction share the same colorscale. The colorscales for the perpendicular direction range are half of the ones for the parallel direction for better visibility of the changes. (e) Time cuts extracted at $h\nu = 1.7$ eV ($\lambda = 730$ nm) for the parallel direction at $T = 300$ K and $T = 15$ K for the two fluence values considered. (f) Same as (e) but for the perpendicular direction. (g) Spectral profiles extracted at $t = 200$ fs for the parallel direction, at $T = 300$ K and $T = 15$ K, for the two fluence values considered. (g) Same as (f) but for the perpendicular direction.

residuals (coherent component only) of the $\Delta R/R(t,h\nu)$ map. The residuals have been computed by subtracting from each time profile at a constant photon energy a fit of the same trace obtained with a double-exponential decay function (convoluted with a Gaussian to account for the finite time resolution), which represents the incoherent response. By stacking the FFT of the residuals, we obtain the map presented in Fig. 3, which displays the evolution of the oscillation

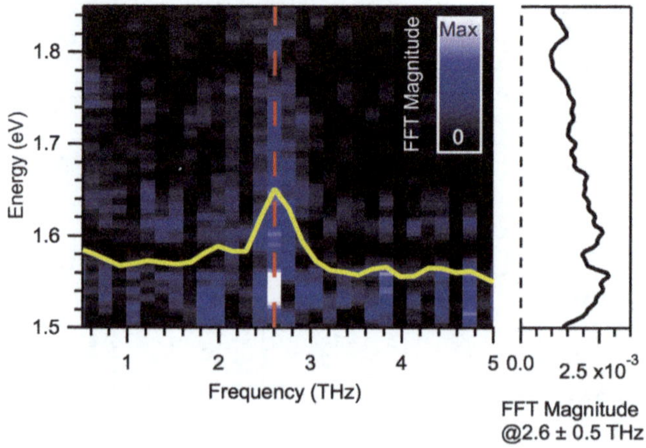

Fig. 3 $I(\omega,h\nu)$ map displaying the FFT of the coherent component extracted from the $\Delta R/R(t,h\nu)$ map acquired at $T = 15$ K and 200 μJ cm^{-2} fluence. The photon energy range 1.5–1.85 eV is considered, where the coherent oscillations associated with the excitation of the amplitude mode are more pronounced. The yellow curve, obtained by integrating the FFT amplitude over the full photon energy range considered, indicates the oscillation frequency. It peaks at 2.6 THz, as marked by the red dashed line. The black curve shows the photon-energy-dependent FFT amplitude, as integrated in a 0.5 THz-wide interval centered at 2.6 THz.

frequency as a function of the probe photon energy. A peak at 2.6 THz is clearly detected, which corresponds to the amplitude mode of the CDW in $(TaSe_4)_2I$. This finding is in agreement with earlier results reported in ref. 25 for $\lambda = 800$ nm. In our case, we gain information about the spectral dependence of the coherent modulation, which extends in a wide spectral range from the near-infrared beyond $\approx$2 eV (*cf.* data shown in Fig. 2c; 200 μJ cm^{-2}).

In order to gain quantitative insights into the microscopic origin of the $\Delta R/R(h\nu)$ profiles, we developed a model for the dielectric function of $(TaSe_4)_2I$ and used a differential approach to find which of its components are modified by photoexcitation. The prerequisite for this analysis is the detailed comprehension of the optical properties at equilibrium. The reflectivity of $(TaSe_4)_2I$ at $T = 300$ K was measured by Berner *et al.*[28] We digitized the reflectivity curves for both the parallel and perpendicular components, which are presented in Fig. 4a. The solid lines are the fit to the reflectivity curves of a Drude–Lorentz model for the dielectric function. The reflectivity at near-normal incidence is computed using the expression $R = |(1-\sqrt{\varepsilon})/(1+\sqrt{\varepsilon})|^2$, where $\varepsilon = \varepsilon(h\nu)$ is the model complex dielectric function. The expression for $\varepsilon(h\nu)$ in terms of the Drude and Lorentz optical oscillators is extensively discussed and reported in ref. 29. The reflectivity curve for the direction parallel to the CDW clearly shows the presence of a plasma edge, which is instead absent in the perpendicular component. Hence, for the dielectric function model, we used the Drude term only for the parallel component. Additionally, a number of Lorentz oscillators were added in order to obtain a satisfactory fit of the reflectivity curves, using the smallest number of oscillators for minimizing the χ^2 of the fit. For the parallel component, four Lorentz oscillators were required; they have central wave-numbers of 12 300 cm^{-1},

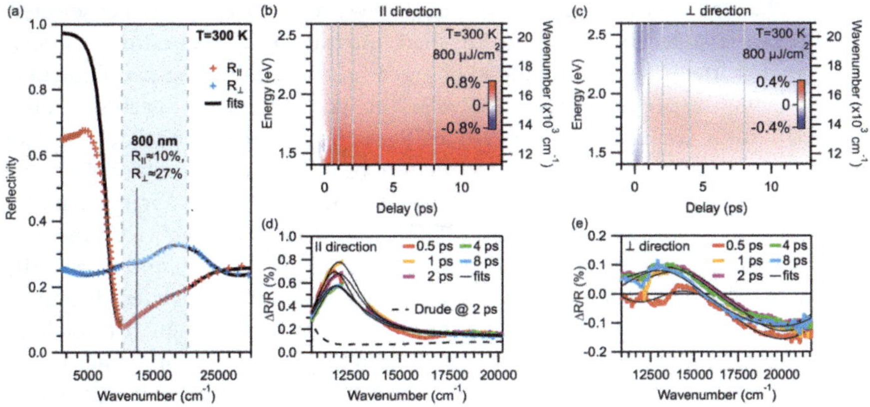

Fig. 4 Analysis of the time-resolved optical spectroscopic data acquired at $T = 300$ K. (a) Reflectivity along the crystal a- and c-axes at equilibrium, reported by Berner *et al.*[28] The spectral range probed by the setup sketched in Fig. 1c is emphasized. The photon energy $h\nu = 1.55$ eV (12 500 cm^{-1}), corresponding to $\lambda = 800$ nm, is marked. This corresponds to the pump photon energy of the setup in Fig. 1c and the probe photon energy of the setup in Fig. 1d. The fits to the two reflectivity curves have been obtained as described in the main text. Panel (b) displays the result of the fit of a Drude–Lorentz model to the data acquired along the parallel direction. The map is obtained by stacking the fits performed at each pump–probe delay. Panel (c) displays the same quantity, but for the perpendicular component. The details of the analysis are described in the main text. The vertical gray lines shown in panels (b and c) mark a few selected time delays at which the fit profiles are extracted and compared to the corresponding spectral profile of the $\Delta R/R(t,h\nu)$ map. The fit profiles (displayed as solid black lines) are reported in panels (d and e) for the parallel and perpendicular components, respectively, while the spectral profiles are displayed with colored curves indicating the time delay. An overall good agreement between data and model is achieved, both in the vicinity of t_0 and at large time delays. For the spectral profiles of the parallel direction, the energy interval (1.5–1.65 eV) has been excluded from the fit because of the difficulty in correcting the probe pulse chirp in the vicinity of $h\nu = 1.55$ eV.

17 000 cm^{-1}, 22 050 cm^{-1} and 27 880 cm^{-1}. For the perpendicular component, a total of six Lorentz oscillators was required, having central wave-numbers of 4500 cm^{-1}, 7000 cm^{-1}, 10 800 cm^{-1}, 15 500 cm^{-1}, 17 000 cm^{-1} and 28 200 cm^{-1}. Once the optical properties at equilibrium were described by a model dielectric function (one for each direction), the time-resolved optical properties were analyzed by computing the $\Delta R/R(h\nu)$ as $\Delta R/R(h\nu) = (R_{\text{neq}}(h\nu) - R_{\text{eq}}(h\nu))/R_{\text{eq}}(h\nu)$, where $R_{\text{eq}}(h\nu)$, referring to the reflectivity at equilibrium, was calculated from the model dielectric function described above, and the non-equilibrium reflectivity $R_{\text{neq}}(h\nu)$ was calculated from a 'copy' of the equilibrium model in which a few parameters were allowed to vary when the quantity $\Delta R/R(h\nu)$ was fitted to the spectral profiles. We applied this analysis to the results obtained at $T = 300$ K, for consistency with the experimental conditions of the equilibrium spectroscopy, for excitation at 800 μJ cm^{-2} fluence. In the analysis, the parallel and perpendicular dielectric functions were considered to be independent. The results are presented in Fig. 4. Panels (b and c) of Fig. 4 show the results of the fitting procedure for the parallel and perpendicular components, respectively. They should be compared to the corresponding datasets reported in panels (a and b) of Fig. 2. In order to

discuss the agreement of the fit to the data, a few profiles are extracted at selected fixed delays, marked by the vertical dashed lines superimposed onto the maps. They are reported in panels (d and e) of Fig. 4 for the parallel and perpendicular components, respectively, together with the corresponding spectral profile extracted from the datasets reported in Fig. 2 at the same time delays. The agreement is satisfactory, both in the vicinity of t_0 and at large time delays, for both components. The fittings have been obtained as follows: for the parallel component, the Drude term (plasma frequency and scattering rate) and the oscillator at 12 300 cm^{-1} (plasma frequency, central frequency and linewidth) were set as free parameters. Their temporal evolution and associated dynamics are shown in panels (a–e) of Fig. 5. For the perpendicular component, the analysis reveals that three Lorentz oscillators are affected by photoexcitation, in particular those at 10 800 cm^{-1}, 15 500 cm^{-1} and 17 000 cm^{-1}. The number of parameters that have to be modified for a satisfactory fit is five, and in particular they are the central frequency and the plasma frequency of the first, the plasma frequency of the second, and the plasma frequency and linewidth of the third. Their time evolution is displayed in panels (f–j) of Fig. 5.

Before approaching the discussion of the dynamics, it is worth noting that some parameters show a non-trivial time-evolution in the vicinity of time-zero, which we associate with the small imperfections in the correction of the supercontinuum pulse chirp. For this reason, we will focus mainly on the discussion of

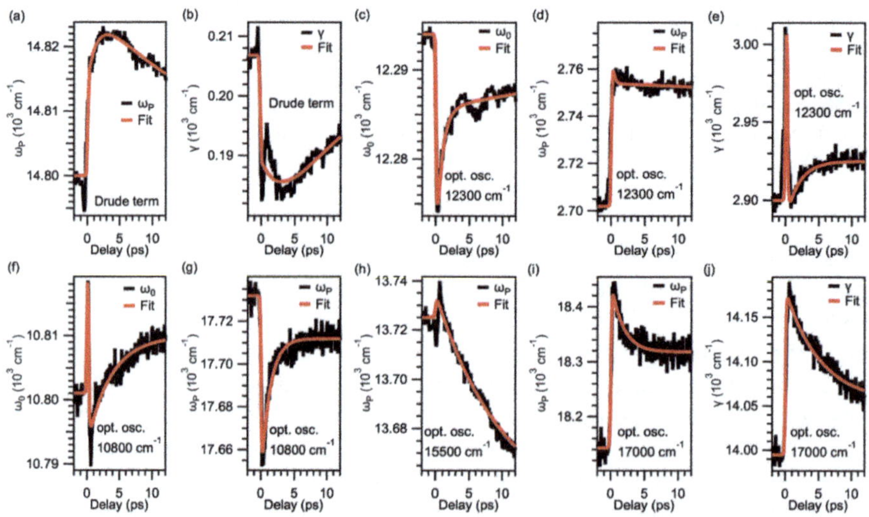

Fig. 5 Dynamics of the parameters of the dielectric function, as obtained from the fitting procedure described in the main text. Panels (a–e) refer to the parameters of the dielectric function for the parallel direction optical properties. Panels (f–j) refer instead to the perpendicular direction optical properties. (a) Plasma frequency of the Drude term. (b) Scattering rate of the Drude term. (c) Central wave-number of the optical oscillator at 12 300 cm^{-1}. (d) Plasma frequency of the optical oscillator at 12 300 cm^{-1}. (e) Linewidth of the optical oscillator at 12 300 cm^{-1}. (f) Central wave-number of the optical oscillator at 10 800 cm^{-1}. (g) Plasma frequency of the optical oscillator at 10 800 cm^{-1}. (h) Plasma frequency of the optical oscillator at 15 500 cm^{-1}. (i) Plasma frequency of the optical oscillator at 17 000 cm^{-1}. (j) Linewidth of the optical oscillator at 17 000 cm^{-1}.

the dynamics on the picosecond timescale. The knowledge of which parameter of the dielectric function is altered by photoexcitation, and how, allows us to understand the effect of photoexcitation at the microscopic level. For the parallel component, photoexcitation affects the free carriers in the conduction band, which are described by the Drude term. In particular, the plasma frequency increases, signaling that extra carriers are injected into the conduction band. This is accompanied by a reduction of the scattering rate. This is a counter-intuitive effect (heating typically broadens the plasma edge, *i.e.*, the scattering rate increases) that requires more detailed investigations in order to determine the reason behind it. The dielectric function analysis allows us to isolate the effect of the modification of the Drude term alone on the $\Delta R/R$ signal. We did this for $t = 2$ ps by picking up the values for the Drude term in the non-equilibrium dielectric function, as provided by the fit. The resulting $\Delta R/R(t = 2\ \text{ps}, h\nu)_{\text{Drude}}$ curve is shown in Fig. 4d. It shows that the contribution of the Drude term to the $\Delta R/R$ in the near-IR/visible spectral range is a nearly-flat positive signal. Hence, we argue that the large, positive signal in the near-IR spectral region is due to the modification of a Lorentz oscillator (namely, at 12 300 cm^{-1} or 1.52 eV) accounting for interband optical transitions. We can speculate on the origin of this optical transition by consulting the electronic band structure of $(TaSe_4)_2I$, as reported in ref. 19 and 21. Despite the fact that detailed calculations of the joint density-of-states (JDOS) for $(TaSe_4)_2I$ are not presently available in the literature, it is clear that optical transitions at photon energies in the order of or larger than $\approx$1.5 eV link the quasi-continuum of nearly-flat bands found below the Fermi level (starting at an energy 0.5–0.6 eV below E_F) to another quasi-continuum of nearly-flat bands located above the Fermi level (starting at an energy 0.6–0.7 eV above E_F). This particular band structure is the reason why, for both the parallel and perpendicular components, there is a large number of optical oscillators at rather high photon energies (overall from 10 800 cm^{-1} to 28 200 cm^{-1}, that is, with photon energies in excess of $\approx$1.3 eV) that links these two sets of bands. The other two oscillators at 4500 cm^{-1} and 7000 cm^{-1}, as determined from the analysis of the perpendicular component of the optical properties, are likely linked to optical transitions from the set of bands below E_F to the states near E_F and from the states at E_F to the set of bands above E_F. Because of the photon-energy range accessible in our time-resolved experiment, we can infer little about the contribution to the $\Delta R/R$ from these two low-energy optical transitions. However, from the analysis of the $\Delta R/R$ along the perpendicular direction, we can learn interesting details. From what was said above, it is not surprising that the only optical oscillator showing a modification of its central wave-number is the one with the lowest central wave-number (ω_0, equal to 10 800 cm^{-1} or 1.34 eV) among the high-energy ones. Indeed, this energy acts as an effective energy-gap between the two sets of bands. Its transient modifications signify a sort of band-gap renormalization. After a first ultrafast increase, ω_0 falls below the initial value, signalling a reduction of the effective band gap (this effect is responsible for the negative $\Delta R/R$ signal observed just after t_0), and then overshoots the initial value on a timescale of a few picoseconds. The fact that the plasma frequencies of three oscillators (10 800 cm^{-1}, 15 500 cm^{-1}, 17 000 cm^{-1}) all vary indicates that photoexcitation triggers charge-transfer processes among different bands.

Finally, we comment on the timescales of the recorded modifications of the parameters, reported in Fig. 5, starting from the parallel direction terms. All the traces have been fitted with one or two exponential decays convoluted with a Gaussian function accounting for the time resolution, as also shown in Fig. 5. After a slow buildup, the modification of the plasma frequency of the Drude term recovers on a timescale of (22 ± 7) ps (the large uncertainty is due to the pump–probe delay range available). The scattering rate also shows a slow-buildup, and recovers its equilibrium value on a slightly shorter timescale, (12 ± 7) ps. The dynamics of the parameters of the 12 300 cm^{-1} interband oscillator are different, and display a faster relaxation followed by a plateau-like decay. The central-frequency relaxes in (900 ± 200) fs (with a second, slower decay of (50 ± 10) ps). The plasma frequency shows a plateau-like behavior only, while the linewidth displays complex dynamics. For what concerns the time-evolution of the perpendicular direction terms, we have already qualitatively described the evolution of the central wave-number, ω_0, of the oscillator at 10 800 cm^{-1}. Quantitatively, after a first transient, which is faster than the time resolution and likely associated with the imperfections in the correction of the probe pulse chirp, ω_0 evolves with an exponential time-constant of (3.75 ± 0.15) ps. The plasma frequencies of the oscillators at 10 800 cm^{-1} and 17 000 cm^{-1} decay with decays of (1.3 ± 0.15) ps and (1.8 ± 0.15) ps, respectively, then they converge to a plateau. The plasma frequency of the oscillator at 15 500 cm^{-1} instead converges to a value smaller than that at equilibrium in ≈40 ps. Finally, the linewidth of the oscillator at 17 000 cm^{-1} shows a time-constant of (5.3 ± 0.15) ps.

3.2 X-ray excitation

After having presented the results obtained for an optical pumping, which can only excite the valence band, we compare them with the non-equilibrium data obtained for FEL excitation, which, in addition to the valence band excitation, enables the derivation of an initial excitation mechanism involving a core-hole state. These experiments were performed with a single-color probe beam at 800 nm, as schematically shown in Fig. 1d. Fig. 6a shows the differential reflectivity data obtained at room temperature along the direction of chain formation (metallic direction), $R_{\parallel}$ (black open circles), and perpendicular to this direction (insulator direction), $R_{\perp}$ (red open circles). The shaded areas mark the standard deviation of the signal for independent measurements. For an absorbed X-ray fluence of 220 μJ cm^{-2}, $R_{\parallel}$ shows an increase in reflectivity of about 2%. Note that the FEL flux was estimated from the nominal transmission of the beamline. The maximum change is reached within the first 2 ps after excitation and is followed by a slow decay. On the contrary, for $R_{\perp}$, we observe a rapid reduction of the reflectivity at 800 nm that reaches its minimal value around 160 fs after excitation and a sign change after *ca.* 1.5 ps. The maximum increase of reflectivity is reached after *ca.* 3 ps for the highest fluences and then decays towards equilibrium on the timescale of several ps. We can compare the general line shape of the transients to former works discussing X-ray induced optical reflectivity for different types of materials.[15] It can be noted that the division of the material response into a fast component (<1 ps) and a second, slower component (several ps) is typically observed in insulating or semiconducting materials. Eckert *et al.*[14] discussed the changes in the refractive index in the case of X-ray excited Si_3N_4 and concluded that the fast response is given by the modification of both the real and imaginary parts of the refractive index, whereas the ps-behavior is dominated by real-index changes. They furthermore observed

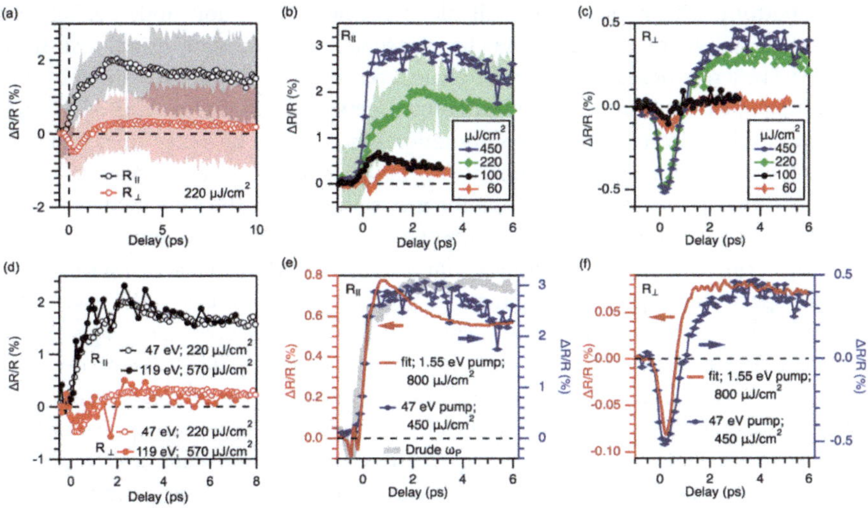

Fig. 6 FEL-induced optical reflectivity changes. (a) Reflectivity changes for $R_{\parallel}$ (black open circles) and $R_{\perp}$ (red open circles) at RT measured for an absorbed fluence of 220 μJ cm^{-2} and an excitation energy of 47 eV. The shaded areas mark the standard deviations of the signal. The FEL flux was estimated from the nominal transmission of the beamline. (b) Fluence series measured for $\Delta R_{\parallel}/R_{\parallel}$ for a pump photon energy of 47 eV. (c) Fluence series measured for $\Delta R_{\perp}/R_{\perp}$ for a pump photon energy of 47 eV. (d) Comparison of the transient changes of $R_{\parallel}$ and $R_{\perp}$, respectively, for two different excitation photon energies. (e and f) Comparisons of the FEL induced reflectivity changes (blue markers) with the changes expected for a probe photon energy of 1.55 eV after excitation with 1.55 eV (red curve) in the parallel and perpendicular directions, respectively. The red traces for the optical excitation data are extracted from the maps shown in Fig. 4b and c. As a guide to the eye, we include the temporal evolution of the Drude plasma frequency (grey shaded curve) in panel (e).

that the sensitivity for probing the fast or slow component depends strongly on the incident angle of the probe beam. In contrast, for metallic materials, such as gold, no long timescale component was observed.[30]

In the case of $R_{\perp}$, we measured along a direction of the $(TaSe_4)_2I$ crystal band structure where it has an energy-gap of *ca.* 1.3 eV,[19] which is lower than the probe photon energy of 1.55 eV. A similar outcome as in our case was shown for GaAs by Gahl *et al.*[13] and Krupin *et al.*[30] In this semiconducting material, they also observed an initial drop in the reflectivity, followed by a zero crossing, and a similar fluence dependence to that which we measure for $(TaSe_4)_2I$ along the *a*-axis, showing a simultaneous decrease of the dip intensity and an overall decrease in reflectivity on the ps scale with decreasing fluence. Gahl *et al.* fitted the time-scale for the initial drop in reflectivity to $\tau = (160 \pm 44)$ fs, which is similar to our observation in the case of $(TaSe_4)_2I$. The interpretation given for the fast drop of the reflectivity in GaAs after X-ray excitation is assigning it as a consequence of a band gap narrowing effect.[15] In the case of band gap narrowing, the number of valence band electrons that can be excited to the conduction band by the probe pulse is enhanced. In this way, the absorption increases and the reflectivity decreases. After the initial population of excited carriers in the conduction band, thermalization of the hot electrons sets in, leading to a filling of the conduction band bottom. The band filling will block further absorption and, hence, the reflectivity will increase again. The characteristic transient of an initial fast drop followed by a net increase of the reflectivity after few ps is, hence, explained by the simultaneous contribution of two different effects: band gap narrowing and band filling. Finally, electron–hole recombination leads to the decrease in reflectivity towards the equilibrium value on longer timescales. As Durbin[15] points out, there are also other possible explanations for the initial dip in the reflectivity besides band gap narrowing, which is a pure effect of the valence/conduction band structure and, hence, an argument that comes directly from a comparison to optical spectroscopy, but does ignore the presence of core holes in the case of X-ray excitation. In the case of the semiconductor GaAs, he describes the significant hole-density induced by X-ray excitation in analogy to semiconductor doping, which would also induce band gap narrowing effects. Furthermore, he mentions effects caused by an interference of the effects that both X-rays and optical pulses have on the modification of the density of excited conduction electrons. Note that the very same shape of the transient after excitation was also observed in the large band-gap insulator Si_3N_4, where the net positive reflectivity on the ps-scale was explained by the formation of metastable electron–hole bound states inside the band gap.[16] For a deeper analysis of our data, one would need to model the modification of the dielectric function due to the X-ray excitation, as was done by Tkaschenko *et al.*[17] for the example of some semiconducting materials using the XTANT code developed at DESY, which is an ongoing field of research.

Also, the shape of the transients measured for $R_{||}$, which should reflect the metallic character of the sample along this direction, resembles best the expected curves for X-ray excited metals, which was described by Durbin[15] for gold. In the case of gold, an abrupt increase of the reflectivity is measured and there is a subsequent decay towards equilibrium within few picoseconds.[15,30] The decay times that we observe for the case of $(TaSe_4)_2I$ are significantly longer, but nonetheless show a strong fluence dependence. The influence of the X-ray fluence on the optical reflectivity changes in gold was not discussed in the former mentioned works and can, hence, not be used for comparison. As Durbin states, the reason for the observed X-ray induced reflectivity changes in the case of gold is still under discussion. In the case of gold, he discusses the inelastic scattering processes of the photoelectrons, which promote d-electrons into the unoccupied parts of the s–p bands, and argues that the Drude reflectivity does not change by increasing the number of conduction electrons.

The measurements shown so far were obtained using an FEL wavelength of 26 nm, which corresponds to a photon energy of ≈ 47 eV. Referring to ultraviolet photoemission studies, with a photon energy of 47 eV, electrons from the Ta 4f and 5p states will be emitted due to direct photoemission, whereas by also using higher photon energies, the I 4d and Se 3d states can be addressed.[31] Furthermore, Othake *et al.*[31] observed that using a photon energy of 48 eV results in an enhanced partial cross-section for the detection of Ta 5d valence electrons in the angle-integrated UPS spectra, which would have an impact on the optical reflectivity. Fig. 6d shows the comparison of photoexcitation with photon energies of 47 eV and 119 eV and comparable amount of photons per pulse. Both transients look very similar. Even though we can address further core levels with the higher photon energy and probably excite less Ta 5d electrons, we do not observe a significant impact on the conduction and valence band structure.

The overall shape of the reflectivity changes observed for X-ray excitation resembles our observations from optical pump–probe spectroscopy. Fig. 6e and f show a direct comparison of the traces of the reflectivity changes along the two directions for excitation with 47 eV photons and an absorbed fluence of 450 μJ cm^{-2} (blue markers) and the extrapolated behavior observed for an excitation energy of 1.55 eV photon energy and an absorbed fluence of 800 μJ cm^{-2} (red traces) for the same probe photon energy. The traces for the 1.55 eV pump case are extracted from the maps shown in Fig. 4b and c. The amplitude of the recorded $\Delta R/R$ signal is much higher in the case of the X-ray pump energy (*cf.* right axes) than for near-IR excitation. A fluence of 450 μJ cm^{-2} at 47 eV corresponds to a number of absorbed photons of 6×10^{13} photons per cm^2 with an attenuation depth of roughly 12 nm. In comparison, the total number of absorbed photons at 1.55 eV for a fluence of 800 μJ cm^{-2} amounts to 3×10^{15} photons per cm^2 with an attenuation depth in the order of 130 nm (this is an average of the values for the parallel and perpendicular direction, which are different in the optical range). We observe maximal changes in the reflectivity of 3% in the parallel direction and ±0.5% in the perpendicular direction for FEL excitation, whereas they are ≤1% and ≤0.01% for near-IR excitation. Since in both cases the probe-beam was hitting the sample at near-normal incidence, we ascribe these variations to the difference in the pump wavelength rather than to experimental geometry effects. Considering that the number of absorbed photons in the case of the near-IR excitation is much larger than in the case of FEL excitation, and that this difference is not equilibrated by the ratio in photon energy, we cannot explain by intuitive arguments the reason why the X-ray excitation triggers larger reflectivity modifications. Even more strikingly, the X-ray induced reflectivity modifications are larger, despite the fact that 90% of the optical-probed volume is not pumped by X-rays because of the different penetration depths. The ability to only probe the pumped layer would, we suppose, further increase the magnitude of the recorded signal, however, a careful estimation of this effect would require a careful simulation of the photoexcitation mechanism. Regarding the dynamics, the rising edge of the initial change in reflectivity is identical for both types of excitations. This is reflected along both measurement directions. Since the core-hole lifetimes are typically in the order of only a few fs, the time-resolution in our experiments does not allow for distinguishing if the pump pulse directly accesses the conduction band, or if secondary processes after a core-hole generation modify the carrier density in the conduction band. Concentrating on the perpendicular direction, we observe a slightly earlier appearing compensation point of the band-gap narrowing and band-filling processes in the case of optical excitation. However, the traces along the perpendicular direction are very similar in both experiments. In contrast to the parallel direction, we see different decay dynamics of the enhanced reflectivity due to the excitation. In the case of X-ray excitation with increasing fluence, we observe a second, slower timescale for reaching the maximal change in reflectivity (*cf.* green curve in Fig. 6b), or a plateau-like behavior (*cf.* blue curve in Fig. 6b) that decays after only a few picoseconds. On the contrary, for valence-band excitation with 1.55 eV photons, we reach the maximal change in reflectivity in less than 1 ps and observe an intermediate initiating decay until it reaches *ca.* 75% of the highest change and saturates. This behavior initially resembles the temporal evolution of the plasma frequency of the Drude term, which was extracted from the optical spectroscopy experiments. As demonstrated by the grey shaded curve in Fig. 6e, the Drude plasma frequency perfectly describes the dynamics of the first 4 ps after FEL excitation. The reflectivity then decreases, which means that some Lorentz oscillators must be affected as well as a consequence of the initial core-hole excitation. In the optical spectroscopy data, the Drude term gives a broadband offset to the reflectivity, whereas the fast dynamics are dominated by the oscillator at 12 300 cm^{-1}, to which we are close to resonance when using a pump photon energy of 1.55 eV. This finding suggests that the core-level excitation affects mainly the free carriers (metallic conduction electrons), which subsequently decay in the conventional way through electron–phonon scattering processes, while the excitation of interband optical transitions is somehow less effective in a metallic environment by secondary processes after a core-level excitation.

We were furthermore interested in investigating if the core-level excitation can couple to the charge density wave, which is present at sample temperatures below 263 K. Optical excitation of the valence band structure leads to the excitation of the amplitude mode (the phonon mode coupled to the structural distortion induced by the CDW), and a faster initial relaxation of the reflectivity change of $R_{\parallel}$, as discussed in Section 3.1. Both effects are not present when exciting with an FEL pulse, as can be seen in the direct comparison of the $\Delta R/R$ transients obtained at RT and at 37 K shown in Fig. 7. One could speculate that this is a consequence of the electron-lattice thermalization taking longer than the period of the amplitude mode, which amounts to *ca.* 400 fs. Ziaja *et al.*[32] developed a model to fit the transients reported by Gahl *et al.*[13] and Krupin *et al.*[30] and extract the electron-lattice thermalization time in GaAs, depending on photon energy and fluence, based on the recorded changes in optical reflectivity. The timescales were found to be in the few-ps range. As we have discussed above, the transients recorded for the perpendicular direction resemble those measured for GaAs, because both samples have a similar spacing between the valence and the conduction band, but the sign-change in the $\Delta R/R$ signal happens earlier. Also, we do not observe a dramatic difference for the cases of near-IR or FEL pumping along the perpendicular direction, which could give a clear hint regarding the differing electron-lattice thermalization timescales. Careful theoretical modelling of the obtained data is required to give a deeper insight into this question.

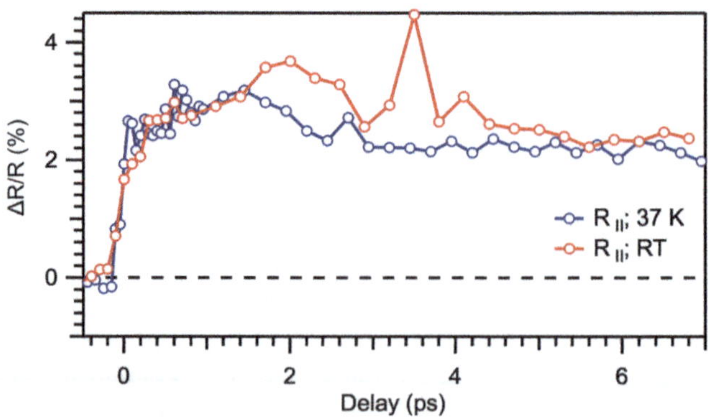

Fig. 7 Comparison of the transient changes of $R_{\parallel}$ above and below the critical temperature for charge density formation for an excitation energy of 119 eV.

4 Conclusions

The comparison of optical reflective changes induced by optical and X-ray excitations for the quasi one-dimensional material $(TaSe_4)_2I$ along its *a*- and *c*-axes showed that the qualitative response along the insulator direction does not depend on the excitation pathway, whereas along the chain direction, the decay mechanisms of the initial reflectivity change are different. Furthermore, we are not able to detect fingerprints of the charge density wave phase in the $\Delta R/R$ signal when using core-level excitations. At the present stage, we cannot conclude whether this is due to the X-rays destroying the CDW or due to their excitation pathway not coupling directly to the CDW. Thus, we think that our results could be of high interest for theoretical studies on X-ray induced optical reflectivity changes, which is a rather new field of investigation.

An interesting future experiment to explore in more detail the variations of the dielectric function after X-ray excitation would be to set up an X-ray pump/optical broadband probe experiment. Data gained from such an experiment would allow for performing an analysis as discussed in Section 3.1 for optical pump–probe spectroscopy and, hence, to identify the modified oscillators after excitation. This would provide insight into the reason for the different de-excitation dynamics we observe for valence- and core-level excitation along the *c*-axis in $(TaSe_4)_2I$. Finally, the study of the influence of photoexcitation *via* different pathways on the non-equilibrium properties of complex materials is an emerging field that will provide new insights about the functional properties of metallic and insulating phases and their possible control.

Author contributions

F. C. designed and performed the time-resolved table top optical spectroscopy experiments. M. T., G. C., M. De C. and F. C. contributed to the analysis of the time-resolved purely optical spectroscopy data. W. B. and F. C. had the experimental lead of the experiments performed at FERMI to study FEL-induced optical reflectivity changes. H. B. grew the samples. G. K. took care of the optical setup at the TIMER endstation. W. B., M. T, M. de C., D. P., D. S., A. G., M. P., D. De A., D. F., E. P., S. P. C., R. M., L. F., F. B. and F. C. performed the experiments on FEL-induced optical reflectivity changes. W. B. analyzed the data gained in the FEL experiments. F. P. contributed to the scientific discussion. W. B. and F. C. wrote the original draft of the paper, which was proofread by all co-authors.

Conflicts of interest

There are no conflicts to declare.

Acknowledgements

We acknowledge the contribution of Simone Peli to the original beam-time proposal, allowing us to perform the experiments on X-ray induced optical reflectivity measurements. Furthermore, we thank Alberto Crepaldi for the fruitful discussions and his support in setting up the collaboration with the sample growers. E. Paltanin acknowledges the contribution of the European

Union's Horizon 2020 research and innovation program under the Marie Sklodowska-Curie grant agreement no. 860553.

Notes and references

1 C. Giannetti, M. Capone, D. Fausti, M. Fabrizio, F. Parmigiani and D. Mihailovic, *Adv. Phys.*, 2016, **65**, 58–238.

2 C. Giannetti, F. Cilento, S. D. Conte, G. Coslovich, G. Ferrini, H. Molegraaf, M. Raichle, R. Liang, H. Eisaki, M. Greven, A. Damascelli, D. van der Marel and F. Parmigiani, *Nat. Commun.*, 2011, **2**, 353.

3 S. Hellmann, T. Rohwer, M. Kalläne, K. Hanff, C. Sohrt, A. Stange, A. Carr, M. M. Murnane, H. C. Kapteyn, L. Kipp, M. Bauer and K. Rossnagel, *Nat. Commun.*, 2012, **3**, 1069.

4 D. T. Payne, P. Barone, L. Benfatto, F. Parmigiani and F. Cilento, 2020, https://arxiv.org/abs/2010.09826.

5 S. Mor, M. Herzog, D. Golež, P. Werner, M. Eckstein, N. Katayama, M. Nohara, H. Takagi, T. Mizokawa, C. Monney and J. Stähler, *Phys. Rev. Lett.*, 2017, **119**, 086401.

6 S. Mor, V. Gosetti, A. Molina-Sánchez, D. Sangalli, S. Achilli, V. F. Agekyan, P. Franceschini, C. Giannetti, L. Sangaletti and S. Pagliara, *Phys. Rev. Res.*, 2021, **3**, 043175.

7 D. Wegkamp and J. Stähler, *Prog. Surf. Sci.*, 2015, **90**, 464–502.

8 G. Jnawali, Y. Xiang, S. M. Linser, I. A. Shojaei, R. Wang, G. Qiu, C. Lian, B. M. Wong, W. Wu, P. D. Ye, Y. Leng, H. E. Jackson and L. M. Smith, *Nat. Commun.*, 2020, **11**, 3991.

9 D. Soranzio, M. Peressi, R. J. Cava, F. Parmigiani and F. Cilento, *Phys. Rev. Res.*, 2019, **1**, 032033.

10 A. Mann, E. Baldini, A. Odeh, A. Magrez, H. Berger and F. Carbone, *Phys. Rev. B*, 2016, **94**, 115122.

11 A. Kogar, A. Zong, P. E. Dolgirev, X. Shen, J. Straquadine, Y.-Q. Bie, X. Wang, T. Rohwer, I.-C. Tung, Y. Yang, R. Li, J. Yang, S. Weathersby, S. Park, M. E. Kozina, E. J. Sie, H. Wen, P. Jarillo-Herrero, I. R. Fisher, X. Wang and N. Gedik, *Nat. Phys.*, 2020, **16**, 159–163.

12 F. Zhou, J. Williams, S. Sun, C. D. Malliakas, M. G. Kanatzidis, A. F. Kemper and C.-Y. Ruan, *Nat. Commun.*, 2021, **12**, 566.

13 C. Gahl, A. Azima, M. Beye, M. Deppe, K. Döbrich, U. Hasslinger, F. Hennies, A. Melnikov, M. Nagasono, A. Pietzsch, M. Wolf, W. Wurth and A. Föhlisch, *Nat. Photonics*, 2008, **2**, 165–169.

14 S. Eckert, M. Beye, A. Pietzsch, W. Quevedo, M. Hantschmann, M. Ochmann, M. Ross, M. P. Minitti, J. J. Turner, S. P. Moeller, W. F. Schlotter, G. L. Dakovski, M. Khalil, N. Huse and A. Föhlisch, *Appl. Phys. Lett.*, 2015, **106**, 061104.

15 S. M. Durbin, *AIP Adv.*, 2012, **2**, 042151.

16 F. Casolari, F. Bencivenga, F. Capotondi, E. Giangrisostomi, M. Manfredda, R. Mincigrucci, E. Pedersoli, E. Principi, C. Masciovecchio and M. Kiskinova, *Appl. Phys. Lett.*, 2014, **104**, 191104.

17 V. Tkachenko, N. Medvedev and B. Ziaja, *Appl. Sci.*, 2016, **6**, 238.

18 P. Monceau, *Adv. Phys.*, 2012, **61**, 325–581.

19 W. Shi, B. J. Wieder, H. L. Meyerheim, Y. Sun, Y. Zhang, Y. Li, L. Shen, Y. Qi, L. Yang, J. Jena, P. Werner, K. Koepernik, S. Parkin, Y. Chen, C. Felser, B. A. Bernevig and Z. Wang, *Nat. Phys.*, 2021, **17**, 381–387.
20 H. Yi, Z. Huang, W. Shi, L. Min, R. Wu, C. M. Polley, R. Zhang, Y.-F. Zhao, L.-J. Zhou, J. Adell, X. Gui, W. Xie, M. H. W. Chan, Z. Mao, Z. Wang, W. Wu and C.-Z. Chang, *Phys. Rev. Res.*, 2021, **3**, 013271.
21 X.-P. Li, K. Deng, B. Fu, Y. Li, D.-S. Ma, J. Han, J. Zhou, S. Zhou and Y. Yao, *Phys. Rev. B*, 2021, **103**, L081402.
22 Z. Wang, M. Saint-Lager, P. Monceau, M. Renard, P. Gressier, A. Meerschaut, L. Guemas and J. Rouxel, *Solid State Commun.*, 1983, **46**, 325–328.
23 J. Gooth, B. Bradlyn, S. Honnali, C. Schindler, N. Kumar, J. Noky, Y. Qi, C. Shekhar, Y. Sun, Z. Wang, B. A. Bernevig and C. Felser, *Nature*, 2019, **575**, 315–319.
24 C. Tournier-Colletta, L. Moreschini, G. Autès, S. Moser, A. Crepaldi, H. Berger, A. L. Walter, K. S. Kim, A. Bostwick, P. Monceau, E. Rotenberg, O. V. Yazyev and M. Grioni, *Phys. Rev. Lett.*, 2013, **110**, 236401.
25 H. Schaefer, M. Koerber, A. Tomeljak, K. Biljaković, H. Berger and J. Demsar, *Eur. Phys. J.: Spec. Top.*, 2013, **222**, 1005–1016.
26 M. Perlangeli, S. Peli, D. Soranzio, D. Puntel, F. Parmigiani and F. Cilento, *Opt. Express*, 2020, **28**, 8819–8829.
27 F. Bencivenga, R. Cucini, F. Capotondi, A. Battistoni, R. Mincigrucci, E. Giangrisostomi, A. Gessini, M. Manfredda, I. P. Nikolov, E. Pedersoli, E. Principi, C. Svetina, P. Parisse, F. Casolari, M. B. Danailov, M. Kiskinova and C. Masciovecchio, *Nature*, 2015, **520**, 205–208.
28 D. Berner, G. Scheiber, A. Gaymann, H. M. Geserich, P. Monceau and F. Lévy, *J. Phys. IV*, 1993, **03**, C2-255–C2-258.
29 F. Wooten, in *Optical Properties of Solids*, ed. F. Wooten, Academic Press, 1972, pp. 42–84.
30 O. Krupin, M. Trigo, W. F. Schlotter, M. Beye, F. Sorgenfrei, J. J. Turner, D. A. Reis, N. Gerken, S. Lee, W. S. Lee, G. Hays, Y. Acremann, B. Abbey, R. Coffee, M. Messerschmidt, S. P. Hau-Riege, G. Lapertot, J. Lüning, P. Heimann, R. Soufli, M. Fernández-Perea, M. Rowen, M. Holmes, S. L. Molodtsov, A. Föhlisch and W. Wurth, *Opt. Express*, 2012, **20**, 11396–11406.
31 K. Ohtake, E. Sato, S. Koshiba, R. Yamamoto, M. Doyama, K. Soda, T. Mori, S. Suga and K. Endo, *Physica B+C*, 1986, **143**, 192–194.
32 B. Ziaja, N. Medvedev, V. Tkachenko, T. Maltezopoulos and W. Wurth, *Sci. Rep.*, 2016, **5**, 18068.

Faraday Discussions

PAPER

Ultrashort and metastable doping of the ZnO surface by photoexcited defects

Lukas Gierster, *[a,b] Sesha Vempati †[b] and Julia Stähler [a,b]

Received 7th February 2022, Accepted 21st March 2022
DOI: 10.1039/d2fd00036a

Shallow donors in semiconductors are known to form impurity bands that induce metallic conduction at sufficient doping densities. The perhaps most direct analogy to such doping in optically excited semiconductors is the photoexcitation of deep electronic defects or dopant levels, creating defect excitons (DX) which may act like shallow donors. In this work, we use time- and angle-resolved photoelectron spectroscopy to observe and characterize DX at the surface of ZnO. The DX are created on a femtosecond timescale upon photoexcitation and have a spatial extent of few nanometers that is confined to the ZnO surface. The localized electronic levels lie at 150 meV below the Fermi energy, very similar to the shallow donor states induced by hydrogen doping [Deinert *et al.*, *Phys. Rev. B: Condens. Matter Mater. Phys.*, 2015, **91**, 235313]. The transient dopants exhibit a multi-step decay ranging from hundreds of picoseconds to 77 μs and even longer. By enhancing the DX density, a Mott transition occurs, enabling the ultrafast metallization of the ZnO surface, which we have described previously [Gierster *et al.*, *Nat. Commun.*, 2021, **12**, 978]. Depending on the defect density, the duration of the photoinduced metallization ranges from picoseconds to μs and longer, corresponding to the decay dynamics of the DX. The metastable lifetime of the DX is consistent with the observation of persistent photoconductivity (PPC) in ZnO reported in the literature [Madel *et al.*, *J. Appl. Phys.*, 2017, **121**, 124301]. In agreement with the theory on PPC [Lany and Zunger, *Phys. Rev. B: Condens. Matter Mater. Phys.*, 2005, **72**, 035215], the deep defects are attributed to oxygen vacancies due to their energetic position in the band gap and their formation by surface photolysis upon UV illumination. We show that the photoexcitation of these defects is analogous to chemical doping and enables the transient control of material properties, such as the electrical conductivity, from ultrafast to metastable timescales. The same mechanism should be at play in other semiconductor compounds with deep defects.

[a]*Humbolt-Universität zu Berlin, Institut für Chemie, Brook-Taylor-Str. 2, 12489 Berlin, Germany. E-mail: lukas.gierster@hu-berlin.de*

[b]*Fritz-Haber-Institut der Max-Planck-Gesellschaft, Abt. Physikalische Chemie, Faradayweg 4–6, 14195 Berlin, Germany*

† Present address: Department of Physics, Indian Institute of Technology Bhilai, Raipur-492015, India.

1 Introduction

Chemical doping of solids can lead to drastic changes in their conduction, optical and magnetic properties.[1] Photoexcitation can induce similar phenomena even on ultrafast timescales, coining the term "photodoping".[2–4] However, photodoping often denotes optical excitation across the band gap of the material,[2,5,6] creating holes in the valence band (VB) and electrons in the conduction band (CB). Such photoexcitation can lead to changes analogous to chemical doping in particular materials, such as Mott insulators.[5,7] In conventional semiconductors and arguably in most materials, however, no true doping effect should occur as photoexcitation generates both charge carrier types – electrons and holes – simultaneously. This is different to chemical doping, which adds either additional electrons to the CB (n-type doping) or additional holes to the VB (p-type doping) from impurity levels. The achievement of a qualitatively similar effect in photoexcited semiconductors requires the exploration of alternative pathways, for example, the photoexcitation across heterolayers which separates the charges.[8] Alternatively, electrons can directly be excited from occupied deep defects, which creates a localized hole and a comparably delocalized electron. Such defect excitons (DX) are the supposed origin of the persistent photoconductivity (PPC) (*i.e.* conductivity that persists for long times, up to days after photoexcitation) observed in many semiconductors.[9] As discussed below, they represent the perhaps most direct analogy to chemical doping.

Fig. 1 illustrates the similarity of a photoexcited deep defect and chemical doping. Electron doping in semiconductors is described by so-called shallow donors.[10] An impurity atom (or lattice defect) with a different valency is introduced into the host. While the system stays charge neutral, an electron can be released from the impurity into the CB, leaving behind a positively charged hole.

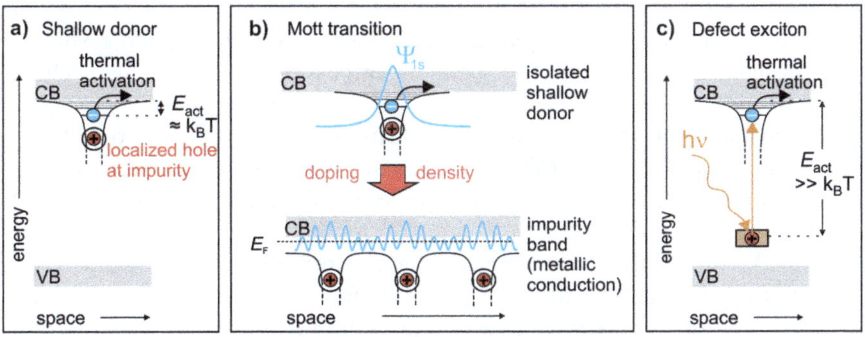

Fig. 1 The defect exciton as a bridging concept between shallow and photoexcited deep donors in a semiconductor. (a) Textbook view of a shallow donor: an impurity or defect with a different valency to that of the host releases an electron into the ZnO CB. The electron is bound in a hydrogenic potential to the positively charged impurity and can be freed by thermal activation with k_BT. (b) Mott transition: shallow donors at low densities lead to thermally activated conduction. At a critical density, a semiconductor-to-metal transition occurs. The electrons delocalize and form a metallic band (impurity band).[1] (c) Deep donors lead to an occupied state within the band gap too far away from the CB for thermal activation. Upon optical excitation, a bound state in the hydrogenic potential of the hole is formed, similar to the shallow donor.

This happens spontaneously and leads to the formation of a hydrogen-like state (extending over several lattice sites) from where electrons can become free carriers by thermal activation with $E_{act} \approx k_BT$, see Fig. 1a. Hence, shallow donors in a low concentration lead to thermally activated conductivity in the semiconductor. Yet, at a sufficient density, the Mott transition occurs, at which electrons delocalize and form a metallic band (impurity band),[1] allowing conduction down to zero K, *cf.* Fig. 1b. If photoexcitation is to have the same effect as n-type chemical doping in semiconductors, it would not only lead to additional electrons in the CB, but in addition these electrons should be first localized and then form a metallic band at a critical density of photodoping. A species that could in principle feature the exact same behaviour as shallow donors is photoexcited deep donors, *i.e.* in-gap states induced by impurities or defects that are too far from the CB edge for thermal activation, *cf.* Fig. 1c. Here, the electron is much more localized at the impurity as compared to the shallow donor, more like a particle in a box than a hydrogen-like state.[10] However, an external stimulus such as optical excitation can produce a DX, *i.e.* a bound electron in the hydrogenic potential of a photoionized impurity. This state is very similar to the shallow donor (*cf.* Fig. 1a and c), as has been pointed out by recent conceptual and theoretical work.[11,12] The question is if this would also lead to a doping effect, *i.e.* photoinduced (thermally activated) conductivity and, eventually, the Mott semiconductor-to-metal transition, in the sense of an exciton Mott transition of DX.

Turning from concepts to real systems, pioneering work by Katsumoto *et al.*[13] showed that the photoexcitation of deep defects can be used to influence the doping density in a GaAs compound, up to reaching a Mott semiconductor-to-metal transition. The system was chemically doped close to the Mott transition and photoexcitation was used to fine-tune the doping density across the Mott density. This is an important step to demonstrate photodoping by DX. However, it remains an open question if nominally undoped semiconductors can also be photodoped by DX up to metallization. Moreover, Katsumoto *et al.*[13] inferred from transport measurements that photoexcited deep defects lead to the Mott transition. The DX themselves, *i.e.* localized electrons bound to holes, and eventually their evolution into an impurity band is impossible to resolve in such an experiment.

DX most likely also play an important role in the conductive properties of ZnO, a wide-band gap (E_{gap} = 3.4 eV) semiconductor, which has been heavily studied due to its potential for optoelectronic applications and photocatalysis. In ZnO, PPC is observed with lifetimes ranging from below seconds to hours and longer,[14–16] and the effect has been attributed theoretically to photoexcited oxygen vacancies which form deep defect levels.[17–19] In fact, photoexcitation could be the long searched for bridge between the notorious (unintentional) n-type doping and oxygen vacancies in ZnO.[18] A metastable doping effect results from the charged defect distorting the lattice locally, forming a polaronic state, which creates an energetic barrier that has to be overcome by thermal excitation prior to electron–hole recombination.[17,18] The theory is supported by experimental findings such as the sensitivity to oxygen exposure[14] and the ability to create PPC with photon energies below the band gap.[14,20] Furthermore, the rise of the conductivity typically never saturates after switching on the illumination, which is attributed to the photogeneration of oxygen vacancies at the ZnO surface,[14,15] an interpretation supported by X-ray photoelectron spectroscopy that evidenced the formation of a Zn-rich surface.[21] It is still being discussed, though, whether photoexcited oxygen vacancies are responsible for both short- and long-lived PPC, or if the photodesorption of adsorbed oxygen as an alternative mechanism is at play.[16] Moreover, even though deep defects such as oxygen vacancies are widely accepted as source of PPC, the mechanism is usually discussed by assuming that ZnO exhibits upward surface band bending in equilibrium, *i.e.* an electron depletion layer with low conductivity.[15,16,22,23] In this picture, upon photoexcitation of deep defects or across the band gap, electrons are driven into the subsurface, while holes accumulate at the surface. This may contribute in two ways to PPC. Firstly, photoexcited electrons in the subsurface may lead to conductivity for rather long times, as the recombination is slowed by the energy barrier resulting from the band bending.[16,22,23] Secondly, the electron–hole separation counteracts the equilibrium band bending (a surface photovoltage effect), eventually leading to flat bands, thereby increasing the surface conductivity with respect to the equilibrium.[23] The question is, though, whether upward band bending is at all required (and an adequate assumption) to explain the PPC effect. DX themselves are already predicted to have long lifetimes. Moreover, if DX are formed at the surface, then the surface electronic structure of ZnO would rather be characterized by downward band bending instead of upward band bending or flat bands. This is well known from doping with shallow donors at semiconductor surfaces, for example due to H "doping" by adsorption at ZnO.[24] Ultimately, at sufficient DX densities, the surface would become metallic.

In our previous work, we showed that, by photoexciting deep defects at its surface, ZnO indeed shows signatures of transient surface doping.[25] The surface is metallized within femtoseconds and returns to the semiconducting state within a sub-ns timescale. Upon tuning through the phase transition by a variation of the laser fluence, the effective mass follows the critical behaviour of Mott semiconductor-to-metal transitions, and we observe transient downward band bending concomitant with metallization, all consistent with photodoping by shallow donor-like DX. In this paper, we extend this work by characterization of the photoexcited defects in more detail. We analyze the lifetimes of DX up to metastable timescales and their influence on the surface electronic structure in direct comparison with chemical doping by H. To do so, we use static as well as time- and angle-resolved photoelectron spectroscopy. This technique allows us to simultaneously measure the work function of ZnO and the degree of localization of photoexcited electronic states, as well as their temporal evolution on ultrafast to metastable timescales across more than 9 orders of magnitude in time. We show that the photoexcitation of ZnO creates a localized electron population below the CB minimum, about 150 meV below the Fermi energy. This feature is isoenergetic with the signature of shallow donor states induced by H adsorption. Monitoring the work function change connected to the electronic population evidences that the photoexcited localized electrons are bound to positive charges at the ZnO surface, forming effective dipoles that affect the surface potential. The electronic population and the positive surface charges grow as the defect density at the sample surface is increased by UV illumination. Furthermore, the photon energy required to excite these states is slightly below the fundamental band gap of ZnO, showing that deep defects <400 meV above the VB maximum are photoexcited. Both findings are consistent with oxygen vacancies as the origin of the deep defects. The DX are created on a femtosecond timescale and exhibit complex decay dynamics ranging from hundreds of picoseconds to significantly longer than the inverse laser repetition rate (5 μs). The lifetime of the metastable portion of the DX is determined to be a few tens of μs using fluence- and repetition rate-dependent measurements. In addition, some DX exhibit even longer lifetimes. The ultrafast excitation of deep defects at sufficient laser fluence can lead to transient metallization of the ZnO surface decaying within hundreds of picoseconds, as described earlier.[25] Complementarily to this, the ZnO surface becomes metastably metallic upon increasing the deep defect density and, hence, the long-lived DX by continuous UV illumination of the surface. These results are in agreement with the phenomenon of PPC in ZnO observed in the literature and show that the effect is due to shallow donor-like DX, which lead to downward band bending and eventually metallize the surface. All in all, our results show that the photoexcitation of deep defects in ZnO is in fact analogous to chemical doping with shallow donors. Such photodoping is expected to also occur in other semiconductor compounds with deep defects and could be the cause of ultrafast phase transitions and PPC in these materials. Hence, this study opens up new pathways for the optical control of semiconductors across multiple timescales.

2 Experimental methods

The experiments are performed on a ZnO single crystal surface with the non-polar (10−10) orientation. The (10−10) surface is thermodynamically the most stable and also prevalent in ZnO nanostructures.[27,28] The sample is prepared and analyzed in a pair of coupled ultrahigh vacuum (UHV) chambers operating at base pressures in the order of 10^{-10} mbar and 10^{-11} mbar, respectively. In the preparation chamber, the ZnO(10−10) surface (MaTecK GmbH) is cleaned by repeated cycles of Ar^+ sputtering (10 min, $p_{Ar} = 2.0 \times 10^{-6}$ mbar, 750 eV at 300 K), succeeded by 30 min annealing in vacuum ($T_{max} = 950$ K, initial heating rate of 30 K min^{-1}) following established procedures.[24,29] All the samples are transferred *in situ* into the analysis chamber, and the experiments are carried out at a sample temperature of 100 K.

The light source for the photoemission measurements is a regenerative amplified femtosecond (50–100 fs pulses) laser system running at a base repetition rate of 200 kHz (PHAROS by Light Conversion), feeding several optical parametric amplifiers (internal OPA, Orpheus-N-2H, Orpheus-N-3H by Light Conversion). The internal OPA provides a photon energy of $h\nu = 1.55$ eV, which is frequency converted to $h\nu = 6.2$–6.3 eV to detect the occupied electronic states at the ZnO surface. Additional laser pulses with $h\nu = 2.5$–4.2 eV are created in the Orpheus-N-2H and 3H. These are required to perform time-resolved pump–probe measurements. Here, two laser pulses (pump and probe) with different photon energies are spatially and temporally overlapped with the help of a delay stage to adjust the optical path length of the probe laser beam. The probe laser pulse with $h\nu = 6.3$ eV is used to monitor the excited state population induced by the pump pulse with photon energies between $h\nu = 2.5$ eV and $h\nu = 4.2$ eV at variable time delays.

The kinetic energy (E_{kin}) and intensity of the photoelectrons are detected using a hemispherical analyzer (PHOIBOS 100, SPECS GmbH) and a fluorescent screen that is photographed with a CCD camera (sensicam qe, PCO AG). The energy resolution of the experiment is 50 meV. The spectra are referenced to the Fermi level E_F of the gold sample holder, which is in electrical contact with the sample. A bias voltage of -1.5 V is applied to the sample with respect to the analyzer, enabling the detection of electrons of zero E_{kin}. The low-energy secondary electron cut-off (E_S) in the photoelectron spectra provides the work function (Φ) of the surface *via*

$$\Phi = h\nu - (E_F - E_S).$$

Note that the secondary electron cutoff is, even for a perfectly homogeneous sample surface, broadened by the energy resolution of the experiment.[30] The work function is therefore extracted from the secondary cutoff *via* fitting the data with a step function convoluted with a Gaussian function that represents the energy broadening.

Besides measuring the electronic binding energies and the work function, the angular spread of the photoelectrons is recorded. This gives access to their distribution as a function of the angular momentum $k_{\|}$ and, hence, the band structure $E(k_{\|})$ of the extended Bloch states within the sample.[31] In the simplest case, such as for ideal metals, $E(k_{\|})$ is a parabola around the center of the Brillouin zone. Electronic states can also be localized to few nm, which leads to a constant energy eigenvalue as a function of the emission angle.[32] Hence, localized and delocalized states can be qualitatively distinguished from the angular distribution of the photoelectron intensity.

A special feature of our laser system is that the base repetition rate can be lowered by a pulse picker to fractions of $R = 200$ kHz (*i.e.* 100 kHz, 50 kHz, 25 kHz…), which is useful for the investigation of photoexcited states with lifetimes exceeding the inverse repetition rate. As such states do not decay completely in between two subsequent laser pulses, a stationary population builds up over the course of many laser excitations, see Fig. 2a. Under the condition that every laser pulse populates as well as depopulates the long-lived states, the maximum population is limited, leading to a stable photostationary state, *cf.* Section 3.4. Note that a photostationary electronic population could easily be misinterpreted as part of the equilibrium electronic band structure of the sample. R-dependent measurements unveil the photostationary state, since, upon lowering R, the pulse-to-pulse distance is increased. For a photostationary state, it means more time to decay and, hence, in case of a lifetime in the order of $1/R$, the photostationary population reduces. In contrast, the intensity of electronic states that are part of the equilibrium band structure stays unaffected. On top of the pure identification of photostationary states, the variation of R can be used to determine their lifetimes using a suitable model, as detailed in Section 3.4.

3 Results

The focus of the present paper is the electronic structure of ZnO around the Fermi level, as this is what determines the conduction properties influenced by doping. However, we firstly note that photoelectron (PE) spectra from the VB region of the as-prepared ZnO(10$-$10) surface shows the topmost VB at -3.2 eV below E_F in our experiments,[25,33] as well as in the literature.[34] In combination with the band gap of 3.4 eV,[35] this means that the CB is 0.2 eV above E_F. Thus, ZnO is n-type doped, as is commonly observed.[36] As discussed in the following Sections 3.1–3.3 for photoexcited ZnO, the n-type doping results (at least partially) from metastable DX at the ZnO surface. Note that we use the term 'metastable' to denote photoexcited states that have a lifetime larger than those typically observed for electronic excitations in semiconductors (*i.e.* picoseconds to nanoseconds). 'Photostationary' refers to photoexcited states that do not decay in between two subsequent laser pulses, leading to an accumulation effect up to reaching a stationary population, as described above.

3.1 Photostationary n-type doping

The central experimental evidence for shallow donor-like DX at the ZnO surface is displayed in Fig. 2b. It shows the photoelectron (PE) spectra of the as-prepared ZnO(10−10) recorded with a photon energy of 6.3 eV and different repetition rates R (5–200 kHz). The photon energy is larger than the work function (4.3–4.4 eV) and leads to direct photoemission of occupied states, *c.f.* the energy level diagram in Fig. 2d. The photoemission intensity is characterized by two features, the secondary electrons with their low energy cut-off and one peak just below E_F with a width of several hundred meV. As shown in the inset, the latter feature is isoenergetic with the shallow dopant state induced by H adsorption at the ZnO surface.[24] However, in contrast to the n-type doping by H, the electronic population is not part of the

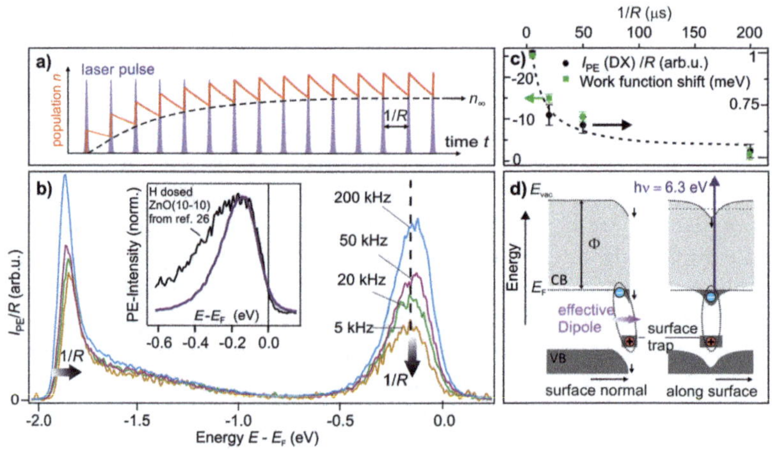

Fig. 2 Central experimental evidence for long-lived defect excitons that act like shallow donors at the ZnO(10−10) surface. (a) Photoexcited states form a photostationary population if they have a lifetime that exceeds the inverse laser repetition rate $1/R$. (b) Repetition-rate dependent PE spectra of as-prepared ZnO(10−10). The intensity of the peak below E_F is reduced as the repetition rate is lowered, revealing that it is a photostationary state. Concomitantly, the secondary edge shifts up. Inset: comparison of the PE spectrum of the photostationary state to a PE spectrum with H adsorbed at the ZnO(10−10) surface. The spectrum is taken from ref. 26, with kind permission from the author. (c) Work function change and intensity of the peak below E_F as a function of $1/R$, extracted from the raw data in (b). The dashed line is a guide for the eye. The data points represent the average from repeating the experiment three times; error bars represent one standard deviation. (d) Interpretation of the observation: optical excitation of deep defects at the ZnO surface creates DX. Left hand side: the complex at the surface has a dipole moment which reduces the work function. Right hand side: the electron is laterally localized due to the hydrogenic potential of the surface defect, as discussed in Section 3.5. The metastable defect excitons are created and detected by laser photons with $h\nu = 6.3$ eV as they form a photostationary state.

equilibrium band structure, but results from the sample being in a photostationary state. This is shown in the following by comparing the PE spectra recorded with different R. In case of a lifetime in the order of $1/R$, the PE intensity due to a photostationary population should reduce upon lowering R, as introduced in Section 2. Note that in order to be able to compare the PE intensity recorded with different R, the same laser fluence (*i.e.* energy per pulse and unit area) is used for all recorded spectra, which are then divided by R, as the measurement integrates over all photoemission events within the camera exposure time.

The PE spectra in Fig. 2b exhibit marked changes upon changing R. Firstly, the PE-intensity of the state below E_F drops by about 50% upon lowering the repetition rate from 200 kHz to 5 kHz. The reduction of the PE intensity with lowering R means that the shallow donor-like electronic population below E_F is photostationary, with a lifetime in the order of the inverse laser repetition rate $1/R$ (5–200 μs). Secondly, at the same time as the PE intensity due to the electron population below E_F drops, lowering R also affects the secondary electron cut-off. The raw data in 2b unveils a systematic upward shift of the secondary electron cut-off when lowering R. The shift of the secondary electron cut-off, *i.e.* the sample work function, together with the PE intensity of the peak below E_F, is shown in Fig. 2c as a function of the temporal distance between two laser pulses $1/R$. The work function is 25 meV lower when the PE-intensity below E_F is at its maximum (at $R = 200$ kHz) with respect to its minimum value (at $R = 5$ kHz). Evidently, the photostationary electron population is connected to an equally long-lived work function reduction. Also, this is in line with H adsorption at the ZnO surface.[24] Both the photostationary electron population with a spectral signature similar to H dopants, as well as the concomitant work function reduction, suggest that the ZnO surface is photostationary n-type doped.

To explain the above observations, metastable free excitons can clearly be excluded, as they are inconsistent with a localization at the surface. Instead, surface-bound electrons and a work function reduction would be readily explained by the formation of DX upon photoexciting deep defects at the ZnO surface: photoexcitation leads to comparably delocalized electrons in the ZnO CB bound to localized hole states at the surface. Thus, an effective dipole is formed, which points outward the surface normal and lowers the work function (left hand side of Fig. 2), as upon the chemical doping of oxide surfaces.[24] Alternatively, the photostationary n-type surface doping could be explained by excitons formed after cross band gap excitation, where the hole is localized at the surface due to polaron formation. Hole polarons with lifetimes up to minutes have been observed in ZnO.[37] What discriminates these two cases is the origin of the photoexcited electrons, which are either VB states or deep defects in the ZnO band gap. Therefore, to identify the DX, we test whether cross-band gap excitation is necessary for creating the photostationary state or if sub-band gap photons are sufficient.

3.2 Photon energy dependency for the identification of the deep defects

Deep defects feature an occupied state within the band gap of semiconductors, *cf.* Fig. 1c. Hence, to photoexcite them, light with a photon energy below the fundamental band gap is sufficient. In order to see whether the photostationary state can be created with below band gap excitation, we perform photon energy-dependent pump–probe experiments. Pump laser pulses with photon energies below and above the band gap are used. The photon energy of the probe laser pulses is kept at 6.2 eV, *i.e.* above the sample work function, to be able to detect population changes below the Fermi energy E_F, *cf.* the energy level diagram in Fig. 2d.

Fig. 3a presents a typical pump–probe experiment, here using a pump photon energy of 3.2 eV. The PE intensity in false colors as a function of pump–probe delay in picoseconds (bottom) and energy (left) is shown. The photostationary signal at negative pump–probe delays was subtracted from the data in order to show the ultrafast dynamics of the photoinduced PE intensity due to the pump laser beam. Upon arrival of the pump laser at time zero, an abrupt rise of the PE intensity below E_F is observed. The ultrafast rise and decay dynamics will be discussed later in Section 3.3. The energy of the pump-induced signal overlaps with the photostationary state observed in the steady state spectra. A part of the pump-induced signal also survives the inverse laser repetition rate, because adding the pump laser beam leads to an increase of the photostationary signal at negative pump–probe delays (not shown). As the photon energy of the pump laser pulse is, at 3.2 eV, significantly below the band gap of ZnO (3.4 eV), this shows that the photostationary state is populated from in-gap states, see the inset in Fig. 3a. It cannot originate from metastable electrons bound to polaronic VB holes at the ZnO surface, because such excitons require at least excitation at the free exciton resonance (3.38 eV at $T = 4$ K; 3.31 eV at $T = 295$ K (ref. 35)).

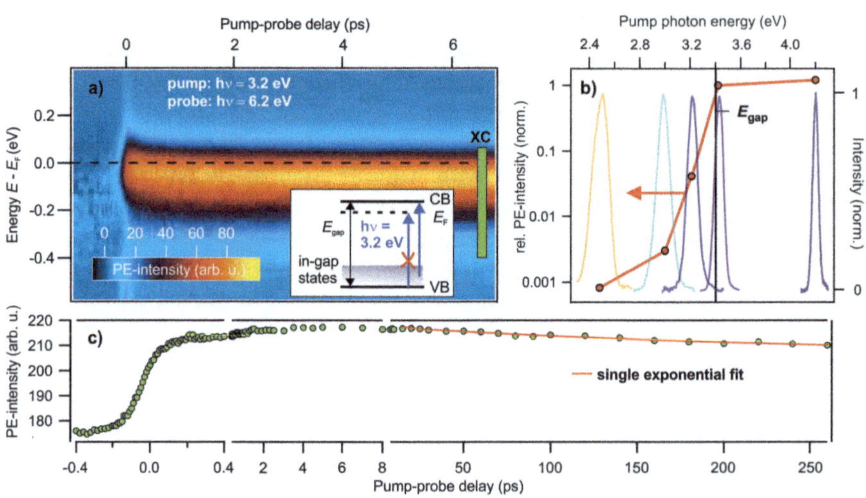

Fig. 3 Photon energy-dependent time-resolved PE measurements using $h\nu = 6.2$ eV as the probe photon energy and pump photon energies between 2.5–4.2 eV. (a) False color representation of the pump-induced population change as a function of the pump–probe delay and energy, with a pump photon energy of $h\nu = 3.2$ eV. The photostationary state was subtracted from the data to show the ultrafast intensity evolution around E_F. Inset: energy level diagram showing that no excitations across the band gap are allowed for $h\nu = 3.2$ eV, but only from in-gap states. (b) Signal increase at positive with respect to negative delays according to eqn (1) for different pump–photon energies. The action spectrum shows that a photon energy of at least 3.0 eV is necessary to populate the state below E_F. The colored lines are the laser spectra of the pump laser pulses to indicate photoexcitation below and above E_{gap}. (c) Pump-induced PE intensity around E_F resulting from integration across the box 'XC' indicated in (a) for a larger delay window up to 260 ps. The red line is a fit with a single exponential decay: $I = I_0\,e^{-\Delta t/\tau_d}$ + offset.

To determine the energetic position of the in-gap states, we tuned the pump photon energy over a wide range, from 2.5 eV to 4.2 eV. Fig. 3b displays (in red) the pump-induced signal increase at −0.3 to 0 eV with respect to E_F evaluated at a positive pump–probe delay of 1 ps with respect to a negative delay of −1 ps. The signal increase is shown on a logarithmic scale *versus* the pump–photon energy and is determined from the non-background subtracted data according to

$$\frac{I(1\ \text{ps}) - I(-1\ \text{ps})}{I(-1\ \text{ps})} \Big/ F_{\text{pump}}. \tag{1}$$

Here, F_{pump} is the pump laser fluence. The action spectrum shows that the efficiency to populate the photostationary state rapidly decreases upon lowering the photon energy. At 3.2 eV, the efficiency is already an order of magnitude lower than at 3.4 eV. For pump energies lower than 3.0 eV, the signal increase almost vanishes. Hence, the in-gap states lie deep within the band gap, less than 400 meV above the VB maximum.

The energetic position of defects is characteristic of their chemical nature, hence the above result may be used to identify the defect species at the ZnO surface. Amongst the native defects, the oxygen vacancy is the only deep donor.[38,39] According to hybrid DFT calculations,[40] the oxygen vacancy in ZnO is located just above the VB edge, at 0.4 eV above the VB maximum or 3.0 eV below the CB minimum. This coincides with the above-determined position of the in-gap states. In fact, the action spectrum in Fig. 4b is also in agreement with the optical absorption of samples with deliberately added oxygen vacancies.[41] This suggests that the origin of the in-gap states is oxygen vacancies. As an alternative to native defects, impurities or defect complexes could be the origin of the deep defects. A Zn vacancy bound to two H impurity atoms was recently predicted to feature a metastable state that could also cause PPC, as could oxygen vacancies.[42] In order to finally identify the chemical nature of the involved defects, the next section shows how the defects are created.

3.3 Relation to UV illumination-induced defects

Defects, impurities and defect complexes can be present in ZnO due to the growth process as well as the preparation of the surface due to sputtering and high temperature annealing.[29,43,44] After the sample preparation, additional defects can

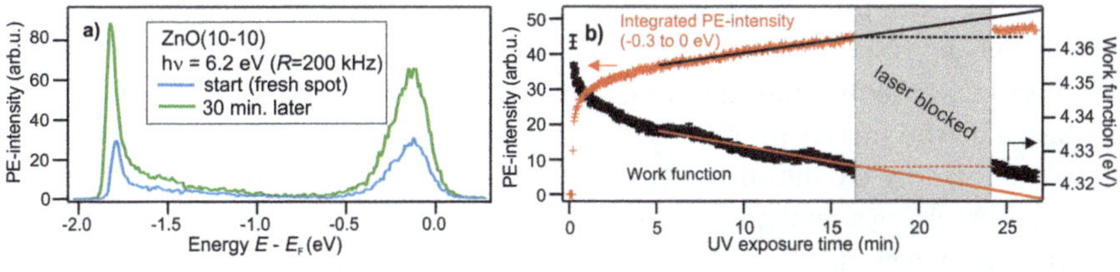

Fig. 4 Change of the PE spectra as a function of UV exposure time. (a) PE spectra recorded with $h\nu = 6.3$ eV directly after addressing a previously not illuminated spot on the as-prepared ZnO surface (blue curve) and after 30 min of continuous UV illumination (green curve). (b) Work function reduction and rise of PE intensity in the energetic region −0.3 to 0 eV with respect to E_F as a function of UV exposure time, extracted from the PE data. The two quantities change concomitantly. The solid and dashed lines are guides for the eye. The grey box indicates that the laser is blocked (during minutes 17–24).

be created by UV illumination; it has been shown by mass and X-ray spectroscopy that oxygen is removed from the ZnO surface upon UV illumination, making the surface more Zn-rich.[21,45] Therefore, to figure out if oxygen vacancies are indeed the origin of the deep defect levels, we investigate if the photoresponse changes as a function of the UV illumination time.

Fig. 4a shows the PE spectra recorded with $h\nu = 6.2$ eV at two different times. The blue spectrum is measured just after preparation of the sample surface and the green one after 30 minutes of UV illumination. The two spectra are clearly very different. The state below E_F has increased by a factor of 2. The secondary edge, *i.e.* the sample work function, has shifted down. Fig. 4b shows these changes in more detail, measured across 30 min. Both spectral features have the same temporal evolution. After an initial non-linear rise, both the PE intensity increase below E_F and the work function drop vary linearly. It is also illustrated that the change is not present when the laser (*i.e.* the UV illumination) is blocked, between minute 17 and 24; there, the work function does not drop further. The PE intensity below E_F increases slightly during this time but clearly not at the same rate as before. The observed change of the PE spectra, *i.e.* the rise of the peak below E_F and the drop of the work function, are readily explained by an increased number of DX as the density of deep defects at the surface is increased by the UV illumination. The small effect that happens in the dark may be due to the adsorption of residual gas present even in the UHV environment, such as H.[46]

An alternative scenario that would explain the growth of the DX signal during UV illumination is the buildup of the population of the photostationary state at a given defect density caused by repeated photoexcitation with subsequent laser pulses (see Fig. 2a). We can differentiate this accumulation effect from the increase of the deep defect density itself by inspecting the DX signal created with an individual laser excitation event and how it changes with UV illumination time. Pump–probe spectroscopy enables us to measure this, as the pump laser pulse also creates DX that decay faster than the inverse laser repetition rate (see the discussion of the ultrafast dynamics in the next section). We find that the number of these fast-decaying DX also increases upon UV illumination and does so with a similar rate as the photostationary population on a timescale of minutes to hours.[33] This unambiguously shows that, on these long timescales, the UV illumination enhances the number of states that can be photoexcited with a single laser pulse, *i.e.* the density of deep defects at the ZnO surface. However, we note that the buildup of the photostationary state due to subsequent laser pulses may be responsible for the initial nonlinear change of the PE spectra that we observe in the first minute in Fig. 4b.

Based on the surface photolysis observed in the literature, the most likely deep defect created by UV illumination is the oxygen vacancy, as opposed to a defect complex involving a Zn vacancy. Hence, due to both their energetic position in the band gap, as well as their creation by UV illumination, we attribute the deep defects to oxygen vacancies.

3.4 Lifetime of the DX

The previous three sections showed that DX are excited at the ZnO surface, as we found (i) a photoexcited electronic population bound to a positive charge at the ZnO surface, which leads to a work function reduction, (ii) it is possible to photoexcite the state with below band gap photons and (iii) the population increases upon enhancing the surface defect density due to UV illumination. The DX are shallow donor-like, as they have a similar spectral signatures to those observed upon doping with H. In this section, we investigate the timescales associated to the DX dynamics.

The pump–probe measurements introduced in Section 3.2 unveil the decay dynamics of the DX on ultrafast timescales. Fig. 3c displays the integrated intensity around E_F. The buildup of the population is a two-step process consisting of a fast rise happening within the time-resolution of the experiment (<100 fs), followed by a slower process with a duration of several picoseconds. Subsequently, the DX start decaying on a timescale of hundreds of picoseconds. This is analogous to the case of using cross band gap excitation with 3.4 eV, which we reported in ref. 25. In line with our previous publication, the fast rise is attributed to the direct excitation of in-gap states at the ZnO surface and the subsequent DX formation, which occurs on screening timescales. The picosecond rise is attributed to the trapping of bulk or sub-surface holes at the surface. As photoexcited electrons are bound to the holes in an excitonic state, this enhances the number of DX at the surface and, hence, the photoemission signal below E_F. This interpretation is also consistent with oxygen vacancies, which are predicted to lie at lower energies in the bulk than at the surface,[47] thus energetically favoring the accumulation of holes in surface states. Eventually, the DX start recombining, which leads to the decay of the PE intensity on a timescale of hundreds of picoseconds. A single exponential fit gives $\tau_d = 250 \pm 80$ ps, also in agreement with our previous work.[25]

However, a significant portion the PE intensity decays more slowly, clearly surpassing the time window accessed in the pump–probe scans (*cf.* Fig. 3c). These DX eventually form the photostationary state as they live longer than the inverse laser repetition rate. In order to determine their lifetime, femtosecond pump–probe spectroscopy cannot be used. Instead, as shown in the following, the lifetime can be determined from the R dependency presented in Section 3.1, combined with an adequate model and supplementary laser fluence-dependent measurements.

An increase of the pulse-to-pulse distance $\Delta T = 1/R$ gives the photoexcited population more time to relax before the next laser pulse arrives (*cf.* Fig. 2a). However, the intensity decay in Fig. 2c does not directly reflect the lifetime of the DX as the laser-induced population and depopulation occurs every time a laser pulse arrives, leading to a complex intensity change of the photostationary state as a function of $1/R$. In the following, we develop the simplest possible model that describes our experimental observations. To do so, we need to consider three contributions to the population dynamics: the photoexcitation of new DX, the depopulation by photoemission from the DX state, and the decay of DX on μs-timescales. Firstly, each laser pulse leads to an increase of the population by αN, where N is the total number of deep defects that can be excited. $\alpha < 1$ is the population probability, which is assumed to be proportional to the laser fluence F, with an unknown factor a, *i.e.*

$$\alpha = aF. \tag{2}$$

Secondly, as the population is probed in PE spectroscopy, electrons are removed from the state each time a laser pulse arrives. This happens with the depopulation probability

$$\beta = bF, \tag{3}$$

where b is again a proportionality factor and $\beta < 1$. The depopulation acts on the already populated states. Hence, the overall population change with the k-th laser pulse is

$$\Delta n_k = \alpha N - \beta n_{k-1}(\Delta T), \tag{4}$$

where $n_{k-1}(\Delta T)$ is the population that remains from the preceding laser excitation events. It is initially zero and then builds up in the course of subsequent laser pulses (*cf.* Fig. 2a). Thirdly, the population decay in between two pulses is assumed to be exponential

$$n_k(t) = n_k^0\, e^{-(t-t_k)/\tau}, \tag{5}$$

where t_k refers to the time of arrival of the laser pulse, leading to the population n_k^0 by the photoexcitation of and PE from DX. For simplicity, the 250 ps decay of the DX is omitted, as it is instantaneous compared to ΔT and does not contribute to the photostationary population. τ is the lifetime of the part of the DX that does not decay in between two subsequent laser pulses and builds up the photostationary state. All of this is illustrated in the inset of Fig. 5a.

n_k^0 is calculated iteratively for each laser pulse. This leads to the sequence shown in Table 1. For the population after the k-th laser pulse, one can identify the geometric series

$$n_k^0 = \alpha N \sum_{j=0}^{k-1} (1-\beta)^j \left(e^{-\Delta T/\tau}\right)^j. \tag{6}$$

For many pulses, $k \to \infty$, so the population becomes[33]

$$n_\infty(\Delta T, F) = \frac{\alpha(F)N}{e^{\Delta T/\tau} + \beta(F) - 1}. \tag{7}$$

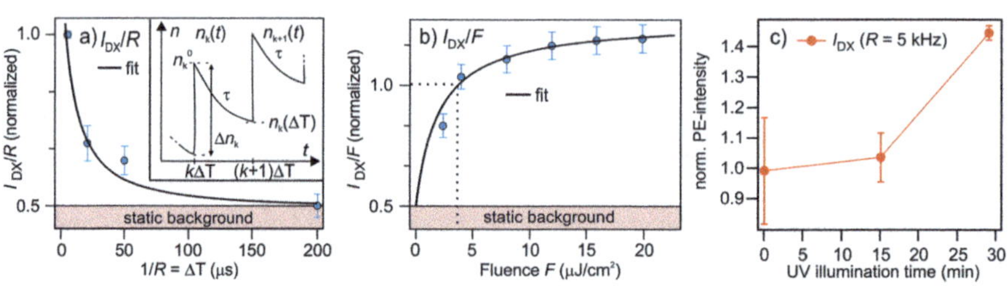

Fig. 5 Measurements and model to determine the lifetime of the photostationary DX. (a) The PE intensity of DX as a function of the inverse laser repetition rate $1/R = \Delta T$, reproduced from 2c. Inset: illustration of the notation used in the model to calculate the photostationary population. (b) The PE intensity of DX as a function of the laser fluence F ($R = 200$ kHz). The black lines in (a) and (b) are global fits with eqn (7) and (8). Both data sets are normalized to the PE intensity at $R = 200$ kHz and the fluence used in the R-dependent measurements, which is $F = 3.6$ μJ cm^{-2}. (c) The PE intensity of DX at $R = 5$ kHz as a function of UV illumination time. The rising trend of I_{DX}(5 kHz) suggests that the 'static background' indicated in (a) and (b) is resulting from the photoexcitation of defects created by UV illumination as well, but with a lifetime $\gg 1/R$ (see text).

Table 1 Sequence for the calculation of the build-up of a photostationary equilibrium. k is the pulse counter, n_k^0 is the population of the long-lived state just after the arrival of the k-th laser pulse and $n_k(\Delta T) = n_k^0\,\mathrm{e}^{-\Delta T/\tau}$ is the population just before the $k+1$-th laser pulse, which depopulates $n_k(\Delta T)$ and populates new DX according to eqn (4)

k	n_k^0	$n_k(\Delta T) = n_k^0\,\mathrm{e}^{-\Delta T/\tau}$
0	0	0 (sample is in equilibrium)
1	$n_0(\Delta T) + \Delta n_1 = \alpha N$	$\alpha N \mathrm{e}^{-\frac{\Delta T}{\tau}}$
2	$n_1(\Delta T) + \Delta n_2 = \alpha N\left(1 + \mathrm{e}^{-\frac{\Delta T}{\tau}}(1-\beta)\right)$	$\alpha N\left(1 + \mathrm{e}^{-\frac{\Delta T}{\tau}}(1-\beta)\right)\mathrm{e}^{-\frac{\Delta T}{\tau}}$
3	...	...

This is the DX population in the photostationary state, which depends on the inverse repetition rate $\Delta T = 1/R$, as well as on the laser fluence F.

Importantly for the comparison to the experiment, b can be directly determined from pump–probe spectroscopy by choosing a pump photon energy above the work function. The DX dynamics monitored with the probe laser pulse are similar to Fig. 3c, where a pump photon energy below the work function is used, but with the crucial difference that the DX signal first drops at time zero, which is due to the depopulation of the DX by photoemission due to the pump laser pulse. The drop in PE intensity can be directly related to b, resulting in $b = 0.034 \pm 0.017$ $\mathrm{cm^2\ \mu J^{-1}}$.[33] The conservatively estimated error bar for b is 50%. In order to fit the detected PE intensities, it should be noted that they are proportional to the number of incident photons per time $R \times F$, *i.e.*

$$I_{\mathrm{DX}} \propto RF(n_\infty + n_0). \tag{8}$$

Here, n_0 is a constant, representing an electronic population that does not show a dependence on F and ΔT. Note that the proportionality constant between the number of electronic states in the sample and the detected number of photoelectrons is not known due the unknown efficiency of electron detection, which can only be estimated. Nevertheless, the formula can be used to describe relative changes of the PE intensity as a function of R and F, enabling the determination of the lifetime τ of the DX, as shown in the following.

The model is now fitted to the experimental data in Fig. 5. Panel (a) shows the PE intensity of the DX divided by R at constant F as a function of $1/R = \Delta T$ and panel (b) shows I_{DX}/F as a function of F for constant R. Qualitatively, a decay and a rise with increasing fluence is observed, respectively, tending towards saturation. The black lines are global fits with eqn (7) and (8), yielding $\tau = 77 \pm 45$ μs and $n_0 = 0.52 \pm 0.1$. Note that the DX population saturates at high fluence due to a balance between population and depopulation, as eqn (7) approaches a constant for large α and β.

n_0 is invariant under F and ΔT and amounts to about half the PE intensity at $1/R = 5$ μs; the PE intensity due to n_0 is indicated as 'static background' in Fig. 5a and b. This shows that either an additional photoexcitation-independent state isoenergetic with the DX is present or that a part of the DX survives much longer than the μs window accessed with the R dependency. For the identification of the source of n_0, it is helpful to recognize that the time-dependent DX population $n_\infty(\Delta T)$ has decayed after 200 μs (*cf.* Fig. 5a) and, therefore, I_{DX}(5 kHz) represents the quasistationary contribution n_0. To identify whether n_0 results from even longer-lived DX or is an entirely new, independent feature, we examine the long-time evolution of I_{DX}(5 kHz) in Fig. 5c. Note that we have shown above that continuous UV illumination leads to a growing number of deep defects and, thus, DX. Fig. 5c clearly indicates that the n_0 population also rises as the sample is exposed to UV light. This rising trend shows that n_0 increases as the surface defect density is enhanced. It, thus, seems highly likely that the quasistationary population n_0 results from DX with lifetimes on ms-timescales or longer. The origin of the different lifetimes of the DX will be discussed in Section 4.

3.5 Metastable metallization

As explained in the Introduction, a crucial property of shallow donor dopants is their influence on the electrical conductivity of semiconductors. At low densities, the electrons are localized at the impurity sites, inducing thermally activated conductivity, while the electrons delocalize at a sufficient density into a metallic band as the Mott semiconductor-to-metal transition occurs. The DX discovered in this work have a similar spectral signature as H dopants, *i.e.* shallow donors. However, do they also lead to metallic conductivity? In our previous publication,[25] we have shown that the surface of ZnO can be reversibly metallized by the photoexcitation of deep defects on an ultrashort timescale. The temporal variation of the density of photoexcited defects leads to the dynamic evolution of localized electrons into a parabolic band, and a subsequent localization within hundreds of picoseconds (with τ_d, the picosecond decay of the DX). Here, we expand these results to the photostationary DX and show that the ZnO surface becomes a metastable metal at sufficient densities of photoexcited deep defects.

Fig. 6a presents the steady-state angle-resolved PE spectrum around E_F, detected with 6.3 eV laser photons. The measurement is recorded after 1.5 hours of UV

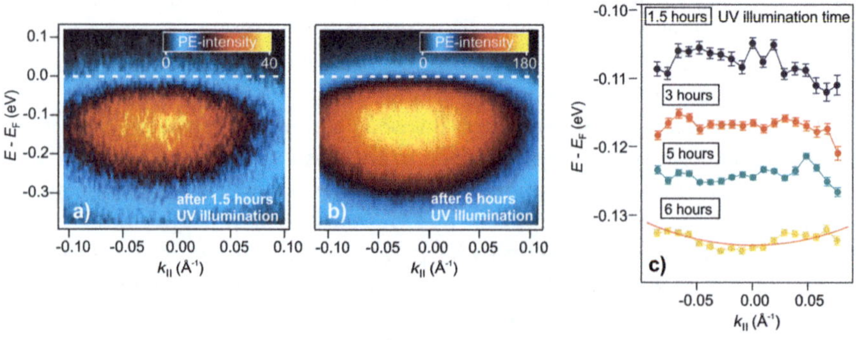

Fig. 6 Metastable metallization of the ZnO surface due to DX. (a) and (b) Angle-resolved PE spectra after 1.5 hours and 6 hours of UV illumination of the as-prepared ZnO(10−10) surface, respectively. After 1.5 hours, the PE intensity shows a dispersionless feature due to localized electrons. (b) After 6 hours, the PE intensity exhibits a more pronounced curvature at lower energies and seems to be cut by E_F, more like a metallic band than localized states. (c) Overview of the evolution of the peak maxima over 6 hours. The traces are vertically offset for clarity. The red line is a free electron-like parabola (eqn (9)).

illumination of the as-prepared sample. The data shows the DX feature *ca.* 100 meV below E_F and with a width of several hundred meV. The angular distribution of this state is symmetric and does not exhibit an apparent curvature, indicative of localized states. Fig. 6b shows the angle-resolved PE spectrum after 6 hours of UV illumination. Within this time span, the PE-intensity from the DX increased by a factor of about 4 (not shown). In contrast to the spectrum after 1.5 hours of UV illumination (Fig. 6a), the PE intensity is now directly adjacent to E_F, appears broadened and assymmetric, *i.e.* it has apparently a more pronounced curvature at lower energies, and seems to be cut by E_F, more like a metallic band than localized states.

To inspect the dispersion change of the DX feature as a function of UV illumination time in more detail, the apparent dispersion is evaluated by determining the position of the peak maximum of the PE intensity at different $k_{\parallel}$. This is done by fitting an asymmetric Gaussian function to the spectra (evaluation procedure not shown, see ref. 33 for details). The so-determined peak maxima are plotted in Fig. 6c for angle-resolved PE spectra recorded after 1.5, 3, 5 and 6 hours of UV illumination. The traces are vertically offset by *ca.* 0.01 eV for clarity. After 1.5 hours, the peak maxima show a slight tendency toward a negative curvature. This has been observed before in the context of localized states and is attributed to the simultaneous sampling of electrons with a different degree of localization and, hence, different binding energies and momentum spread.[32] For longer UV illumination times, the dispersion starts to flatten and evolve into a parabolic dispersion with positive curvature. The red line on top of the trace after 6 hours is the plot of a free electron-like parabola,

$$E(k) = \frac{\hbar^2 k^2}{2m_{\text{eff}}}, \tag{9}$$

with a heavy effective electron mass of $m_{\text{eff}} = 10m_e$, where m_e is the free electron mass. The apparent dispersion of the peak maxima can be described well by this parabola. This indicates that the electrons are no longer localized after 6 hours of UV exposure, but form a free electron-like band with a large effective mass, *i.e.* the sample has become metallic.

We expect that even longer UV illumination times lead to more shallow donor-like DX and, in analogy to the density-dependent evolution on ultrafast timescales,[25] to a lowering of the effective mass. Eventually, the effective mass should follow the critical behaviour for a shallow donor Mott semiconductor-to-metal transition.

4 Discussion

4.1 DX–shallow donor analogy

Can photoexcitation truly act like chemical doping? Chemical doping by shallow donors features additional electrons which are first localized (leading to thermally activated conduction) and then form a metallic impurity band as the doping density crosses the Mott limit. These two conditions must be met by photoexcitation in order to justify the term "photodoping". From the conceptual comparison given in the Introduction, it is expected that DX might act like shallow donors, however, so far, it was unknown to what extent the analogy holds true, as photodoping by DX was previously only inferred from transport measurements for a chemically pre-doped sample by Katsumoto *et al.*[13] Using photoelectron spectroscopy, we show above that ZnO exhibits a photoinduced localized electronic state below E_F and unveil – by the associated work function reduction – that it is bound to a positive surface charge. The state can be excited with laser photon energies slightly below the fundamental band gap and grows as the defect density at the surface is increased. Hence, the positive surface charge is located at a deep defect at the surface. In other words, we directly observe DX. The effect of the photoexcitation is by all means comparable to chemical doping of the ZnO surface with shallow donors: firstly, the spectral signature of the electronic state below the CB (at 150 meV below E_F) overlaps with that of adsorbed H, and the work function reduction is also in line with the latter. Secondly, the electronic state is first localized and then evolves into a metallic band at a sufficient density of DX. Remarkably, we find that the metallization can be performed on two different timescales by controlling the DX density: For low defect densities, *i.e.* short UV illumination times after sample preparation, the metastable (μs to ms and above) component of the DX is localized. We observe this for all photoexcitation fluences (not shown), consistent with the finding that the photostationary population saturates as a function of laser fluence (*cf.* Section 3.4). This is the initial state for the ultrafast metallization described previously,[25] where an individual strong laser pulse enhances the DX density above the Mott limit. The system returns to the semiconducting state within τ_d, the ultrashort decay of the DX, *i.e.* much faster than the inverse laser repetition rate. Beyond varying the laser fluence, the DX population is enhanced by the increase of the defect density due to continuous UV illumination. This also means an increase of the photostationary population, *cf.* eqn (7). At a sufficient defect density, the photostationary DX population renders the ZnO surface a metastable (photostationary) metal. Hence, the ZnO surface always becomes metallic when the DX density crosses the Mott limit, caused by an exciton Mott transition of DX. It is noteworthy that on both timescales (metastable as well as ultrashort), the Fermi energy E_F of the metallic band equals the equilibrium E_F (*cf.* Fig. 6b and ref. 25). Although photoexcitation drives the system into a non-equilibrium state, the resulting phase transition is equivalent to the equilibrium metallization by chemical doping. We suspect that this is due to the surface character of the photodoping and results from equilibration of the surface Fermi level with that of bulk ZnO. As charge accumulation layers of oxides can be as thin as 1 nm,[24,48] charge exchange between the metallic state at the surface and the bulk happens on ultrafast timescales. In conclusion, the spectroscopic similarity to H doping, as well as the photoinduced semiconductor-to-metal transition in a nominally undoped ZnO sample, illustrates that DX are truly equivalent to shallow donors. This "photodoping" allows us to control the conduction properties of ZnO over at least 6 orders of magnitude in time, from ps to μs and even longer timescales.

4.2 Relation to PPC

Persistent photoconductivity has been predicted theoretically to arise from photoexcited oxygen vacancies, creating shallow donor states.[17] Although oxygen vacancies are widely accepted to be the source of PPC, many experimental studies explain the effect within the framework of equilibrium upward band bending. Moreover, as PPC lifetimes range from below seconds to hours and longer, it is still being discussed whether it is due to a single mechanism, *i.e.* photoexcited oxygen vacancies, or whether the photodesorption of adsorbed oxygen as alternative mechanism is at play.[16]

The photodoping of the ZnO surface due to photoexcited oxygen vacancies reported in this work means that there must be photoinduced conductivity lasting as long as the DX live. We identified a species decaying with $\tau_d = 250$ ps and two metastable species, one decaying within 77 μs and one exceeding the μs regime. Within the theory of photoexcited oxygen vacancies,[17,19,38] the different lifetimes of the DX can be explained as follows: the ultrafast decay $\tau_d = 250$ ps is likely to be attributed to direct recombination of photoexcited electron–hole pairs prior to forming the polaronic state, which prolongs the lifetime.[17,38] Regarding the two metastable species, the theory unanimously predicts that the oxygen vacancies can be singly and doubly charged. Both charging states deform the lattice and thereby suppress electron–hole recombination, which leads to metastable states.[17,19] Therefore, we tentatively assign the metastable species to the singly and doubly photoionized oxygen vacancies, respectively.

While our results directly demonstrate that the Mott transition of DX leads to photoconductivity on sub-second timescales, DX-driven PPC at longer times can be inferred indirectly, as outlined in the following. Our experiments indicate a long-lived, metastable part of the DX population that does not decay on μs timescales. These long lifetimes can be rationalized by considering that the defect-induced, metallic impurity band lies below the sample's Fermi energy. This means that, on the one hand, any depopulation process beyond electron–hole recombination (*e.g.* thermal activation to unoccupied levels) will eventually be compensated by repopulation, as the electronic level will persist as long as the hole polaron remains. Electron–hole recombination, on the other hand, is highly unlikely, because no vertical transitions are available in conformational space due to the strong lattice deformation.[17] Conceptually, this can be viewed in a Marcus-type picture for electron transfer,[49] where the reorganization energy stored in the environment of the hole is so large that there is no free energy gain upon recombination with the electron. Such trapping of electrons below E_F can easily lead to lifetimes of minutes, hours, or even days. Although we cannot determine this lifetime experimentally and, therefore, cannot provide direct evidence for PPC on these timescales, we would like to point out that the action spectrum in Fig. 3b for the surface metallization is nearly identical to the photon energy dependence of PPC creation in various samples with lifetimes up to hours and longer.[14,16,20] In other words, as the minimum photon energies to create PPC or DX are 3.0–3.15 eV, consistent with the photoexcitation of oxygen vacancies, as discussed in Section 3.3, it seems highly likely that DX are also responsible for PPC on these long timescales. This conclusion is in agreement with the previous assignment by ref. 14, 20 and in opposition to ref. 16 which considered sub-band gap photoexcitation so close to the fundamental band gap as insufficient proof and only attributed the short-lived PPC to photoexcited oxygen vacancies. We believe that our experiments show clear evidence for photoexcited oxygen vacancies as the origin of PPC on timescales below seconds as well as on hours and longer, possibly due to single and double photoionization. The oxygen vacancies are created by UV illumination at the ZnO surface and are subsequently photoexcited into metastable shallow donor-like states, as predicted theoretically.

Beyond associating the fast and slow timescales of PPC with the photoexcitation of oxygen vacancies, our study, importantly, resolves that PPC due to DX must be understood by their shallow donor behaviour. As ZnO is photodoped at the surface, this leads to downward surface band bending and eventually to a metallic surface. The auxiliary condition of upward band bending at ZnO surfaces is not necessary to explain PPC. On the contrary, as DX may already be excited under daylight illumination, ZnO surfaces are likely to always exhibit downward band bending. The literature suggests that the photoinduced downward shift of the surface potential of ZnO, inferred from a downward shift of the core levels[22] and the work function,[21] is evidence for equilibrium upward band bending. In this picture, upward surface band bending leads to a separation of the charges, and a transient polarization field at the surface mitigates the band bending (surface photovoltage). Our work shows that this circuitous view of equilibrium surface band bending abolished by photoexcitation can be replaced by the plain photoexcitation of DX, as these provide one single, simple explanation of all the phenomena discussed above: DX induce downward surface band bending, also known as photovoltage inversion,[50] which goes along with a downward shift of the surface potential. They also undergo a Mott transition and render the ZnO surface metallic on ultrafast and macroscopic timescales.

The above conclusion that ZnO surfaces tend to have downward band bending has important implications for ZnO devices, such as ZnO-based hybrid solar cells, for example. In these solar cells, a current is generated after the injection of photogenerated electrons from an organic adlayer into ZnO. The assumption of upward band bending led to the conclusion that ZnO is a good substrate as electrons would be driven away from the interface, would have long lifetimes and could thus contribute to the current.[22,51] Downward surface band bending due to DX as observed in this work might actually have the opposite effect, *i.e.* the trapping of injected charges at the interface and hindering of charge extraction. This is consistent with reports of ZnO-based hybrid solar cells that become less efficient upon photoexcitation with UV light, which was attributed to UV illumination-induced surface states.[52,53] The suppression of ZnO deep defect states due to surface modification by adlayers was found to increase the solar cell performance.[52]

The insights gained in this work are thus quite universally relevant to applications of ZnO, as the DX influence the electronic surface properties, inducing conductivity and downward band bending. We expect that DX are excited whenever ZnO is exposed to UV light and affect the electronic properties even very long after excitation.

5 Summary

In summary, static as well as time- and angle-resolved PE spectroscopy of ZnO(10−10) under UHV conditions allowed us to reveal the presence of photoexcited excitons with the hole at a deep defect site at the surface. The DX lead to a similar spectroscopic signature as that produced by chemical doping with H and cause a Mott semiconductor-to-metal transition at sufficient densities. Depending on the initial defect densities, the metallization can be short-lived or metastable, following the complex decay dynamics of the DX, which range from hundreds of ps to μs and even longer. The Fermi level in the surface metallic phase equals the equilibrium E_F on ultrashort to metastable timescales, *i.e.* photoexcitation acts like chemical doping in thermodynamic equilibrium. The deep defects are created by UV illumination and feature an (occupied) energy level at <400 meV above the VB edge. These results are consistent with photoconductivity measurements of ZnO reported in the literature and strongly support that both short- and long-lived (persistent) photoconductivity arise due to photoexcited oxygen vacancies at the ZnO surface, which behave like shallow donors. The transient surface doping leads to downward band bending, as opposed to the usually assumed upward surface band bending in photoexcited ZnO.[15,16,22,23] To the extent that we can judge with PE spectroscopy, we conclude that the conceptual analogy between shallow donors and optically excited DX holds true, ranging from quasi-equilibrium metallization and critical behaviour on ultrafast timescales to metastable n-type doping at the ZnO surface. Beyond these fundamental research aspects, our results are also relevant to ZnO-based devices. DX may be created without knowledge, simply by daylight exposure, leading to photoinduced conductivity and surface band bending with effects, for example, on ZnO solar cells. In addition, the intentional photoexcitation of DX can be used for the optical control of the conductive properties of ZnO even on ultrafast timescales. Lastly, we expect that photodoping by DX can also be reached in other semiconductor compounds with deep defect states, including complex materials.[54]

Author contributions

L. G. and S. V. did the experiments, L. G. analyzed the data, L. G. and J. S. wrote the manuscript. J. S. supervised the work. All authors contributed to the discussion and interpretation of the results.

Conflicts of interest

There are no conflicts to declare.

Acknowledgements

This work was funded by the Deutsche Forschungsgemeinschaft (DFG, German Research Foundation)-Project-ID 182087777-SFB 951. S. V. would like to thank the Max Planck Research Society for a Max Planck Postdoctoral Fellowship. The authors thank J.-C. Deinert for providing the reference spectrum (H dosed ZnO) in the inset of Fig. 2b.

Notes and references

1 N. F. Mott, *Semiconductor-to-metal Transitions*, Taylor & Francis, London, 1st edn, 1974.

2 G. Yu, C. H. Lee, A. J. Heeger, N. Herron and E. M. McCarron, *Phys. Rev. Lett.*, 1991, **67**, 2581–2584.

3 D. N. Basov, R. D. Averitt, D. Van Der Marel, M. Dressel and K. Haule, *Rev. Mod. Phys.*, 2011, **83**, 471–541.

4 C. Giannetti, M. Capone, D. Fausti, M. Fabrizio, F. Parmigiani and D. Mihailovic, *Adv. Phys.*, 2016, **65**, 58–238.

5 A. Cavalleri, M. Rini, H. H. Chong, S. Fourmaux, T. E. Glover, P. A. Heimann, J. C. Kieffer and R. W. Schoenlein, *Phys. Rev. Lett.*, 2005, **95**, 237401.

6 D. Wegkamp, M. Herzog, L. Xian, M. Gatti, P. Cudazzo, C. L. McGahan, R. E. Marvel, R. F. Haglund, A. Rubio, M. Wolf and J. Stähler, *Phys. Rev. Lett.*, 2014, **113**, 216401.
7 S. Iwai, M. Ono, A. Maeda, H. Matsuzaki, H. Kishida, H. Okamoto and Y. Tokura, *Phys. Rev. Lett.*, 2003, **91**, 057401.
8 J. Quereda, T. S. Ghiasi, C. H. Van Der Wal and B. J. Van Wees, *2D Materials*, 2019, **6**, 025040.
9 L. V. Scalvi and C. F. Bueno, *J. Phys. D: Appl. Phys.*, 2020, **53**, 033001.
10 P. A. Cox, *The Electronic Structure and Chemistry of Solids*, Clarendon Press, 1st edn, 1987.
11 Y. Zhang, *Chin. Phys. B*, 2018, **27**, 117103.
12 Y. Zhang and J. Wang, *Phys. Rev. B: Condens. Matter Mater. Phys.*, 2014, **90**, 155201.
13 S. Katsumoto, F. Komori, N. Sano and S. Kobayashi, *J. Phys. Soc. Jpn.*, 1987, **56**, 2259–2262.
14 R. J. Collins and D. G. Thomas, *Phys. Rev.*, 1958, **112**, 388.
15 J. Bao, I. Shalish, Z. Su, R. Gurwitz, F. Capasso, X. Wang and Z. Ren, *Nanoscale Res. Lett.*, 2011, **6**, 404.
16 M. Madel, F. Huber, R. Mueller, B. Amann, M. Dickel, Y. Xie and K. Thonke, *J. Appl. Phys.*, 2017, **121**, 124301.
17 S. Lany and A. Zunger, *Phys. Rev. B: Condens. Matter Mater. Phys.*, 2005, **72**, 035215.
18 S. Lany and A. Zunger, *Phys. Rev. Lett.*, 2007, **98**, 045501.
19 A. Janotti and C. G. Van De Walle, *Appl. Phys. Lett.*, 2005, **87**, 122102.
20 P. Liu, G. She, Z. Liao, Y. Wang, Z. Wang, W. Shi, S.-T. Lee and D. Chen, *Appl. Phys. Lett.*, 2009, **94**, 063120.
21 R. Gurwitz, R. Cohen and I. Shalish, *J. Appl. Phys.*, 2014, **115**, 033701.
22 B. F. Spencer, D. M. Graham, S. J. O. Hardman, E. A. Seddon, M. J. Cliffe, K. L. Syres, A. G. Thomas, S. K. Stubbs, F. Sirotti, M. G. Silly, P. F. Kirkham, A. R. Kumarasinghe, G. J. Hirst, A. J. Moss, S. F. Hill, D. A. Shaw, S. Chattopadhyay and W. R. Flavell, *Phys. Rev. B: Condens. Matter Mater. Phys.*, 2013, **88**, 195301.
23 P. M. Wojcik, L. D. Bastatas, N. Rajabi, P. V. Bakharev and D. N. McIlroy, *Nanotechnology*, 2021, **32**, 035202.
24 J.-C. Deinert, O. T. Hofmann, M. Meyer, P. Rinke and J. Stähler, *Phys. Rev. B: Condens. Matter Mater. Phys.*, 2015, **91**, 235313.
25 L. Gierster, S. Vempati and J. Stähler, *Nat. Commun.*, 2021, **12**, 978.
26 J.-C. Deinert, PhD thesis, Technische Universität Berlin, 2016.
27 C. Wöll, *Prog. Surf. Sci.*, 2007, **82**, 55–120.
28 S.-H. Na and C.-H. Park, *J. Korean Phys. Soc.*, 2009, **54**(9(2)), 867–872.
29 U. Diebold, L. V. Koplitz and O. Dulub, *Appl. Surf. Sci.*, 2004, **237**, 336–342.
30 J. Stähler and P. Rinke, *Chem. Phys.*, 2016, **485–486**, 149–165.
31 A. Damascelli, *Phys. Scr.*, 2004, **T109**, 61–74.
32 U. Bovensiepen, C. Gahl and M. Wolf, *J. Phys. Chem. B*, 2003, **107**, 8706–8715.
33 L. Gierster, PhD thesis, Technische Universität Berlin, 2021.
34 K. Ozawa, K. Sawada, Y. Shirotori and K. Edamoto, *Phys. Rev. B: Condens. Matter Mater. Phys.*, 2003, **68**, 125417.
35 C. Klingshirn, J. Fallert, H. Zhou, J. Sartor, C. Thiele, F. Maier-Flaig, D. Schneider and H. Kalt, *Phys. Status Solidi B*, 2010, **247**, 1424–1447.

36 C. Klingshirn, *Semiconductor Optics, Springer Series in Materials Science*, 4th edn, 2012.
37 H. Sezen, H. Shang, F. Bebensee, C. Yang, M. Buchholz, A. Nefedov, S. Heissler, C. Carbogno, M. Scheffler, P. Rinke and C. Wöll, *Nat. Commun.*, 2015, **6**, 6901.
38 A. Janotti and C. G. Van de Walle, *Rep. Prog. Phys.*, 2009, **72**, 126501.
39 K. Ellmer and A. Bikowski, *J. Phys. D: Appl. Phys.*, 2016, **49**, 413002.
40 C. H. Patterson, *Phys. Rev. B: Condens. Matter Mater. Phys.*, 2006, **74**, 144432.
41 J. Wang, Z. Wang, B. Huang, Y. Ma, Y. Liu, X. Qin, X. Zhang and Y. Dai, *ACS Appl. Mater. Interfaces*, 2012, **4**, 4024–4030.
42 Y. Kang, H. H. Nahm and S. Han, *Sci. Rep.*, 2016, **6**, 35148.
43 W. Göpel and U. Lampe, *Phys. Rev. B: Condens. Matter Mater. Phys.*, 1980, **22**, 12.
44 S. Vempati, J. Mitra and P. Dawson, *Nanoscale Res. Lett.*, 2012, **7**, 470.
45 J. Cunningham, E. Finn and N. Samman, *Faraday Discuss. Chem. Soc.*, 1974, **58**, 160–174.
46 J.-C. Deinert, D. Wegkamp, M. Meyer, C. Richter, M. Wolf and J. Stähler, *Phys. Rev. Lett.*, 2014, **113**, 057602.
47 K. K. Korir, A. Catellani and G. Cicero, *J. Appl. Phys.*, 2016, **120**, 125301.
48 S. McKeown Walker, A. De La Torre, F. Y. Bruno, A. Tamai, T. K. Kim, M. Hoesch, M. Shi, M. S. Bahramy, P. D. King and F. Baumberger, *Phys. Rev. Lett.*, 2014, **113**, 177601.
49 R. A. Marcus, *J. Chem. Phys.*, 1956, **24**, 966–978.
50 L. Kronik and Y. Shapira, *Surf. Sci. Rep.*, 1999, **37**, 1–206.
51 B. F. Spencer, M. J. Cliffe, D. M. Graham, S. J. Hardman, E. A. Seddon, K. L. Syres, A. G. Thomas, F. Sirotti, M. G. Silly, J. Akhtar, P. O'Brien, S. M. Fairclough, J. M. Smith, S. Chattopadhyay and W. R. Flavell, *Faraday Discuss.*, 2014, **171**, 275–298.
52 B. Wu, Z. Wu, Q. Yang, F. Zhu, T. W. Ng, C. S. Lee, S. H. Cheung and S. K. So, *ACS Appl. Mater. Interfaces*, 2016, **8**, 14717–14724.
53 Y. Kim and B. Park, *Nanoscale*, 2018, **10**, 20377–20383.
54 W. Li, J. Shi, K. H. Zhang and J. L. Macmanus-Driscoll, *Mater. Horiz.*, 2020, **7**, 2832–2859.

Faraday Discussions

ROYAL SOCIETY OF CHEMISTRY

DISCUSSIONS

Material science: ultrafast transformation, electron–phonon coupling, multi-scale aspects: general discussion

Martina Basini, Lara Benfatto, Yorrick Boeije, Wibke Bronsch, Lin X. Chen, Eric Collet, Aswathy V. Girija, Hamoon Hedayat, Kyoko Ishizaka, Shinichiro Iwai, J. Olof Johansson, Steven L. Johnson, Kota Katsumi, Shinya Koshihara, Jan Michael Rost, Habib Rostami, Julia Stähler and Mattia Udina

DOI: 10.1039/d2fd90035a

Martina Basini opened discussion of the introductory lecture by Steven L. Johnson (https://doi.org/10.1039/d2fd00098a): Please clarify the case of an electronic excitation that is able to change the statistical distribution of the phonon population.

Steven Johnson answered: I assume this question refers to the phenomenon of "phonon squeezing", which is experimentally seen as an oscillation of the variance of the generalized coordinate of one or more vibrational modes. This is in contrast to coherent phonons, which are typically identified with oscillations of the expectation value of the generalized coordinate.

When a harmonic crystalline material is in thermal equilibrium, each of the vibrational modes can (in a quantum mechanical picture) be represented as consisting of an admixture of number states, with a Bose–Einstein distribution. The expectation value of each mode coordinate is zero and independent of mode frequency Ω and temperature T; the variance of the coordinate, however, depends on both Ω and T. In the low temperature limit, the variance is dominated by the ground-state variance, which is inversely proportional to Ω^2. One way to see the "squeezing" phenomenon is to drive a sudden change in the value of Ω by using an ultrashort light pulse to change the electronic state of the system. If the electronic state change couples to the vibrational modes *via* terms in the Hamiltonian that are quadratic in the mode coordinates, the electronic state change can trigger a nearly instantaneous change in the frequency Ω. After this is done, the admixture of vibrational states that characterized the initial conditions are no longer eigenstates of the modified vibrational Hamiltonian, and will, therefore, evolve in time. This results in oscillations of the variance of the mode coordinate at a frequency of 2Ω. More mathematical detail and descriptions of experiments that have observed this can be found in several references, *e.g.*, ref. 1 or 2. At high temperatures, one can also use a classical framework to describe the phenomenon accurately, and, in fact, this is the regime where most experiments are actually done.

It is maybe also important here to note that squeezing indicates a very non-thermal vibrational state, and so talking about a "statistical distribution of the phonon population" is not really the right picture since the vibrational state is not accurately represented as an admixture of number states.

1 S. L. Johnson, P. Beaud, E. Vorobeva, C. J. Milne, É. D. Murray, S. Fahy and G. Ingold, *Phys. Rev. Lett.*, 2009, **102**, 175503.
2 M. Trigo, M. Fuchs, J. Chen, M. P. Jiang, M. Cammarata, S. Fahy, D. M. Fritz, K. Gaffney, S. Ghimire, A. Higginbotham, S. L. Johnson, M. E. Kozina, J. Larsson, H. Lemke, A. M. Lindenberg, G. Ndabashimiye, F. Quirin, K. Sokolowski-Tinten, C. Uher, G. Wang, J. S. Wark, D. Zhu and D. A. Reis, *Nat. Phys.*, 2013, **9**, 790.

Mattia Udina commented: You mentioned that Impulsive Stimulated Raman Scattering (ISRS) and Displacive Excitation of Coherent Phonons (DECP) can be seen as two different regimes of the same excitation process. Nonetheless, while we have clear experimental evidence for the excitation of lower-symmetry modes *via* ISRS, it seems that only fully symmetric (A_{1g}) modes can be generated *via* DECP. Why?

Steven Johnson replied: The key difference between fully symmetric modes and other vibrational modes is the lifetime of the electronic state population that couples to a displacement of the minimum of the potential energy surface along the mode coordinate. For fully symmetric modes, the magnitude and sign of the electron–phonon coupling that leads to this displacement typically varies only a little along a given band. In this situation, fast intraband carrier–carrier scattering processes that occur at very high excitation levels do not significantly change the vibrational mode dynamics.

For non-fully symmetric vibrational modes, only excited electronic states with the same symmetry as the mode can cause a displacement along the mode coordinate. These states are typically selected by the polarization of the pump pulse relative to the crystal structure. For large excitation levels, carrier–carrier scattering will cause these lower-symmetry electronic states to rapidly depopulate on a time scale much faster than the phonon period, resulting in a quasi-thermal population of electronic states with the full symmetry of the crystal. Thus one sees only an "impulsive" force on these modes, rather than the persistent displacement often seen for fully-symmetric modes.

Shinichiro Iwai asked: What do you think about the role of electronic coherence in coherent phonon generation and the structural phase transition driven by it (in the real excitation case)? Although an electronic coherence only survives for several fs to several tens of fs, the initial processes of the coherent phonon generation may be affected by it, in particular for the strong electron–phonon coupling case. In that sense, I guess that the pulse duration being shorter or longer than the electronic coherence time may be an important problem. Can we expect coherent control of the coherent phonon/structural change if the pulse width is shorter than the electronic coherence time?

Steven Johnson replied: In principle, the key element for the electron–phonon coupling is the lifetime of the state (or states) that couple to the vibration. I do not believe that the phase coherence of the electronic state usually has a direct impact on the vibrational dynamics, since the frequency of the electronic transition is typically many orders of magnitude higher than that of the vibrational mode. This may not be the case when the electronic transition has an energy comparable to that of the phonon, for example, when exciting very low-energy transitions with THz light.

The electronic coherence time probably plays a more prominent role if one tries to exploit the interference between two or more electronic excitations (*e.g.* with time-shaped pulses); in this case, the excited state population can be significantly modified. This can in turn change the amplitude of the coherent phonon.

Jan Rost remarked: Short pulses excite a lot of levels, at least transiently. Could this lead to a situation where the materials' dynamics are ruled by non-linear (semi-)classical mechanics, therefore making it susceptible to methods in control theory to steer the system to a desired state?

Steven Johnson answered: I think so, although the dynamics would likely depend strongly on the system choice, as well as the pulses themselves. This would be a very fruitful area to explore in the future.

Yorrick Boeije communicated: In my understanding, phonons activated in the displacive and impulsive limits are the equivalent of excited-state and ground-state vibrational coherence in molecules. For molecular systems, these may be easily distinguished with optical pump–probe spectroscopy when the excited-state absorption (ESA) and stimulated emission (SE) are spectrally separated from the ground-state bleach (GSB). How do you distinguish between impulsive and displacive phonons with pump–probe spectroscopy in solids, where usually there is no clear separation of excited-state (ESA and SE) and ground-state transitions (GSB)?

Steven Johnson communicated in reply: This can be very tricky. Often this distinction is attempted by trying to accurately measure the phase of the oscillation, which is nominally cosine-like for the displacive limit, and sine-like in the impulsive limit. In practice it can be extremely difficult to measure this accurately, and so there is often ambiguity. In some cases the ambiguity can be partially lifted with theory, maybe in combination with wavelength-dependent studies to see if the phase depends on the pump wavelength (going into or out of resonance with particular electronic transitions). The most direct experimental method is time-resolved X-ray or electron diffraction, where one can directly see if the structural dynamics occur around a displaced structure or oscillate around the initial equilibrium structure.

Lin Chen asked: Would you please comment on the effect of excess energy in the photoexcitation of solids mentioned in your introductory lecture (https://doi.org/10.1039/d2fd00098a)? What exactly does this imply?

Steven Johnson answered: I was referring to the fact that, for many (or even most) photoexcitation schemes, a large fraction of the energy deposited into a material *via* electronic state transitions is often distributed across all degrees of freedom in a quasi-thermal distribution. This can be viewed as a consequence of the large energy difference between the ground and initially-excited state *versus* the energy difference between the ground and metastable state. Although the ergodic hypothesis is only partially satisfied at finite times after excitation, control schemes are still constrained to some extent by the second law of thermodynamics.

Lin Chen added: We have observed transient optical absorption features that can be reproduced by heating of the solid sample upon photoexcitation. Some of the old literature reports transient absorption (TA) or X-ray transient absorption (XTA) features of photoexcitation that could be mostly from heating. Thus, it is always important to distinguish between the thermal effects and electronic transition effects in solid samples under light illumination, especially with excess photon energies.

Steven Johnson opened a general discussion of the paper by Kyoko Ishizaka: You report a time resolution of approximately 2 ps and a spatial resolution of 15 μm for the confocal microscope. These choices are a compromise due to the fact that you require a certain minimum charge per pulse in order to get a sufficient signal-to-noise ratio within a time frame where you do not see significant alignment drifts. You also mention that it might be possible to use pulse compression techniques to either increase the number of electrons per pulse or to improve spatial or temporal resolution (https://doi.org/10.1039/d2fd00062h). What do you think would be the potential gains from pulse compression? Would you prefer to optimize the spatial resolution, the temporal resolution, or the acquisition time?

Kyoko Ishizaka responded: First of all, let me confirm that our spatial resolution is currently about 15 nm, not 15 μm. Regarding the pulse compression, we have not yet made any estimation, but we would like to try some simulations. At present, since we are focusing on the nanoscale spatiotemporal dynamics appearing in the materials, we are trying to improve the spatial resolution from ~15 nm to a few-nm scale, by reducing the system's fluctuations, as well as the acquisition time.

Habib Rostami asked: I have a question about the mechanism of the light-induced strain. With an intense laser, we may also heat up the sample locally, and, therefore, due to a thermal expansion, a strain can be generated. Can't it be just a laser-induced thermalization effect in the THz frequency, instead of being a coherent acoustic phonon excitation? How relevant is the laser-induced thermalization in your experiment?

Kyoko Ishizaka answered: You are right that in the case of the Si/W interface as presented in our manuscript, the laser-induced thermalization is the first trigger to cause the strain. Of course, in some more complex materials (*e.g.* charge density wave (CDW) materials as presented in ref. 1), more non-trivial light–lattice interactions seemingly exist. I understand your comment is the matter of the definition of "coherent acoustic phonon". In our work, we refer to the "coherent acoustic phonons" as demonstrated in some previous works, such as ref. 2, where several mechanisms, including thermoelasticity, are raised as the origin.

1 A. Nakamura, T. Shimojima, Y. Chiashi, M. Kamitani, H. Sakai, S. Ishiwata, H. Li and K. Ishizaka, *Nano Lett.*, 2020, **20**(7), 4932–4938.
2 P. Ruello and V. E. Gusev, *Ultrasonics*, 2015, **56**, 21–35.

Aswathy V. Girija queried: Would having a better time resolution help to distinguish between the purely optically-induced strains and the photothermally-induced strains?

Kyoko Ishizaka replied: Yes. At present, the ~2 ps time resolution is quite comparable to the timescale of the photothermal effect in many materials. If we want to distinguish between the different (faster) mechanisms of the photo-induced strains, we should pursue the better time resolution, which requires further effort.

Aswathy V. Girija asked: In this system, would how the tungsten is bound to the silicon affect the mechanically-induced strains? Or is this not relevant for this work?

Kyoko Ishizaka replied: In our present case, the tungsten is deposited on silicon. How the tungsten is bound to the silicon may affect the strains generated at the interface, but we have not confirmed it experimentally.

Shinichiro Iwai queried: Just from a technical viewpoint, I would like to ask you what determines the time resolution?

Do you have an idea for improving it?

I agree that the time resolution is determined by the space-charge effect. Do you have a trade-off between the spatial and time resolutions? Are they related or independent?

Kyoko Ishizaka answered: As you mentioned, the time resolution is limited by the space-charge effect. To eliminate the space-charge effect, the number of electrons per pulse should be suppressed, which requires a long acquisition time. At present, the long acquisition time (>days) severely affects the spatial resolution, due to the noise and fluctuations inherent in our measurement setup. Thus, practically, the spatial and time resolutions are related problems.

Eric Collet asked: In ferroelastic materials there are two types of strain: volume strain and symmetry-breaking strain. Is it possible to use your five-dimensional ultrafast electron microscopy techniques to distinguish the out-of-equilibrium dynamics of both types of strain? This would open the way to distinguish/decouple symmetry-breaking and non-symmetry-breaking lattice distortion driven by optical excitation.

Kyoko Ishizaka replied: Thank you for your question. Yes, that is indeed what we are working on right now. By using five-dimensional convergent beam electron diffraction-scanning transmission electron microscopy (5D-CBED-STEM), and by detecting and analyzing the responses of different multiple diffraction indices, we can extract the different components of the displacement gradient tensor, *e.g.*, *xx*, *zz*, *zx* components, quantitatively.

Lara Benfatto opened the discussion of a paper by Wibke Bronsch: You modeled the equilibrium reflectivity data with a Drude term and Lorentz oscillators (https://doi.org/10.1039/d2fd00019a). I noticed that the fit fails in reproducing the low-energy part, which is actually beside the range of frequencies you are addressing with the supercontinuum probe pulse. Can you comment on the failure of the fit for the low-energy Drude part?

Wibke Bronsch responded: This question refers to the equilibrium reflectivity data published in ref. 1, which we are showing together with a fit based on the Drude–Lorentz model in Fig. 4a of our article (https://doi.org/10.1039/d2fd00019a). We fitted the sharp decrease in reflectivity around 8000 cm^{-1} (1 eV) with a Drude term. Since the Drude term describes intraband excitations, one would expect no absorption to be happening at energies lower than the plasma edge (as shown by the evolution of the fit (black curve)). Instead, the data from Berner *et al.*[1] show a clearly reduced reflectivity at energies below the Drude edge. We agree that this behavior is unexpected and unusual. Berner *et al.*[1] were interpreting the deviation from the behavior of a free electron gas with the existence of a charge density wave phase without long-range order for temperatures above 263 K, which is also indicated as a pseudogap.[2] This means that the sample is never purely metallic along the chain direction. Given the fact that, in our laboratories, we do not have the possibility to repeat the equilibrium measurements and that the range of the fit we are using for the analysis of the time-resolved data (see the blue shadowed area in Fig. 4a of our article (https://doi.org/10.1039/d2fd00019a)) won't be influenced by modifications in the low-energy range, we cannot further contribute to the discussion of what is causing the absorption of photons in the range up to 0.7 eV. We are also confident that these effects are not altering the results presented in our article. However, this issue is worthy of further investigation.

1 D. Berner, G. Scheiber, A. Gaymann, H. M. Geserich, P. Monceau and F. Lévy, *J. Phys. IV*, 1993, **3**, C2-255–C2-258.
2 N. Shannon and R. Joynt, *Solid State Commun.*, 2000, **8**, 411–415.

Lara Benfatto said: In your experiment, you can see oscillations of the amplitude mode by pumping with optical light, but not by pumping with X-rays. In the former case, the oscillations disappear when the fluency increases. Is this an indication of CDW melting at large optical fluency, which only occurs with X-rays? In the case of optical pumping, can you determine the amplitude of the oscillations as a function of the fluency to determine where they disappear?

Wibke Bronsch answered: This question refers to the fluence-dependent data acquired for optical pumping, which we show in Fig. 2c and e of our paper in comparison to the data acquired in the case of X-ray pumping shown in Fig. 7 (https://doi.org/10.1039/d2fd00019a). We confirm that it is possible to melt the CDW by pumping with a large optical fluence. A detailed study of the fluence dependence of the $\Delta R/R$ signal on $(TaSe_4)_2I$ is reported in ref. 1, and for this reason was not repeated by us. In the dataset (Fig. 5 in ref. 1), it is shown that the CDW can be quenched upon intense photoexcitation. However, weak oscillations connected to the CDW amplitude mode persist up to the highest fluence reported (1.5 mJ cm^{-2}), which is a rather high value. For this reason, the authors could not determine the value of the critical fluence. Hence, photoexcitation can indeed quench the CDW order, but the complete vaporization happens at very high fluences, and this should also ensure that the CDW is not (completely) vaporized in the X-ray pump experiment, in which the fluence used amounted to 570 μJ cm^{-2}.

1 H. Schaefer, M. Koerber, A. Tomeljak, K. Biljaković, H. Berger and J. Demsar, *Eur. Phys. J. Spec. Top.*, 2013, **222**, 1005–1016.

Lara Benfatto commented: You mentioned that the reflectivity changes observed when you pump with X-ray light can be ascribed to band-gap narrowing. Can you give us a hint why such a band-gap narrowing effect is happening?

Wibke Bronsch replied: The characteristic shape of the transient reflectivity change indicating a band-gap renormalization is observed, not only for the X-ray excitation, but also for the excitation with 800 nm. From the analysis of the optical data, our fit gave a modification of the oscillator with central wavenumber 10 800 cm^{-1} (or 1.34 eV), which represents the transition between the sets of valence and conduction bands along the crystal *a*-axis. The modification of this oscillator disturbs also the reflectivity at the probe wavelength of 800 nm in our X-ray pump experiments. The modification of the central wavenumber, however, is very small. Indeed, the effect is only in the meV range. The origin of band-gap renormalization effects in semiconductors is usually related to many-body screening effects of the produced electron–hole pair due to photoexcitation.

When pumping with X-rays, we are not only perturbing the core levels, but also the valence bands. On the basis of the data we have, we cannot conclude if the excitation of core levels gives an additional contribution to the band-gap narrowing.

Hamoon Hedayat remarked: In the X-ray pump experiments, the charge density waves (CDW) modulation is not observed in the $\Delta R/R$ signal, while it is observed when using the 800 nm pump wavelength. What can be the possible mechanism behind this difference and its relation to the CDW? The maximum change in $\Delta R/R$ using the X-ray pump is about one order of magnitude larger than with the 800 nm excitation. Can you explain why? Can it be related to the different photoexcitation densities due to different photon energies? Can you comment on what we can understand from this combined experiment? Thanks.

Wibke Bronsch answered: Hamoon Hedayat is addressing a number of puzzling questions. From the analysis of the changes in the dielectric function after optical excitation, we were concluding that the excitation of the valence band structure with low photon energies couples to low-energy Lorentz oscillators on a fs-timescale, contributing significantly to the observed dynamics. This component in the dynamics seems to be absent in the case of X-ray excitation, so one could investigate the role of interband excitations for coupling to the phonon modes related to the CDW. However, we should not forget that, not only does the excitation cross section for valence band excitations differ for the different pump photon energies, but there might also be effects due to the presence of core holes and following secondary processes, as well as due to different excitation densities. Surprisingly, we are observing larger changes in the reflectivity when exciting with X-rays, even though the excitation density should be less dependent on a smaller number of impinging photons and a washing out of the signal magnitude as X-rays probe larger unpumped areas (the penetration depth of the exciting X-rays is roughly 10 times larger than the penetration depth of the near-infrared probe). Our experiment hence shows that the X-ray excitation has a huge effect on the optical properties of the sample and we hope to open a discussion for further investigations to enlighten us about the underlying processes, and perform calculations in order to simulate the excitation mechanism.

J. Olof Johansson queried: How confident are you about the time resolution in the X-ray experiment? Could it be that the oscillations are washed out due to problems with jitter, *etc.*?

Wibke Bronsch answered: The typical time resolution obtained at the EIS-TIMER endstation, FERMI, Elettra-Sincrotrone Trieste, where we were performing the experiments, is in the order of 150 fs. The beamline has been used for a number of experiments so far, and the time resolution is a parameter which is under control. See, for example, the data reported in ref. 1. Moreover, the sharp-rising edge measured at the temporal overlap of the pump and probe pulses is also a clear indication that there were no unusual issues in the experiment with washing out the time resolution.

1 F. Bencivenga, R. Cucini, F. Capotondi, A. Battistoni, R. Mincigrucci, E. Giangrisostomi, A. Gessini, M. Manfredda, I. P. Nikolov, E. Pedersoli, E. Principi, C. Svetina, P. Parisse, F. Casolari, M. B. Danailov, M. Kiskinova and C. Masciovecchio, *Nature*, 2015, **520**, 205–208.

J. Olof Johansson asked: Do you expect to populate different states associated with the Lorentz oscillators if you varied the pump-pulse polarization direction with respect to the crystal orientation?

Wibke Bronsch replied: We agree that, for resonant excitations, the polarization of the exciting light has to match the orientation of the transition dipole moment of a specific interband transition. In our experiments, we are working with p-polarized pump pulses that are impinging on the sample surface at an angle, which means that we can excite not only out-of-plane, but also in-plane orbital components. Furthermore, we are not pumping resonantly to any of the identified Lorentz oscillator energies. Since we have no detailed knowledge about the symmetries of specific bands, we cannot speculate further on the effects of the pump-pulse polarization, but we expect only minor effects. However, we agree that for experiments pumping specific transitions resonantly, the chosen pump polarization can play a major role and give insight into the symmetries of the involved bands.

Kota Katsumi remarked: The decay time of the pump-induced change in the reflectivity looks very long. Can you explain why the decay time is so long? I would expect it to relax very quickly if it probes the relaxation of the photo-excited electrons.

Wibke Bronsch replied: This question refers to the fact that the differential reflectivity data we are showing display a component with a long lifetime, independent of the sample temperature and the crystal direction along which we are probing. In contrast to photoemission spectroscopy techniques that directly probe the photo-excited electrons from specific energy levels, meaning the band-structure population, optical spectroscopy probes the polarization of the material due to the interaction with light. A long-lasting change in the reflectivity of the sample after excitation, therefore, does not necessarily mean that photo-excited electrons are populating an excited state for a long time (as in the case of, *e.g.*, exciton or polaron formation, or trapping at impurity sites), but could also indicate a modification in the electronic structure that accompanies a lattice distortion, for example. Another possibility could be the formation of metastable states, as recently reported in ref. 1 and 2 for other materials. However, to date there is no evidence for metastable state formation in $(TaSe_4)_2I$. On the other hand, since the long-lasting dynamics are observed both with pulsed sources at 250 kHz (table-top laser) and 25 Hz (FEL), we can exclude heat accumulation as the cause of this effect (the average pump power differs by orders of magnitude in the two experiments). Hence, although we don't have a final proof for the origin of such dynamics, we would point toward a photo-induced modification (expansion) of the lattice.

1 L. Fanfarillo, D. Kopić, A. Sterzi, G. Manzoni, A. Crepaldi, D. T. Payne, W. Bronsch, V. Tsurkan, D. Croitori, J. Deisenhofer, F. Parmigiani, M. Capone and F. Cilento, *J. Electron Spectrosc. Relat. Phenom.*, 2021, **250**, 147090.
2 L. X. Yang, G. Rohde, Y. J. Chen, W. J. Shi, Z. K. Liu, F. Chen, Y. L. Chen, K. Rossnagel and M. Bauer, *J. Electron Spectrosc. Relat. Phenom.*, 2021, **250**, 147085.

Steven Johnson commented: The overall idea of comparing the non-equilibrium dynamics in charge density wave systems (or really any system) for very different pump-wavelength excitations is very interesting, but I think quite difficult, as acknowledged in this very nice work. One item in particular that I think needs to be emphasized due to its importance is the fact that the different pump wavelengths used here have very different absorption lengths (in this case, a factor of 10). This implies that, even for similar fluences, the per-volume initial excitation density is very different. For this reason, I would suspect that a likely explanation of why the amplitude mode oscillations are not seen for the vacuum ultraviolet (VUV)/X-ray excitation is indeed that the excitation density is high enough to melt the CDW within its absorption depth, as you already discuss in the paper (https://doi.org/10.1039/d2fd00019a).

Another interesting and related point: these excitation profiles may not be time-independent. For example, ballistic or diffusive carrier transport may cause rapid transport of energy into the bulk. These transport effects may be responsible for some of the time-dependent changes observed here.

I think that, in general, it might be difficult to see the effects from the excitation of core levels in such systems, since the core hole created by the initial excitation lives only for a few femtoseconds at best. Most of the energy of the pump I would expect to be rapidly transferred to general heating of the electronic system *via* cascading inelastic electron scattering. An ideal experiment to tease out any such effects would need to directly compare pump wavelengths with nearly identical absorption lengths, with one tuned to a specific core-level excitation. The dynamics that you might see will likely be impulsive (*versus* displacive) in nature due to the highly short-lived nature of the core holes.

Wibke Bronsch answered: We are thankful for this comment, which covers a number of different aspects and suggests pathways for further studies. Indeed, the use of very different wavelengths goes hand-in-hand with the strongly differing penetration depths (10 to 15 nm in the case of the X-rays; 100 nm in the case of 1.55 eV pumping) of the pump-beam, whereas the probing depth remained the same in the performed experiments. In the case of $E_{\text{pump}} = 1.55$ eV, pump and probe have comparable attenuation lengths, which means that we are roughly probing the pumped volume of the sample. In contrast, in the case of X-ray excitation, we are probing also a large part of the unpumped regions. This means that, in the case of the data shown for X-ray pumping, only ~10% of the probed volume contributes to a change in the reflectivity signal. However, averaged over the total amount of reflected light, we observe an effective reflectivity change of 3% (parallel direction) and 0.5% (perpendicular direction) due to pumping. This means that if we were able to probe only the pumped regions, the effect would be even ~10 times larger. Looking at the depth-dependent excitation density, under the given conditions of our experiment, in the surface, most layers are absorbing ~three times the number of optical photons than X-ray photons. Computing the total number of photons absorbed for excitation with 1.55 eV and 47 eV photons through all layers, we are reaching 3×10^{15} photons per cm^2 and 6×10^{13} photons per cm^2, respectively. This means that the total number of absorbed photons is about 50 times larger for optical excitation. Still, the recorded change in reflectivity is much larger in the case of the X-ray excitation. Hence, we can conclude that likely it is not the number of absorbed photons that determines the magnitude of the reflectivity change, even when one invokes the 30-times higher photon energy. However, when discussing the effect on the optical reflectivity changes, it is not the number of created core-holes, but rather the resulting impact on the carrier density in the valence and conduction bands that is decisive. Whereas one absorbed photon of 1.55 eV will create one hole in the valence band and one electron occupying the conduction band, an absorbed photon of 47 eV could either directly emit electrons from the valence bands or from deeper core levels, and subsequently modify the carrier density in the valence band structure due to, *e.g.*, the Auger effect. We expect that, due to secondary processes, one absorbed photon of 47 eV can result in multiple changes in the carrier density in the valence band structure. However, we are not able to quantify this. We agree that it would definitely be interesting to look for effects due to smaller variations in the pump wavelength in order to disentangle the role of different excitation and secondary processes. Furthermore, the role of transport effects is an interesting question, and exploration could lead to further insights.

Lin Chen asked: Do you expect any structural change on the timescale that you detect, which could modify the band structure? We have conducted a study on hematite and observed a band-structure change caused by the lattice-structure change.[1] Thus, the lattice nuclear structural change could cause the band-structure modification.

Wibke Bronsch responded: The techniques used in this work offer no direct access to the optically-induced structural changes of the sample. Anyhow, the observation of the long-lasting changes in the reflectivity of the sample might suggest a coupling of the optical excitation to the lattice under the form of a transient structural modification. It is generally true that a lattice modification leads to the modification of the electronic band structure (see, for example, ref. 2), and this is particularly true in CDW materials where the charge density modulation and the periodic-lattice distortion are intertwined. However, given the experimental conditions (pump photon energy, fluence) of our experiment, we tend to exclude a direct excitation of the lattice; rather, a possible lattice distortion could arise as a consequence of the photo-induced redistribution of charges in the system. In conclusion, a possible technique to study optically-induced changes in the lattice structure could be time-resolved X-ray diffraction. To our knowledge, such experiments have not yet been performed on this sample system.

1 D. Hayes, R. G. Hadt, J. D. Emery, A. A. Cordones, A. B. Martinson, M. L. Shelby, K. A. Fransted, P. D. Dahlberg, J. Hong, X. Zhang, Q. Kong, R. W. Schoenlein and L. X. Chen, *Energ. Env. Sci.*, 2016, **9**, 3754–3769.
2 D. Soranzio, M. Peressi, R. J. Cava, F. Parmigiani and F. Cilento, *Phys. Rev. Res.*, 2019, **1**, 032033(R).

Aswathy V. Girija remarked: As to future prospects, it is suggested that experiments with an X-ray pump and visible probe would be useful. Are there practical challenges in that? Furthermore, would integrating this spectroscopy with transient microscopy provide more understanding on the evolution of charge density waves? Which of these would be more relevant?

Wibke Bronsch answered: Experiments where an FEL X-ray pump is coupled to broadband detection of a supercontinuum light pulse are especially challenging because of the signal statistics that can be achieved at low repetition rates (50 Hz), that only allow us to measure very large transients. We recognize that this is only a technical issue and experiments in this sense have been designed and tested, however, the setups are still not ready for routine use. Once the above issues are mitigated, an X-ray pump/broadband optical probe experiment would be extremely powerful and would allow us to perform the same analysis of transient changes in the dielectric function as for the data obtained with the optical pump–probe experiment presented in Fig. 2 of the article (https://doi.org/10.1039/d2fd00019a). We could hence directly compare differences in the differential reflectivity maps and draw conclusions on the different coupling to the Drude and the Lorentz oscillators for the different pump wavelengths.

Transient absorption microscopy offers the possibility to study spatially-resolved coherent phonon excitations. This experiment would add complementary information to the studies we were performing so far, but it would not directly contribute to the analysis of changes in the dielectric function in the optical range in order to compare the impact of X-ray and optical excitations.

Yorrick Boeije asked: Have you identified the 2.6 THz phonon mode that is coupled to the charge density wave? If so, do you understand how this particular mode is excited, either through an impulsive-limited or displacive-limited mechanism?

Wibke Bronsch responded: The identified phonon mode of 2.6 THz agrees with the CDW-related phonon mode reported by ref. 1. In the case of displacive excitation, the phonon couples to the excited state. With respect to the phonon coordinate, the excited state is shifted with respect to the ground state. The atoms in the lattice start oscillating and move towards new central positions. Hence, at t_0, we are observing maximal intensity. The oscillation of the intensity is following a cosine function. In contrast, in the case of impulsive-stimulated Raman scattering processes, the phonon couples to the ground state. The oscillation as a function of pump–probe delay follows a sinusoidal function, instead of a cosine function. However, in real cases on complex materials, it has been shown that the phase of the oscillation can be influenced by a number of factors that can even lead to large variations of the initial phase from the two ideal cases. Moreover, it has been shown that the two cases are two limits of a more general theory.[2] However, in the case of strongly absorptive materials, like $(TaSe_4)_2I$, the main mechanism for coherent phonon generation is the DECP.

1 H. Schaefer, M. Koerber, A. Tomeljak, K. Biljaković, H. Berger and J. Demsar, *Eur. Phys. J. Spec. Top.*, 2013, **222**, 1005–1016
2 T. E. Stevens, J. Kuhl and R. Merlin, *Phys. Rev. B*, 2002, **65**, 44304.

Jan Rost opened a general discussion of the paper by Julia Stähler: You emphasized the analogy between chemical and photoexcitation doping – is there a difference given that the angular momentum character appears to be more restricted for photoexcitation doping due to the strict selection rules?

Julia Stähler replied: Our comparison between chemical and photoexcitation doping is based on considering shallow donors as localized excitons, which are spontaneously formed in semiconductors. With photoexcitation, we produce a localized exciton by photoexcitation of a deep defect level. The electron that is bound in the Coulomb potential of the hole is very much like a shallow donor. We show that they behave the same, and can cause metallization of the semiconductor above a critical density. There might, of course, be differences in the exciton states. Typically, in textbooks, one assumes that electrons are in the 1s state of the exciton when interpreted as shallow donors. However, for semiconductors, this transition might be forbidden due to the dipole selection rules (as, *e.g.*, the famous case of CuO_2, which has a direct forbidden energy gap[1]). We, unfortunately, do not know the transition dipole moment for the excitation from the oxygen vacancy to the ZnO conduction band. Generally, if states with higher angular momentum, *e.g.*, p-states, are excited preferentially, their spatial extension will be different compared to the s-state. This will modify the Mott density, *i.e.*, the critical density at which metallization occurs. This could affect the early stage after photoexcitation (*i.e.*, on a femto- to picosecond timescale), while the exciton relaxes to the 1s ground state on the longer term.

1 T. Kazimierczuk, D. Fröhlich, S. Scheel, H. Stolz and M. Bayer, *Nature*, 2014, **514**, 343–347.

Shinya Koshihara asked: Is it possible to analyze/show the crystal deformation combining with an exciton (hole) to form a defect exciton (DX) center?

Julia Stähler communicated in reply: The crystal deformation (polaron), which is in fact expected from theory, should lead to a lowering of the electron energies detected in our photoemission measurements. In the early time delays after the excitation of the defects, we indeed observe a small energy relaxation of the state below the Fermi energy, *cf.* the first 2 ps in Fig. 3a of the paper (https://doi.org/10.1039/d2fd00036a)). However, it is difficult to attribute this to polaron formation as, on a similar timescale, a downward shift of the band structure occurs due to surface charging by hole trapping at the surface.[1] To answer your question: No, we can not directly characterize the crystal deformation with our experiments, unfortunately. This would require theoretical input or a structural probe, ideally with spatial resolution.

1 L. Gierster, S. Vempati and J. Stähler, *Nat. Commun.*, 2021, **12**, 978.

(**Eric Collet** communicated: In Fig. 6(c) of your paper (https://doi.org/10.1039/d2fd00036a), there is a kind of dispersion, which appears after 6 h of UV exposure. Is this kind of free-electron-like band due to interactions between defect excitons, or is it the large number of defect excitons that is responsible for the appearance of this band?

Julia Stähler communicated in reply: From our understanding, in the case of a bare density enhancement of non-interacting defect excitons, their spectroscopic signatures would be identical and just add up, leading to a pure intensity increase. However, when you enhance the defect exciton density sufficiently, such that the quasiparticles start interacting, the situation changes. At a critical density, the Mott transition occurs between the defect excitons. This is commonly assumed to happen when the Bohr radii of the quasiparticles start overlapping.

As mentioned in the paper (https://doi.org/10.1039/d2fd00036a), we observe an analogous dispersion change upon exciting defect excitons in the ultrafast time domain[1] where the number of defect excitons can be tuned by pump-laser fluence, *i.e.*, more easily than in the photostationary state. The effective mass of the free electron-like band decreases further as the excitation density (*i.e.*, the number of excited deep defects) is increased, following the critical behaviour expected theoretically for the Mott transition and observed for doped semiconductors. A second important argument for the impurity band formation (*i.e.*, metallization) accessed in the ultrafast time domain is that the electrons start to follow Fermi–Dirac statistics at the very same density as the parabolic band appears, as is evidenced by the hot electron gas cooling in the ultrafast time domain (Fig. 1e in ref. 1).

1 L. Gierster, S. Vempati and J. Stähler, *Nat. Commun.*, 2021, **12**, 978.

Conflicts of interest

There are no conflicts to declare.

Faraday Discussions

PAPER

Quantum dynamics simulations of the thermal and light-induced high-spin to low-spin relaxation in $Fe(bpy)_3$ and $Fe(mtz)_6$†

Marc Alías-Rodríguez,*[ab] Miquel Huix-Rotllant[b] and Coen de Graaf[ac]

Received 31st January 2022, Accepted 21st February 2022
DOI: 10.1039/d2fd00027j

First row transition metal complexes with d^4 to d^7 electronic configurations exhibit spin-crossover (SCO), which can be induced by external stimuli, such as temperature, pressure and light. The low-spin to high-spin transition has been widely studied, but very little is known about the reverse process. Here, we present a theoretical study of thermal and light-induced high-to-low spin crossover in prototypical Fe(II) complexes. The lifetime of the high-spin state in the thermal process is determined using Fermi's golden rule. With this methodology, we have accurately computed the transfer rate of the HS state thermal relaxation at several time scales (from sub-nanosecond to a few seconds) in two different iron complexes. The use of quasi-degenerate perturbation theory (QDPT2) in the analysis of the LS–HS spin–orbit coupling has allowed us to identify $^3T_1$ as the main intermediate state coupling the LS and HS states. The light-induced process has been studied using wavepacket quantum dynamics along the main vibrational coordinates (one symmetric and two asymmetric Fe–N stretchings). The study suggests that after the initial excitation from the $^5T_{2g}$ to the ^5E_g state, the population is transferred back to a vibrationally hot $^5T_{2g}$ state, from which a small amount of the population is transferred to the $^1A_{1g}$ state *via* the intermediate $^3T_{1g}$. Most of the population remains trapped in the HS state at the time scale of the simulation.

Introduction

Spin-crossover (SCO) is a molecular phenomenon wherein the spin state of the system changes because of an external stimulus, such as pressure, temperature, light irradiation or the influence of a magnetic field. This process commonly takes place from the diamagnetic thermally stable low-spin state to the paramagnetic

[a]*Departament de Química Física i Inorgànica, Universitat Rovira i Virgili, Marcellí Domingo, 143007 Tarragona, Catalunya, Spain*
[b]*Aix-Marseille Univ, CNRS, ICR, Marseille, France. E-mail: marc.alias-rodriguez@univ-amu.fr*
[c]*ICREA, Passeig Lluís Companys 23, Barcelona, Spain*

† Electronic supplementary information (ESI) available. See DOI: 10.1039/d2fd00027j

meta-stable high-spin state. Although SCO can be observed in some molecules,[1,2] the phenomena is typically appreciated in d^4 to d^7 first row transition metal complexes.[3–5]

Among the different transition metal complexes exhibiting spin-crossover, Fe(II) complexes have probably been the most widely studied ones. Fe(II) has a d^6 valence configuration and usually presents a (quasi-)octahedral coordination.[6–9] In ferrous complexes, the magnetization varies between the diamagnetic singlet electronic state ($S = 0$) and the paramagnetic quintet electronic state ($S = 2$).

The magnetic bistability gives these complexes a wide range of possible applications, such as thermochromic paints, molecular switches or nanophotonic devices, to name a few.[10] For technological uses, low and high spin states must show a reasonable lifetime in the working temperatures, ideally room temperature. The lifetimes of meta-stable high-spin states strongly depend on their adiabatic energies, which makes the lifetime vary from a few picoseconds to several seconds or even days.

Hauser *et al.* described the thermal relaxation of the high-spin state as a tunneling process, which does not need to overcome the classical barrier.[11] They approximated the quantum mechanical expression for the rate constant in a tunneling process; assuming that the LS–HS spin–orbit coupling has a fixed value of 150 cm^{-1}, the vibrational frequency along the Fe–N symmetric stretching is 250 cm^{-1} for both the HS and LS states and these states are only displaced along this mode at a constant distance $\Delta r_{HL} \approx 0.2$ Å. The so-obtained analytical expression to determine the rate of the high-spin to low-spin transition depends only on the adiabatic energy difference between these states, and is known as the energy gap rule. This approach was validated for isotropic complexes, for which the reaction coordinate is mainly a totally symmetric Fe–N breathing mode, such as $[Fe(bpy)_3]^{2+}$.

The light-driven mechanism for the formation of the HS state in SCO complexes, (commonly known as light-induced excited spin state trapping, LIESST) has been largely studied in Fe(II) complexes using $[Fe(bpy)_3]^{2+}$ as a prototype complex. Auböck *et al.* determined the direct transfer from the metal to ligand charge transfer (^{1,3}MLCT) band to the ^{5}T state in the sub-50 fs time scale through ultrafast pump-probe spectroscopy.[12] Zhang *et al.* based on X-ray emission spectroscopy concluded that the population of the ^{5}T state was in two sequential steps, ^{1,3}MLCT → ^{3}T in 150 fs and ^{3}T → ^{5}T in 70 fs.[13] From a theoretical perspective, one of us has performed several studies based on Fermi's golden rule. The deactivation path *via* the ^{3}MC states was confirmed in the symmetric structures,[14] but the presence of geometrical distortions re-opened the direct transfer from the ^{1,3}MLCT band, making both paths competitive.[15] Recently, Pápai published a study about the LIESST in $[Fe(NCH)_6]^{2+}$, where only MC states play a role, based on quantum dynamic simulations.[16] He constructed a vibronic Hamiltonian along the three main modes (Fe–N stretchings) and determined an ultrafast first ISC $^1T_{1g} \rightarrow {}^3T_{2g}$ in 100 fs and a second slower ISC to the ^{5}T.

Despite the significant amount of LIESST studies, of which only a few are mentioned in the above paragraph, the number of studies on the reverse process is low. Marino *et al.* studied the reverse-LIESST after excitation to the ^5E_g in $[Zn_{1-x}Fe_x(ptz)_6](BF_4)_2$, where ptz = 1-propyltetrazole, using ultrafast pump–probe

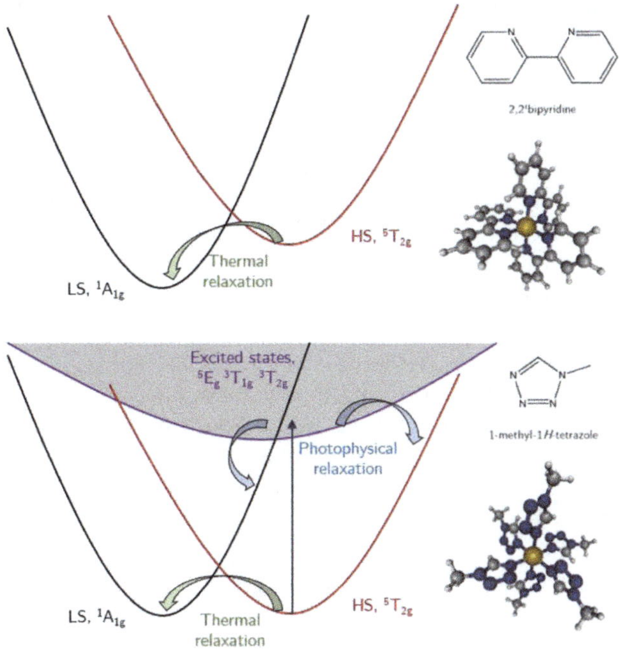

Fig. 1 Schematic diagram of the main electronic states involved in the thermal HS → LS transition in $[Fe(bpy)_3]^{2+}$ (top) and in the thermal and photoinduced relaxation in $[Fe(mtz)_6]^{2+}$ (bottom).

spectroscopy.[17] Following the time-resolved excited-state absorption (ESA), they established that two different transitions are involved in the process, whose rates were determined by a double exponential fitting. The first one was attributed to the $^5E_g \rightarrow {}^3T_{1g}$ in 1.7 ps ISC and the second to the $^3T_{1g} \rightarrow {}^1A_{1g}$ in 39 ps. The quantum yield was determined to be 0.1 for the reverse-LIESST and 0.8 for the LIESST in $[Fe(ptz)_6](BF_4)_2$ at 10 K, with a branching ratio $^3T_1 \rightarrow {}^5T_2/{}^3T_1 \rightarrow {}^1A_1$ of 4 : 1

In the current study, we simulated the thermal and light-induced conversion from the high-spin to low-spin in $[Fe(bpy)_3]^{2+}$, where bpy = 2,2′-bipyridine, and $[Fe(mtz)_6]^{2+}$, where mtz = 1-methyl-1*H*-tetrazole. Two different processes have been studied, the HS → LS thermal relaxation and the photophysical process upon the absorption of a photon from the HS ($^5T_{2g} \rightarrow E_g$), *i.e.* the reverse-LIESST (Fig. 1). The study of the thermal process is based on Fermi's golden rule and the photo process has been performed using wave-packet quantum dynamics with a vibronic model Hamiltonian.

Thermal relaxation

The HS → LS direct transfer has been studied in two Fe(II) complexes with significantly different HS–LS relative energies. $[Fe(bpy)_3]^{2+}$ is a commonly studied complex with a short-living HS state and $[Fe(mtz)_6]^{2+}$ is either a paramagnetic or diamagnetic complex depending on the temperature. The intersystem crossing rate has been calculated using Fermi's golden rule within the Condon approximation, which allows us to express the rate as a direct product between the vibrational overlap and the (effective) spin–orbit coupling.

$$k_{\mathrm{ISC}}^{\mathrm{a}\rightarrow\mathrm{b}} = \frac{2\pi}{\Omega_{\mathrm{a}}}\sum_{\alpha}\sum_{\gamma}\left|\left\langle\psi_{\mathrm{a}}{}^{\alpha}\middle|\hat{H}_{\mathrm{SO}}\middle|\psi_{\mathrm{b}}{}^{\gamma}\right\rangle\right|^{2}\sum_{j=0,k}\left|\langle\{\nu_{\mathrm{a}j}\}||\{\nu_{\mathrm{b}k}\}\rangle\right|^{2}\delta\left(E_{\mathrm{a}j}-E_{\mathrm{b}k}\right). \quad (1)$$

Herein, $\psi_a{}^{\alpha}$ and $\psi_b{}^{\gamma}$ are the spin-free electronic wave functions where the a and b labels refer to electronic states of different multiplicities and α and γ are, respectively, the MS-components of the initial and final states. j and k are the vibronic levels, E_{aj} and E_{bk} are, respectively, the energies of electronic state a in vibrational level j and electronic state b in vibrational level k, $E_{bk} = E_b + (k+1)\hbar\omega_k/2$. Ω_a is the degeneracy of the initial state and $\langle\{\nu_{aj}\}||\{\nu_{bk}\}\rangle$ is the two-center Franck–Condon integral. Note that this is the vibrationally cold approximation and the vibrational level of the initial state is fixed, $j = 0$.

The effect of temperature is taken into account by the introduction of the Boltzmann distribution in the initial vibronic states,

$$k_{\mathrm{ISC}}^{\mathrm{a}\rightarrow\mathrm{b}} = \frac{2\pi}{Z\Omega_{\mathrm{a}}}\sum_{\alpha}\sum_{\gamma}\left|\left\langle\psi_{\mathrm{a}}{}^{\alpha}\middle|\hat{H}_{\mathrm{SO}}\middle|\psi_{\mathrm{b}}{}^{\gamma}\right\rangle\right|^{2}\times\sum_{j,k}\mathrm{e}^{-\beta E_j}\left|\langle\{\nu_{\mathrm{a}j}\}||\{\nu_{\mathrm{b}k}\}\rangle\right|^{2}\delta\left(E_{\mathrm{a}j}-E_{\mathrm{b}k}\right) \quad (2)$$

where $Z = \sum_{j}\mathrm{e}^{-\beta E_j}$ is the canonical partition function for the vibrational term in the initial state, E_j is the vibrational energy of the level j and β is $1/(k_BT)$, in which k_B is the Boltzmann constant.

We mainly use the time-dependent approach, obtained after Fourier transforming the δ function in eqn (1), implemented in the VIBES program.[19] Then, the rate is determined analytically and the number of grid points and the spacing between them to perform the discrete Fourier transform are selected upon convergence of the correlation and cumulant functions. In one case, we use the time-independent approach for comparison. η, the threshold that determines the number of vibronic states included, is chosen to be 1000 cm^{-1} after some testing of its influence. The basic ingredients for these calculations of the rates are: (i) the vibrational normal modes and frequencies of the states to account for the vibrational overlap; (ii) the adiabatic electronic energy difference between the states, and (iii) the modulus of the (effective) spin–orbit among the different spin components of each state.

The vibrational modes and frequencies are computed after a geometry optimization for the lowest singlet (LS) and quintet (HS) states using the TurboMole package.[20] The Perdew–Burke–Ernzerhof (PBE0) hybrid functional is used in combination with the triple-ζ valence polarized Gaussian-type basis set, def2-TZVP. Two-electron integrals are approximated by the resolution of identity (RI) using the auxiliary basis def2-universal-JFIT. The hessian matrix is calculated through analytical second derivatives and no symmetry restrictions are imposed during the optimization procedure.

The B3LYP adiabatic energies are unfortunately not accurate enough in Fe(II) complexes to be used for further analysis. Therefore, we determine the adiabatic energies with CASPT2 and shift the DFT potentials accordingly. Unfortunately, the DFT minimum considerably differs from the CASPT2 one and we carry out a relaxed scan along the main reaction coordinate, the Fe–N symmetric stretching. The geometries at each point are optimised through a constrained optimization at the B3LYP/def2-SVP level. C_2 symmetry is imposed in the case of

$[Fe(bpy)_3]^{2+}$ and no symmetry restrictions are applied in $[Fe(mtz)_6]^{2+}$. The Fe–N distances for the LS and HS states in the relaxed scan minima are in good agreement with the (XMS-)CASPT2 fully optimized geometries by Finney *et al.*[21] Further information about the relaxed scan can be found in the ESI.†

CASPT2 calculations are performed using (Open)Molcas.[22] The basis sets used are ano-rcc type with the following contractions: (7s6p5d4f3g2h) for the Fe, (4s3p1d) for N bonded with Fe, (3s2p) for C and the other N and (2s) for H atoms. Cholesky decomposition for the two-electron integrals is used to speed up the calculations. The reference wave function is a CASSCF state-averaged (SA) wave function, including the four lowest states for the singlets, the six lowest for the triplets and the five lowest for the quintets. The active space is formed by 10 electrons distributed over 12 orbitals; three more orbitals are added to those calculations that include MLCT states. The molecular orbitals for these systems are well-established[23,24] and include two σ orbitals, mainly corresponding to a small bonding interaction between ligands and Fe, the three 3d (t_{2g}) orbitals of Fe, the two 3d (e_g) orbitals of Fe with a small anti-bonding contribution with respect to the ligands, and an additional set of five 4d orbitals on Fe to introduce radial electron correlation.[25] The influence of the MLCT states has also been studied, expanding the active space by including three π* orbitals corresponding to the aromatic system of the ligands. The perturbational treatment is carried out in all the electrons except the deep core ones, Fe (1s2s2p), C and N (1s), and an imaginary shift of $0.15E_\mathrm{h}$ is used.

The spin–orbit coupling is calculated using Molcas[22] in the singlet–quintet minimum energy crossing point geometry and in several conformations of the metadynamics for the LS and HS in $[Fe(bpy)_3]^{2+}$. The settings for CASPT2 calculations are the same as for the adiabatic energy differences. The direct coupling is determined using the effective one-electron Douglas–Kroll–Hess (DKH) Hamiltonian. The off-diagonal spin–orbit matrix elements are calculated by state interaction[26] using the CASSCF wave functions and the diagonal matrix elements are replaced by the CASPT2 energies. The direct coupling between states with $\Delta S = 2$, such as singlet and quintet states, is strictly zero using the standard effective one-electron spin–orbit operators, such as the DKH Hamiltonian. However, these states may have a non-vanishing effective coupling through intermediate states, which are states with non-zero direct coupling simultaneously with both states. In this case, triplet states can couple singlet and quintet states simultaneously. The effective Hamiltonian is constructed using the orthonormalised projections of SO wave functions, $\tilde{\Psi}_k^{\perp}$, and their respective eigenvalues, E_k, using Bloch's formula.

$$H_{ij}^{\mathrm{SO,eff}} = \langle \psi_i | \hat{H}^{\mathrm{SO,eff}} | \psi_j \rangle = \langle \psi_i | \left[\sum_{k \in S_0} \left| \tilde{\psi}_k^{\perp} \right\rangle E_k \left\langle \tilde{\psi}_k^{\perp} \right| \right] | \psi_j \rangle. \tag{3}$$

Herein, $\tilde{\Psi}_k^{\perp}$ is the des Cloizeaux orthonormalised spin–orbit wave function expressed on the basis of the spin-free states $\{\psi\}$ spanning a smaller subspace S_0, and E_k is the spin–orbit energy.

The effective Hamiltonian theory is an accurate and elegant way to determine effective couplings but it is difficult to extract information about the intermediate states that allow the coupling. For this reason, the weight of the intermediate triplet states in the singlet-quintet coupling was studied using second order

Table 1 Influence of the intermediate states in the singlet–quintet spin–orbit coupling for $[Fe(bpy)_3]^{2+}$. All the couplings and energies are expressed in cm^{-1} [a]

Conf.	ψ_μ	$H^{SO}_{S\mu}$	$H^{SO}_{\mu Q}$	$\gamma_{S\mu Q}$	H^{SO}_{QDPT}	H^{SO}_{eff}	ΔE
ISC	$^3T_{1g}$	529	284	20.8	21.8	32.7	−186
	$^3T_{2g}$	55	232	1.0			
ISC*	$^3T_{1g}$	548	449	27.9	29.0	29.9	−317
	$^3T_{2g}$	56	248	1.0			
	^{3}MLCT	25	58	0.03			
1a	$^3T_{1g}$	526	252	40.3	42.3	30.1	2615
	$^3T_{2g}$	92	291	1.9			
2a	$^3T_{1g}$	516	464	32.2	34.4	28.2	696
	$^3T_{2g}$	123	275	2.2			
1b	$^3T_{1g}$	524	465	19.4	20.6	43.0	−5623
	$^3T_{2g}$	70	294	1.1			
2b	$^3T_{1g}$	519	452	23.8	25.4	37.0	−3168
	$^3T_{2g}$	99	273	1.6			

[a] $\Delta E = E(\text{HS}) - E(\text{LS})$ in cm^{-1}, ISC*: CAS(10,15); active space formed by 10 electrons distributed over 15 orbitals. $\gamma_{S\mu Q}$ is the 2nd order perturbative correction from state μ. $H^{SO}_{S\mu}$ and $H^{SO}_{\mu Q}$ are the average matrix elements $\langle\psi_S|\hat{H}^{SO}|\psi_\mu\rangle$ and $\langle\psi_\mu|\hat{H}^{SO}|\psi_Q\rangle$, respectively. The labels a and b refer to conformations from the singlet and quintet metadynamics simulations, respectively.

quasi-degenerate perturbation theory (QDPT2) in different conformations. QDPT2 allows us to decompose the spin–orbit coupling as a sum over the different intermediate states.

$$H_{ij}^{SO,QDPT2} = \left\langle\psi_i\middle|\hat{H}^{SO}\middle|\psi_j\right\rangle + \sum_{\mu\neq i,j}\frac{\left\langle\psi_i\middle|\hat{H}^{SO}\middle|\psi_\mu\right\rangle\left\langle\psi_\mu\middle|\hat{H}^{SO}\middle|\psi_j\right\rangle}{E_\mu - E_j}. \quad (4)$$

Here, $\{\psi\}$ are the spin-free states and $\{E\}$ the spin-free energies. ψ_i and ψ_j are, respectively, the singlet and quintet states and ψ_μ the intermediate state.

Table 2 Singlet quintet effective spin–orbit coupling in cm^{-1} at different conformations and its corresponding lifetime using different approaches in $[Fe(bpy)_3]^{2+}$ [c]

Sys.	Conf.	$H^{SO,eff}$	$T = 0$ K, τ/ns	$T = 300$ K, τ/ns	$T = 0$ K, τ_{FC}
bpy	ISC	32.7	77.9	2.14	—
	ISC(10,15)	29.9	92.8	2.54	—
	1a	30.1	91.7	2.51	—
	2a	28.2	104.7	2.87	—
	Avg.[a]	28.1	106.1	2.9	—
	1b	43.0	45.0	1.24	—
	2b	37.0	60.7	1.66	—
	Avg.[b]	37.5	61.1	1.67	—
mtz	ISC	31.4	8.9 s	—	~10 years

[a] Average of the ten structures extracted from the singlet metadynamics. [b] Average of the ten structures extracted from the quintet metadynamics. [c] The labels a and b refer to conformations from the singlet and quintet simulations, respectively.

The results in Table 1 show that $^3T_{1g}$, the lowest set of ^{3}MC states, has the largest influence in the coupling of the singlet and quintet states. This can be easily rationalised because it is the intermediate state closest in energy to the singlet and quintet and has strong direct couplings with both states. In addition, the decomposition also showed an almost negligible influence of the ^{3}MLCT states in the coupling of the $^1A_{1g}$ and $^5T_{2g}$ states.

Table 1 also shows discrepancies between the effective Hamiltonian theory and the quasi-degenerate perturbation theory. These differences are mainly associated with two intrinsic characteristics of the perturbative method used. The first is that QDPT2 loses accuracy when the initial and final states have larger energy differences. Eqn (4) shows that the larger the difference between E_i and E_j, the larger the difference becomes between H_{ij} and H_{ji}. The relative similarity of the couplings with both methods in the ISC geometry where singlet and quintet are quasi degenerate, in comparison with the disagreement in most of the conformations where the energy difference is larger, helps to corroborate this observation. The other shortcoming is that in QDPT2, the interactions are expanded up to second order and only the interactions through one intermediate state are accounted for, while all higher order terms are taken into account in the effective Hamiltonian theory.

The influence of the thermal disorder in the spin–orbit coupling and, therefore, in the intersystem crossing rates was also studied. Previously, it was demonstrated by one of us that the geometrical distortions have a strong effect on the size of the coupling, especially when MLCT states are involved. These changes in the couplings changed the rates of some of the intersystem crossings by a few orders of magnitude in Fe(II) polypyridinal complexes.[15] However, in the case of the ^{1}MC–^{5}MC coupling, the effect of geometrical distortions is almost negligible, giving in all the conformations the same order of magnitude for the couplings and, therefore, the rates (or lifetimes) (Table 2).

The results show that the spin–orbit coupling strength is almost identical for both complexes. However, the HS–LS vibrational overlap is completely different in the two complexes, which is responsible for the different HS lifetimes in these complexes. Temperature is also an important factor; the rates at 300 K are one order of magnitude faster than those at 0 K using the vibrationally cold approximation. The HS state lifetime in $[Fe(bpy)_3]^{2+}$ computed with the methodology described here is in good agreement with the 700 ps determined by Consani *et al.* for the same complex in aqueous solution,[27] while the very slow ISC rate for $[Fe(mtz)_6]^{2+}$ is indicative of a stable HS spin, as observed in experiments.[18] Fermi's golden rule methodology is presented as an appropriate method to determine the order of magnitude of the HS state lifetime in Fe(II) complexes.

Photophysical relaxation

The reverse-LIESST is studied from the ^5E_g state, after an excitation from the high spin state ($^5T_{2g} \rightarrow {}^5E_g$). The process is studied in $[Fe(mtz)_6]^{2+}$ because of the large HS state lifetime in this complex and the presence of experimental studies in an analogous complex, $[Fe(ptz)_6]^{2+}$.[17] In the high-spin minimum geometry, the ^{5}MLCT states are clearly higher in energy than the ^{5}MC states, therefore the former are highly unlikely to play a role in the deactivation and only the latter are included in the study. The reverse-LIESST was, in a first attempt, studied using

Fermi's golden rule, as a follow-up of the satisfactory results reported in the literature for the singlet–quintet conversion. The only energetically favourable intersystem crossing transition from the ^5E_g state to the triplets is the $^5E_g \rightarrow {}^3T_{1g}$. However, the rate for this transition using Fermi's golden rule is about a few milliseconds, several orders of magnitude slower than the experimentally reported value of 1.7 ps. For this reason, the process is studied using wave-packet quantum dynamic simulations with a vibronic model Hamiltonian (eqn (7)).

The simulations are carried out with the multi-configuration time-dependent Hartree (MCTDH) method (Heidelberg package, version 8.5.13).[28] The basis of the method is to solve the time-dependent Schrödinger equation using the following *ansatz* for the nuclear wave function:

$$\Psi(q_1, q_2, \ldots, q_p, t) = \sum_{j_1=1}^{n_1} \cdots \sum_{j_p=1}^{n_p} A_{j_1 \ldots j_p}(t) \prod_{k}^{p} \varphi_{j_k}^{(k)}(q_k, t). \tag{5}$$

Herein, A_j are the expansion coefficients and φ_j are the single-particle functions (SPFs), which are linear combinations of the primitive functions. In this case, the primitive functions take the shape of the n_j's lowest solutions for an harmonic oscillator. The multi-set formalism is employed, where a different wave function is written for each state

$$\Psi^{\mathrm{m}}(q_1, q_2, \ldots, q_p, t) = \sum_{\alpha=1}^{\sigma} \Psi^{(\alpha)} |\alpha\rangle \tag{6}$$

where σ is the number of electronic states. We have included 5 SPFs for each state along each mode and 151 primitive functions for each mode.

The vibronic coupling model Hamiltonian takes the following shape:

$$\begin{aligned} &\hat{H}(\{q\}) = \\ &\sum_{S=0}^{2} \sum_{n=1}^{N_S} \sum_{\alpha=-|S|}^{|S|} \left[\left(T_N(\{q\}) + V_n^S(\{q\}) \right) |n_{S,\alpha}\rangle\langle n_{S,\alpha}| + \sum_{n' \neq n}^{N_S} W_{n,n',S}^{\mathrm{NA}}(\{q\}) |n_{S,\alpha}\rangle\langle n'_{S,\alpha}| \right] \\ &+ \sum_{S=0}^{2} \sum_{n=1}^{N_S} \sum_{\alpha=-|S|}^{|S|} \sum_{S'=1}^{2} \sum_{n'=1}^{N_{S'}} \sum_{\alpha'=-|S'|}^{|S'|} \left[W_{nS\alpha,n'S'\alpha'}^{\mathrm{SO}} |n_{S,\alpha}\rangle\langle n_{S',\alpha'}| + W_{n'S'\alpha',nS\alpha}^{\mathrm{SO},*} |n'_{S',\alpha'}\rangle\langle n_{S,\alpha}| \right]. \end{aligned} \tag{7}$$

Herein, S is the spin quantum number, N_S is the number of states with S spin quantum number and α is the M_S component. $T_N(\{q\})$ is the nuclear kinetic energy, $V_n^S(\{q\})$ is the diabatic potential, $W_{n,n',S}^{\mathrm{NA}}(\{q\})$ is the non-adiabatic coupling and $W_{nS\alpha,n'S'\alpha'}^{\mathrm{SO}}$ and its hermitian conjugate $W_{nS\alpha,n'S'\alpha'}^{\mathrm{SO},*}$ are the spin–orbit coupling terms. Note that in our model, spin–orbit is coordinate independent, while the other terms are coordinate dependent.

The nuclear kinetic energy is defined (in atomic units) as

$$T_{\mathrm{N}} = -\sum_{i}^{N_{\mathrm{vib}}} \frac{\omega_i}{2} \frac{\partial^2}{\partial {q_i}^2} \tag{8}$$

where ω_i is the frequency of the HS state in mode i and N_{vib} is the number of vibrations included in the model Hamiltonian.

The diabatic potential is defined using a quartic polynomial expression:

$$V_n^S = E_n^S + \sum_i^{N_{\text{vib}}} \sum_{j=1}^{4} \frac{1}{j!} k_{j,i}^{n,S} q_i^j. \tag{9}$$

Here E_n^S is the diabatic energy for the state n with spin S at the Franck–Condon region and $k_{j,i}^{n,S}$ is the jth expansion term along the vibrational mode i for this state.

The non-adiabatic coupling term is described as

$$W_{n,n',S}^{\text{NA}} = \sum_i^{N_{\text{vib}}} \lambda_i^{n,n',S} q_i \tag{10}$$

where $\lambda_i^{n,n',S}$ is a linear-expansion coefficient for the non-adiabatic coupling between states n and n' with spin S along mode i.

The spin–orbit coupling term is defined as follows

$$W_{nS\alpha,n'S'\alpha'}^{\text{SO}} = \left\langle n_{S\alpha} \middle| \hat{H}^{\text{SO}} \middle| n'_{S'\alpha'} \right\rangle \tag{11}$$

where $\hat{H}^{\text{SO}}$ is the DKH Hamiltonian, $\langle n_{S\alpha}|$ is the state n with spin S and $M_S\alpha$, and $|n_{S\alpha}\rangle$ is the state n' with spin S' and $M_S\alpha'$.

The model is constructed by the expansion along the most representative modes of the HS state optimized at the PBE0/def2-TZVP level and includes the electronic states $^1A_{1g}$, $^1T_{1g}$, $^3T_{1g}$, $^3T_{2g}$, $^5T_{2g}$ and ^5E_g, which generates a model of 47 states, taking into account the spatial and spin symmetries. The terms for the Hamiltonian are obtained through a fitting procedure of the XMS-CASPT2 energies. These calculations are carried out using an active space of 10 electrons distributed over 12 molecular orbitals. The perturbative treatment of the electron correlation is applied to all electrons except the deep core ones, Fe (1s2s2p) and C, N (1s), and the 30 highest orbitals. An imaginary level shift of $0.25E_h$ is applied to avoid intruder state problems. From the adiabatic energies along the main modes, we obtain the diabatic energies as a quartic polynomial and the non-adiabatic couplings between states of the same M_S value within the linear-expansion approximation. The off-diagonal terms between states of different M_S are included with the spin–orbit coupling at the DFT optimised geometry. The nuclear kinetic energy operators in mass-frequency-weighted coordinates are added using the frequencies of the HS state at the DFT level. The simulations are initialised after a relaxation of the wave-packet in the $^5T_{2g}$ state, which starts the dynamics at the minimum energy geometry of the HS state, and the wave-packet is distributed over the different components of the ^5E_g state according to the oscillator strengths. Further (numerical) information about the couplings is given in the ESI.†

The selected normal modes are chosen based on those modes with the largest reorganization energy connecting the minima of the different states involved in the process. The difference between the optimized geometries for each state at the DFT level is projected into the HS state normal modes, following the expression

$$q_k^{\text{A}} = {m_k}^{1/2} \sum_l L_{kl}^{\text{HS}} {m_l}^{1/2} (\mathbf{R}_{\text{A},l} - \mathbf{R}_{\text{HS},l}). \tag{12}$$

Herein, q^{A} is the normal mass-frequency-weighted Cartesian coordinate expanding the geometrical differences between state A and the HS state, projected

Table 3 Coordinates q^A of the main electronic states involved in reversed LIESST projected on the HS state normal modes. The values of q^A are given in parentheses, those modes with absolute values larger than 1 are reported here. The modes in the reaction coordinate are in bold

	Included modes	Excluded modes
$^1A_{1g}$–$^5T_{2g}$	**42** (−8.95), **30** (−1.99) **10** (1.72), **39** (1.63)	—
$^3T_{1g}$–$^5T_{2g}$	**42** (−4.28), 31 (2.50) 33 (2.34)	29 (−2.05), 165 (1.79) 11 (−1.11)
^5E_g–$^5T_{2g}$	**42** (3.89), 31 (−3.42) 33 (−1.36), **10** (−1.28)	29 (2.97), 11 (2.00) 21 (1.31), 28 (−1.17)

in the basis of HS state reference normal modes. m_l is the mass of atom l, $\mathbf{R}_A$ and $\mathbf{R}_{HS}$ are the Cartesian geometries for states A and the HS state, respectively, and L^{HS} is the matrix that contains the eigenvectors of the HS state normal modes (Table 3).

According to this analysis, there are three main modes representing the reverse-LIESST. These modes have Fe–N stretching character, q_{42} is the symmetric motion and q_{31} and q_{33} correspond to Fe–N asymmetric stretching vibrations, which are especially important for Jahn–Teller active states. This model with three Fe–N modes coincides with the one proposed by Pápai in his study of the LIESST for $[Fe(NHC)_6]^{2+}$. In our model, we include the two asymmetric modes q_{31} and q_{33}, and the reaction coordinate mode (q_{rc}) that connects HS and LS geometries, which is a combined mode described as:

$$q_{rc} = 0.95q_{42} + 0.21q_{30} - 0.18q_{10} - 0.17q_{39}. \tag{13}$$

This reaction coordinate mode describes the Fe–N symmetric breathing (q_{rc}) but is slightly assisted for N–Fe–N bending modes (q_{30}, q_{10}, q_{39}), which aids the adaptation of the N–Fe–N angles to each Fe–N distance.

The reaction coordinate mode does not break the pseudo-O_h symmetry. In this mode, the triple degeneracy of $^1T_{1g}$, $^3T_{1g}$, $^3T_{2g}$ and $^5T_{2g}$ and the double degeneracy of ^5E_g is maintained. The Fe–N asymmetric stretching (q_{31}) is described as a compression of the four Fe–N distances for the N in the equatorial plane with an elongation of the two Fe–N distances for the axial N. This breaks the pseudo-O_h symmetry to a pseudo-D_{4h}, which breaks the triply degenerate states into doubly degenerate states and removes the degeneracy of the ^5E_g state. In the other asymmetric stretching (q_{33}), the axial N remains in place, while the two equatorial N come closer and recede in pairs. After this motion, the pseudo-O_h symmetry is broken to a pseudo-D_{2h} that removes the degeneracy for all the states (Fig. 2).

The efficiency of an intersystem crossing depends on the energy difference between a pair of states and the size of their coupling. In this complex, the energetically accessible intersystem crossing from the ^5E_g is the $^5E_g \rightarrow {}^3T_{1g}$. Near the absorption geometry, that is the $^5T_{2g}$ minimum, $^3T_{1g}$ is slightly lower in energy than ^5E_g, therefore, a crossing between these states is found after the absorption. However, the gradients to the ^5E_g minimum are high and the system will evolve to this region where the energy gap between the quintet and the triplet is large and

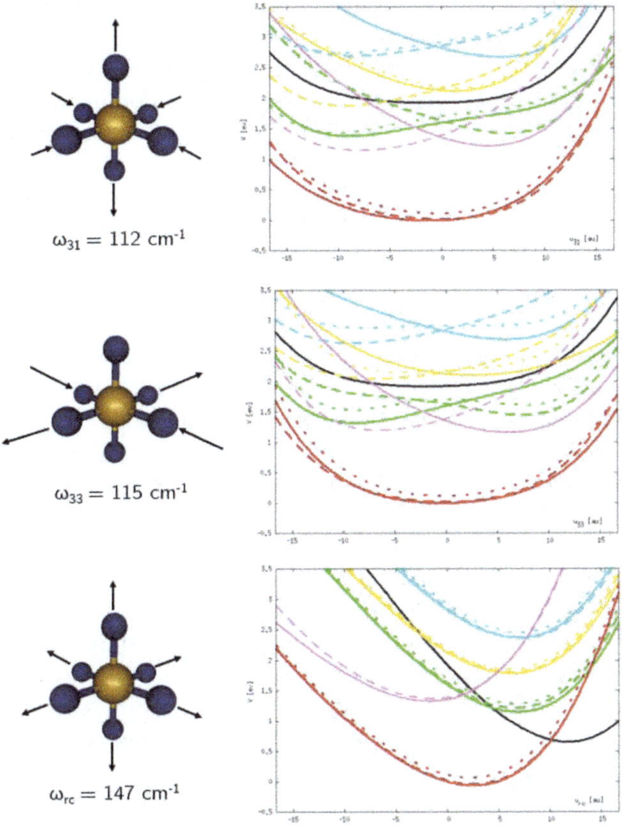

Fig. 2 Schematic of the motions and diabatic potentials along the modes q_{31}, q_{33} and q_{rc}. $^1A_{1g}$ is in black, $^1T_{1g}$ in cyan, $^3T_{1g}$ in green, $^3T_{2g}$ in gold, $^5T_{2g}$ in red and ^5E_g in violet. The different components are represented by solid, dashed and dotted lines.

the efficiency of the ISC is small. The transition to the $^3T_{2g}$ is discarded because these states are above the ^5E_g states in the absorption region.

Although the energy gap does not seem to favour the ISC transfer, this could still be efficient when there exists a large coupling between the states. This is not the case for the ^5E_g–$^3T_{1g}$ coupling, which is about 10 cm^{-1} in average at the $^5T_{2g}$ DFT minimum. The small coupling can be easily rationalized observing the two electrons of difference in the main configurations of the $^5E_g(t_{2g}^3e_g^3)$ and $^3T_{1g}(t_{2g}^5e_g^1)$, which are strictly uncoupled in the usual effective one-electron spin–orbit operator.

The minimum of the ^5E_g is energetically far away from the other states; ^5E_g is lower in energy than $^3T_{1g}$ and above $^5T_{2g}$, well separated from both. Despite the large energy gap with these other states, the nuclear kinetic energy may bring the wave-packet to regions with nearly degenerate $^5T_{2g}$–^5E_g, in a sloped conical intersection. Unfortunately, this intersection could not be determined because of the lack of an accurate computational method to determine precisely at the same time geometries (for the minima of states and minimum energy crossings) and energies at an affordable computational cost. The plausibility of the crossing in the quintets is also not favoured by the energy difference between the states in the regions around the ^5E_g minimum, but at least there is a strong coupling with values larger than 250 cm^{-1}. This coupling is strong, as expected for the one electron of difference between the states in their dominant configurations ($t_{2g}^4e_g^2$ and $t_{2g}^3e_g^3$). These ingredients make us think that the most favourable transition is $^5E_g \rightarrow {}^5T_{2g}$.

The evolution of the diabatic population (Fig. 3) shows that almost half of the population of the ^5E_g is rapidly transferred to the $^5T_{2g}$ state. Then, part of the remaining population in the ^5E_g evolves to the $^5T_{2g}$ when the wave-packet visits regions with a small energy gap between both states. From this state, part of the population is gradually transferred to the $^3T_{1g}$, above 10% in 2 ps, from which the LS may be populated. This last event is not appreciated in our dynamics because it happens in a larger time scale than our simulation time.

Our results indicate a faster $^5E_g \rightarrow {}^5T_{2g}$ transition than that experimentally observed. This unexpectedly fast rate is probably because of the limitations of the model Hamiltonian. The complete vibrational space is not represented but it is reduced to the three main coordinates. Moreover, the diabatic coupling terms between modes are not included in the model. These terms may be especially important in the description of the Jahn–Teller modes. The lack of these terms gives stronger gradients in the model and the wave-packet may rapidly visit regions with small $^5T_{2g}$–^5E_g energy gaps with high efficient population transfer. Unfortunately, the inclusion of these terms dramatically increases the computational cost of these calculations. In addition, the spin–orbit coupling is fixed at the HS geometry and the variation along coordinates, vibronic spin–orbit coupling, is not included in our model.

The quintet–quintet transition observed in the wave-packet quantum dynamics is also determined by Fermi's golden rule. The calculated lifetime for the ^5E_g at 0 K is 2.3 ps, in good agreement with the first transition determined by Marino *et al.* in $Fe(ptz)_6$.[17] Then, calculations at different levels of theory point in

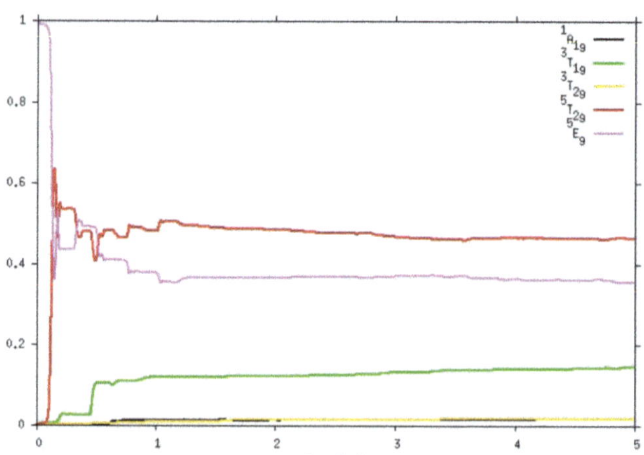

Fig. 3 Time-evolution population of the diabatic states along the dynamics. The states are represented by the following colours: $^1A_{1g}$: black, $^1T_{1g}$: cyan, $^3T_{1g}$: green, $^1A_{1g}$: gold, $^5T_{1g}$: red and ^5E_g: violet.

the same direction. The first transition into the reverse-LIESST process is the relaxation to the $^5T_{2g}$. After that, in a longer time scale, part of the population is transferred from the vibrationally hot HS state to the intermediate $^3T_{1g}$ and, finally, the $^1A_{1g}$.

Conclusions

To conclude, we have applied two different methodologies to study the thermal and photophysical conversions from the high-spin to the low-spin state. The order of magnitude of the HS state lifetime in the thermal relaxation process has been determined satisfactorily using Fermi's golden rule. The influence of the $^3T_{1g}$ as the main intermediate state coupling HS and LS has been demonstrated, while the role of the ^{3}MLCT states in this coupling can be neglected. This rate theory has been used to determine the HS state lifetime in Fe(II) complexes with nearly O_h symmetry. This may be also tested to determine the HS state lifetime in complexes breaking the symmetry, such as $[Fe(terpy)_2]^{2+}$, where the empirical energy gap rule is not valid.[11] The determination of the adiabatic energies is the key step in this process and could be done using an appropriate DFT functionals or with the recent algorithms to account for the (XMS-)CASPT2 (ref. 29 and 30) optimization implemented in quantum codes such as BAGEL.

The calculations indicate that the intersystem crossing from the ^5E_g to the triplets in the reverse-LIESST is not favourable. The mechanism that we propose, according to the quantum dynamics, is $^5E_g \rightarrow {}^5T^*_{2g} \rightarrow {}^3T_{1g} \rightarrow {}^1A_{1g}$. The intersystem crossing is produced from the vibrationally hot $^5T^*_{2g}$, which is formally a thermal relaxation at an effective high temperature. This path could explain the limitation of the results observed by Fermi's golden rule, when it should work fine, because it happens from the minimum of the excited state due to the rate of the transition. This also would explain the two different rates reported in the reverse-LIESST and the quantum yield reported by Marino *et al.*[17] According to the LIESST, the rate for the $^3T_{1g} \rightarrow {}^5T_{2g}/{}^3T_{1g} \rightarrow {}^1A_{1g}$ should be about 1.5 ps and the branching ratio 4 : 1. This does not compare with the 37 ps and the quantum yield of 0.1 associated with these transitions. For this reason and based on our results, we propose an alternative mechanism from the vibrationally hot $^5T^*_{2g}$ state. Our results do not match completely with the experimentally reported rates, which may possibly be associated with a limitation in our model Hamiltonian. New improvements in the generation model Hamiltonian, such as the inclusion of diabatic coupling terms or vibronic spin–orbit, may be explored.

Conflicts of interest

The authors declare no conflicts of interest.

Acknowledgements

MHR acknowledges financial support by the Agence Nationale pour la Recherche through the project MULTICROSS (ANR-19-CE29-0018). MAR and CdG acknowledge the Spanish ministry of Science and Innovation (Project PID2020-113187GB-I00). The project leading to this publication has received funding from the

Excellence Initiative of Aix-Marseille Université – A*Midex, a French "Investissement d'Avenir" programme, AMX-21-ERC-01, and from the funding "ERC Booster" from Région Sud Provence-Alpes-Côte d'Azur.

Notes and references

1 M. R. Geraskina, A. T. Buck and A. H. Winter, *J. Org. Chem.*, 2014, **79**, 7723–7727.

2 W.-W. Wu, S.-G. Wu, Y.-C. Chen, G.-Z. Huang, B.-H. Lyu, Z.-P. Ni and M.-L. Tong, *Chem. Commun.*, 2020, **56**, 4551–4554.

3 I. Krivokapic, M. Zerara, M. L. Daku, A. Vargas, C. Enachescu, C. Ambrus, P. Tregenna-Piggott, N. Amstutz, E. Krausz and A. Hauser, *Coord. Chem. Rev.*, 2007, **251**, 364–378.

4 G. G. Morgan, K. D. Murnaghan, H. Muller-Bunz, V. McKee and C. J. Harding, *Angew. Chem., Int. Ed.*, 2006, **45**, 7192–7195.

5 D. J. Harding, P. Harding and W. Phonsri, *Coord. Chem. Rev.*, 2016, **313**, 38–61.

6 O. I. Kucheriv, I. O. Fritsky and I. A. Gural'skiy, *Inorg. Chim. Acta*, 2021, **521**, 120303.

7 A. Gaspar, M. Seredyuk and P. Gütlich, *J. Mol. Struct.*, 2009, **924–926**, 9–19.

8 A. Arroyave, A. Lennartson, A. Dragulescu-Andrasi, K. S. Pedersen, S. Piligkos, S. A. Stoian, S. M. Greer, C. Pak, O. Hietsoi, H. Phan, S. Hill, C. J. McKenzie and M. Shatruk, *Inorg. Chem.*, 2016, **55**, 5904–5913.

9 S. Zerdane, M. Cammarata, O. Iasco, M.-L. Boillot and E. Collet, *J. Chem. Phys.*, 2019, **151**, 171101.

10 G. Molnár, L. Salmon, W. Nicolazzi, F. Terki and A. Bousseksou, *J. Mater. Chem. C*, 2014, **2**, 1360.

11 A. Hauser, C. Enachescu, M. L. Daku, A. Vargas and N. Amstutz, *Coord. Chem. Rev.*, 2006, **250**, 1642–1652.

12 G. Auböck and M. Chergui, *Nat. Chem.*, 2015, **7**, 629–633.

13 W. Zhang, R. Alonso-Mori, U. Bergmann, C. Bressler, M. Chollet, A. Galler, W. Gawelda, R. G. Hadt, R. W. Hartsock, T. Kroll, K. S. Kjaer, K. Kubiček, H. T. Lemke, H. W. Liang, D. A. Meyer, M. M. Nielsen, C. Purser, J. S. Robinson, E. I. Solomon, Z. Sun, D. Sokaras, T. B. van Driel, G. Vankó, T.-C. Weng, D. Zhu and K. J. Gaffney, *Nature*, 2014, **509**, 345–348.

14 C. Sousa, C. de Graaf, A. Rudavskyi, R. Broer, J. Tatchen, M. Etinski and C. M. Marian, *Chem.–Eur. J.*, 2013, **19**, 17541–17551.

15 C. Sousa, M. Llunell, A. Domingo and C. de Graaf, *Phys. Chem. Chem. Phys.*, 2018, **20**, 2351–2355.

16 M. Pápai, *Inorg. Chem.*, 2021, **60**, 13950–13954.

17 A. Marino, P. Chakraborty, M. Servol, M. Lorenc, E. Collet and A. Hauser, *Angew. Chem., Int. Ed.*, 2014, **53**, 3863–3867.

18 A. Hauser, in *Spin Crossover in TransitionMetal Compounds II. Topics in Current Chemistry*, ed. P. Gütlich and H. A. Goodwin, Springer Berlin, Heidelberg, Germany, 2004, ch. Light-Induced Spin Crossover and the High-Spin → Low-Spin Relaxation, pp. 155–198.

19 M. Etinski, J. Tatchen and C. M. Marian, *J. Chem. Phys.*, 2011, **134**, 154105.

20 TURBOMOLE V6.4.2 2012, a development of University of Karlsruhe and Forschungszentrum Karlsruhe GmbH, 1989-2007, TURBOMOLE GmbH, since 2007; available fromhttp://www.turbomole.com.

21 B. A. Finney, S. R. Chowdhury, C. Kirkvold and B. Vlaisavljevich, *Phys. Chem. Chem. Phys.*, 2022, **24**, 1390–1398.
22 I. F. Galván, M. Vacher, A. Alavi, C. Angeli, F. Aquilante, J. Autschbach, J. J. Bao, S. I. Bokarev, N. A. Bogdanov, R. K. Carlson, L. F. Chibotaru, J. Creutzberg, N. Dattani, M. G. Delcey, S. S. Dong, A. Dreuw, L. Freitag, L. M. Frutos, L. Gagliardi, F. Gendron, A. Giussani, L. González, G. Grell, M. Guo, C. E. Hoyer, M. Johansson, S. Keller, S. Knecht, G. Kovačević, E. Källman, G. Li Manni, M. Lundberg, Y. Ma, S. Mai, J. P. Malhado, P. A. Malmqvist, P. Marquetand, S. A. Mewes, J. Norell, M. Olivucci, M. Oppel, Q. M. Phung, K. Pierloot, F. Plasser, M. Reiher, A. M. Sand, I. Schapiro, P. Sharma, C. J. Stein, L. K. Sørensen, D. G. Truhlar, M. Ugandi, L. Ungur, A. Valentini, S. Vancoillie, V. Veryazov, O. Weser, T. A. Wesołowski, P.-O. Widmark, S. Wouters, A. Zech, J. P. Zobel and R. Lindh, *J. Chem. Theory Comput.*, 2019, **15**, 5925–5964.
23 K. Pierloot and S. Vancoillie, *J. Chem. Phys.*, 2006, **125**, 124303.
24 B. Ordejón, C. de Graaf and C. Sousa, *J. Am. Chem. Soc.*, 2008, **130**, 13961–13968.
25 K. Andersson and B. O. Roos, *Chem. Phys. Lett.*, 1992, **191**, 507–514.
26 B. O. Roos and P.-Å. Malmqvist, *Phys. Chem. Chem. Phys.*, 2004, **6**, 2919–2927.
27 C. Consani, M. Prémont-Schwarz, A. ElNahhas, C. Bressler, F. van Mourik, A. Cannizzo and M. Chergui, *Angew. Chem.*, 2009, **121**, 7320–7323.
28 H. D. Meyer and O. Vendrell, http://mctdh.uni-hd.de/, 2013.
29 M. K. MacLeod and T. Shiozaki, *J. Chem. Phys.*, 2015, **142**, 051103.
30 B. Vlaisavljevich and T. Shiozak, *J. Chem. Theory Comput.*, 2016, **12**, 3781–3787.

Faraday Discussions

ROYAL SOCIETY OF CHEMISTRY

PAPER

Quantum entanglement control of electron–phonon systems by light irradiation

Kunio Ishida, *[a] Hiroaki Matsueda[b] and Akira Kamada[c]

Received 11th January 2022, Accepted 5th April 2022
DOI: 10.1039/d2fd00007e

We numerically study the dynamics of quantum entanglement between interacting electron–phonon and qubit–spin systems under photoirradiation, employing a model of multiple spins and boson modes. The interplay of the antiferromagnetic exchange and electron–phonon interactions provides us with a phase diagram, wherein each phase is characterized by the ground state property of the electron–phonon system. Light irradiation of the electron–phonon system facilitates the generation of quantum entanglement, according to the spin configuration and the phonon state in the ground state. Analyses using the quantum mutual information and the singular values of the reduced density matrix indicate that the quantum mechanical effect of the irradiated light appears in the state of the material.

1 Introduction

The generation of quantum entanglement is key to realizing various quantum information devices. The irradiation with light induces quantum entanglement between systems without direct interaction.[1,2] Theoretically, the irradiation with classical light of remote systems corresponds to "local operations and classical communication" (LOCC) in the terminology of quantum information science,[3] and no entanglement generation is obtained between them. Hence, the quantum mechanical nature of light is essential in understanding the entanglement generation mechanism.[4] However, considering that quantum entanglement plays an important role in quantum many-body physics,[5,6] the coherent dynamics of remote systems is an appropriate reference when studying general quantum many-body states under photoirradiation. In other words, it provides an insight into the photoinduced entanglement in interacting many-body systems.

[a]*School of Engineering and Center for Optical Research and Education, Utsunomiya University, Utsunomiya, Tochigi 321-8585, Japan. E-mail: ishd_kn@cc.utsunomiya-u.ac.jp; Fax: +81 28 689 6101; Tel: +81 28 689 6101*
[b]*Department of Applied Physics, Graduate School of Engineering, Tohoku University, Sendai 980-8579, Japan*
[c]*Graduate School of Regional Development and Creativity, Utsunomiya University, Utsunomiya, Tochigi 321-8585, Japan*

Because recent advancements in ultrashort pulse laser technology have enabled the observation of the time evolution of quantum mechanical states in the coherent regime,[7–9] the dynamical control of entanglement by an external light field is an intriguing problem that is crucial in the study of light–matter interaction and quantum information processing. In particular, a previous experimental study on diamond crystals[1] demonstrated that the Raman scattering process contributes to creating phonon entanglement between spatially separated samples. Although the entanglement is evaluated after a certain measurement process in previous studies, it is expected that entangled states are formed during the irradiation by control light. Hence, provided we have information on transient entanglement before measurement, we can obtain a clue to determine appropriate quantities or methods and design a method to control the entanglement properties. In other words, if the transient behavior of entanglement is elucidated, it will assist us to design/select appropriate ways of projecting the subspace that are realized as required.

This study aims to investigate the dynamics of entanglement generation by light irradiation. In Section 2, we study a two-site electron–phonon system in which the electrons are coupled by exchange interactions. A method is required to study the quantum dynamics in the coherent regime, which implies, at first, that the quantization of incident/scattered light is also required. If we treat the electromagnetic field classically, no additional entanglement is generated by light, *i.e.*, only the internal interaction of the material determines the entanglement. We infer that the simultaneous treatment of Maxwell and Schrödinger equations cannot address the aforementioned challenges. Hence, this situation renders it difficult to calculate the dynamics of the system, since the degrees of freedom (DoF) of the entire electron–phonon–photon system are significantly large. We aim to construct a simple model system that contains the essential physics of the phonon entanglement generation in this study.

In addition, a method is required to evaluate the entanglement with the time evolution of the system. In this study, we considered multipartite systems comprising electrons, phonons, and photons, and we determined that the quantum mutual information is appropriate to discuss the dynamical behavior of the entanglement generation.[4] Furthermore, we focused on the phonon properties by dividing the entire electron–phonon–photon state $|\Phi\rangle$ into phonon $|\eta_n\rangle$ and electron–photon $|\theta_n\rangle$ states using the Schmidt decomposition as $|\Phi\rangle = \sum_n \lambda_n |\eta_n\rangle |\theta_n\rangle$. Numerical calculations on the time-evolution of the singular values λ_n and $|\eta_n\rangle$ demonstrate how light irradiation induces the entanglement between the electrons or phonons on different sites. The calculated results also indicate that the quantum mechanical nature of light should be considered even in directly interacting systems to design quantum control methods in the transient regime.

As the quantum nature of photons is important in the coherent dynamics of materials, we consider that the quantum state of light is regarded as a parameter of coherent control. In Section 3, we study the dynamics of remote qubit–spin systems to demonstrate that entangled photons contribute to generate inter-system entanglement. Numerical calculations show that the entanglement between distant qubit–spin systems is generated by entangled photons, which indicates that the validity of the qubit–spin–photon system as an entanglement storage system.

2 Electron–phonon systems with exchange interactions

Photoinduced entanglement is theoretically[4] and experimentally[1,2] derived in remote systems. When each material system in a previous study[4] is interpreted as a molecule or a unit cell of solids, the electron–phonon system becomes part of a crystal, and the intersite interaction is responsible for the generation of quantum entanglement. In this case, irradiated photons affect the quantum state of the material by modulating the entanglement properties, which is discussed in the present study.

2.1 Hamiltonian of the system

As a simple model of interacting electron–phonon sites, we consider localized electrons with exchange interactions, where each electron also interacts with the Einstein phonons at each site. When all of the systems interact with multimode photons, the Hamiltonian is described by

$$\begin{aligned}\mathcal{H} = &\sum_{i=1}^{3} \Omega_i c_i^\dagger c_i \\ &+ \sum_{j=1}^{2} \left[\omega a_j^\dagger a_j + \sigma_x^j \sum_{i=1}^{3} \mu_i \left(c_i^\dagger + c_i\right) \right. \\ &\left. + \hat{n}_j \left\{ \nu \left(a_j^\dagger + a_j \right) + \varepsilon \right\} \right] + J \vec{\sigma}_1 \cdot \vec{\sigma}_2, \end{aligned} \tag{1}$$

where a_j and c_i denote the annihilation operators of optical phonons in the j-th site and the photons of the i-th mode, respectively. σ_ρ^j ($\rho = x,y,z$) describes the Pauli matrix that operates on the electronic states of the j-th material site, $|\downarrow\rangle_j$(ground state), and $|\uparrow\rangle_j$(excited state). $\hat{n}_j = (\sigma_z^j + 1)/2$ represents the excited state electronic population of the j-th site, and μ_i denotes the dipole interaction between electrons and photons. Fig. 1 presents a schematic view of the model.

We take $\hbar = 1$ together with the parameter values $\omega = 1$, $\mu_i = 0.5(i = 1, 2, 3)$, and $\varepsilon = 13.5$. When $\hbar\omega$ is ~150 meV, the light absorption of the system is observed in the near infrared or visible region.

The number of irradiated photons is much smaller than that in an optical pulse due to the limitation of the numerical calculation. We choose $\mu_i = 0.5$ ($i = 1$, 2, 3) in order to make the effect of light irradiation clear under such a weak excitation condition. As for the electron–phonon coupling ν, it lies between 0 and 6 in various materials.[10,11] In the present study, it is set to 3.5, which is a strong coupling case. Because Raman scattering is a primary process in the excitation of electron–phonon systems, three-photon modes corresponding to the pump mode ($\Omega_1 = 13.5$), Stokes mode ($\Omega_2 = 12.5$), and anti-Stokes mode ($\Omega_3 = 14.5$) are considered.

The initial states of the photons are set to the coherent state $|\alpha_i\rangle = \mathrm{e}^{-|\alpha_i|^2/2} \sum_{n=0}^{\infty} \frac{\alpha_i^n}{\sqrt{n!}} |n\rangle_i$ for the i-th photon mode. In this case, the light field

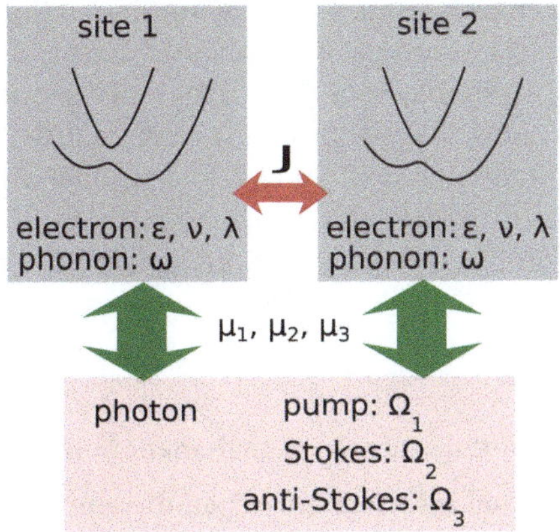

Fig. 1 Schematic diagram of the model of coupled electron–phonon systems.

is considered to be confined in a one-dimensional cavity, though the details of its structure are not discussed in this paper. We set $\alpha_1 = 5$ and $\alpha_2 = \alpha_3 = -2.5$ to describe a train of optical pulses. The expectation value for the electric field of the incident optical pulse is described by $E(t) \propto \sum_{n=1}^{3} i\langle \alpha_n \, e^{-i\omega_n t} | (c_n - c_n^\dagger) | \alpha_n \, e^{-i\omega_n t} \rangle \propto \sum_{n=1}^{3} \alpha_n \sin \omega_n t$, and illustrated in Fig. 2.

In this study, we consider the dynamics of structural change accompanied by a nonadiabatic transition, *e.g.*, photoisomerization; however, we do not restrict ourselves to discussing molecular systems. We also stress that the proposed model covers various types of photoinduced phenomena, *e.g.*, coherent phonon dynamics,[12] conical intersection dynamics,[13] and photoinduced nucleation in a coherent regime.[14]

The time-dependent Schrödinger equation for Hamiltonian (1) was numerically solved to obtain the wavefunction $|\Phi(t)\rangle$. The initial condition is given by $|\Phi(0)\rangle = |\alpha_1\rangle \otimes |\alpha_2\rangle \otimes |\alpha_3\rangle \otimes |G\rangle$, where $|\alpha_i\rangle$ denotes a coherent state for the i-th photon mode and $|G\rangle$ represents the ground state of the electron–phonon system.

Because the phonons in sites 1 and 2 comprise a subsystem of the entire electron–phonon–photon system, we must consider multipartite entanglement to study phonon entanglement. Hence, we discuss an appropriate measure for phonon entanglement. Because the concurrence or tangle is only applicable to the evaluation of multi-qubit entanglement,[15–17] we instead discuss the quantitative

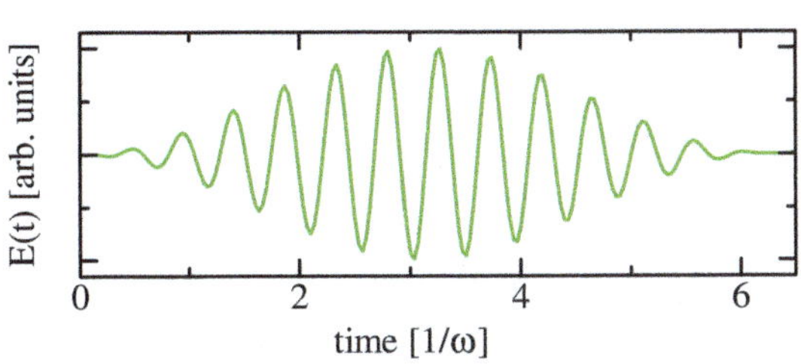

Fig. 2 Electric field amplitude of the incident optical pulse.

properties of phonon entanglement by the quantum mutual information, which is interpreted as the relative entropy of the reduced density matrix for the phonon subspace to its closest separable state.[18,19] Because it is also an order parameter in quantum phase transitions,[20] it is appropriate to discuss the structural dynamics concomitant with photoexcitations. We also calculate the dynamics of the singular values obtained by the Schmidt decomposition of the time-dependent wavefunction, and then discuss their relationship with the measurement of the phonon states.

2.2 Evaluation method of quantum entanglement between subsystems

2.2.1 Quantum mutual information. When the entire system is divided into three subsystems, A, B, and C, the quantum mutual information between A and B is defined by

$$I_{\mathrm{M}} = S_{\mathrm{A}} + S_{\mathrm{B}} - S_{\mathrm{A}\otimes\mathrm{B}}, \tag{2}$$

where S_α denotes the von Neumann entropy of the reduced density matrix on subsystem α. In addition, the wavefunction of the entire system is expressed as

$$|\Phi\rangle = \sum_{lmnk'\sigma\sigma'} p_{lmnkk'\sigma\sigma'} |l\rangle|m\rangle|n\rangle|\sigma\sigma' kk'\rangle, \tag{3}$$

where $|l\rangle$, $|m\rangle$, and $|n\rangle$ denote the Fock state for the photons of each mode. $|k\sigma\rangle$ represents the k phonon state in the electronic level σ $(=\uparrow, \downarrow)$. When we calculate the phonon entropy, A and B correspond to the phonon states in sites 1 and 2, respectively. In this case, C denotes the residual DoF, including photons, electrons, and the irrelevant phonons. Hence, the reduced density matrices for the S_{A}, S_{B}, and $S_{\mathrm{A}\otimes\mathrm{B}}$ calculations are given by

$$\rho_{\mathrm{A}} = \mathrm{Tr}_{\mathrm{B}\otimes\mathrm{C}}|\Phi\rangle\langle\Phi| = \sum_{k,k'} q_{kk'}|k\rangle\langle k'|, \tag{4}$$

$$\rho_{\mathrm{B}} = \mathrm{Tr}_{\mathrm{A}\otimes\mathrm{C}}|\Phi\rangle\langle\Phi| = \sum_{k,k'} r_{kk'}|k\rangle\langle k'|, \tag{5}$$

and

$$\rho_{\mathrm{A}\otimes\mathrm{B}} = \mathrm{Tr}_{\mathrm{C}}|\Phi\rangle\langle\Phi| = \sum_{k,k',j,j'} s_{kk'jj'}|kj\rangle\langle k'j'|, \tag{6}$$

respectively, where

$$q_{kk'} = \sum_{lmnj\sigma\sigma'} \overline{p}_{lmnkj\sigma\sigma'} p_{lmnk'j\sigma\sigma'}, \tag{7}$$

$$r_{kk'} = \sum_{lmnj\sigma\sigma'} \overline{p}_{lmnjk\sigma\sigma'} p_{lmnjk'\sigma\sigma'}, \tag{8}$$

and

$$s_{kk'jj'} = \sum_{lmn\sigma\sigma'} \overline{p}_{lmnkj\sigma\sigma'} p_{lmnk'j'\sigma\sigma'}. \tag{9}$$

$\bar{x}$ denotes the complex conjugate of x and $\mathrm{Tr_X}$ represents the partial trace over the basis of site X.

When we calculate the electron entropy, A and B correspond to the electronic states in sites 1 and 2, respectively. Therefore, C denotes the residual DoF including photons, phonons, and the irrelevant electrons. The reduced density matrices for the S_A, S_B, and $S_{\mathrm{A}\otimes\mathrm{B}}$ calculations are given by

$$\rho_\mathrm{A} = \mathrm{Tr}_{\mathrm{B}\otimes\mathrm{C}}|\Phi\rangle\langle\Phi| = \sum_{\sigma,\sigma'} u_{\sigma\sigma'}|\sigma\rangle\langle\sigma'|, \tag{10}$$

$$\rho_\mathrm{B} = \mathrm{Tr}_{\mathrm{A}\otimes\mathrm{C}}|\Phi\rangle\langle\Phi| = \sum_{\sigma,\sigma'} v_{\sigma\sigma'}|\sigma\rangle\langle\sigma k'|, \tag{11}$$

and

$$\rho_{\mathrm{A}\otimes\mathrm{B}} = \mathrm{Tr}_\mathrm{C}|\Phi\rangle\langle\Phi| = \sum_{\sigma,\sigma',\tau,\tau'} w_{\sigma\sigma'\tau\tau'}|\sigma\tau\rangle\langle\sigma'\tau'|, \tag{12}$$

respectively, where

$$u_{\sigma\sigma'} = \sum_{lmnkk'\tau} \bar{p}_{lmnkk'\sigma\tau} p_{lmnkk'\sigma'\tau}, \tag{13}$$

$$v_{\sigma\sigma'} = \sum_{lmnkk'\tau} \bar{p}_{lmnkk'\tau\sigma} p_{lmnkk'\tau\sigma'}, \tag{14}$$

and

$$w_{\sigma\sigma'\tau\tau'} = \sum_{lmnkk'} \bar{p}_{lmnkk'\sigma\tau} p_{lmnkk'\sigma'\tau'}. \tag{15}$$

The quantum mutual information is obtained by

$$I_\mathrm{M} = -\mathrm{Tr}\rho_\mathrm{A} \ln \rho_\mathrm{A} - \mathrm{Tr}\rho_\mathrm{B} \ln \rho_\mathrm{B} + \mathrm{Tr}\rho_{\mathrm{A}\otimes\mathrm{B}} \ln \rho_{\mathrm{A}\otimes\mathrm{B}}. \tag{16}$$

2.2.2 Singular value decomposition on time-dependent wavefunctions. The numerically obtained wavefunction $|\Phi(t)\rangle$ comprises two sites and three photon systems. The entanglement between partial DoFs (phonon–phonon or electron–electron) is evaluated by the quantum mutual information I_M, while the dynamical behavior of composite modes corresponds to the time-dependence of I_M.

To clarify the details of the quantum mechanical state at each moment, it is beneficial to perform the Schmidt decomposition to $|\Phi(t)\rangle$ for the focused DoF. Because we are interested in the phonon–phonon entanglement, it is appropriate to decompose $|\Phi\rangle$ into

$$|\Phi\rangle = \sum_n \lambda_n |\eta_n\rangle|\theta_n\rangle, \tag{17}$$

where $|\eta_n\rangle$ belongs to the phonon space and $|\theta_n\rangle$ is in the space of the other DoF. They are complete orthonormal systems of the phonon states and the electron–photon states, *i.e.*, $\langle\eta_m|\eta_n\rangle = \langle\theta_m|\theta_n\rangle = \delta_{mn}$. λ_n denotes the singular value of the coefficient matrix of $|\Phi\rangle$.

By tracing out the photon and electron DoF in the density matrix $\rho = |\Phi\rangle\langle\Phi|$, we obtain a reduced density matrix Ξ as

$$\Xi = \sum_n \lambda_n{}^2 |\eta_n\rangle\langle\eta_n|. \tag{18}$$

Eqn (18) demonstrates that $|\eta_n\rangle$ corresponds to experimentally observed phonon states, provided the experiments are designed to detect the phonons in both systems simultaneously.

Because $\mathcal{M}_n \equiv |\eta_n\rangle\langle\eta_n|$ satisfies the closure relation $\sum_n \mathcal{M}_n = 1$, they construct a set of projection-valued measures (PVM).[3] When we consider a measurement of the phonon states corresponding to $\mathcal{M}_n$, we expect to obtain a single $|\eta_n\rangle$ with probability $\lambda_n{}^2$. Hence, though the accumulated data in pump–probe experiments, for example, bring the averaged properties of the phonon states, it is important to discuss the entanglement properties of individual singular vectors $|\eta_n\rangle$ and reveal the capacity of the phonon states as an entanglement source in the present configuration. If $|\eta_n\rangle$ corresponds to experimentally observed phonon states, it is beneficial to discuss each component $|\eta_n\rangle$, as they are to be detected at each datum of the pump–probe detection.

2.3 Calculated results

2.3.1 Effect of exchange interaction. Fig. 3 presents the I_M between phonons for $\nu = 3.5$ and $J = 0, -0.5$. In these cases, the ground state of the electron–phonon system is given by

$$|\Phi(t=0)\rangle = |\downarrow\downarrow 00\rangle, \tag{19}$$

and the phonon entanglement increases by the irradiation of photons in both cases. Although the exchange interaction triggers the increase in I_M, we stress that the value of I_M is of the same order between the two cases, which implies that the photoinduced entanglement cannot be neglected even in interacting electron–phonon systems.

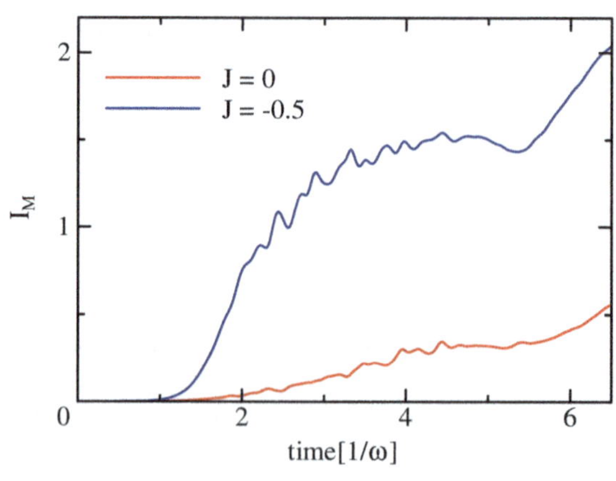

Fig. 3 I_M for $\nu = 3.5$ and $J = 0, -0.5$.

The ground state is described by eqn (19) for the ferromagnetic case ($J < 0$), while the phonon entanglement increases monotonically as the value of $|J|$ increases.

2.3.2 Ground-state phase diagram on the J–ν plane. When an antiferromagnetic exchange interaction is present between electrons, the electronic state prefers to be a spin singlet $\propto |\uparrow\downarrow\rangle - |\downarrow\uparrow\rangle$. When J is small, a spin triplet $\propto |\downarrow\downarrow\rangle$ becomes more preferable, because the spin flip from $|\downarrow\rangle$ to $|\uparrow\rangle$ costs energy by ε. However, when ν is large, we obtain energy gain by lattice deformation for the wavepacket on the excited-state potential energy surface and, thus, a spin singlet recovers to be the ground state. This feature is intuitively understood by regarding the lattice deformation x as a classical variable. In this case, the total spin of electrons is a good quantum number, and the potential energy for the electrons is given by

$$U(x) = \omega^2 x^2 + \sqrt{2}\nu x + \varepsilon - 3J \geq -\frac{\nu^2}{\omega} + \varepsilon - 3J, \tag{20}$$

for a spin singlet, and

$$U(x) = J, \tag{21}$$

for a spin triplet. Hence, the boundary between the spin triplet and the spin singlet on the J–ν plane is given by

$$\omega(4J - \varepsilon) + \nu^2 = 0. \tag{22}$$

The above consideration allows us to draw a phase diagram that characterizes the ground state of the material system. For this purpose, we numerically diagonalize $\mathscr{H}_0$, varying the values of J and ν, and calculate I_M between the electron and the phonon at site 1 to distinguish the lattice-spin states from each other. Fig. 4 presents the phase diagram on the J–ν plane, where the values of I_M are depicted by colors. The boundary line is illustrated in the figure, which shifts from that calculated by eqn (22). This difference demonstrates the effect of the quantum nature of phonons.

Precisely, Fig. 4 indicates the boundary between phase 1 $\left(\sum_j \sigma_z^j = -2\right)$ and phase 2 $\left(\sum_j \sigma_z^j = 0\right)$.

Fig. 4 also shows that I_M gradually changes in phase 2. It is important to consider that the ground state comprises the electrons in either $|\uparrow\rangle$ or $|\downarrow\rangle$, including the phonons in the coherent state. Hence, it is important to understand this case by considering a variational wavefunction

$$|\Psi_\nu(\alpha)\rangle = \frac{1}{\sqrt{2}}|\uparrow\downarrow\alpha 0\rangle - |\downarrow\uparrow 0\alpha\rangle, \tag{23}$$

where $|\alpha\rangle$ denotes the coherent state of the phonon. This state demonstrates that the ground state is neither a spin singlet nor a spin triplet due to the dynamical Jahn–Teller effect.[21] In fact, the Schmidt decomposition of eqn (23) is given by

$$|\Psi_\nu(\alpha)\rangle = \sqrt{\frac{1 + e^{-|\alpha|^2/2}}{2}}|S\rangle|p_+\rangle + \sqrt{\frac{1 - e^{-|\alpha|^2/2}}{2}}|T\rangle|p_-\rangle, \tag{24}$$

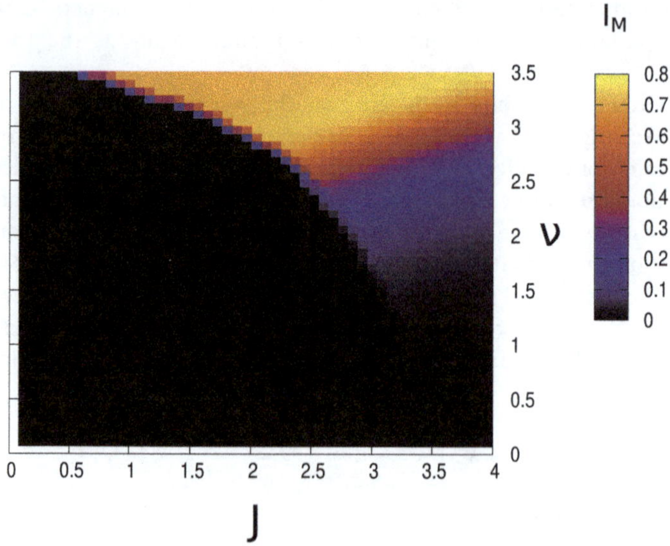

Fig. 4 Phase diagram of the system characterized by the ground state property depicted by I_M values between phonons on sites 1 and 2.

where

$$|S\rangle = \frac{|\uparrow\downarrow\rangle - |\downarrow\uparrow\rangle}{\sqrt{2}} \tag{25}$$

$$|T\rangle = \frac{|\uparrow\downarrow\rangle + |\downarrow\uparrow\rangle}{\sqrt{2}} \tag{26}$$

$$|p_\pm\rangle = \frac{|\alpha 0\rangle \pm |0\alpha\rangle}{\sqrt{2(1 \pm \mathrm{e}^{|\alpha|^2})}}. \tag{27}$$

Hence, $|\psi_\nu\rangle$ is a spin triplet for $\alpha \to \infty$, and a spin singlet for $\alpha \to 0$. The electron and the phonons are decomposable in the above two limits; they are entangled otherwise.

The numerically obtained wavefunction in phase 2 is well-approximated by eqn (23) by varying the value of α, *i.e.*, the Jahn–Teller effect continuously diminishes as α decreases. Therefore, for example, $I_M = 0$ for $\nu = 0$, which indicates that the spin singlet state without lattice deformation is the ground state.

2.3.3 Dynamical behavior of the wavefunction under photoirradiation. After irradiation with an optical pulse, the spin configuration will be switched to another, depending on the values of the parameters. Fig. 5 presents the number of phonons per site $n_p(t) = \langle \Phi(t)|a_1^\dagger a_1|\Phi(t)\rangle$ and I_M between the phonons and that between the electrons in different sites, as functions of time, where the values of J and ν are presented in the left panel of the figures. We also calculated the corresponding properties when the irradiated light is considered a classical electromagnetic field.

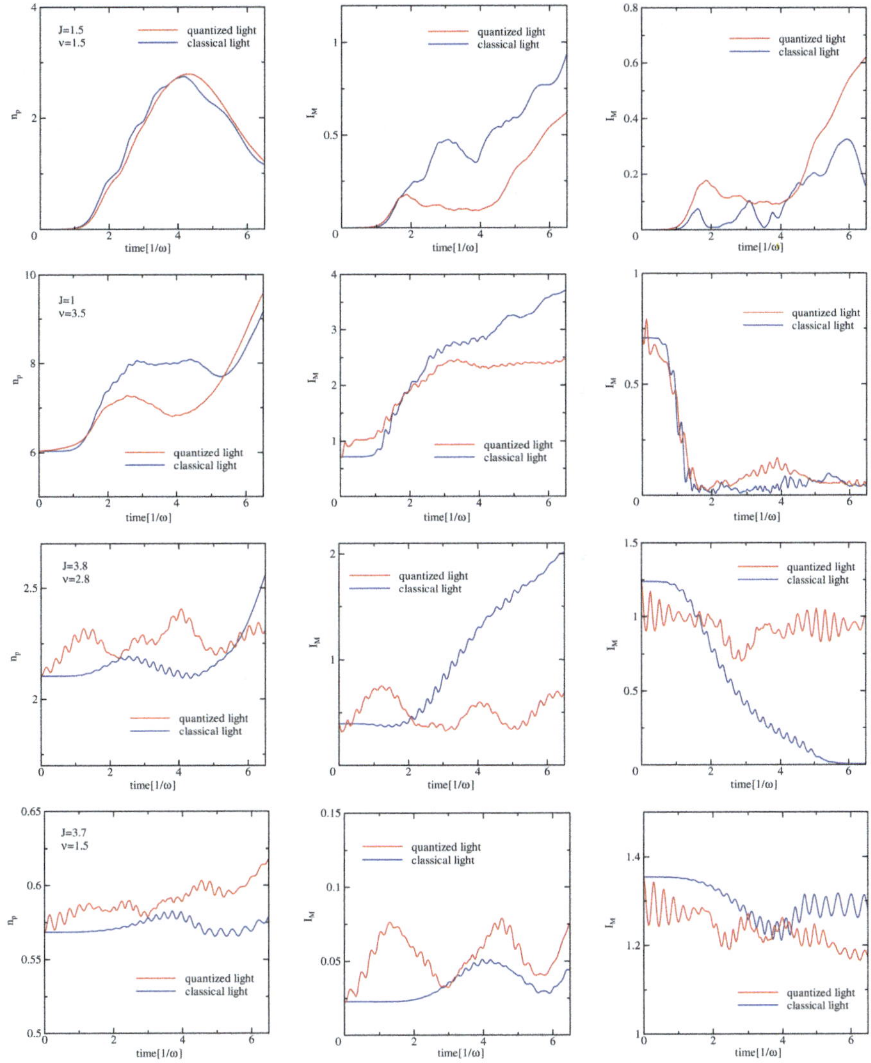

Fig. 5 From left to right: n_p, I_M between phonons on sites 1 and 2, and I_M between electrons on sites 1 and 2. From top to bottom: $(J, \nu) = (1.5, 1.5)$, $(1, 3.5)$, $(3.8, 2.8)$, and $(3.7, 1.5)$.

In the three cases ($(J, \nu) = (1.5, 1.5)$, $(1, 3.5)$, and $(3.7, 1.5)$), $n_p(t)$ and I_M exhibit similar behaviors to those with classical light. These results indicate that the effect of the quantum nature of light is limited in these cases, which implies that the quantum fluctuation is not large in the ground state. However, when $(J, \nu) = (3.8, 2.8)$, the difference between the two simulations is substantial. In particular, the I_M value between electrons vanishes when the light is treated classically, while it remains finite when the quantum nature of the light is considered. These values are in the area of the phase diagram (Fig. 4) where I_M rapidly changes and, thus, the quantum fluctuation in the ground state is larger than the other points on the J–ν plane. Because the quantized light generates additional entanglement

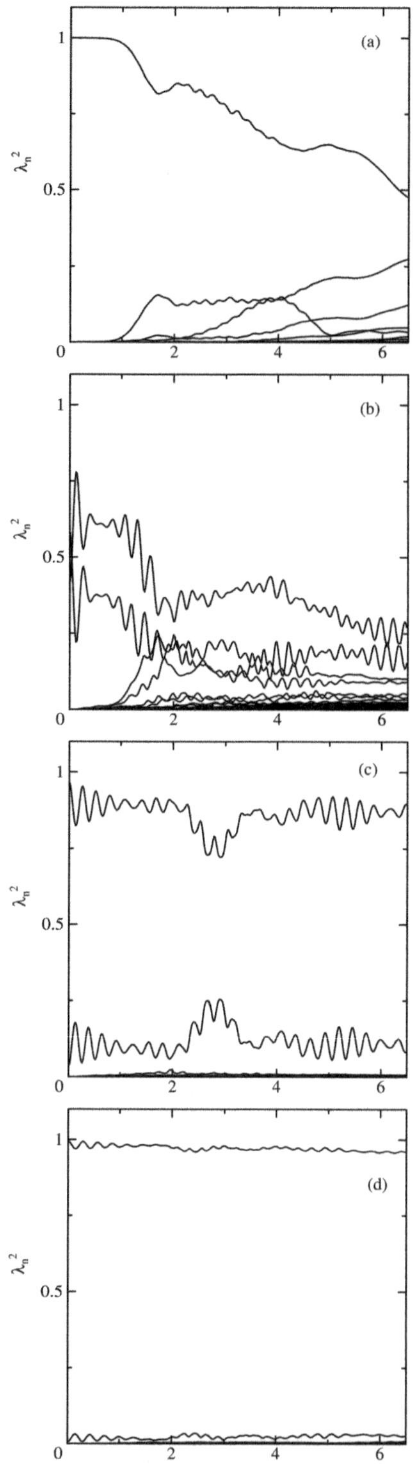

Fig. 6 Squares of the singular values $\lambda_n{}^2$ for the reduced density matrix, where the entire system is divided into phonon and electron–photon systems. (J, ν) = (a) (1.5, 1.5), (b) (1, 3.5), (c) (3.8, 2.8), and (d) (3.7, 1.5).

between the electronic/phonon DoF, we consider that the fluctuations triggered by the photoinduced entanglement affect the dynamics of the system, which should be considered particularly when the ground state property is fluctuating. In other words, these results indicate that the quantum fluctuation relevant to the photoinduced entanglement plays an important role in the transient dynamics, particularly near the photoinduced structural or electronic change. Hence, the quantum nature of the incident light is relevant to the transient dynamics, and is therefore a control parameter for coherent control.

We also calculated the singular values λ_n obtained by dividing the entire system into phonon and electron–photon systems. Fig. 6 presents ${\lambda_n}^2$ as functions of time for $(J, \nu) = (1.5, 1.5)$, $(1, 3.5)$, $(3.8, 2.8)$, and $(3.7, 1.5)$. The number of lines in each figure of Fig. 6 corresponds to the rank of the reduced density matrix, which represents the strength of the entanglement between the material and photons. Therefore, Fig. 6 demonstrates that the material–photon entanglement is large when the switching of the spin configurations occurs ($(J, \nu) = (1.5, 1.5)$, and $(1, 3.5)$). When the material DoFs are measured in these cases, the singular vectors characterized by the projection operator $\mathcal{M}_n$ are obtained, which implies that multiple quantum mechanical states are provided as a result. In contrast, the other cases correspond to fewer resultant states after the measurement, although they differ from those obtained by classical light. In summary, we obtained different patterns for the quantum mechanical effect of light as J and ν are varied in the phase diagram.

3 Spin chains irradiated by entangled photons

As the quantum state of light plays an important role in the physical properties of photoirradiated systems, we study another configuration of material–photon systems in which two remote systems are irradiated by entangled photons. In this case, we regard the quantum state of the incident light as a parameter for quantum state control. For this purpose, we consider a pair of two-level systems (qubits) coupled with antiferromagnetic spin chains instead of phonons. They are irradiated by the outputs of a beam splitter which has quantum entanglement corresponding to the input states, and modify the entanglement properties between the qubit–spin system.

3.1 Hamiltonian of the system

According to the photon quantum mechanics of beam splitters,[22] the annihilation operators of the input (b_0, b_1) and the output (c_1, c_2) of a beam splitter satisfy the following relation

$$c_1 = \frac{1}{\sqrt{2}}(b_0 + ib_1) \tag{28}$$

$$c_2 = \frac{1}{\sqrt{2}}(ib_0 + b_1). \tag{29}$$

When the input states on port 0 and 1 are the vacuum and a coherent state $|\alpha\rangle$, both of the output states on port 1 and 2 are the coherent states $|\alpha/\sqrt{2}\rangle$ and $|i\alpha/\sqrt{2}\rangle$,[22] respectively, and, thus, they are not entangled. In this case, no entanglement generation takes place between the spin chains, since the light irradiation corresponds to LOCC, *i.e.*, the quantum mechanical nature of the light is irrelevant.

Although it is not realistic to set the input state on port 1 to a Fock state, it is suggested to study the role of the quantum state of light. When the light input is given by $|N\rangle\rangle \equiv= (b_1^\dagger)^N/N!|0\rangle\rangle$, the light output is described by the Fock state constructed by c_i as

$$\frac{\left(c_0^\dagger - ic_1^\dagger\right)^N}{2^{N/2}N!}|0\rangle|0\rangle = \sum_{n=0}^{N}(-i)^{N-n}\sqrt{\frac{N!}{2^N n!(N-n)!}}|n\rangle|N-n\rangle, \tag{30}$$

which are entangled with each other.

As shown in Fig. 7, c_i is coupled with each spin chain and the Hamiltonian of the interacting spin chain and photons is described by

$$\begin{aligned}\mathcal{H}_s = &\sum_{i=1}^{2}\left(\varepsilon\sigma_{i0}^z + J\sum_{k=1}^{n}\vec{\sigma}_{ik}\cdot\vec{\sigma}_{ik+1}\right) + \Omega\left(c_1^\dagger + c_2^\dagger\right)\\ &+\mu\sum_{i=1}^{2}\left(\sigma_{i0}^+ c_n + \sigma_{i0}^- c_n^\dagger\right),\end{aligned} \tag{31}$$

where $\sigma_{i0}^\pm = (\sigma_{i0}^x \pm i\sigma_{i0}^y)/2$. We note that only σ_{i0} is coupled with light and has an energy gap between $|\downarrow\rangle$ and $|\uparrow\rangle$. σ_{ik} ($k \geq 1$) interacts with the nearest neighbor spins, which shows that each spin chain comprises a two-level system (qubit) and an antiferromagnetic spin. Thus, the qubits are excited by the photons and the excitation of the attached spins follows. The entanglement between different chains is generated by a certain input light of the beam splitter.

The values of the parameters are: $\varepsilon = \Omega = J = \mu = 1$, and the initial state of the time-dependent Schrödinger equation for the Hamiltonian (31) is the tensor product of the ground state of the spin chains and the output of the beam splitter. The numerically obtained wavefunction is denoted by $|\Psi\rangle$.

3.2 Calculated results

The interchain entanglement is evaluated by the quantum mutual information between two chains defined by

$$I_M^C = S_{c1} + S_{c2} - S_{c12}, \tag{32}$$

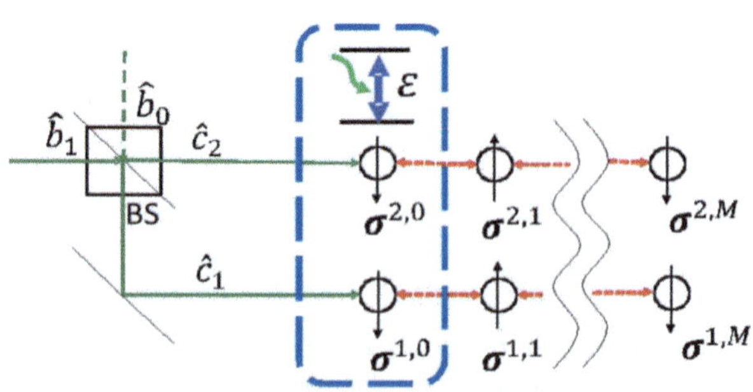

Fig. 7 A schematic diagram of the model of qubit–spin systems and a beam splitter-based optical system.

where S_{c1}, S_{c2}, and S_{c12} denote the entanglement entropy obtained by tracing out the DoFs for chain 1, chain 2, and both chains, respectively. We also calculated the spin correlation functions $\langle\Psi|\vec{\sigma}_{10}\cdot\vec{\sigma}_{11}|\Psi\rangle$ and $\langle\Psi|\vec{\sigma}_{11}\cdot\vec{\sigma}_{12}|\Psi\rangle$. The calculated results are shown in Fig. 8(b), where the light input of the beam splitter is a coherent state $|\alpha\rangle$ and a Fock state $|N\rangle$. As for the number of photons, we set $\alpha^2 = N = 50$.

I_M^C rapidly increases after photoirradiation starts, and its oscillatory behavior reflects the Rabi oscillation of σ_{10} for the Fock state input. However, due to the aforementioned property of the coherent state, no entanglement generation is observed for the coherent state input. On the contrary, the spin correlations coincide with each other for the two types of input, which implies that the quantum state of light is irrelevant to the internal properties of each chain.

Hence, photoirradiation makes a pair of coupled spins act as an entanglement storage, even when they are distant from each other. When we consider generating entanglement between distant materials by light, the light intensity is an important parameter. Precisely, the decoherence during light propagation disturbs the entanglement generation. Therefore, it is preferable to irradiate the materials with as intense a light as possible for effective entanglement generation. We discuss the strength of the interchain entanglement as a function of the number of photons. Fig. 9 shows the I_M^C for $t = 8$ as a function of N, which shows that I_M^C decreases as the number of photons increases.

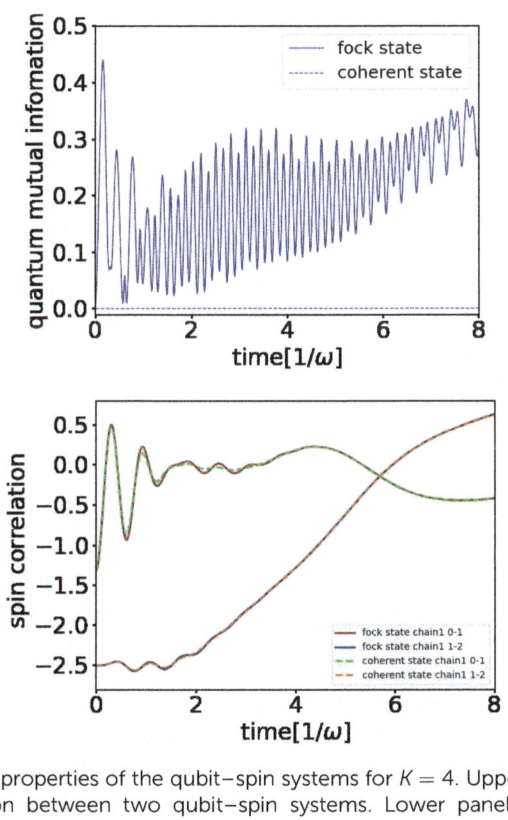

Fig. 8 Dynamical properties of the qubit–spin systems for $K = 4$. Upper panel: quantum mutual information between two qubit–spin systems. Lower panel: spin correlation functions $\langle\Psi|\vec{\sigma}_{10}\cdot\vec{\sigma}_{11}|\Psi\rangle$ and $\langle\Psi|\vec{\sigma}_{11}\cdot\vec{\sigma}_{12}|\Psi\rangle$.

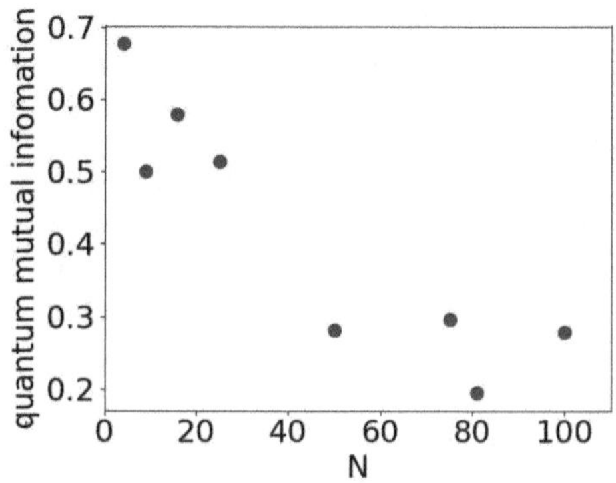

Fig. 9 I_M^C for $t = 8$ as a function of the number of photons N. The input of the beam splitter is a Fock state.

Since the output of the beam splitter is described by eqn (30), the entanglement entropy of the two output modes is

$$S_f(N) = -\sum_{n=0}^{N} \frac{N!}{2^N n!(N-n)!} \ln \frac{N!}{2^N n!(N-n)!}, \tag{33}$$

which decreases as N_p increases. This contrast in the behaviors of I_M^C and S_f is reminiscent of the monogamy of entanglement,[23] although I_M^C remains finite even when N_p becomes large as far as the quantum nature of the photons is taken into account. Thus, the irradiation of entangled photons on remote material systems is suitable for effective entanglement storage.

4 Conclusions and outlook

We studied the role of quantum nature of light on the entanglement generation in multiple electronic systems.

For the coupled electron–phonon systems, we calculated the phonon entanglement generation dynamics in two-site electron–phonon systems with exchange interactions. Employing a model of coupled electron–phonon–photon systems, we inferred that the quantum mutual information for phonons elucidates the dynamics of phonon entanglement generation, which depends on the ground state of the electron–phonon system. Precisely, the interplay of the antiferromagnetic exchange and electron–phonon interactions provides us with a phase diagram, in which each phase is characterized by the ground state property of the electron–phonon system. Light irradiation of the electron–phonon system results in the generation of quantum entanglement. We compared two simulation methods corresponding to the classical and quantum treatments of light. Although they both provided a similar results in general, remarkable differences were observed in some cases. In particular, when the I_M value between the phonons in different sites rapidly changes in the J–ν plane, the change in I_M is suppressed, which indicates that the photoinduced entanglement affects the quantum mechanical state of the material to maintain the spin configuration.

Furthermore, we focused on the phonon properties by dividing the entire electron–phonon–photon state $|\Phi\rangle$ into phonon $|\eta_n\rangle$ and electron–photon $|\theta_n\rangle$

states by the Schmidt decomposition. Numerical calculations on the time-evolution of the singular values λ_n indicate that the measurement of the phonon states corresponding to $\mathcal{M}_n = |\eta_n\rangle\langle\eta_n|$ will provide a clue for understanding the quantum mechanical effect of the irradiated photons.

We also studied the dynamics of coupled qubit–spin systems irradiated by entangled photons to demonstrate that the quantum nature of photons explicitly appears when the quantum state of photons plays an important role. The entanglement between them is recovered by the irradiation of the entangled photons, even if it is consumed at each site. Thus, the remote systems are applicable to use as rechargeable entanglement storage. Furthermore, we found that advanced light sources, *e.g.*, single photon sources, are not required to realize such entanglement storage, as the relationship between the number of photons and the quantum mutual information shows. Although the detail of the relationship between the quantum state of the photons and the strength of the entanglement is not revealed in the present study, it is particularly important to study the effect of more realistic light input, *e.g.*, the squeezed state of light. We stress, however, that the results of this paper open up a possibility of using the quantum state of light as a parameter for the coherent control of materials' states.

The measurement methods of the entanglement properties in this study have not been clarified so far. We, however, mention that slight modulation of the quantum state of the scattered light gives us a clue as to the quantum mechanical correlation (entanglement) between electrons and/or phonons. Analysis by the Heisenberg equation of motion is a powerful tool to extract the relevant operators for the generation of the entanglement, which shows that the quantization of light is also essential to derive the probe signal of the entanglement.[4] The details of the relationship between the probe signals and the quantum states of the materials are left for future study.

Conflicts of interest

There are no conflicts to declare.

Acknowledgements

K. I. is grateful to K. G. Nakamura, N. Aoyagi, and F. Sunaga for fruitful discussion. This work was partly supported by JSPS KAKENHI Grant Number JP21H01752 and the Collaborative Research Project of Laboratory for Materials and Structures, Institute of Innovative Research, Tokyo Institute of Technology, Japan. Numerical calculations were performed on the computer resources provided under the category of General Projects by the Research Institute for Information Technology, Kyushu University, Japan.

Notes and references

1 K. C. Lee, M. R. Sprague, B. J. Sussman, J. Nunn, N. K. Langford, X.-M. Jin, T. Champion, P. Michelberger, K. F. Reim, D. England, D. Jaksch and I. A. Walmsley, *Science*, 2011, **334**, 1253–1256.
2 F. Borjans, X. G. Croot, X. Mi, M. J. Gullans and J. R. Petta, *Nature*, 2020, **577**, 195–198.

3 M. A. Nielsen and I. L. Chuang, *Quantum Computation and Quantum Information*, Cambridge University Press, Cambridge, 2010.

4 K. Ishida and H. Matsueda, *J. Phys. Soc. Jpn.*, 2021, **90**, 104714.

5 L. Amico, R. Fazio, A. Osterloh and V. Vedral, *Rev. Mod. Phys.*, 2008, **80**, 517–576.

6 L. Mazza, D. Rossini, R. Fazio and M. Endres, *New J. Phys.*, 2015, **17**, 013015.

7 M. Gallart, M. Ziegler, O. Crégut, E. Feltin, J.-F. Carlin, R. Butté, N. Grandjean, B. Hönerlage and P. Gilliot, *Phys. Rev. B*, 2017, **96**, 041303(R).

8 T. Ghosh, S. Aharon, A. Shpatz, L. Etgar and S. Ruhman, *ACS Nano*, 2018, **12**, 5719–5725.

9 M. Chergui and E. Collet, *Chem. Rev.*, 2017, **117**, 11025–11065.

10 A. Hoffmann, B. K. Meyer, and E. Malguth, in *Zinc Oxide: From Fundamental Properties Towards Novel Applications*, Springer, Berlin, Heidelberg, 2010, vol. 120, Springer Series in Materials Science, p. 233.

11 A. Y. Freidzon, R. R. Valiev and A. A. Berezhnoy, *RSC Adv.*, 2014, **4**, 42054–42065.

12 K. G. Nakamura, K. Ohya, H. Takahashi, T. Tsuruta, H. Sasaki, S. Uozumi, K. Norimatsu, M. Kitajima, Y. Shikano and Y. Kayanuma, *Phys. Rev. B*, 2016, **94**, 024303.

13 F. Bouakline, *Chem. Phys.*, 2014, **442**, 31–40.

14 K. Ishida and K. Nasu, *Phys. Rev. Lett.*, 2008, **100**, 116403.

15 T. E. Tessier, I. H. Deutsch, A. Delgado and I. Fuentes-Guridi, *Phys. Rev. A: At., Mol., Opt. Phys.*, 2003, **68**, 062316.

16 Z. X. Man, Y. J. Xia and N. B. An, *Eur. Phys. J. D*, 2009, **53**, 229–236.

17 S. Agarwal, S. M. Hashemi Rafsanjani and J. H. Eberly, *Phys. Rev. A: At., Mol., Opt. Phys.*, 2012, **85**, 043815(R).

18 B. Groisman, S. Popescu and A. Winter, *Phys. Rev. A: At., Mol., Opt. Phys.*, 2005, **72**, 032317.

19 A. Kumar, *Phys. Rev. A*, 2017, **96**, 012332.

20 M. Vojta, *Rep. Prog. Phys.*, 2003, **66**, 2069–2110.

21 J. C. Slonczewski, *Phys. Rev.*, 1963, **131**, 1596–1610.

22 C. H. Holbrow, E. Galvez and M. E. Parks, *Am. J. Phys.*, 2002, **70**, 260–265.

23 V. Coffman, J. Kundu and W. K. Wootters, *Phys. Rev. A: At., Mol., Opt. Phys.*, 2000, **61**, 052306.

Faraday Discussions

ROYAL SOCIETY OF CHEMISTRY

PAPER

Quench–drive spectroscopy of cuprates

Matteo Puviani * and Dirk Manske

Received 17th January 2022, Accepted 2nd March 2022
DOI: 10.1039/d2fd00010e

Cuprates are d-wave superconductors which exhibit a rich phase diagram: they are characterized by superconducting fluctuations even above the critical temperature, and thermal disorder can reduce or suppress the phase coherence. However, photoexcitation can have the opposite effect: recent experiments have shown an increasing phase coherence in optimally doped BSCCO with mid-infrared driving. Time-resolved terahertz spectroscopies are powerful techniques to excite and probe the non-equilibrium states of superconductors, directly addressing collective modes, such as amplitude (Higgs) oscillations. In this work, we calculate the full time evolution of the current generated by a cuprate with a quench–drive spectroscopy setup. Analyzing the response in Fourier space with respect to both the real time and the quench–drive delay time, we look for the signature of a transient modulation of higher harmonics, as well as the Higgs mode, in order to characterize the ground state phase. In particular, this approach can provide a smoking gun for induced or increased phase coherence when applied to the pseudogap phase. These results can pave the way for future experimental schemes to characterize and study superconductors alongside incoherent phases and phase transitions, including induced and transient superconductivity.

1. Introduction

High-temperature superconductors have attracted much of the attention in the research on superconductivity since their discovery. In recent years, experiments on high-temperature superconductors have revealed the presence of signatures of superconducting fluctuations above the critical temperature T_c. This behaviour has been attributed to incoherent or pre-formed Cooper pairs, leading to third-harmonic generation (THG) and enhancement of the reflectivity change in a pump–probe experimental configuration.[1–3]

Cuprates are the prototypical example of high-temperature superconductors, exhibiting a complex phase diagram as a function of doping and temperature, with a superconducting dome enriched by charge-density wave and pseudogap phases.[4] In fact, optimally doped Y-Bi2212 exhibits a superconducting phase below $T_c = 97$ K, and a pseudogap phase above T_c and up to the temperature $T^* \approx 135$ K.[1,5–8] It has been argued that the pseudogap phase on top of the

Max Planck Institute for Solid State Research, 70569 Stuttgart, Germany. E-mail: m.puviani@fkf.mpg.de

superconducting dome at optimal doping in unconventional superconductors is driven by the loss of phase coherence between the Cooper pairs, rather than the softening or vanishing of the pairing strength.[9–11] Therefore, in the pseudogap phase, the electrons are still paired, but their local phase is different from the global phase of the order parameter, which is lowered as a consequence.

A variety of non-equilibrium experiments on cuprates have indicated the importance and the interaction between collective modes, such as the amplitude (Higgs) mode and Josephson plasmon; different setups have been used to investigate this, from high-harmonic generation to pump–probe spectroscopy, to reflectivity measurements.[2,3,12–14] Moreover, it has been recently shown that THz pulses in the mid-infrared region can dynamically enhance the phase coherence of Cooper pairs in optimally doped cuprates, which is lowered by thermal disorder in equilibrium conditions.[7]

In this paper, we go beyond previous works of pump–probe spectroscopy on cuprates, applying the quench–drive spectroscopy setup[15–18] (see Fig. 1) recently extended to study different features of superconductors,[19] to superconductors with anisotropic d-wave order parameters. In this configuration, a first few-cycle THz broadband pulse (quenching, red in Fig. 1) impinges on the material, followed after an adjustable delay time, Δt, by a multi-cycle THz narrow-band (asymmetric) driving pulse (driving, green in Fig. 1). Then, the generated current, which is experimentally addressed by measuring the change of the transmitted electric field or the nonlinear optical conductivity,[20,21] can be analyzed in transmission in real time after filtering out the linear response directly proportional to the incoming pulses. Quench–drive spectroscopy has been shown to be a versatile and powerful tool to systematically analyze the superconducting response, as well as characterizing the signatures of collective modes, possibly enhancing the overall measured signal.[19] This method goes beyond standard pump–probe spectroscopy, where a short-time pulse drives the system out of equilibrium, followed by a subsequent weak and short-time pulse which probes the system (*e.g.*, current, reflectivity).[22,23] In a quench–drive setup, in fact, both the short-time pulse and the long-time drive can perturb the system, and the non-fixed relative geometry of the two pulses allows to shift the quench–

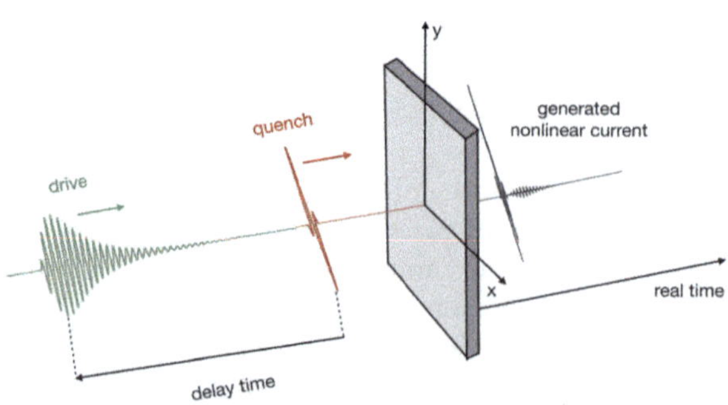

Fig. 1 Quench–drive setup. The figure shows the scheme of the quench–drive setup used here: the quenching pulse (red) impinges on the material (grey). Then, after a delay time (Δt, measured from the maximum peak of the quench to the maximum peak of the drive) the driving pulse also interacts with the material; the nonlinear current output is then generated and can be detected in real time.

drive time delay. This provides an extended number of possible configurations, allowing for the quenching pulse to overlap with the driving pulse , or even acting on the driven superconductor after the longer pulse. The analysis of the generated nonlinear current in both time and Fourier space, for example, including not only the real evolution but also the time-dependence of the quenching and driving relative time, allows us to obtain a richer signal than the usual harmonic response of the non-equilibrium material. In fact, the application of this spectroscopic technique to conventional clean s-wave superconductors has demonstrated that it can provide the nonlinear current signal due to the quasiparticles' and collective modes' excitations, visible as distinctive features in the two-dimensional time and frequency plots, respectively.[19] Alongside the high-harmonic generation, both transient excitation and dynamical modulation of the generated harmonics are visible, and can be theoretically described by solving the real-time Heisenberg's equation of motion, or interpreted with a diagrammatic approach.

However, in quench–drive spectroscopy, the real and delay times are not fully independent, since they both refer to the driving pulse, and are not able to catch the decoherence processes. This mainly differs from other three-pulse techniques, like pump–dump–probe, pump–push–probe or pump–repump–probe mechanisms, which have been widely used to study molecular excitations and transient states for decades.[24–27] In pump–repump–probe spectroscopy, for example, the first pulse creates a macroscopic polarization which decays due to dephasing, then the second one induces a population of the excited state, while the probe converts it back to a coherent polarized state.[28] An extension of our scheme to a real two-dimensional coherent spectroscopy, with two independent times and probing of decoherence processes, will be the object of a future work.

In the present work, we numerically solve the Heisenberg's equation of motion, derived within the pseudospin formalism,[29,30] in order to describe the superconducting state, and we support our interpretations and results with the derivation of the nonlinear susceptibility by means of a diagrammatic approach.[31]

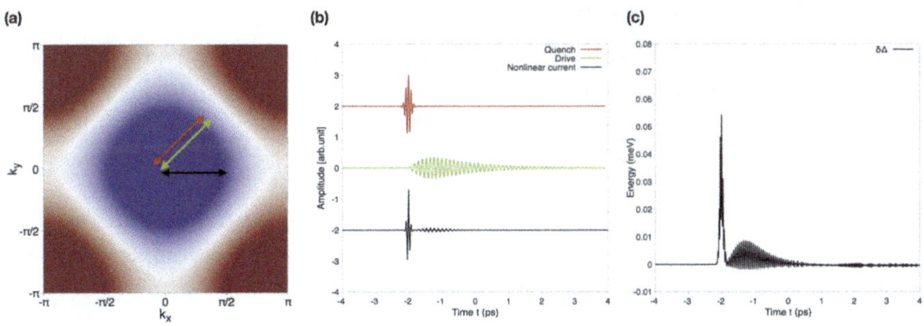

Fig. 2 Quench–drive spectroscopy. (a) Band structure in the first Brillouin zone used to reproduce the optimally doped Bi2212; the red and green arrows represent the direction of linear polarization of the quenching and driving pulses, respectively. The black arrow indicates the direction of the measured current, *i.e.* along the *x* axis. (b) The figure shows the external vector potential of the quenching (red) and the driving (green) pulses for an example case with fixed quench–drive delay time, as well as the nonlinear current response (black). (c) Time-dependent superconducting order parameter variation, $\delta\Delta(t) = \Delta(t) - \Delta(0)$, due to the quenching and driving pulses in (b).

Moreover, in order to treat the pseudogap phase, characterized by incoherent pairs, we extend the pseudospin model, artificially adding a phase to the Cooper pairs,[1] and solving the corresponding equations to obtain the time-evolution of the order parameter and the generated nonlinear current (see Fig. 2(b) and (c)).

The paper is organized as follows: in Section 2, we describe the theoretical models that we used. Starting from the pseudospin approach to solve the equations of motion in order to calculate the time-dependent superconducting gap and the generated nonlinear current, we extend it in order to be able to describe incoherent pairs. In addition, we perform the same calculations for the superconducting state using a diagrammatic approach, deriving the nonlinear susceptibility responsible for the measured response. In Section 3, we show the results of the numerical experiments for the quench–drive setup on a cuprate: we analyze the two dimensional plots in the time and frequency domains, detecting the presence of higher harmonics and signatures of quasiparticles' and the Higgs mode. Then, we repeat the calculation in the presence of incoherent pairs: we show that even a moderate incoherence, which only slightly reduces the superconducting gap, can suppress the high-harmonic generation in the nonlinear current, as well as the quasiparticles' and amplitude mode's response. Finally, we provide a summary and an outlook on future applications and perspectives of quench–drive spectroscopy in Section 4.

2. Theoretical background

In this section, we formulate the theoretical approach used to investigate the current generated by a clean (high-temperature) superconductor subject to an external field. We first develop the pseudospin model for a general superconductor, then we extend it in order to be able to describe incoherent pairs. Finally, we show a diagrammatic approach, where we derive the nonlinear susceptibility used to obtain the nonlinear current, disentangling the quasiparticles' and Higgs' contributions.

2.1 Pseudospin model approach

In order to describe the superconducting phase of a material, we adopt the BCS model expressed by the mean field Hamiltonian:

$$H_{\mathrm{BCS}} = \sum_{\mathbf{k},\sigma} \varepsilon_{\mathbf{k}} \hat{c}^{\dagger}_{\mathbf{k},\sigma} \hat{c}_{\mathbf{k},\sigma} - \sum_{\mathbf{k}} \Delta_{\mathbf{k}} \hat{c}^{\dagger}_{\mathbf{k},\uparrow} \hat{c}^{\dagger}_{-\mathbf{k},\downarrow} + \text{h.c.}, \tag{1}$$

where $\varepsilon_{\mathbf{k}}$ is the electronic band dispersion and $\Delta_{\mathbf{k}}$ the momentum-dependent superconducting order parameter. The latter is described by a complex number which satisfies the gap equation:

$$\Delta_{\mathbf{k}} = \sum_{\mathbf{k}'} V_{\mathbf{k},\mathbf{k}'} \left\langle \hat{c}_{-\mathbf{k}',\downarrow} \hat{c}_{\mathbf{k}',\uparrow} \right\rangle, \tag{2}$$

with $V_{\mathbf{k},\mathbf{k}'}$ being the (momentum-dependent) pairing interaction. It can be factorized as $V_{\mathbf{k},\mathbf{k}'} = V f_{\mathbf{k}} f_{\mathbf{k}'}$, with $f_{\mathbf{k}}$ being the form factor of the superconducting order parameter: for s-wave pairing, $f_{\mathbf{k}} = 1$, while for d-wave pairing, $f_{\mathbf{k}}^{d_{x^2-y^2}} = (\cos k_x - \cos k_y)/2$. Therefore, it follows from eqn (2) that the gap function itself can be factorized as $\Delta_{\mathbf{k}} = \Delta_0 f_{\mathbf{k}}$.

We write the BCS Hamiltonian using the pseudospin formalism, as in ref. 30 and 32, namely

$$\hat{H}_{\mathrm{BCS}} = \sum_{\mathbf{k}} \mathbf{b}_{\mathbf{k}} \hat{\sigma}_{\mathbf{k}}, \tag{3}$$

with the pseudospin vector

$$\hat{\sigma}_{\mathbf{k}} = \frac{1}{2} \hat{\Psi}_{\mathbf{k}}^{\dagger} \tau \hat{\Psi}_{\mathbf{k}}, \tag{4}$$

which is defined in Nambu–Gor'kov space, with spinor $\hat{\Psi}_{\mathbf{k}}^{\dagger} = (\hat{c}_{\mathbf{k},\uparrow}^{\dagger} \hat{c}_{-\mathbf{k},\downarrow})$ and the Pauli matrices $\tau = (\tau_1, \tau_2, \tau_3)$. The pseudo-magnetic field is defined by the vector

$$\mathbf{b}_{\mathbf{k}} = (-\Delta' f_{\mathbf{k}}, -\Delta'' f_{\mathbf{k}}, \varepsilon_{\mathbf{k}}), \tag{5}$$

with $\varepsilon_{\mathbf{k}} = \xi_{\mathbf{k}} - \mu$, $\xi_{\mathbf{k}}$ being the fermionic band dispersion and μ the chemical potential.

In the presence of an external gauge field represented by the vector potential $\mathbf{A}(t)$ coupling to the electrons, the pseudospin changes in time according to

$$\boldsymbol{\sigma}_{\mathbf{k}}(t) = \boldsymbol{\sigma}_{\mathbf{k}}(0) + \delta\boldsymbol{\sigma}_{\mathbf{k}}(t), \tag{6}$$

with $\delta\boldsymbol{\sigma}_{\mathbf{k}}(t) = (x_{\mathbf{k}}(t), y_{\mathbf{k}}(t), z_{\mathbf{k}}(t))$. The external electromagnetic field is included in the pseudo-magnetic field by means of the minimal substitution $\mathbf{k} \to \mathbf{k} - e\mathbf{A}(t)$ in the fermionic energy, resulting in

$$\mathbf{b}_{\mathbf{k}}(t) = (-\Delta'(t) f_{\mathbf{k}}, -\Delta''(t) f_{\mathbf{k}}, \varepsilon_{\mathbf{k}-e\mathbf{A}(t)} + \varepsilon_{\mathbf{k}+e\mathbf{A}(t)}). \tag{7}$$

The Heisenberg equation of motion for the pseudospin can be written in the Bloch form, $\partial_t \boldsymbol{\sigma}_{\mathbf{k}} = 2\mathbf{b}_{\mathbf{k}} \times \boldsymbol{\sigma}_{\mathbf{k}}$, providing the set of differential equations

$$\begin{cases} \partial_t x(t) = -(\varepsilon_{\mathbf{k}-e\mathbf{A}} + \varepsilon_{\mathbf{k}+e\mathbf{A}}) y(t) - \dfrac{f_{\mathbf{k}}}{E_{\mathbf{k}}} \varepsilon_{\mathbf{k}} \delta\Delta''(t) \\ \qquad + 2\delta\Delta''(t) f_{\mathbf{k}} z(t), \\ \partial_t y(t) = 2\varepsilon_{\mathbf{k}} x(t) + 2(\Delta + \delta\Delta'(t)) f_{\mathbf{k}} z(t) \\ \qquad - \delta\Delta' f_{\mathbf{k}} \dfrac{\varepsilon_{\mathbf{k}}}{E_{\mathbf{k}}} + \dfrac{\Delta f_{\mathbf{k}}}{2E_{\mathbf{k}}} (\varepsilon_{\mathbf{k}-e\mathbf{A}} + \varepsilon_{\mathbf{k}+e\mathbf{A}} - 2\varepsilon_{\mathbf{k}}), \\ \partial_t z(t) = -2\Delta f_{\mathbf{k}} y(t) - \dfrac{\Delta {f_{\mathbf{k}}}^2}{E_{\mathbf{k}}} \delta\Delta''(t) - 2\delta\Delta''(t) f_{\mathbf{k}} x(t). \end{cases} \tag{8}$$

Here, for simplicity of calculations and without loss of generality, we assumed a real order parameter, therefore $\Delta''(t = -\infty) = 0$ at equilibrium, so that $y(-\infty) = 0$.

Moreover, in order to describe a quench–drive experiment, we choose the appropriate total vector potential $\mathbf{A}(t) = \mathbf{A}_{\mathrm{q}}(t) + \mathbf{A}_{\mathrm{d}}(t) = \bar{\mathbf{A}}_{\mathrm{q}}(t - t_{\mathrm{q}}) + \bar{\mathbf{A}}_{\mathrm{d}}(t - t_{\mathrm{d}})$, where $\mathbf{A}_{\mathrm{q}}(t)$ is the quenching pulse centered at time $t = t_{\mathrm{q}}$ and $\mathbf{A}_{\mathrm{d}}(t)$ is the driving field centered at $t = t_{\mathrm{d}}$. The expression we used for the modulus of the quenching pulse is a Gaussian-modulated wave,

$$\overline{\mathbf{A}}_{\mathrm{q}}(t - t_{\mathrm{q}}) = \mathbf{A}_{\mathrm{q}}\, \mathrm{e}^{-(t-t_{\mathrm{q}})^2/{\tau_{\mathrm{q}}}^2} \cos(\omega_{\mathrm{q}}(t - t_{\mathrm{q}})), \tag{9}$$

while for the driving we adopted an asymmetric pulse (see Fig. 2(b)) in order to induce an effective quenching of the superconductor at the initial time:

$$\overline{\mathbf{A}}_{\mathrm{d}}(t)=\begin{cases}\mathbf{A}_{\mathrm{d}}\dfrac{\sin(\omega_{\mathrm{d}}(t-t_{\mathrm{d}}))}{1+(t-t_{\mathrm{d}})^{2}}(t-t_{\mathrm{d}})\mathrm{e}^{-(t-t_{\mathrm{d}})/\tau_{\mathrm{d}}^{2}}, & \text{for } t\geq t_{\mathrm{d}},\\ 0, \quad \text{for } t<t_{\mathrm{d}}.\end{cases} \tag{10}$$

By introducing the quench–drive time-delay $\Delta t = t_{\mathrm{d}} - t_{\mathrm{q}}$ and choosing $t_{\mathrm{d}} = 0$, we can rewrite $\mathbf{A}(t) = \bar{\mathbf{A}}_{\mathrm{q}}(t + \Delta t) + \bar{\mathbf{A}}_{\mathrm{d}}(t)$. Therefore, the expressions in eqn (6)–(8) depend on both t and Δt.

The solution of eqn (8) provides the full time-dependent pseudospin, from which the time-dependent order parameter $\Delta(t)$ can be calculated, namely

$$\Delta_{\mathbf{k}}(t)=Vf_{\mathbf{k}}\sum_{\mathbf{k}'}f_{\mathbf{k}'}\left(\sigma_{\mathbf{k},x}(t)-i\sigma_{\mathbf{k},y}(t)\right), \tag{11}$$

where $\sigma_{\mathbf{k},x}(t)$ ($\sigma_{\mathbf{k},y}(t)$) is the $x(y)$ time-dependent component of the full pseudospin. The current generated by the superconductor in this quench–drive setup is given by the expression

$$\mathbf{j}(t,\Delta t)=e\sum_{\mathbf{k}}\mathbf{v}_{k-e\mathbf{A}(t,\Delta t)}\left\langle\hat{c}^{\dagger}_{\mathbf{k},\uparrow}\hat{c}_{\mathbf{k},\uparrow}+\hat{c}^{\dagger}_{\mathbf{k},\downarrow}\hat{c}_{\mathbf{k},\downarrow}\right\rangle(t,\Delta t), \tag{12}$$

where the electron band velocity is calculated *via* $\mathbf{v}_{\mathbf{k}-e\mathbf{A}(t,\Delta t)} = \nabla_{\mathbf{k}}\varepsilon_{\mathbf{k}-e\mathbf{A}(t,\Delta t)}$. In particular, if we consider measuring the current generated along the x direction, it can calculated by the expression

$$j_x(t,\Delta t)=e\sum_{\mathbf{k}}\frac{\partial\varepsilon_{\mathbf{k}-e\mathbf{A}(t,\Delta t)}}{\partial k_x}\langle\hat{n}_{\mathbf{k}}(t,\Delta t)\rangle. \tag{13}$$

However, since the fundamental harmonic is dominant in this regime, while the superconducting features are visible in the nonlinear response, we are interested in the lowest order nonlinear current contribution. The first non-vanishing nonlinear term generated by the driving pulse is the third order component, which reads

$$\mathbf{j}^{(3)}(t,\Delta t)=-2e^{2}\sum_{\mathbf{k}}\sum_{i=x,y,z}\mathbf{A}(t,\Delta t)\cdot\mathbf{r}_i(\partial_{k_i}\mathbf{v}_{\mathbf{k}})z_{\mathbf{k}}(t,\Delta t). \tag{14}$$

Here, $\mathbf{r}_i$ is the unit vector in the direction of axis $i = x, y, z$, and $\mathbf{A}(t, \Delta t)$ is the total vector potential. The non-equilibrium term of the third component of the pseudospin, $z_{\mathbf{k}}(t, \Delta t)$, contains a quadratic dependence on the full vector potential, A^2, and is characterized by oscillations with frequencies $2\omega_{\mathrm{q}}$, $2\omega_{\mathrm{d}}$ and $\omega_{\mathrm{q}} \pm \omega_{\mathrm{d}}$, as well as 2Δ, due to the quasiparticles' and amplitude mode's excitation. The Fourier transform of the nonlinear current with respect to both the real time t and the quench–drive delay time Δt, $j^{(3)}(\omega, \omega_{\Delta t})$, provides the spectrum of the generated harmonics.

By identifying the factor depending on the derivative of the velocity and the direction of the external field with $C_x(\mathbf{k})$, we can write the x component of the third harmonic generated current as

$$j_x^{(3)}(\omega = 3\omega_\mathrm{d}, \omega_{\Delta t} = 0) = -2e^2 A_\mathrm{d} \sum_\mathbf{k} C_x(\mathbf{k}) z_\mathbf{k}(\omega = 2\omega_\mathrm{d}, \omega_{\Delta t} = 0) \\ - 2e^2 A_\mathrm{d} \sum_\mathbf{k} C_x(\mathbf{k}) z_\mathbf{k}(\omega = 4\omega_\mathrm{d}, \omega_{\Delta t} = 0), \tag{15}$$

where the two terms provide the contribution with a sum- and difference-frequency mechanism with the external driving field, respectively. We used ω as the Fourier transform of the real-time variable t, and $\omega_{\Delta t}$ for the Fourier transform of the quench–drive delay time Δt. Analogously, the third order contribution to the fundamental harmonic is given by

$$j_x^{(3)}(\omega = \omega_\mathrm{d}, \omega_{\Delta t} = 0) = -2e^2 A_\mathrm{d} \sum_\mathbf{k} C_x(\mathbf{k}) z_\mathbf{k}(\omega = -2\omega_\mathrm{d}, \omega_{\Delta t} = 0). \tag{16}$$

Moreover, since the third component of the pseudospin has a (non-resonant) peak in the frequency domain at $\omega = 2\Delta$, it is possible to obtain other local maxima for the generated nonlinear current at $\omega = 2\Delta \pm \omega_\mathrm{d}$:

$$j_x^{(3)}(2\Delta \pm \omega_\mathrm{d}, \omega_{\Delta t} = 0) = -2e^2 A_\mathrm{d} \sum_\mathbf{k} C_x(\mathbf{k}) z_\mathbf{k}(\omega = 2\Delta, \omega_{\Delta t} = 0). \tag{17}$$

In addition to these contributions, other terms which involve the quenching pulse are present, such as – among others – the nonlinear current term,

$$j_x^{(3)}(\omega_\mathrm{d}, \omega_{\Delta t} = 2\omega_\mathrm{q}) = -2e^2 A_\mathrm{d} \sum_\mathbf{k} C_x(\mathbf{k}) z_\mathbf{k}(\omega = 0, \omega_{\Delta t} = 2\omega_\mathrm{q}). \tag{18}$$

This expression involves a sum-frequency process of two photons of the quenching pulse, each with frequency ω_q, embedded in the third component of the pseudospin, $z_\mathbf{k}$, thus, the dependence $z_\mathbf{k}(\omega_{\Delta t} = 2\omega_\mathrm{q})$, and it also implicitly depends quadratically on the amplitude of the quenching pulse, ${A_\mathrm{q}}^2$.

2.2 Extended pseudospin model for incoherent pairs

We now want to describe a state with a superconducting instability and characterized by the presence of pre-formed incoherent pairs, which can be identified as the origin of the pseudogap phase; in this regard, as in ref. 1, we use a new artificial equilibrium superconducting state obtained by adding a random momentum-dependent phase $\phi_\mathbf{k}$ to the original Cooper pairs' state. Therefore, the strength of the pairing potential remains unchanged, as well as the number of total Cooper pairs, while the superconducting order parameter decreases due to the reduced coherence. According to the maximum angle ϕ_max, which defines the range of the random phase $\phi_\mathbf{k}$, with $\phi_\mathbf{k} \in [-\phi_\mathrm{max}, +\phi_\mathrm{max}]$, we are able to describe different conditions of the material, from the pure superconducting phase for $\phi_\mathrm{max} = 0$ to the complete loss of coherence for $\phi_\mathrm{max} = \pi$.

We define the gap of the pure superconducting state $\Delta_\mathbf{k}^{(0)} = \Delta_0^{(0)} f_\mathbf{k}$, obtained from the pure BCS gap equation, and the superconducting order parameter in the presence of incoherent pairs of the pseudogap phase as $\hat{\chi}_\mathbf{k}^{(\phi)} = \hat{\chi}^{(\phi)} f_\mathbf{k}$, such that

$$\tilde{\Delta}^{(\phi)} = V f_\mathbf{k} \sum_{\mathbf{k}'} {f_{\mathbf{k}'}}^2 \frac{\Delta_0^{(0)}}{2E_{\mathbf{k}'}^{(0)}} e^{i\phi_{\mathbf{k}'}}, \tag{19}$$

where V is the same pairing strength of the original state, and with the quasiparticles' energy $E_{\mathbf{k}}^{(0)} = \sqrt{\varepsilon_{\mathbf{k}}{}^2 + (\varDelta_0^{(0)} f_{\mathbf{k}})^2}$. The superconducting gap in the new equilibrium state can be written using the pseudospin formalism

$$\tilde{\varDelta}_{\mathbf{k}}{}^{(\phi)} = V f_{\mathbf{k}} \sum_{\mathbf{k}'} f_{\mathbf{k}'} \left(\tilde{\sigma}_{\mathbf{k}',x} - i \tilde{\sigma}_{\mathbf{k}',y} \right), \tag{20}$$

where we have introduced the equilibrium pseudospin components

$$\begin{cases} \tilde{\sigma}_{\mathbf{k},x} = \sigma_{\mathbf{k},x} \cos \phi_{\mathbf{k}} = f_{\mathbf{k}} \dfrac{\varDelta_0^{(0)} \cos \phi_{\mathbf{k}}}{2E_{\mathbf{k}}^{(0)}}, \\ \tilde{\sigma}_{\mathbf{k},y} = \sigma_{\mathbf{k},y} \sin \phi_{\mathbf{k}} = -f_{\mathbf{k}} \dfrac{\varDelta_0^{(0)} \sin \phi_{\mathbf{k}}}{2E_{\mathbf{k}}^{(0)}}. \end{cases} \tag{21}$$

Analogously to the derivation in Section 2.1 for the original superconducting phase, we can obtain the Heisenberg's equation of motion

$$\partial_t \tilde{\sigma}_{\mathbf{k}} = \tilde{\mathbf{b}} \times \tilde{\sigma}_{\mathbf{k}}, \tag{22}$$

with the new pseudomagnetic field defined as

$$\tilde{\mathbf{b}} = (-\hat{\chi}' f_{\mathbf{k}}, -\hat{\chi}'' f_{\mathbf{k}}, \varepsilon_{\mathbf{k}}). \tag{23}$$

The solution of eqn (22) provides the time-dependent value of the pseudospin $\tilde{\sigma}_{\mathbf{k}}$, from which the evolution of the new order parameter $\hat{\chi}^{(\phi)}$ can be obtained. However, we notice that the complex order parameter can be written as

$$\hat{\chi}^{(\phi)} = |\hat{\chi}^{(\phi)}| \mathrm{e}^{i\theta}, \tag{24}$$

where θ is the global phase of the superconducting gap, which differs from the local phase of the Cooper pairs in momentum space, $\phi_{\mathbf{k}}$. As a consequence, the gap equation is not self-consistent anymore (see eqn (19)) and the value of the gap is subject to some time-dependent noise due to the phase incoherence of the preformed pairs.

2.3 Quasi-equilibrium nonlinear susceptibility

In this section, we tackle the problem of a quenched–driven clean superconductor by means of a diagrammatic approach to calculate the nonlinear susceptibility; this is provided only as a tool to interpret the numerical results obtained *via* the pseudospin model.

Starting from the BCS Hamiltonian in eqn (1), we add the interaction of the external field that we treat perturbatively, which can be expressed as a sum of the Feynman diagrams around a quasi-equilibrium condition. In particular, the current generated by the perturbed superconductor in the quasi-equilibrium condition satisfies a proportionality relation with the generalized density–density susceptibility, namely $j(\omega) \sim \chi_{\gamma\gamma}(\omega') A^3(\omega - \omega')$, with $\chi_{\gamma\gamma}(\omega) = \langle \tilde{\rho}\tilde{\rho} \rangle$ and $\tilde{\rho} = \sum_{\mathbf{k}} \gamma_{\mathbf{k}} \langle \hat{c}_{\mathbf{k}}^{\dagger} \hat{c}_{\mathbf{k}} \rangle$.

Since we are considering a clean superconductor, the lowest non-vanishing order of the nonlinear response is provided by the third-order nonlinear susceptibility $\chi_{\gamma\gamma}^{(3)}(\omega)$,[31] where $\gamma_{\mathbf{k}}$ is the diamagnetic light–matter (vertex) interaction strength, which can be written in the effective mass approximation as $\gamma_{\mathbf{k}} = \sum_{i,j=k_x,k_y} \partial_{ij}{}^2 \varepsilon_{\mathbf{k}}$.[33,34]

The pure quasiparticles' diamagnetic response is given by the bare density–density susceptibility, which in Nambu notation within the Matsubara formalism reads

$$\chi_{\gamma\gamma}{}^{(3)}(i\nu_m) = T\sum_{\mathbf{k},i\omega_n} \gamma_{\mathbf{k}}{}^2 \mathrm{Tr}\left[\hat{G}(\mathbf{k}, i\omega_n)\tau_3 \hat{G}(\mathbf{k}, i\omega_n + \nu_m)\tau_3\right], \tag{25}$$

in the limit of the light momentum $\mathbf{q} \to 0$. Here, the Matsubara complex frequency is $i\omega_n$, and T is the temperature. The Nambu–Green's function in matricial form is given by

$$\hat{G}(\mathbf{k}, i\omega_n) = \frac{1}{(i\omega_n)^2 - E_{\mathbf{k}}{}^2} \begin{pmatrix} i\omega_n + \varepsilon_{\mathbf{k}} & -\Delta_{\mathbf{k}} \\ -\Delta_{\mathbf{k}}^* & i\omega_n - \varepsilon_{\mathbf{k}} \end{pmatrix}. \tag{26}$$

This function can be expressed in its spectral form as

$$\hat{G}(\mathbf{k}, i\omega_n) = -\frac{1}{\pi}\int_{-\infty}^{+\infty} \mathrm{d}\omega \frac{\hat{G}''(\mathbf{k}, \omega + i\delta)}{i\omega_n - \omega}, \tag{27}$$

where the imaginary part of the Green's function in real frequency is

$$\hat{G}''(k, \omega) = -\frac{\pi}{2E_{\mathbf{k}}} \begin{pmatrix} \omega + \varepsilon_{\mathbf{k}} & \Delta_{\mathbf{k}} \\ \Delta_{\mathbf{k}}^* & \omega - \varepsilon_{\mathbf{k}} \end{pmatrix} \cdot [\delta(\omega - E_{\mathbf{k}}) - \delta(\omega + E_{\mathbf{k}})]. \tag{28}$$

Therefore, the expression in eqn (25) can be solved, obtaining

$$\chi_{\gamma\gamma}{}^{(3)}(\omega) = -\sum_{\mathbf{k}} \gamma_{\mathbf{k}}{}^2 \frac{\Delta_{\mathbf{k}}{}^2}{E_{\mathbf{k}}{}^2} \frac{\tanh(\beta E_{\mathbf{k}}/2)}{(2E_{\mathbf{k}} + \omega + i\delta)(2E_{\mathbf{k}} - \omega - i\delta)}. \tag{29}$$

This is the pure quasiparticles' contribution, obtained by neglecting the oscillations of the order parameter for small perturbations. The Higgs propagator is obtained by the random phase approximation (RPA) summation of the pairing interaction in the amplitude channel:

$$D_{\mathrm{Higgs}}(\mathbf{q}, \omega) = -V/2 - V/2\chi_{ff}(\mathbf{q}, \omega)D_{\mathrm{Higgs}}(\mathbf{q}, \omega), \tag{30}$$

which provides the expression

$$D_{\mathrm{Higgs}}(\mathbf{q}, \omega) = \frac{-1}{2/V + \chi_{ff}(\mathbf{q}, \omega)}, \tag{31}$$

where $\chi_{ff}(\mathbf{q}, \omega)$ is given by

$$\chi_{ff}(\mathbf{q}, \omega) = T\sum_{\mathbf{k},i\omega_n} f_{\mathbf{k}}{}^2 \mathrm{Tr}\left[\tau_1 \hat{G}(\mathbf{k}, i\omega_n)\tau_1 \hat{G}(\mathbf{k}, i\omega_n + i\nu_m)\right]\Big|_{i\nu_m \to \omega + i\delta}. \tag{32}$$

We can include in the pure quasiparticles' Raman response in eqn (25) the Higgs contribution given by the propagator in eqn (31), obtaining the full Raman response

$$\chi_{\gamma\Gamma}(i\nu_m) = T\sum_{\mathbf{k},i\omega_n}\gamma_{\mathbf{k}}\mathrm{Tr}\left[\hat{G}(\mathbf{k},i\omega_n)\tau_3\hat{G}(\mathbf{k},i\omega_n+i\nu_m)\hat{\Gamma}(\mathbf{k},i\nu_m)\right], \tag{33}$$

where $\hat{\Gamma}(\mathbf{k}, i\nu_m)$ is the vertex matrix, which contains the corrections due to the Higgs mode. In the RPA, we can identify it as

$$\hat{\Gamma}(\mathbf{k},i\nu_m) = \gamma_{\mathbf{k}}\tau_3 - \frac{V}{2}\tau_1 f_{\mathbf{k}} T\sum_{\mathbf{k}',i\omega_n} f_{\mathbf{k}'}\mathrm{Tr}\left[\tau_1\hat{G}(\mathbf{k}',i\omega_n+i\nu_m)\hat{\Gamma}(\mathbf{k}',i\omega_n)\hat{G}(\mathbf{k}',i\omega_n)\right]. \tag{34}$$

Additional corrections can be included in this vertex with different forms.

The linearized equations of motion obtained by removing higher-order terms in eqn (8) provide the same result of the nonlinear susceptibility calculated from the diagrammatic contributions of the density–density response including the RPA summation of the amplitude mode, responsible for the value of $\delta\Delta(\omega)$. The real part of the oscillation of the order parameter, namely the amplitude (Higgs) mode, reads in frequency space

$$\delta\Delta'(\omega) = -2\frac{\sum_{\mathbf{k}'}\frac{\gamma_{\mathbf{k}'}{f_{\mathbf{k}'}}^2\varepsilon_{\mathbf{k}'}\Delta_{\max}}{E_{\mathbf{k}'}\left(4{E_{\mathbf{k}'}}^2-\omega^2\right)}}{1/V-\sum_{\mathbf{k}'}\frac{2{\varepsilon_{\mathbf{k}'}}^2{f_{\mathbf{k}'}}^2}{E_{\mathbf{k}'}\left(4{E_{\mathbf{k}'}}^2-\omega^2\right)}}. \tag{35}$$

The vertex in eqn (34) after analytic continuation of the complex Matsubara frequency can be written, including explicitly the amplitude mode in eqn (35), as $\hat{\Gamma}(\mathbf{k}, \omega) = \gamma_{\mathbf{k}}\tau_3 + \delta\Delta'(\omega)f_{\mathbf{k}}\tau_1$.

The expression of the vertex factor $\gamma_{\mathbf{k}}$ depends on the symmetry of the response which is measured; however, if we consider the experimental configurations with linearly polarized light along high-symmetry crystallographic directions, we can replace the factor $\gamma_{\mathbf{k}}$ with the tensor $\gamma_{ij}(\mathbf{k}) = {\partial_{ij}}^2\varepsilon_{\mathbf{k}}$, and define the corresponding susceptibility from eqn (25) as follows:

$$\chi_{ijlm}{}^{(3)}(i\nu_m) = T\sum_{\mathbf{k},i\omega_n}\gamma_{ij}(\mathbf{k})\gamma_{lm}(\mathbf{k})\mathrm{Tr}\left[\hat{G}(\mathbf{k},i\omega_n)\tau_3\hat{G}(\mathbf{k},i\omega_n+\nu_m)\tau_3\right]. \tag{36}$$

For example, for quenching and driving pulses with nonzero components along both the x and y directions, the third-order nonlinear response current generated along the x axis will be given by

$$j_x{}^{(3)}(\omega') \propto \sum_{i,j=x,y}\chi_{ijxx}{}^{(3)}(\omega)A_i(\omega_1)A_j(\omega_2)\delta(\omega'-\omega-\omega_1-\omega_2), \tag{37}$$

where ω' is obtained by energy conservation and the vector potential $A_{i,j}$ is the i, $j = x, y$-component of the total vector potential, given by the sum of the quenching and driving pulses.

We notice that, in contrast to the solution of the Heisenberg's equations of motion which are valid on a general basis, the diagrammatic approach is valid for small perturbations of the superconducting order parameter Δ. Therefore, when the intensity of the external pulses is such that the gap is significantly enhanced or suppressed, so that a new (transient) equilibrium value of the gap Δ' is reached, the quasi-equilibrium susceptibility calculation fails to catch all the features of the corresponding nonlinear response, requiring a full non-equilibrium calculation.

Moreover, this approach cannot be easily applied when pre-formed pairs or superconducting islands are present in the material, such as in the pseudogap phase of a cuprate, giving rise to long-range incoherence of the Cooper pairs.

3. Numerical results

We now present the results obtained from the numerical implementation of the time-dependent Bloch equations and expressions for the current derived from the pseudospin model in Sections 2.1 and 2.2. For the calculations, we used the electronic band dispersion $\varepsilon_{\mathbf{k}} = -2t(\cos k_x + \cos k_y + \mu/2)$, where the wave vector's components are expressed in units of the lattice constant a. We used the values of $t = 125$ meV for the nearest-neighbour hopping energy and chemical potential $\mu = -0.2$, in units of t, in order to obtain an electron occupation $n = 0.9$, as in ref. 1; the corresponding band structure is shown in Fig. 2(a). For the d-wave order parameter with symmetry $\Delta_{\mathbf{k}} = \Delta_{\max}(\cos k_{\mathbf{k}} - \cos k_y)/2$, we adopted the value $\Delta_{\max} = 31$ meV; the calculations were performed with a summation over the full Brillouin zone with a homogeneous square sampling and a total number of k points $N_{\mathbf{k}} = 10^6$. For the time-dependent evolution, we used a time-step of $\delta t = 3 \times 10^{-4}$ ps, and for the quench–drive delay $\delta\Delta t = 2.5 \times 10^{-2}$ ps.

For the pulses, we used a Gaussian-shaped few-cycle qeuenching and an asymmetric long-duration driving pulse, as expressed in eqn (9) and (10) and shown in Fig. 2(b), linearly polarized along the (1, 1) direction (Fig. 2(a)), with parameters ${\tau_q}^2 = 0.01$ ps^2 and ${\tau_d}^2 = 5$ ps^2, respectively. The maximum intensity used for each pulse is provided for the corresponding vector potential in units of $\hbar/(ea)$, where e is the electron charge and a the lattice constant; the conversion to the value of the electric field for each frequency is provided in Table 1. Table 2 reports the conversion of each frequency (in THz) of the pulses used for the calculations to the energy scale (in meV).

3.1 Quench–drive response of the superconducting state

We first focus on the response of a cuprate in its superconducting phase, at optimal doping; in particular, we first analyze the features of the current response as a function of both the real-time evolution and the quench–drive delay time. Then, we investigate the effect of both the quenching and driving pulses on the superconducting order parameter and its amplitude oscillations.

3.1.1 Nonlinear current generation. We consider a quench–drive setup with both pulses linearly polarized along the diagonal direction (1, 1), and we analyze the nonlinear current generated along the x axis, with the geometry shown in Fig. 2(a). Here, we used a quenching pulse with frequency $\omega_q = 7.80$ THz ($\ll \Delta_{\max}$) and amplitude of the vector potential $A_q = 0.8$, corresponding to an electric field $E_{\max} = 76.1$ kV cm^{-1}.

Table 1 This table shows the conversion of some values of the vector potential maximum intensity, A_{max}, at a given central frequency, as used in the calculations, into the maximum value of the corresponding electric field, E_{max}. Here, the calculation of E_{max} has been done considering the value of the lattice parameter $a = 5.4$ Å

Frequency (THz)	A_{max}	E_{max} (kV cm^{-1})
4.57	0.4	22.3
5.09	1.6	99.3
7.16	0.2	17.5
7.16	0.4	35.0
7.80	0.8	76.1
8.28	0.4	40.4
11.14	0.8	109
12.73	0.8	124

Table 2 This table shows the conversion of the frequency values of the pulses used for the numerical calculations from THz to the energy scale in meV

Frequency (THz)	Energy (meV)
4.57	18.90
5.09	21.05
7.16	29.61
7.80	32.26
8.28	34.24
11.14	46.07
12.73	52.65

In Fig. 3(a), we show the nonlinear current $j^{NL}(t, \Delta t)$ generated by the superconductor as a function of the real evolution time t and for different quench–drive delay times, labeled as Δt. The diagonal line represents the current induced by the quenching pulse, while the vertical signal is due to the asymmetric driving field which sets in at time $t = -2$ ps. The triangular area for $0 \geq t \geq \Delta t$, for $\Delta t \in [-2, 0]$ ps is characterized by a response due to the overlap (wave mixing) of both the quenching and driving pulses.

A Fourier analysis of this plot can provide information about the harmonics present in the current itself. In Fig. 3(b), we present the same results as a function of the frequency ω, obtained by Fourier transforming the time evolution, t, and the delay time, Δt. We can clearly distinguish the fundamental and the third generated harmonic, which are modulated by the delay time as the quenching is swiped with respect to the driving field. This is in accordance with previous theoretical findings on conventional s-wave superconductors.[19] However, we also observe that faded modulated responses appear for $\Delta t > -2$ ps, both at frequencies slightly higher than ω_d and $3\omega_d$, respectively; they also show a modulation in the delay time, as with the other two aforementioned harmonics. Since their appearance and intensity modulation match the time at which the driving field sets in, these features can be attributed to the wave mixing pattern due to the overlap in time of the quenching and the driving fields.

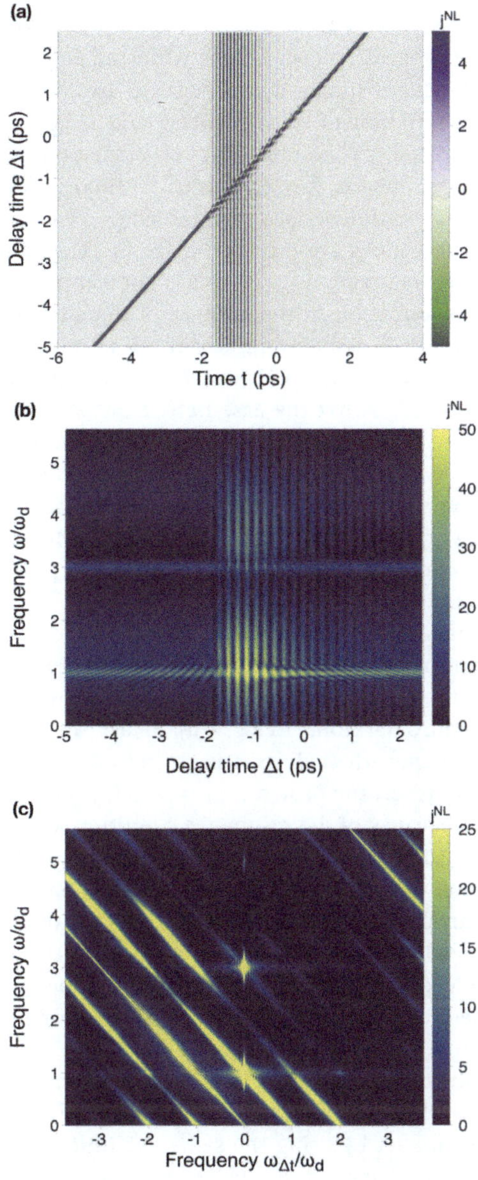

Fig. 3 Two-dimensional plots of the generated nonlinear current. (a) Nonlinear current response, $j,^{NL}$ as a function of real time, t, and quench–drive delay time, Δt. The quenching frequency is $\omega_q = 7.80$ THz, the driving frequency $\omega_d = 5.09$ THz, while the maximum gap value $\Delta_{max} = 31$ meV. The maximum intensities of the pulses used were $A_q = 0.8$ and $A_d = 1.6$, respectively. The diagonal line represents the current due to the quenching pulse, while the vertical lines correspond to the oscillations induced by the asymmetric driving field, which sets in at $t = -2$ ps. (b) The figure corresponds to the same in (a), where the horizontal axis of the real time t has been Fourier transformed to ω, whose values are referred to as the drive frequency ω_d. (c) 2D Fourier transform of the signal in (a), representing the nonlinear current as a function of ω and $\omega_{\Delta t}$.

In Fig. 3(c), the full two-dimensional Fourier transform of the nonlinear current is shown, with the frequency scales referring to the driving frequency ω_d. The bright spots along the vertical axis at $\omega_{\Delta t} = 0$ correspond to the static harmonics generated by the driving field only, while the diagonal stripes are given by wave mixing of the quenching and the driving pulses.

3.1.2 Transient excitation of quasiparticles and Higgs. In order to disentangle the main features, it is useful to separately extract one-dimensional tracks along high-interest directions in this 2D spectrum. In Fig. 4(a), we highlight the spectral intensity of the nonlinear current obtained by a vertical cut at $\omega_{\Delta t} = 0$ in Fig. 3(c). The frequency values are provided there in THz. However, despite the value of the absorption frequency $\omega_{\Delta t} = 0$, this is not completely equivalent to the case of a driven superconductor in the absence of the quenching pulse. In fact, a peak at $\omega \approx 10$ THz $= 2\omega_d$ shows up; a second harmonic generation (SHG), however, cannot be generated in a driven material without breaking spatial inversion symmetry, which is not the case here. However, the presence of such SHG can be interpreted as a transient signal due to the overlap of the quenching and the driving pulses; as can be seen in Fig. 3(c), in fact, the weak peak at $\omega = 2\omega_d$, $\omega_{\Delta t} = 0$ is the result of the sideband contribution of the diagonal feature given by wave mixing of the external field's components. Moreover, in addition to the fundamental and third harmonics, a fifth harmonic is present, in accordance with what has been previously reported in literature.[35]

In Fig. 4(b), we present the spectrum of the generated third harmonic with respect to the absorption frequency, $\omega_{\Delta t}$; this corresponds to tracking the values at $\omega = 3\omega_d$, parallel to the horizontal axis, or, equivalently, to analyze the modulation spectrum of the third harmonic in Fig. 3(b). In fact, the third harmonic signal is modulated by the quench–drive delay time, with intensity peaks at frequencies of multiples of ω_d. However, the peak at $\omega_{\Delta t} \approx 15$ THz is particularly sharp with respect to the others because of the resonance condition with the quasiparticles' and Higgs energies $2\Delta_{max}$ being $\Delta_{max} = 31$ meV ≈ 7.5 THz. Therefore, the dynamical modulation of the third harmonic is here enhanced by the transient excitation of quasiparticles and the Higgs mode, the latter of which is responsible for the amplitude oscillation of the order parameter, which will be discussed in the next section. Indeed, the enhancement of the broadband quenching fundamental signal by these two contributions can also be detected in Fig. 4(c), where the track of $\omega = -\omega_{\Delta t} + \omega_d$ from Fig. 5(c) is shown.

3.1.3 Superconducting order parameter and amplitude mode. Additional information can be obtained by investigating the behaviour of the superconducting order parameter by using the same two-times plots, similarly to the nonlinear current analysis. Indeed, as shown in Sections 2.1 and 2.3, the order parameter oscillates due to the pseudospins' precession in momentum space, and this collective amplitude (Higgs) mode contributes to the enhancement of the nonlinear response, shown in the previous section.

In Fig. 5 we compare the gap variation, $\delta\Delta$, for different parameters of the external fields, plotting them as a function of the time evolution, t, and the delay time of the quenching and driving pulses, Δt.

In Fig. 5(a), we use a quenching pulse with frequency $\omega_q = 8.28$ THz and intensity $A_q = 0.4$ and a driving pulse with $\omega_d = 12.73$ THz and $A_d = 0.8$. The oscillations of the order parameter have an amplitude $\delta\Delta_{max} < 1$ meV; they are overall positive for the duration of the quenching pulse and the driving pulse, but

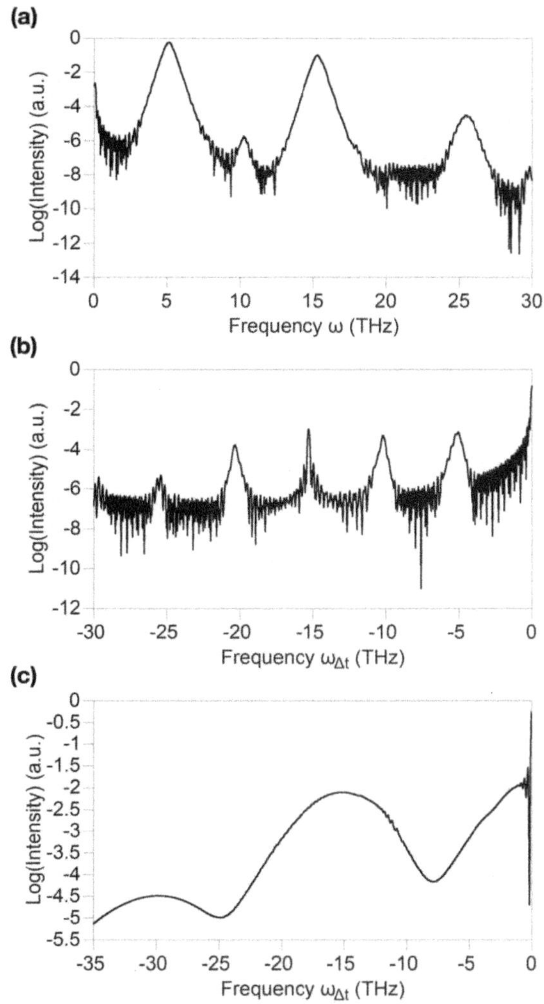

Fig. 4 One dimensional plots of the nonlinear current. Plots of the nonlinear current j^{NL} extracted from Fig. 3(c) along some relevant directions. (a) Current spectrum for $\omega_{\Delta t=0}$ as a function of ω, corresponding to a vertical cut in the two-dimensional Fourier spectrum. (b) Spectrum of the modulation of the third harmonic (TH), corresponding to a cut at $\omega = 3\omega_d$ parallel to the x axis. (c) Current intensity for $\omega = -\omega_{\Delta t} + \omega_d$, corresponding to the diagonal feature in Fig. 6(a), starting from the fundamental harmonic at $\omega = \omega_q$, $\omega_{\Delta t} = 0$.

they become negative at longer times until a new equilibrium value of the gap (smaller than the initial one) is reached. In Fig. 5(b), the frequencies of the pulses are set as $\omega_q = 5.09$ and $\omega_d = 12.73$, while their intensities are $A_q = 0.8$ and $A_d = 1.6$, respectively. Due to the higher fluence of both pulses, the oscillations of the gap are higher (up to 4–5 meV) and the effect of the pump becomes stronger; even during the time overlap of the quenching and driving pulses, in fact, the gap is suppressed and characterized by an intensity modulation in the delay time, Δt.

As shown in Fig. 5(c), we reduced the quenching frequency to $\omega_q = 7.80$ THz, obtaining the same parameters of the calculations of the previous section on the

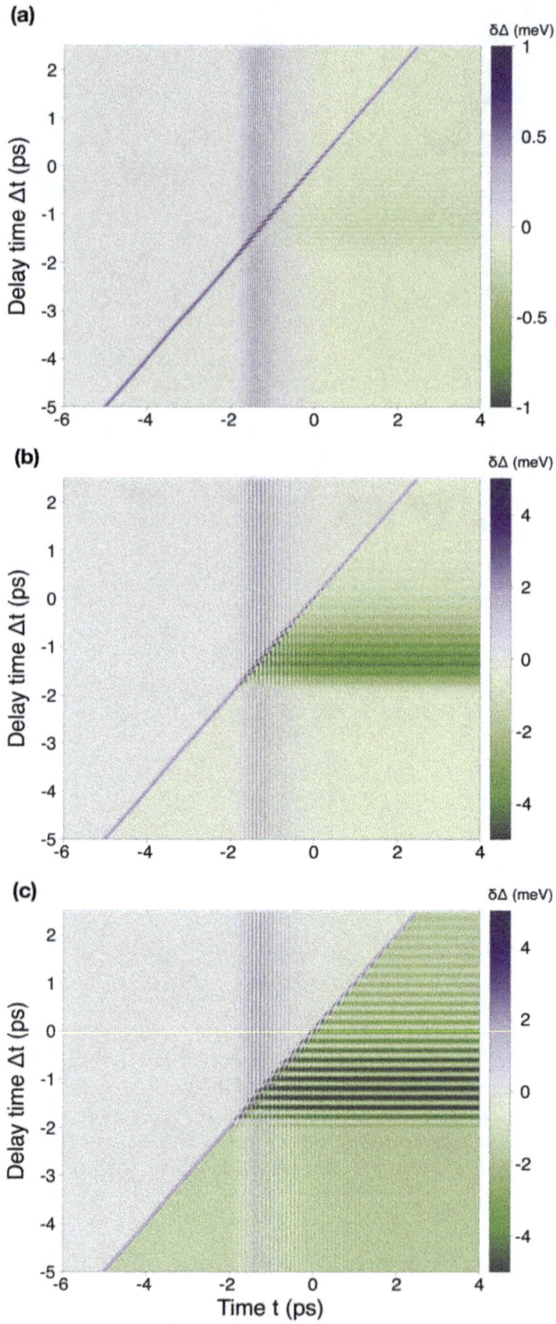

Fig. 5 Two-dimensional plots of the amplitude mode. The plots represent the amplitude mode, *i.e.* the variation of the order parameter $\delta\Delta(t, \Delta t) = \Delta(t, \Delta t) - \Delta_{max}$, as a function of real time, *t*, and quench–drive delay time, Δt, for different quenching and driving pulses. (a) Quenching frequency $\omega_q = 8.28$ THz, intensity $A_q = 0.4$, driving with $\omega_d = 12.73$ THz and $A_d = 0.8$. (b) Pulse frequencies $\omega_q = 5.09$ and $\omega_d = 12.73$, intensities $A_q = 0.8$ and $A_d = 1.6$. (c) The same parameters as in (b) were used, except for the quenching frequency $\omega_q = 7.80$ THz.

nonlinear current. The suppression of the gap by the quenching pulse is enhanced here due to the approaching of the resonance condition $\omega_q \approx \Delta_{max}$, and the intensity modulation in Δt for $t > 2$ ps, namely when the quenching and the driving pulses overlap, is increased.

Therefore, we can conclude that higher fluences are associated, for higher frequencies, with a suppression of the gap and an enhancement of its oscillations at the same time, which leads to a more intense amplitude mode and a sizeable response in the nonlinear current. On the contrary, for lower frequencies, the gap is transiently enhanced during both the quenching and the driving, in accordance with previous results.[1]

3.2 Quench–drive response of incoherent pairs

Similarly to the study of the pure superconducting phase of cuprates of the previous section, we now focus on the response of the same cuprate structure in the presence of reduced phase coherent giving rise to pre-formed incoherent Cooper pairs, as described in Section 2.2. In particular, for our calculations, we adopted a random phase within the range $[-\pi/2, \pi/2]$, which decreases the initial gap value of $\Delta_{max} = 31$ meV to $\hat{\chi}^{(\phi)} = 27.92$ meV. We also used the same duration and shape parameters for both the quenching and the (asymmetric) driving pulses.

3.2.1 Nonlinear current generation. We study here the nonlinear current generated by the superconductor with incoherent pairs under the application of a quenching pulse and a driving pulse, with the same setup described before. In particular, we focus directly on the two-dimensional Fourier spectrum of the current, since all the other features in the time domain are qualitatively similar to those of the pure superconducting case.

In Fig. 6(a), we show the result for $A_q = 0.4$, $\omega_q = 7.16$ THz, $A_d = 0.8$ and $\omega_d = 11.14$ THz. While a third harmonic generated by the driving field is still present, its intensity is strongly suppressed (notice the log scale used). This is confirmed also by the one dimensional track in Fig. 7(a), which shows a 6 orders of magnitude difference between the first and third generated harmonics.

In Fig. 7(b), the spectrum of the TH modulation is shown. While the $2\omega_d$ and $4\omega_d$ are still present, the peak at 2Δ is suppressed, in contrast to the result of the pure superconducting state (Fig. 4(b)). The same conclusion can be reached from Fig. 7(c), which shows the diagonal track of Fig. 6(a), corresponding to $\omega = -\omega_{\Delta t} + \omega_d$, where the broadband signal of the quench pulse is visible, which is not enhanced here by the quasiparticles' and Higgs resonance, in contrast to the pure superconducting case.

If we further decrease the intensity of the pulses, we obtain a complete suppression of the third harmonic with respect to the fundamental, which is shown in Fig. 6(b), where we used $A_q = 0.2$, $\omega_q = 7.16$ THz, $A_d = 0.4$ and $\omega_d = 4.57$ THz. In this case, even approaching the resonance condition $\omega_d \approx \hat{\chi}^{(\phi)}$, the third harmonic is not present, as well as the dynamical modulation of the harmonics.

3.2.2 Superconducting order parameter and amplitude mode. We now analyze the time-dependent response of the superconducting order to the quench–drive setup in the same conditions and with corresponding parameters of the nonlinear current to those discussed above. Indeed, the interpretation of

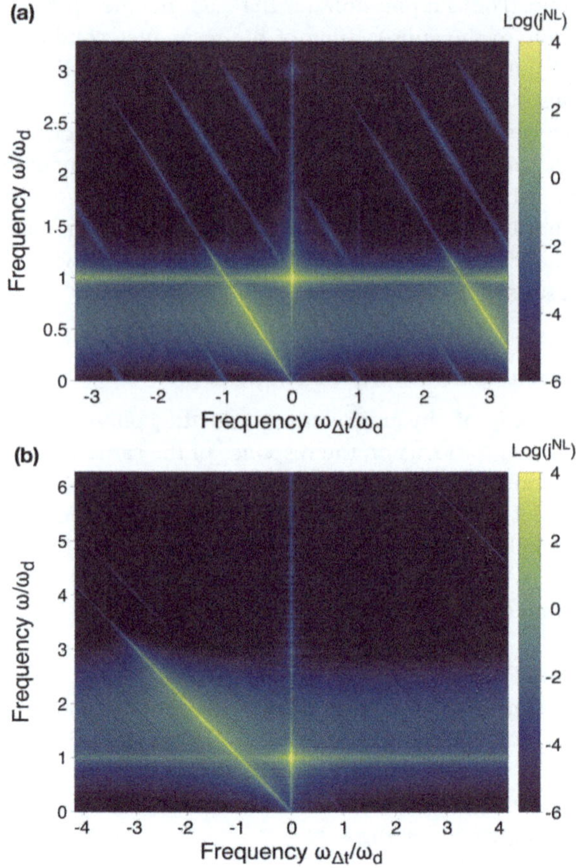

Fig. 6 Two-dimensional plots in Fourier space of the generated nonlinear current. (a) Nonlinear current (log scale) generated by the superconductor with quenching intensity $A_q = 0.4$ and frequency $\omega_q = 7.16$ THz, driving intensity $A_d = 0.8$ and frequency $\omega_d = 11.14$ THz. (b) Nonlinear current (log scale) generated by the superconductor with quenching intensity $A_q = 0.2$ and frequency $\omega_q = 7.16$ THz, driving intensity $A_d = 0.4$ and frequency $\omega_d = 4.57$ THz.

the previous results of the nonlinear current can be supported by the study of the amplitude mode of the superconducting order parameter.

In Fig. 8, we present the plots of the time-dependent change of the superconducting order parameter $\delta\Delta(t, \Delta t)$, corresponding to the same setup conditions of the results in Fig. 6. In particular, we observe that stronger quenching and driving pulses enhance the gap for the duration of the perturbation itself, except when the quenching overlaps in time with the driving field, *i.e.* $t \in [-2, 0]$ ps (see Fig. 8(a)). On the contrary, weak pulses do not significantly enhance the superconducting gap, nor activate amplitude oscillations (Fig. 8(b)). As deduced from the nonlinear current response described in the previous section, the amplitude mode is suppressed by the presence of incoherence in the Cooper pairs; physically, this can be understood by the fact that instead of one coherent amplitude mode with a definite phase, we have a dispersion of modes with different phases

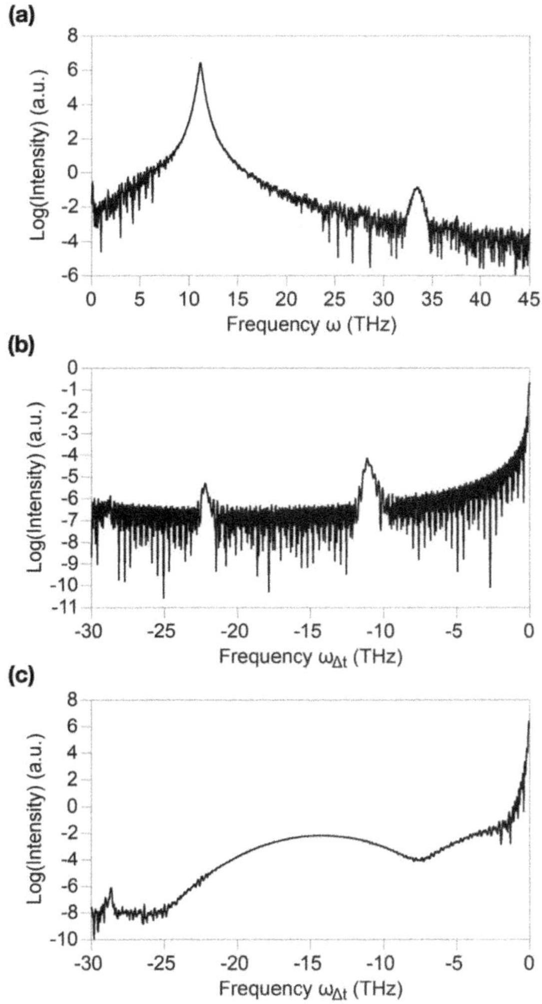

Fig. 7 One dimensional plots of the nonlinear current. These plots represent the nonlinear current j^{NL} in log scale extracted from Fig. 6(a) along some relevant directions. (a) Current spectrum for $\omega_{\Delta t=0}$ as a function of ω, corresponding to a vertical cut in the two-dimensional Fourier spectrum. (b) Spectrum of the modulation of the third harmonic, corresponding to a cut at $\omega = 3\omega_d$ parallel to the x axis. (c) Current intensity for $\omega = -\omega_{\Delta t} + \omega_d$, corresponding to the diagonal feature in Fig. 6(a) starting from the fundamental harmonic at $\omega = \omega_q$, $\omega_{\Delta t} = 0$.

and signs, which tend to cancel each other out. Therefore, unless strong pulses increase the superconducting gap and, hence, the phase coherence, the Higgs and quasiparticles' modes are suppressed.

4. Conclusions

In this work, we have studied the quench–drive spectroscopy response of cuprate superconductors, which are interesting for their anisotropic d-wave superconducting order parameter and their rich phase diagram. In this work, we have

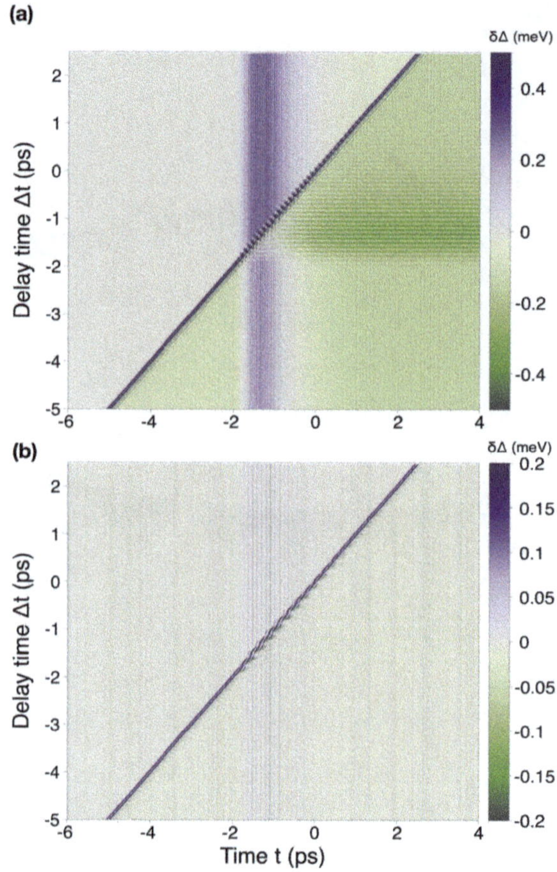

Fig. 8 Comparison of the time-dependent order parameter. Plot of the time-dependent gap change $\delta\Delta(t, \Delta t)$, referred to as the equilibrium value. The results shown in panels (a and b) are obtained with the same parameters of the corresponding plots in Fig. 6, respectively.

explored two different situations: the pure superconducting state and the phase with incoherent pre-formed Cooper pairs, which can reproduce the results of the pseudogap phase, or even originate from other conditions. With this method, we have moved a step forward with respect to previous experiments and calculations, where only a pump (short quenching or longer driving) of cuprates was considered. In fact, in the common pump–probe configuration, only the real time evolution of the system driven out of equilibrium by the pump was tracked by the probe and analyzed. On the contrary, we have shown that in the quench–drive setup, both the short-time and the long-time pulses can act as out-of-equilibrium drivers on the material, affecting both the order parameter and the nonlinear current response. Moreover, this setup has an additional temporal degree of freedom with respect to the pump–probe setup, namely the quench–drive delay time is continuously swiped in order to gain insight into the dependence of the response on the absorption frequency $\omega_{\Delta t}$. This has allowed us to investigate the high-harmonic generation, the modulation of such harmonics in the quench–

drive delay time and the role of the amplitude mode in the features of the nonlinear current. In particular, we have detected a fifth harmonic (visible at $\omega = 5\omega_d \approx 25$ THz) generated by the superconductor in some conditions, as well as a peak at the energy 2Δ, which can be attributed to quasiparticles and the amplitude (Higgs) mode, as shown by the analysis of the fluence and frequency dependence of the change of the order parameter.

Furthermore, we have applied quench–drive spectroscopy to a cuprate with incoherent pre-formed pairs, artificially adding a random noise phase to the Cooper pairs in momentum space. The results have shown that, even for moderate incoherence, the decrease of the superconducting gap is accompanied by the partial or total suppression of the third harmonic response, as well as the amplitude mode contribution to the nonlinear current. These results can be experimentally addressed and tested by the measurement of the transmitted electric field or the nonlinear optical conductivity.

A more detailed and systematic analysis of the response of incoherent pairs and the amplitude mode of their superconducting gap will be the subject of a future work. We further speculate that an extended configuration, such as a pump–pump–probe scheme, could be used to add an independent time degree of freedom on the real time evolution and the pump–probe delay time, as already used to study molecular excitations and semiconductors.[28,36]

Conflicts of interest

There are no conflicts of interest to declare.

Acknowledgements

Fruitful discussions with P. M. Bonetti, R. Haenel, S. Kaiser, M.-J. Kim, M. Monaco and D. Vilardi are thankfully acknowledged. Open Access funding provided by the Max Planck Society.

Notes and references

1 F. Giusti, A. Marciniak, F. Randi, G. Sparapassi, F. Boschini, H. Eisaki, M. Greven, A. Damascelli, A. Avella and D. Fausti, *Phys. Rev. Lett.*, 2019, **122**, 067002.

2 H. Chu, M.-J. Kim, K. Katsumi, S. Kovalev, R. D. Dawson, L. Schwarz, N. Yoshikawa, G. Kim, D. Putzky, Z. Z. Li, *et al.*, *Nat. Commun.*, 2020, **11**, 1–6.

3 H. Chu, S. Kovalev, Z. X. Wang, L. Schwarz, T. Dong, L. Feng, R. Haenel, M.-J. Kim, P. Shabestari, H. L. Phuong, K. Honasoge, R. D. Dawson, D. Putzky, G. Kim, M. Puviani, M. Chen, N. Awari, A. N. Ponomaryov, I. Ilyakov, M. Bluschke, F. Boschini, M. Zonno, S. Zhdanovich, M. Na, G. Christiani, G. Logvenov, D. J. Jones, A. Damascelli, M. Minola, B. Keimer, D. Manske, N. Wang, J.-C. Deinert and S. Kaiser, *Fano Interference of the Higgs Mode in Cuprate High-T_c Superconductors*, 2021.

4 B. Keimer, S. A. Kivelson, M. R. Norman, S. Uchida and J. Zaanen, *Nature*, 2015, **518**, 179–186.

5 W. Zhang, C. L. Smallwood, C. Jozwiak, T. L. Miller, Y. Yoshida, H. Eisaki, D.-H. Lee and A. Lanzara, *Phys. Rev. B: Condens. Matter Mater. Phys.*, 2013, **88**, 245132.

6 E. Baldini, A. Mann, B. P. P. Mallett, C. Arrell, F. van Mourik, T. Wolf, D. Mihailovic, J. L. Tallon, C. Bernhard, J. Lorenzana and F. Carbone, *Phys. Rev. B*, 2017, **95**, 024501.

7 F. Giusti, A. Montanaro, A. Marciniak, F. Randi, F. Boschini, F. Glerean, G. Jarc, H. Eisaki, M. Greven, A. Damascelli, A. Avella and D. Fausti, *Phys. Rev. B*, 2021, **104**, 125121.

8 A. K. Abdullah Kaplan, *Corpus J. Case Rep.*, 2021, **2**, 1–2.

9 L. Li, Y. Wang, S. Komiya, S. Ono, Y. Ando, G. D. Gu and N. P. Ong, *Phys. Rev. B: Condens. Matter Mater. Phys.*, 2010, **81**, 054510.

10 F. Yang and M. W. Wu, *Phys. Rev. B*, 2021, **104**, 214510.

11 D. K. Singh, S. Kadge, Y. Bang and P. Majumdar, *Fermi Arcs and Pseudogap Phase in a Minimal Microscopic Model of D-Wave Superconductivity*, 2021.

12 R. Shimano and N. Tsuji, *Annu. Rev. Condens. Matter Phys.*, 2020, **11**, 103–124.

13 M. Buzzi, G. Jotzu, A. Cavalleri, J. I. Cirac, E. A. Demler, B. I. Halperin, M. D. Lukin, T. Shi, Y. Wang and D. Podolsky, *Phys. Rev. X*, 2021, **11**, 011055.

14 F. Gabriele, M. Udina and L. Benfatto, *Nat. Commun.*, 2021, **12**, 752.

15 M. Woerner, W. Kuehn, P. Bowlan, K. Reimann and T. Elsaesser, *New J. Phys.*, 2013, **15**, 025039.

16 J. Lu, X. Li, H. Y. Hwang, B. K. Ofori-Okai, T. Kurihara, T. Suemoto and K. A. Nelson, *Phys. Rev. Lett.*, 2017, **118**, 207204.

17 Y. Wan and N. P. Armitage, *Phys. Rev. Lett.*, 2019, **122**, 257401.

18 F. Mahmood, D. Chaudhuri, S. Gopalakrishnan, R. Nandkishore and N. P. Armitage, *Nat. Phys.*, 2021, **17**, 627–631.

19 M. Puviani, R. Haenel and D. Manske, *Transient Excitation of Higgs and High-Harmonic Generation in Superconductors with Quench-Drive Spectroscopy*, 2021.

20 I. Katayama, H. Aoki, J. Takeda, H. Shimosato, M. Ashida, R. Kinjo, I. Kawayama, M. Tonouchi, M. Nagai and K. Tanaka, *Phys. Rev. Lett.*, 2012, **108**, 097401.

21 J. Demsar, *J. Low Temp. Phys.*, 2020, **201**, 676–709.

22 K. Katsumi, N. Tsuji, Y. I. Hamada, R. Matsunaga, J. Schneeloch, R. D. Zhong, G. D. Gu, H. Aoki, Y. Gallais and R. Shimano, *Phys. Rev. Lett.*, 2018, **120**, 117001.

23 W. Li, C. Zhang, X. Wang, J. Chakhalian and M. Xiao, *J. Magn. Magn. Mater.*, 2015, **376**, 29–39.

24 F. Gai, J. C. McDonald and P. A. Anfinrud, *J. Am. Chem. Soc.*, 1997, **119**, 6201–6202.

25 L. J. G. W. van Wilderen, I. P. Clark, M. Towrie and J. J. van Thor, *J. Phys. Chem. B*, 2009, **113**, 16354–16364.

26 A. E. Fitzpatrick, C. N. Lincoln, L. J. G. W. van Wilderen and J. J. van Thor, *J. Phys. Chem. B*, 2012, **116**, 1077–1088.

27 T. W. Kee, *J. Phys. Chem. Lett.*, 2014, **5**, 3231–3240.

28 C. Giannetti, *Il Nuovo Cimento C*, 2016, **39**, 1–10.

29 R. Matsunaga, N. Tsuji, H. Fujita, A. Sugioka, K. Makise, Y. Uzawa, H. Terai, Z. Wang, H. Aoki and R. Shimano, *Science*, 2014, **345**, 1145–1149.

30 N. Tsuji and H. Aoki, *Phys. Rev. B: Condens. Matter Mater. Phys.*, 2015, **92**, 064508.

31 T. Cea, C. Castellani and L. Benfatto, *Phys. Rev. B*, 2016, **93**, 180507.
32 L. Schwarz and D. Manske, *Phys. Rev. B*, 2020, **101**, 184519.
33 T. P. Devereaux, D. Einzel, B. Stadlober, R. Hackl, D. H. Leach and J. J. Neumeier, *Phys. Rev. Lett.*, 1994, **72**, 396–399.
34 T. P. Devereaux and R. Hackl, *Rev. Mod. Phys.*, 2007, **79**, 175–233.
35 Z. X. Wang, J. R. Xue, H. K. Shi, X. Q. Jia, T. Lin, L. Y. Shi, T. Dong, F. Wang and N. L. Wang, *Transient Higgs Oscillations and High-Order Nonlinear Light-Higgs Coupling in Terahertz-Wave driven NbN Superconductor*, 2021.
36 S. T. Cundiff and S. Mukamel, *Phys. Today*, 2013, **66**, 44–49.

Faraday Discussions

PAPER

Signatures of coherent vibronic exciton dynamics and conformational control in the two-dimensional electronic spectroscopy of conjugated polymers

Dominik Brey,[a] Robert Binder,[a] Rocco Martinazzo[bc] and Irene Burghardt*[a]

Received 20th January 2022, Accepted 22nd February 2022
DOI: 10.1039/d2fd00014h

Two-dimensional electronic spectroscopy (2DES) signals for homo-oligomer J-aggregates are computed, with a focus on the role of structural change induced by low-frequency torsional modes, along with quasi-stationary trapping effects induced by high-frequency polaronic modes. To this end, a model system is derived from an *ab initio* parametrized site-based Hamiltonian for oligothiophenes [Binder *et al.*, *Phys. Rev. Lett.*, 2018, **120**, 227401]. To obtain a compact representation, we introduce a collective lattice mode whose vibronic coupling depends nonlinearly on the exciton density. As a result, an N-site model with a single polaronic mode and a single torsional mode is obtained. Furthermore, a quantum–classical treatment is employed where the torsional mode is treated within a mean-field Ehrenfest/Langevin approximation. 2D electronic spectra are computed using the equation-of-motion phase-matching approach (EOM-PMA) within a wavefunction description. It is seen that the spectra combine the vibronic fine structure, due to the polaronic mode, and a dynamic Stokes shift, due to torsional relaxation. The signatures of the coherent effects and adiabatic evolution in the 2DES signals are discussed.

1 Introduction

Conjugated polymer materials are multi-chromophoric systems[1–4] whose molecular properties determine the efficiency of the energy and charge transfer processes that eventually relate to device efficiencies in organic electronics.[5,6] The electronic properties primarily define the accessible transfer pathways and mechanisms of electronic delocalization, such that excitonic and charge-separated states can be described in terms of Frenkel-type[7,8] or Merrifield-

[a]*Institute of Physical and Theoretical Chemistry, Goethe University Frankfurt, Max-von-Laue-Str. 7, 60438 Frankfurt am Main, Germany. E-mail: burghardt@chemie.uni-frankfurt.de; Fax: +49 69 798 29709*
[b]*Department of Chemistry, Università degli Studi di Milano, Via Golgi 19, 20133 Milano, Italy*
[c]*Istituto di Scienze e Tecnologie Molecolari, CNR, Via Golgi 19, 20133 Milano, Italy*

type[9,10] Hamiltonians in an electron–hole quasi-particle basis. However, electron-vibration (vibronic) coupling plays a prominent role in excitonic processes in conjugated polymer materials and influences the elementary transport events.[11,12] The relevant vibrations include high-frequency modes which are prone to induce polaronic effects, and low-frequency modes that are typically taken to generate disorder and, hence, lead to localization of the excitonic wavefunction. These soft, low-frequency modes, however, also play a dynamical role by generating fluctuations that drive exciton transport, either in a near-adiabatic fashion or else involving non-adiabatic pathways.[12–14] Upon photo-excitation, a pronounced dynamical Stokes shift has been associated with the low-frequency modes and excited-state planarization of typical conjugated polymeric systems.[15,16] The interplay between the electronic transfer events and the interaction with these different types of vibrations also determines to what extent electronically or vibronically coherent processes are observed.

In order to unravel the complex vibronic dynamics of such materials, diverse time-resolved spectroscopies offer complementary insights, ranging from time-resolved absorption and emission spectroscopies to time-resolved Raman spectroscopy,[3] single-molecule spectroscopy,[2] coherent multi-dimensional spectroscopies[17,18] to very recent time-resolved optical microscopies.[19,20] Specifically, two-dimensional electronic spectroscopy (2DES)[17,18,21–24] has proven to be a highly useful tool that has brought insights into the role of localization[25,26] and coherent transients associated with intra-chain exciton migration[22] and polaron generation.[23] However, spectral congestion and the coexistence of electronic and vibronic features may hamper the interpretation of spectroscopic results.[27,28] Also, 2DES provides a frequency–frequency map that only indirectly yields information, *e.g.*, about the spatial extension and migration of excitons. For these reasons, higher-order spectroscopies have been proposed, which include additional exciton–exciton interactions that reveal exciton migration in an unequivocal fashion.[29]

The present contribution focuses on the 2DES approach and, specifically, the inclusion of different types of vibrational modes, as discussed above. To this end, we will use the model for exciton migration at zero temperature that was developed in ref. 30 and is here further reduced to a description in terms of an effective polaronic high-frequency mode and a torsional mode. The latter triggers a microscopic exciton migration event as a function of a conjugation break that gradually gives way to planarization. This model is employed in the framework of the equation-of-motion phase-matching approach (EOM-PMA)[31–33] that is implemented here for an N-site lattice coupled to the abovementioned polaronic high-frequency mode and a torsional mode. The latter is described within a mean-field Ehrenfest approximation, which has proven acceptable for the exciton dynamics in question.[34] As a result, a hybrid quantum–classical framework is obtained, where an N-site system and the polaronic high-frequency mode are treated in an accurate grid-based scheme (here, using a split-operator implementation[35,36]) which is, in turn, coupled to a slow torsional mode *via* Ehrenfest dynamics.[37,38]

Despite the simplicity of the present set-up, some key features can be highlighted that have been shown to occur in a more realistic, full quantum treatment involving tens to hundreds of modes:[13,14,30,39] Notably, (i) coherent electronic transients can occur following photoexcitation, (ii) these are rapidly damped due to polaronic dephasing (internal conversion), and (iii) on a longer time scale, torsional motion determines a displaced local exciton ground state (LEGS).[40,41] The net effect is, therefore, an exciton migration step induced by a single torsional fluctuation event. The generalization to temperature-dependent fluctuations and the resulting diffusive exciton motion has been discussed in detail in ref. 13 and 14.

In the following, we first describe the molecular Hamiltonian that is employed (Section 2), followed by a description of our implementation of the EOM-PMA scheme (Section 3). Following this, the quantum–classical propagation (Section 4) and the results of the spectroscopic calculations (Section 5) are presented and discussed in detail. Finally, Section 6 concludes.

2 Quantum–classical exciton dynamics

In this study, we map a full microscopic reference Hamiltonian, developed in our earlier work,[30,42] to a reduced-dimensional, self-consistent quantum–classical representation that is convenient for use in the context of the 2DES computations reported in the following sections. Briefly, the reference Hamiltonian describes Frenkel excitons in an *N*-site oligomer, *e.g.*, of oligothiophene (OT) or oligo-(*p*-phenylene-vinylene) (OPV) types, including different types of high-frequency modes which are anharmonically coupled to the excitonic subsystem, along with one or more torsional modes. The parametrization was obtained from electronic structure calculations that were subsequently mapped to a generalized Frenkel–Holstein Hamiltonian.[43] In our previous treatments,[13,14,30] a fully correlated time propagation using the multi-layer multiconfiguration time-dependent Hartree method[44,45] was employed. However, we also showed that the torsional modes can be reasonably well described by a mean-field approximation,[46] and also by classical-limit Ehrenfest dynamics.[34]

In the present context, several approximations are made while keeping the key features of the dynamics. First, a quantum–classical description is employed,[37,38] where the classical subspace corresponds to the torsional modes and the quantum subspace comprises the electronic states and high-frequency modes. Second, the latter are subsumed in an effective "polaronic mode" which replaces the site-local high-frequency modes. The polaronic mode couples to the excitonic subsystem *via* a non-linear vibronic interaction term that depends on the time-dependent density itself. As a result, this mode represents a collective high-frequency vibration that travels with the time-evolving exciton. Experimental observations[11,23] and simulations in the full system dimensionality[12,14,30,47] indicate that such collective high-frequency motions are indeed a signature of quasi-stationary trapping effects. Differently from approaches chosen in previous works in the context of the Davydov model,[48] the polaronic mode is treated quantum mechanically, since this mode likely gives rise to an observable vibronic fine structure.

2.1 Reduced-dimensional Frenkel–Holstein Hamiltonian

The molecular Hamiltonian ($\hat{H}_{\mathrm{M}}$) represents a vibronic lattice model for intra-chain exciton transport, close to those employed in our previous studies.[13,14,30] The Hamiltonian is written in the generic form,

$$\hat{H}_{\mathrm{M}} = \hat{H}_{\mathrm{el}} + \hat{H}_{\mathrm{vib}} + \hat{H}_{\mathrm{el\text{-}vib}} \tag{1}$$

where $\hat{H}_M$ includes an electronic part ($\hat{H}_{el}$), a vibrational part ($\hat{H}_{vib}$), and the electronic–vibrational (vibronic) interaction ($\hat{H}_{el\text{-}vib}$). The electronic part comprises the electronic ground state Hamiltonian ($\hat{H}_G$) and an excitonic lattice Hamiltonian ($\hat{H}_{XT}$) which is restricted to the single-excited manifold (noting that a more complete analysis would include the double-excited manifold),

$$\hat{H}_{el} = \hat{H}_G + \hat{H}_{XT} \tag{2}$$

where

$$\hat{H}_G = c_G|0\rangle\langle 0| \tag{3}$$

and

$$\hat{H}_{XT} = c_E \sum_{n=1}^{N} |n\rangle\langle n| + w_0 \sum_{n=1}^{N-1} (|n\rangle\langle n+1| + \text{h.c.}) \tag{4}$$

where $N = 20$ in the simulations reported below. Here, the ground state and single-excited configurations $|n\rangle$ are defined in a direct product space spanned by N monomers, *i.e.*, $|0\rangle = \prod_{j=1}^{N}|G^{(j)}\rangle$ and $|n\rangle = \prod_{j\neq n}^{N}|G^{(j)}\rangle \otimes |E^{(n)}\rangle$ where $|G^{(j)}\rangle$ and $|E^{(j)}\rangle$ denote the monomer ground and excited states, respectively. In eqn (3) and (4), $c_G = N\varepsilon_G$ where ε_G is the ground-state monomer energy, and $c_E = (N-1)\varepsilon_G + \varepsilon_E$, where ε_E is the monomer excited state energy. Furthermore, w_0 is the excitonic coupling which is restricted to nearest-neighbor interactions. The excitonic coupling takes a negative value (see Table 1), in line with the description of the oligomer as a J-aggregate,[12,49] with head-to-tail aligned monomer transition dipole moments.

Next, the zeroth-order vibrational Hamiltonian includes a single torsional mode $\hat{\theta}$ and an effective polaronic mode $\hat{X}$,

$$\begin{aligned}\hat{H}_{vib} &= \hat{H}_\theta + \hat{H}_X \\ &= \left(\frac{1}{2I_\theta}\hat{p}_\theta^2 + \hat{V}_G(\hat{\theta})\right) + \left(\frac{1}{2}\hat{P}_X^2 + \frac{1}{2}\Omega^2\hat{X}^2\right)\end{aligned} \tag{5}$$

where I_θ is the torsional moment of inertia, and mass-weighted coordinates are used for the $\hat{X}$ mode with harmonic frequency Ω (see Table 1). The anharmonic ground-state torsional potential $\hat{V}_G$ is defined as a cosine series based upon electronic structure calculations, as described in ref. 30 and 43, see eqn (9) and Table 2.

As in ref. 30, we consider a torsional mode localized at the junction between two neighboring sites $n = n_J^-$ and $n + 1 = n_J^+$ of an N-site lattice, as illustrated in Fig. 1a. In practice, the mode will be placed at the center of the lattice, such that two symmetric sub-lattices arise.

Table 1 Parameters of the molecular Hamiltonian $\hat{H}_M$; all values are given in atomic units

c_G	c_E	w_0	
0	9.08×10^{-2}	-1.15×10^{-2}	
Ω	κ	I_θ	γ
7.18×10^{-3}	1.23×10^{-3}	3.13×10^{-5}	2.4189×10^{-4}

Table 2 Torsion (θ) monomer potential and potential-type coupling parameters. All values are given in atomic units. See eqn (9) for the cosine series form of the fit function

$c_{G,0}$	$c_{G,1}$	$c_{G,2}$	$c_{G,3}$
1.01×10^{-3}	1.40×10^{-4}	-1.12×10^{-3}	1.06×10^{-4}
$c_{G,4}$	$c_{G,5}$	$c_{G,6}$	
3.52×10^{-4}	8.83×10^{-6}	3.87×10^{-5}	
$c_{E,0}$	$c_{E,1}$	$c_{E,2}$	$c_{E,3}$
2.13×10^{-2}	-4.50×10^{-3}	-3.15×10^{-2}	4.68×10^{-3}
$c_{E,4}$	$c_{E,5}$	$c_{E,6}$	
3.75×10^{-3}	-5.77×10^{-4}	-4.85×10^{-5}	
$c_{w,0}$	$c_{w,1}$	$c_{w,2}$	$c_{w,3}$
-0.6×10^{-3}	-9.28×10^{-5}	6.79×10^{-4}	1.30×10^{-4}
$c_{w,4}$	$c_{w,5}$	$c_{w,6}$	
2.09×10^{-5}	-2.03×10^{-4}	-1.28×10^{-4}	

The electronic–vibrational coupling for these two types of modes,

$$\hat{H}_{\text{el-vib}} = (\hat{H}_{\theta,\text{el}}^{\text{on-site}} + \hat{H}_{\theta,\text{el}}^{\text{coup}}) + \hat{H}_{X,\text{el}}, \tag{6}$$

includes diagonal and off-diagonal torsional vibronic contributions at the junction,

$$\hat{H}_{\theta,\text{el}}^{\text{on-site}} = (\hat{V}_{\text{E}}(\hat{\theta}) - \hat{V}_{\text{G}}(\hat{\theta}))\,(|n_{\text{J}}^{-}\rangle\langle n_{\text{J}}^{-}| + |n_{\text{J}}^{+}\rangle\langle n_{\text{J}}^{+}|) \tag{7}$$

and

$$\hat{H}_{\theta,\text{el}}^{\text{coup}} = \hat{w}(\hat{\theta})\,(|n_{\text{J}}^{-}\rangle\langle n_{\text{J}}^{+}| + |n_{\text{J}}^{+}\rangle\langle n_{\text{J}}^{-}|), \tag{8}$$

where the excited-state potential, $\hat{V}_{\text{E}}$, and ground-state potential, $\hat{V}_{\text{G}}$, are junction potentials[43] that are further defined as sums and differences of monomer potentials, *i.e.*, $\hat{V}_{\text{G}} = 2\hat{v}_{\text{G}}$ and $\hat{V}_{\text{E}} = \hat{v}_{\text{E}} + \hat{v}_{\text{G}}$. The latter, along with the coupling $\hat{w}$, are given by a cosine series,[30,43] see Table 2,

$$\begin{aligned} \hat{v}_{\text{G/E}}(\hat{\theta}) &= \sum_{j=0}^{6} c_{\text{G/E},j} \cos(j\hat{\theta}), \quad 90^\circ \le \theta \le 270^\circ \\ \hat{w}(\hat{\theta}) &= \sum_{j=0}^{6} c_{w,j} \cos(j\hat{\theta}). \end{aligned} \tag{9}$$

These potentials were generated by a mapping procedure[43] which converts electronic structure calculations for an OT-*n* oligomer – here, obtained using the second-order algebraic diagrammatic construction (ADC(2))[50] scheme, as well as

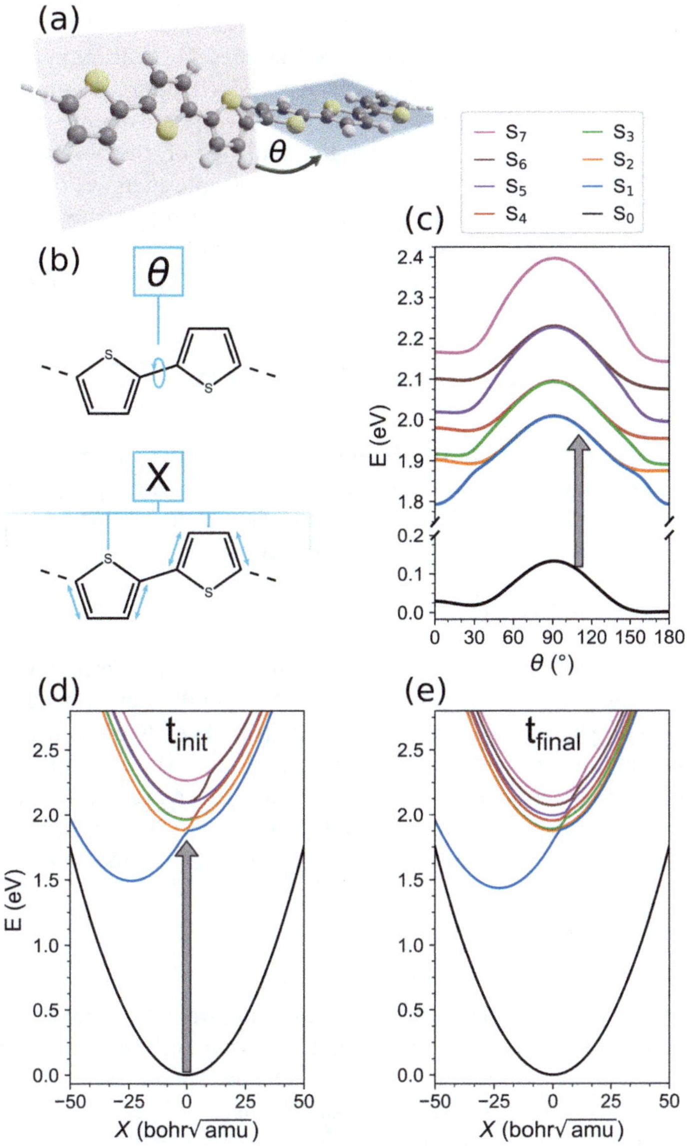

Fig. 1 (a) Schematic representation of a defect geometry of an oligothiophene type oligomer, relating to a torsion-induced conjugation break. (b) Schematic representation of the two relevant modes, where the torsional mode (θ) acts at the center of the lattice, while the polaronic mode (X) features a time-dependent interaction with all lattice sites, according to eqn (10). (c) Adiabatic PESs for the torsional mode (θ). (d) Snapshot of the time-dependent PESs of the polaronic mode (X) immediately after photoexcitation (at time t_{init}, taken at the center of the first pulse) at the defect geometry. (e) Snapshot of the time-dependent PESs of the polaronic mode at the final time (t_{final}, about 700 fs after the first pulse) of the dynamics (see Fig. 3d), where a planar geometry is reached. The vertical arrows in panels (c) and (d) indicate photoexcitation at the defect geometry, taken as $\theta_0 = 110°$ in the dynamical calculations.

time-dependent density functional theory (TDDFT)[51] – to a nearest-neighbor coupled representation of the Hamiltonian. For the purpose of the present analysis, the coupling $\hat{w}(\hat{\theta})$ was rescaled by a factor of 0.25 as compared with the potentials reported in ref. 30, such that the present realization should be considered a model system (in fact, the reduced coupling, around −0.3 eV, remains comparatively strong but permits us to define a dynamics where electronic *vs.* vibronic effects can be clearly distinguished on the observation timescale, as explained below). For this reason, the electronic energy c_E of eqn (4) was also adapted, such as to yield excitation energies that match the experimental values for poly(3-hexylthiophene) (P3HT) type systems.[23,52]

Finally, the vibronic coupling for the polaronic mode takes a nonlinear, time-dependent form,

$$\hat{H}_{X,\mathrm{el}}(t) = \kappa \hat{X} \sum_{n,n'=1}^{N} \left|\rho_{nn'}^{\mathrm{XT}}\right|(t)|n\rangle\langle n'| \tag{10}$$

with a density-dependent vibronic interaction. As mentioned above, this mode should be understood as a collective lattice motion whose coupling to the individual sites depends on the instantaneous excitonic density. In eqn (10), the elements of the electronic density matrix are defined as

$$\rho_{nn'}^{\mathrm{XT}}(t) = (P_{\mathrm{XT}}(t))^{-1}\mathrm{Tr}\{|n'\rangle\langle n|\Psi(t)\rangle\langle\Psi(t)|\} \tag{11}$$

where Tr refers to the trace over the full system (electronic and vibrational coordinates) and $P_{\mathrm{XT}}(t) = \sum_{n=1}^{N} P_n(t)$ denotes the overall population of the excitonic manifold, where $P_n(t) = \mathrm{Tr}\{|n\rangle\langle n|\Psi(t)\rangle\langle\Psi(t)|\}$. The effective interaction of eqn (10) introduces a time-dependent nonlinearity into the equation of motion for the molecular system. Similar formulations have been previously employed in the context of exciton Hamiltonians, notably in the Davydov variant.[48] However, these formulations do not usually refer to a collective polaronic mode.

2.2 Adiabatic potential energy surfaces

While the site-based diabatic representation is a natural setting for the excitonic Hamiltonian, and it captures the spatial evolution of excitonic states, the frequency domain description inherent in 2DES corresponds more closely to the adiabatic representation. Hence, Fig. 1 shows slices of the adiabatic potential energy surfaces (PESs) as a function of θ and X, obtained by diagonalizing the diabatic matrix potential with elements $\hat{V}_{nn'} = \hat{V}_0 + \langle n|\hat{H}_{\mathrm{el}} + \hat{H}_{\mathrm{el\text{-}vib}}|n'\rangle$, where $\hat{V}_0 = \hat{V}_G(\hat{\theta}) + 1/2\Omega^2\hat{X}^2$. Due to the nonlinear vibronic interaction of the collective X mode, which depends on the instantaneous electronic density, according to eqn (10), the adiabatic potential is time-dependent, such that representative snapshots at the initial and final time of the propagation are shown.

Based on these potential energy surfaces, the photoinduced process can be understood in terms of the low-lying adiabatic states, notably the S_1 and S_2 states, with an admixture of the S_3 state. In line with the J-aggregate model of the oligomer,[12,30,49] the S_1 state is the bright, nodeless state in the absence of the torsional defect. In the presence of the conjugation break, a superposition of S_1 and S_2 corresponds to the bright state on the left or right sub-lattice.

The following features can be inferred from Fig. 1: first, the S_1 and S_2 states exhibit a strong sensitivity to the torsional degree of freedom (see panel c); they are near-degenerate around the twisted geometry ($\theta = 90°$), while a gap opens towards the planar geometry. As a result, planarization in the S_1 state is a dominant dynamical pathway. Meanwhile, the collective polaronic mode X induces quasi-stationary trapping effects, due to a marked vibronic displacement of the S_1 state (Fig. 1d and e). More precisely, the X mode exhibits a time-dependent PES due to the density-dependent vibronic interaction of eqn (10). In Fig. 1d and e, two snapshots are shown, first for the defect geometry immediately after photoexcitation (panel d), and second, for the final time of the dynamics (panel e) where the system has planarized (see Fig. 3d). At the initial time (panel d), two sub-manifolds of excitonic states are observed: (i) a set of interacting PESs which exhibit curve crossings; these correspond to the moiety where the initial excitonic density is located, (ii) second, a set of parallel harmonic PESs which do not exhibit any curve-crossings; these correspond to the moiety which does not accommodate any initial exciton density. At the final time (panel e), the excitonic density is delocalized at the planar geometry such that the full set of more densely spaced states takes part in nonadiabatic interactions. For both snapshots, the lowest-energy excitonic (S_1) state undergoes a marked polaronic stabilization, around 0.4 eV. The S_n, $n \geq 2$ states are prone to internal conversion towards S_1, due to the observed curve crossings. This corresponds to the observations of ref. 13, 14 and 30 for a more complete, high-dimensional Hamiltonian representing the dynamics.

2.3 Quantum–classical approximation

While the vibronic lattice Hamiltonian described above could be implemented in a full quantum mechanical framework, similarly to our earlier work,[13,14,30,39] we choose here a hybrid quantum–classical approach where the torsional mode is treated in a mean-field Ehrenfest setting.[37,38] This is motivated by our analysis of ref. 34 showing that the Ehrenfest approximation produces quite good results for the present system, assuming that the dynamics within the excitonic J-aggregate manifold remain largely adiabatic. In this previous work, the Ehrenfest approximation was employed for both torsional and high-frequency modes, whereas it is restricted to the torsional mode θ in the present context.

Hence, we construct multi-state wavefunctions of the following type,

$$|\Psi(X,t)\rangle = \psi_G(X,t)|0\rangle + \sum_{n=1}^{N} \psi_n(X,t)|n\rangle \tag{12}$$

which are computed in a Fourier grid representation (see Section 3.2 for further details) and carry a dependence on the time-evolving torsion $\theta(t)$ *via* the interaction terms of eqn (7) and (8). The torsional mode, in turn, follows a classical-limit Ehrenfest–Langevin evolution,[34]

$$\begin{aligned} \dot{\theta} &= \frac{p_\theta}{I_\theta} \\ \dot{p}_\theta &= -\frac{\partial H_\theta}{\partial \theta} - \frac{\partial \bar{H}_{\theta,\text{el}}(\theta)}{\partial \theta} - \gamma\dot{\theta} + \xi(t) \end{aligned} \tag{13}$$

where $\bar{H}_{\theta,\text{el}}$ is the mean-field interaction potential derived from eqn (7) and (8),

$$\bar{H}_{\theta,\mathrm{el}}(\theta) = \sum_n \sum_{n'} \rho_{nn'}^{\mathrm{XT}} \langle n' | \left(\hat{H}_{\theta,\mathrm{el}}^{\text{on-site}}(\theta) + \hat{H}_{\theta,\mathrm{el}}^{\mathrm{coup}}(\theta) \right) | n \rangle. \tag{14}$$

Here, $\rho_{nn'}^{\mathrm{XT}}$ are elements of the reduced electronic density matrix for the excitonic manifold as introduced in eqn (11). Similar Ehrenfest–Langevin dynamics, including the effects of friction (associated with the friction coefficient γ) and fluctuations (associated with the stochastic force $\xi(t)$), were introduced in ref. 34 and compared with a full quantum mechanical level of treatment at finite temperature, see also ref. 13 and 14. Here, we consider a classical-like zero-temperature case where the torsion experiences damping through the friction coefficient γ (see Table 1), in the absence of fluctuations.

3 2DES computation

In the remainder of this paper, we will examine how the key features of the dynamics resulting from the above exciton Hamiltonian translate to 2DES signals.[17,18,21–24] The 2DES pulse sequence, sketched in Fig. 2, describes a third-order, or four-wave mixing (4WM) experiment, like the simpler two-pulse pump-probe (PP) experiment. In 2DES, the first pulse of the PP experiment is essentially replaced by two pulses with a time delay, τ, denoting the coherence time, which enables frequency resolution of the excitation. Similar to the PP experiment, this is followed by the waiting (or population) time, T, between the second and third pulse. Signal detection occurs in the time t following the third pulse. For a single spectrum, the signal is detected for a fixed waiting time T and varying coherence time τ. Fourier transformation in the τ and t domains leads to a two-dimensional frequency–frequency correlation map. This map yields information on the couplings between different electronic states, through the emergence of cross peaks, as well as their optical accessibility through the relative intensities of the peaks. Spectra recorded for different waiting times, T, reveal characteristic beatings of the signals.

Time propagation is carried out based on the EOM-PMA method, which has been introduced in ref. 31–33 in order to efficiently compute the third-order polarization relevant for 2DES, and specifically three-pulse photon echo experiments. The EOM-PMA method can be viewed as intermediate between perturbative response function approaches and a non-perturbative propagation approach which requires additional post-propagation processing in order to select the relevant phase matching conditions. In the EOM-PMA approach, a propagation of several

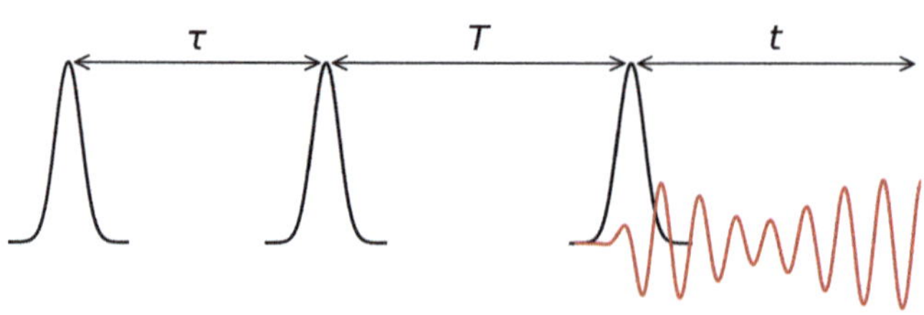

Fig. 2 Schematic representation of the 2DES pulse sequence. The coherence time (τ) is followed by the waiting time (T) and the detection time (t). Pulses with a typical duration of 5 fs full width at half maximum (FWHM) are used in our simulations. The resulting third-order polarization $\mathscr{P}^{(3)}(t)$ is indicated in red.

auxiliary wavefunctions[32,33] (or density matrices[31]) is carried out, which comply with the relevant phase matching conditions. In the present simulations, we focus on the wavefunction implementation, as detailed in ref. 32 and 33.

3.1 EOM-PMA formulation

Considering the full Hamiltonian, which combines the molecular Hamiltonian $\hat{H}_M$ of eqn (1)–(11) with the interaction with the electromagnetic field ($\hat{H}_F$),

$$\hat{H} = \hat{H}_M + \hat{H}_F, \tag{15}$$

the interaction takes the following form, here for $N_P = 3$ pulses, in the rotating wave approximation (RWA),

$$\hat{H}_F(t) = -\sum_{j=1}^{N_P} \hat{h}_j^F(t)\exp(-i\mathbf{k}_j\mathbf{r}) + \text{h.c.} \tag{16}$$

where

$$\hat{h}_j^F(t) = \hat{\mu}\lambda_j E_j(t - \tau_j)\exp(i\omega_j(t - \tau_j)). \tag{17}$$

Here, $\hat{\mu}$ is the transition dipole operator, $\mathbf{k}_j$ denotes the wave vector, λ_j the amplitude, E_j the dimensionless pulse envelope, τ_j the time-domain center of the pulse, and ω_j the carrier frequency.

The following convention will be used for the pulse arrival times,

$$\tau_1 = -T - \tau,\ \tau_2 = -T,\ \tau_3 = 0, \tag{18}$$

where τ is the coherence time and T is the waiting time.

In the following, we will consider rephasing conditions, *i.e.*, the wavevector of the resulting polarization is selected according to the condition $\mathbf{k}_{PE} = -\mathbf{k}_1 + \mathbf{k}_2 + \mathbf{k}_3$, such that a photon echo (PE) experiment is considered.[31]

As mentioned above, only the single-excited manifold will be included in the present context, such that the transition dipole operator $\hat{\mu}$ takes the form

$$\hat{\mu} = \sum_{n=1}^{N} \mu_n |0\rangle\langle n|, \quad \hat{\mu}^\dagger = \sum_{n=1}^{N} \mu_n |n\rangle\langle 0| \tag{19}$$

where μ_n corresponds to the nth monomer transition dipole, which generates the Frenkel configuration, $|n\rangle$, from the ground state, $|0\rangle$. In the present context, we disregard any coordinate dependence of the transition dipole operator. Below, we will either assume that all transition dipole moments μ_n are identical, or else a tailored distribution of μ_n values is assumed, for example to match the initial excitation of a partially delocalized exciton on one or the other side of a conjugation break.

As shown in ref. 32 and 33, the relevant polarization can be obtained from a set of auxiliary wavefunctions which are propagated in parallel and yield the third-order polarization as

$$\begin{aligned}\mathscr{P}^{(3)}(t) = &\left\langle \overline{\Psi}_{12^\dagger 3^\dagger}(t) \middle| \hat{\mu} \middle| \Psi_{12^\dagger 3^\dagger}(t) \right\rangle \\ &- \left\langle \overline{\Psi}_{12^\dagger}(t) \middle| \hat{\mu} \middle| \Psi_{12^\dagger}(t) \right\rangle - \left\langle \overline{\Psi}_{13^\dagger}(t) \middle| \hat{\mu} \middle| \Psi_{13^\dagger}(t) \right\rangle\end{aligned} \tag{20}$$

where the subscript indices of the ket and bra wavefunctions indicate which pulses, according to eqn (16), these auxiliary wavefunctions experience. The bra wavefunctions are marked by an overbar for clarity.

The equations of motion are given as follows[33] for the ket auxiliary wavefunctions,

$$i\frac{\partial}{\partial t}|\psi_{12^\dagger}\rangle = \left(\hat{H}_{\mathrm{M}} - \hat{h}_1^{\mathrm{F}}(t) - \hat{h}_2^{\mathrm{F}\dagger}(t)\right)|\psi_{12^\dagger}\rangle$$

$$i\frac{\partial}{\partial t}|\psi_{13^\dagger}\rangle = \left(\hat{H}_{\mathrm{M}} - \hat{h}_1^{\mathrm{F}}(t) - \hat{h}_3^{\mathrm{F}\dagger}(t)\right)|\psi_{13^\dagger}\rangle$$

$$i\frac{\partial}{\partial t}|\psi_{12^\dagger 3^\dagger}\rangle = \left(\hat{H}_{\mathrm{M}} - \hat{h}_1^{\mathrm{F}}(t) - \hat{h}_2^{\mathrm{F}\dagger}(t) - \hat{h}_3^{\mathrm{F}\dagger}(t)\right)|\psi_{12^\dagger 3^\dagger}\rangle \tag{21}$$

where the $\hat{h}_j^{\mathrm{F}}$ components relating to the different pulses are defined in eqn (16) and (17). Analogously, one obtains for the bra wavefunctions,

$$-i\frac{\partial}{\partial t}\langle\overline{\psi}_{12^\dagger}| = \langle\overline{\psi}_{12^\dagger}|\left(\hat{H}_{\mathrm{M}} - \hat{h}_1^{\mathrm{F}}(t) - \hat{h}_2^{\mathrm{F}\dagger}(t)\right)$$

$$-i\frac{\partial}{\partial t}\langle\overline{\psi}_{13^\dagger}| = \langle\overline{\psi}_{13^\dagger}|\left(\hat{H}_{\mathrm{M}} - \hat{h}_1^{\mathrm{F}}(t) - \hat{h}_3^{\mathrm{F}\dagger}(t)\right)$$

$$-i\frac{\partial}{\partial t}\langle\overline{\psi}_{12^\dagger 3^\dagger}| = \langle\overline{\psi}_{12^\dagger 3^\dagger}|\left(\hat{H}_{\mathrm{M}} - \hat{h}_1^{\mathrm{F}}(t) - \hat{h}_2^{\mathrm{F}\dagger}(t) - \hat{h}_3^{\mathrm{F}\dagger}(t)\right). \tag{22}$$

While we employed the above equations in the presented form, we point out that only five combinations of the auxiliary wavefunctions are in principle needed, since the coupling to higher electronic states (notably the double-excited excitonic manifold) is disregarded.[33]

3.2 Implementation of the split-operator scheme

In view of the time dependence of the Hamiltonian, a split-operator propagation scheme[35,36] is well suited for the time propagation of the quantum mechanical subsystem, *i.e.*, the above auxiliary wavefunctions for $(N + 1)$ electronic states as a function of the polaronic $\hat{X}$ mode. Within the Ehrenfest set-up outlined above, the quantum subsystem is coupled with the classical-limit torsional mode θ within a mean-field description, see Section 2.3.

Our implementation of the split-operator scheme uses a standard Fourier grid representation and has been implemented for a direct-product grid accommodating a number of vibrational modes and electronic states, such that the present case of a single quantum mode $\hat{X}$ is a simple limiting situation. The Strang splitting[53] is employed to approximate the full propagator by a sequence of short-time propagators for the kinetic *vs.* potential parts, *i.e.*, $\hat{\boldsymbol{U}}(\tau) = \hat{\boldsymbol{U}}_V(\tau/2)\hat{\boldsymbol{U}}_T(\tau)\hat{\boldsymbol{U}}_V(\tau/2)$, with $\hat{\boldsymbol{U}}_T(\tau) = \exp(-i\hat{T}\tau/\hbar)\mathbf{1}$ and $\hat{\boldsymbol{U}}_V(\tau) = \exp(-i\hat{\boldsymbol{V}}\tau/\hbar)$, noting that the kinetic energy $\hat{T}$ is electronically diagonal, while $\hat{\boldsymbol{V}}$ represents a matrix potential in the site-based diabatic representation. Exponentiation of the matrix potential requires a diagonalization of $\hat{\boldsymbol{V}}$, which can be carried out in advance if the potential is time-independent. The potential part is further split up into the molecular potential and the interaction with the fields, again using a Strang splitting. For the interaction with the electromagnetic field, the propagator is approximated by a Taylor expansion. In cases where the molecular potential contains nonlinearities – as is the case for the vibronic interaction with the polaronic mode – the N-state matrix potential $\hat{V}$ is diagonalized at time intervals that are chosen to be larger than the propagation step (*e.g.*, for a typical propagation step of 5×10^{-3} fs, the matrix potential is diagonalized at every 20th step, *i.e.*, at intervals of 0.1 fs).

In practice, this basic, numerically exact propagation scheme is restricted to a few vibrational modes, such that different methods are required if a large number of modes are to be propagated, notably the multiconfigurational schemes employed in our previous work[13,14,30] and related tensor train methods,[54] or else modular path integral methods,[55] coherent state methods like the Davydov approach[56] and multiconfiguration Ehrenfest approaches,[57,58] or the hierarchical equations of motion (HEOM) method.[59,60]

4 Time evolution of the photogenerated superposition state

Similarly to our studies in ref. 30 and 39, a photogenerated coherent superposition state is considered, which corresponds to a partially delocalized exciton delimited by an initial conjugation break. This type of photogenerated state has also been denoted spectroscopic unit.[61] For a conjugation break at the center of the lattice, the initial state corresponds to a superposition of S_1 and S_2 states, with a smaller admixture of S_3.[30] The removal of the conjugation break induced by a planarization of the torsion in the excited state leads to a spatial displacement of the exciton – *i.e.*, a microscopic exciton migration event. While the present perspective is restricted to $T = 0$ K, we showed in ref. 13 and 14 that the basic picture carries over to torsional motions under thermal conditions.

In Fig. 3, the time evolution of the initial coherent superposition state is illustrated for the case of purely electronic dynamics (panel (b)), the case of electronic and torsional dynamics (panel (c)), and for the full dynamics involving the θ and X modes (panel (d)). The relevant torsional dynamics are illustrated in panel (a), showing that the torsional evolution starts at $\theta = 110°$ and leads to planarization ($\theta \sim 180°$) within $\sim$500 fs. Clearly, the three realizations of the dynamics differ significantly, and the electronic dynamics (panel (b)) are not representative of the evolution in the presence of the (θ, X) modes.

On the observation time scale, the purely electronic dynamics (panel (b)) do not induce any spatial displacement of the exciton; however, on a longer time scale, a periodic exchange between left *vs.* right localized states would be observed. The high-frequency oscillations likely stem from beatings between the predominant S_1/S_2 superposition states and the S_3 state. Including the torsional motion (panel (b)) changes the picture, since the torsional dynamics leads to a gradual widening of the S_1–S_2 gap (see Fig. 1c) and induces coherent oscillations of the excitonic density between the left/right localized exciton states when a threshold value $\theta \sim 160°$ is reached. Continued electronic wavepacket motion is seen as the torsion planarizes. Finally, including both θ and X modes (panel (d)) again has a striking effect, since the fast electronic oscillations subside after $\sim$300 fs and give way to a smoother electronic density that first delocalizes across both moieties (around 200–300 fs) and then relaxes and strongly contracts at the center

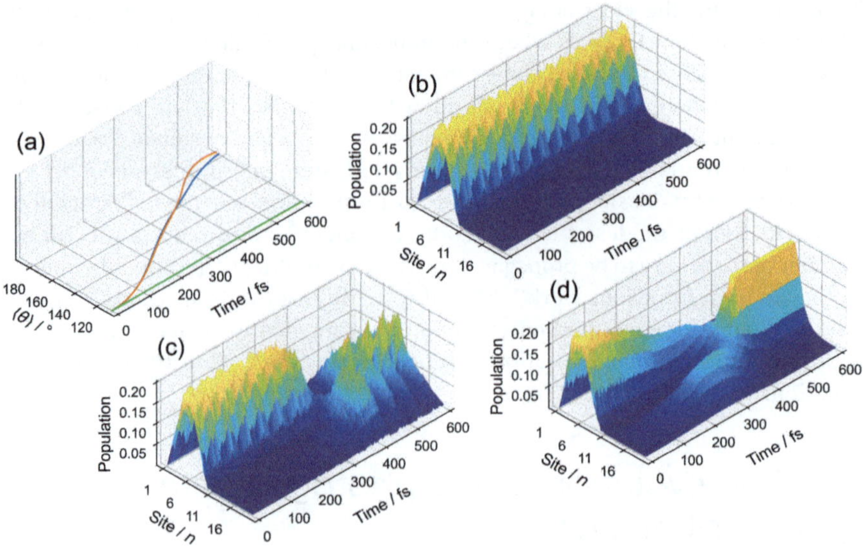

Fig. 3 Time evolution following the initial excitation pulse acting on the oligomer exhibiting a conjugation break, such that the photogenerated state is a partially delocalized exciton extending across sites $n = 1$–10. (a) Torsional evolution, indicating a gradual relaxation of the conjugation break, from $\theta_0 = 110°$ to the final planar conformation ($\theta = 180°$). (b) Time-evolving exciton density for purely electronic dynamics. (c) Exciton density for dynamics including the torsional mode θ. (d) Exciton density for the full dynamics including the torsional (θ) and polaronic (X) modes (note the cut-off of the population above 0.2). In panel (a), the torsional evolution is distinguished for the electronic dynamics of (b) with a persistent conjugation break (green line), the electronic-plus-torsional evolution of (c) (blue line), and the full dynamics of (d) (orange line).

of the lattice. Both features indicate that the S_1/S_2 superposition state decays towards S_1, and a quasi-stationary exciton–polaron, or local exciton ground state (LEGS), is formed.

Clearly, the dynamics are driven by both the torsion and the high-frequency polaronic mode. The torsional mode induces a change of the S_1/S_2 energy gap, accompanying a change in the electronic state character;[30] this process is near-adiabatic. The polaronic mode, in turn, is essential in inducing electronic relaxation – or internal conversion – resulting in a LEGS at the center of the lattice. The LEGS is electronically relaxed but may be accompanied by persistent vibronic oscillations due to the high-frequency polaronic mode, since vibrational relaxation is expected to set in on a longer time scale. Indeed, experimental observations provide evidence for such long-lived vibronic oscillations.[23,24,52] In the present simulations, the polaronic X mode does not undergo damping, such that pronounced oscillatory features are observed which strongly impact the 2D electronic spectra, as shown below. A more detailed, high-dimensional simulation set-up[30,39] shows that relaxation effects are to be expected, even at $T = 0$ K. However, even though the dynamics of panel (d) are generated by a simplified Hamiltonian, they do bear the key features of the high-dimensional quantum dynamical simulations reported earlier for OT and OPV oligomers.[30,39]

5 Dynamical signatures in 2DES

Next, we study the signature of the observed dynamics in the 2D frequency maps obtained by 2DES. Several effects are expected to become manifest: first, a gradual energetic relaxation due to the downhill process in the S_1 state as a result of torsional planarization, and second, vibrational fine structure should appear when the X mode is included; such splittings are indeed observed in the experimental 2D electronic spectra of P3HT and related materials.[23,52] In our simulations, the X mode does not undergo damping, as mentioned above, as vibrational relaxation times for these high-frequency modes are longer than our observation time. Third, cross-peak signatures of electronic coupling between S_1/S_2 and S_3 could be observable in the 2DES maps.

In all 2DES calculations reported below, the initial pulse generates a superposition state, as shown in Fig. 3. To this end, we assume that the initial monomer transition dipole moments of eqn (19) are non-zero for the "left" side of the lattice (*i.e.*, sites 1–10). At later times, relevant for the second and third pulses, all transition dipole moments are active and assumed to be equal. A more refined procedure would tailor the initial transition dipole moments to the bright exciton state at the chosen torsional geometry.

In the following, we first report on the 2D electronic spectra for dynamics which only involve the torsional mode (as shown in Fig. 3c), followed by the analysis of the full dynamics (as shown in Fig. 3d). The pulse duration for the 2DES sequence of Fig. 2 was chosen to be 5 fs (FWHM), for a Gaussian-shaped envelope. Within the present RWA treatment, the carrier frequency and the excited-state energies are reduced by the reference energy c_E, *i.e.*, the monomer excited-state energy, see eqn (4) (however, c_E is added again in the following figures and discussion).

5.1 Torsional dynamics in 2DES maps

Fig. 4 reports the 2DES maps obtained for dynamics which only involve the torsional mode θ, in the absence of the polaronic mode, X, corresponding to the situation shown in Fig. 3c. Fig. 4a illustrates the adiabatic energies relative to the ground state as a function of time, showing the energy relaxation that takes place due to the torsional planarization. These energies are in line with the PESs of Fig. 1a. The time scale of planarization matches Fig. 3a, *i.e.*, the torsion reaches the planar geometry within 400–600 fs. Fig. 4a also illustrates the duration of the coherence time ($\tau = 250$ fs), followed by the detection interval (again $t = 250$ fs, shown by yellow shading) for selected values of the intermediate waiting time ($T = 0$, 100, 200 and 350 fs).

Fig. 4b shows 2DES maps for the indicated values of the mixing time T. At $T = 0$, a prominent diagonal peak is visible, whose position corresponds to the S_1 (and S_2) states at the initial, twisted geometry. This peak shows a deformation, which indicates that a frequency change as a function of the torsional evolution is taking place during the detection interval (see panel a). Moving to the final panel, at $T = 350$ fs, we discern a cross-peak whose frequency corresponds to ($\omega_\tau = 1.88$ eV, $\omega_t = 1.79$ eV), such that the S_1 (and S_2) frequency during the coherence time is correlated with the S_1 frequency of the planarized system during the detection time. The vertical distance between the diagonal peak at $T = 0$ fs and the off-diagonal cross-

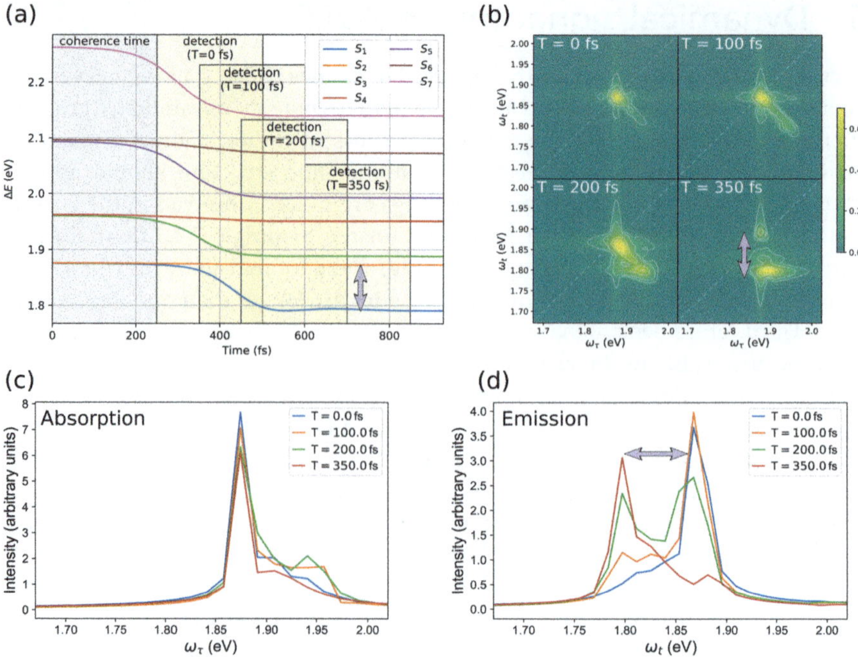

Fig. 4 2D electronic spectra for the dynamics including the torsional mode; absolute values are shown. The carrier frequencies of all pulses were fixed at $\omega_j = 1.87$ eV, $j = 1, \ldots, 3$. (a) Evolution of the adiabatic energies of the S_n, $n = 1, \ldots, 7$, states relative to the ground state (S_0), *i.e.*, $\Delta E = E_{S_n} - E_{S_0}$, during the coherence time ($\tau = 250$ fs), variable mixing times (T) and the detection time ($t = 250$ fs), whose start is shifted depending on T, as indicated by the yellow-shaded areas. The vertical arrow indicates the reorganization energy of the torsional mode, around $\Delta = 0.08$ eV. (b) 2DES signals for different waiting times, see the discussion in the text. At $T = 350$ fs, planarisation has occurred, and a cross-peak exhibits a red shift consistent with the reorganisation energy shown in (a). (c) Absorption spectrum obtained by integration over the ω_t domain. (d) Emission spectrum obtained by integration over the ω_τ domain, highlighting the red shift shown in (b).

peak at $T = 350$ fs corresponds to the torsional reorganization energy, indicated by an arrow, generating a spectral red shift of around 0.08 eV.

Besides this expected red-shift feature, another observation relates to the anti-diagonal features that start appearing at $T = 0$ fs and become pronounced at $T = 100$ fs and $T = 200$ fs. This feature is likely associated with a cross-peak whose frequency $\omega_\tau \sim 1.95$ eV corresponds to the S_3 (and S_4) states at the initial twisted geometry, whereas the frequency ω_t ~1.82 eV to ~1.79 eV corresponds to the decreasing energy of S_1 with increasing planarization of the torsion. At the same time, a cross-peak at ($\omega_\tau \sim 1.88$ eV, $\omega_t \sim 1.89$ eV) starts to emerge and becomes more pronounced with increasing waiting time. At the planarized geometry, S_3 has stabilized to $\omega_t \sim 1.89$ eV and can be detected due to a non-zero transition dipole moment at $\theta = 180°$ (in fact, of equal magnitude as the transition dipole moment of S_1). Therefore, the feature at ($\omega_\tau \sim 1.88$ eV, $\omega_t \sim 1.89$ eV; upper peak in the spectrum for $T = 350$ fs) could relate to a cross-peak between S_1/S_2 (twisted) and S_3 (planar).

Fig. 4c and d show the integrated absorption and emission spectra which have been obtained by integration over the ω_t domain (in the case of the absorption spectrum) and by integration over the ω_τ domain (in the case of the emission spectrum). Here, the red shift due to torsional planarization, and its gradual emergence between $T = 0$ fs and $T = 350$ fs, is clearly visible. This provides a precise map of the dynamical process, in excellent agreement with the energetics of Fig. 4a.

5.2 Torsional/polaronic dynamics in 2DES maps

Fig. 5 shows 2DES maps for the full dynamics, comprising the torsional mode θ and the polaronic mode X. Here, the vibronic fine structure associated with the X mode becomes a dominant feature, as has been shown experimentally.[23,24,52] Fig. 5a illustrates 2DES signals for different mixing times, where an initial time spacing of 10 fs was chosen to capture the rapid changes as a function of the X mode, whose periodicity lies around 20 fs. At these early times, the vibronic structure in the ω_τ domain is the most visible feature; for longer mixing times ($T =$ 160 fs, $T = 350$ fs), the spectral structures become more complex. As in the case of torsional dynamics addressed in the preceding section, it is likely that the observed vibronic fine structure essentially relates to the S_1 state.

Fig. 5b compares the integrated absorption spectrum for different waiting times T, obtained by integration over the ω_t domain, with the Fourier transform of the wavepacket autocorrelation function $C(t) = \langle \Psi(0)|\Psi(t)\rangle$. It is seen that the vibronic progression observed in the 2D spectra is in very good agreement with the reference spectrum.

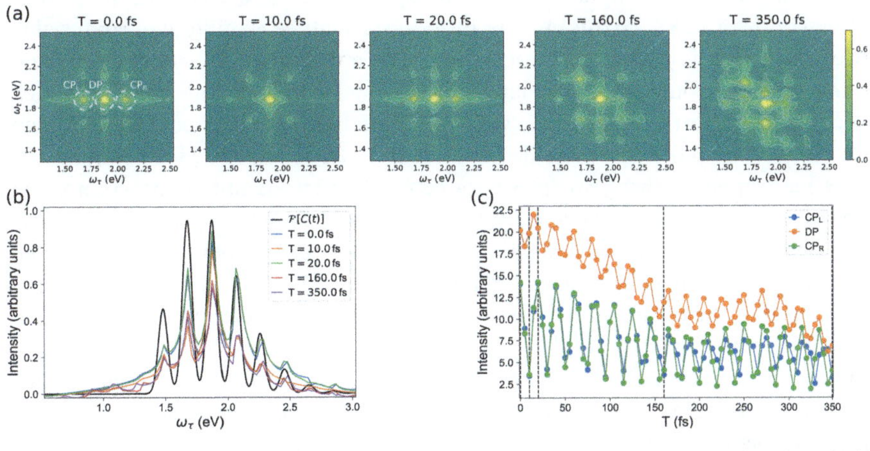

Fig. 5 2D electronic spectra for the dynamics including the torsional and polaronic X modes; absolute values are shown. The carrier frequencies of all pulses were fixed at $\omega_j =$ 1.87 eV, $j = 1, \ldots, 3$. (a) 2DES signals for different waiting times, see the discussion in the text. The individual spectra were rescaled. (b) Integrated absorption spectrum as a function of ω_τ, obtained by integration over the ω_t domain, as compared with the absorption spectrum generated by Fourier transformation of the wavepacket autocorrelation function $C(t) = \langle \Psi(0)|\Psi(t)\rangle$, shifted by the zero-point energy. All spectra are suitably rescaled. (c) Dependence of selected diagonal/cross peaks on the waiting time, up to $T = 350$ fs. Vertical dashed lines indicate the waiting times illustrated in the snapshots of panel (a). The chosen diagonal peak (DP) and cross peaks (CP_L and CP_R) are indicated in the left panel of (a) for $T = 0$ fs.

The wavepacket motion associated with the superposition of these vibronic levels translates to a pronounced T-dependence of the spectral features, as exemplified in Fig. 5c for a diagonal peak (DP) and two cross-peaks (CP_L, CP_R). These selected peaks are indicated in panel (a) (leftmost panel) of the figure. The most prominent feature is the frequency of the polaronic mode around 0.48 fs^{-1} (or ~21 fs). Fourier transformation suggests that additional, minor high-frequency features could play a role. These time dependent signatures are very similar to experimental 2DES observations for P3HT and related materials,[23,52] suggesting that the description in terms of a collective high-frequency mode matches the observed dynamics.

As for longer mixing times ($T > 150$ fs), the spectral features become increasingly complex. One might conjecture that the torsional red shift of ~0.08 eV demonstrated in Fig. 4 is observable, even though this is combined with the polaronic shift induced by the X mode. The latter should have a value of around 0.39 eV from our simulations, and is reduced by about 0.04 eV when moving from the twisted geometry to the planarized geometry (see Fig. 1d and e). A more detailed analysis of specific frequency ranges, using variable pulse durations and carrier frequencies, is necessary to elucidate these features.

6 Discussion and conclusions

Exciton dynamics in molecular materials are known to be strongly influenced by vibronic interactions with two types of modes – high-frequency modes, which give rise to trapping effects, and low-frequency soft modes, which induce conformational changes that influence the spatial distribution of the excitonic density and, hence, induce exciton migration. Here, we explored the signatures of these vibronic interactions in 2DES. To this end, we introduced a lattice Hamiltonian comprising two characteristic vibrational modes – a soft, anharmonic torsional mode and a collective polaronic mode. The latter is associated with a vibronic coupling which depends nonlinearly on the time-evolving exciton density. The model Hamiltonian was derived from our more detailed treatment of ref. 14 and 30, where simulations were performed for first principles parametrized lattice Hamiltonians, comprising several hundred modes. Specifically, we focused here on a situation where a single torsional mode induces a conjugation break in a J-type oligomer, and the observed dynamics at $T = 0$ K relate to the removal of this conjugation break, *i.e.*, the planarization of the chain and concomitant migration of the exciton to the center of the lattice.[30,39] That is, we single out a specific, microscopic exciton migration event induced by conformational change. Despite the simplicity of the present model, the salient features of the dynamics are faithfully reproduced; notably, exciton migration is found to be a near-adiabatic process following the torsion, and the high-frequency polaronic mode stabilizes the exciton in the local exciton ground state (LEGS). Importantly, the latter effect is not due to vibrational relaxation of the polaronic mode (which occurs on a longer time scale), but rather to nonadiabatic electronic relaxation, *i.e.*, internal conversion mediated by this mode.

Since exciton migration is an energetically downhill process in the present system, a spectral red shift is expected in the 2DES map. This is borne out by an analysis of the dynamics involving exclusively the torsional mode. In a real polymer system, we expect a pronounced red shift due to the concerted response of multiple modes, as shown experimentally.[16] At the same time, additional features are visible in our simulations that relate to cross peaks between the S_1/S_2 and S_3 states. When the full Hamiltonian is considered, including the X mode, coherent vibronic effects related to the characteristic periodicity ~20 fs of this mode become a dominant feature, as observed experimentally for P3HT and related systems.[23,52] These experiments underscore that the quasi-stationary dynamics of a collective, polaronic mode is indeed a generic feature of ultrafast photoinduced dynamics in molecular materials.

The present approach can be extended in multiple ways. Specifically, the present quantum–classical set-up is entirely suitable to explore temperature-dependent dynamics.[13,14] Furthermore, various intermediate steps can be chosen between the present simplified representation of the dynamics, based on two modes, and high-dimensional simulation strategies that we employed earlier, based upon efficient multiconfigurational propagation schemes.[13,14,30] Also, related computations of 2DES signals have been carried out using various other methodologies, notably the HEOM,[62] the Davydov approach,[63] and tensor-train methods.[54] Using these various methods, an accurate picture can be developed of the emergence of collective vibronic effects from local, microscopic interactions. Despite the "single-molecule" perspective of the present simulation strategy, we anticipate that the observed effects are observable in the ultrafast dynamics of photoexcited ensembles, where collective, correlated motions play an important role.[16]

Even though the scope of 2DES is limited regarding the characterization of exciton transport, more sophisticated higher-order spectroscopies have been introduced[29] that directly inform about exciton diffusion *via* exciton–exciton annihilation events. The present set-up can be generalized to these spectroscopies to elucidate how soft mode fluctuations influence the observed diffusion process. Furthermore, more refined spectroscopic approaches for the observation of excitonic coherence and its dependence on such low-frequency modes can be pursued.[64]

Finally, the observation of largely adiabatic exciton evolution guided by slow fluctuations of soft modes can be used as a starting point for contemplating the active control of exciton dynamics due to soft modes. The targeted activation of such modes could be employed to optimize transport properties, *e.g.*, in highly ordered materials. Controlled dynamical disorder could, hence, be turned into a strategy to guide quasi-particle motion and optimize functionalities.

Conflicts of interest

There are no conflicts to declare.

Acknowledgements

We thank Dassia Egorova and Manel Mondelo-Martell for useful discussions. Funding by the Deutsche Forschungsgemeinschaft (project BU-1032-2) is gratefully acknowledged.

Notes and references

1 A. Thiessen, J. Vogelsang, T. Adachi, F. Steiner, D. Vanden Bout and J. M. Lupton, *Proc. Natl. Acad. Sci. U. S. A.*, 2013, **110**, E3550.

2 D. Raithel, S. Baderschneider, T. B. de Queiroz, R. Lohwasser, J. Köhler, M. Thelakkat, S. Kümmel and R. Hildner, *Macromolecules*, 2016, **49**, 9553.

3 A. E. Bragg, W. Yu, J. Zhou and T. Magnanelli, *J. Phys. Chem. Lett.*, 2016, **7**, 3990.

4 L. Simine and P. J. Rossky, *J. Phys. Chem. Lett.*, 2017, **8**, 1752.

5 A. Köhler and H. Bässler, *Electronic Processes in Organic Semiconductors*, Wiley-VCH Verlag, Weinheim, Germany, 2015.

6 W. Barford, *Electronic and Optical Properties of Conjugated Polymers*, Clarendon, Oxford, 2nd edn, 2013.

7 J. Frenkel, *Phys. Rev.*, 1931, **37**, 17.

8 V. May and O. Kühn, *Charge and Energy Transfer Dynamics in Molecular Systems*, VCH-Wiley, 3rd edn, 2011.

9 R. E. Merrifield, *J. Chem. Phys.*, 1961, **34**, 1835.

10 M. Hoffmann, K. Schmidt, T. Fritz, T. Hasche, V. M. Agranovich and K. Leo, *Chem. Phys.*, 2000, **258**, 73.

11 A. De Sio and C. Lienau, *Phys. Chem. Chem. Phys.*, 2017, **19**, 18813.

12 W. Popp, D. Brey, R. Binder and I. Burghardt, *Annu. Rev. Phys. Chem.*, 2021, **72**, 591.

13 F. D. Maiolo, D. Brey, R. Binder and I. Burghardt, *J. Chem. Phys.*, 2020, **153**, 184107.

14 R. Binder and I. Burghardt, *Faraday Discuss.*, 2020, **221**, 406.

15 S. Westenhoff, W. Beenken, R. H. Friend, N. C. Greenham, A. Yartsev and V. Sundström, *Phys. Rev. Lett.*, 2006, **97**, 166804.

16 N. P. Wells and D. A. Blank, *Phys. Rev. Lett.*, 2008, **100**, 086403.

17 S. Mukamel, *Annu. Rev. Phys. Chem.*, 2000, **51**, 691.

18 T. Brixner, R. Hildner, J. Köhler, C. Lambert and F. Würthner, *Adv. Energy Mater.*, 2017, 7, 1700236.

19 N. S. Ginsberg and W. A. Tisdale, *Annu. Rev. Phys. Chem.*, 2020, **71**, 1.

20 T. Zhu, J. M. Snaider, L. Yuan and L. Huang, *Annu. Rev. Phys. Chem.*, 2019, **70**, 219.

21 N. S. Ginsberg, Y. C. Cheng and G. R. Fleming, *Acc. Chem. Res.*, 2009, **42**, 1352.

22 E. Collini and G. D. Scholes, *Science*, 2009, **323**, 369.

23 A. De Sio, F. F. Troiani, M. Maiuri, J. Réhault, E. Sommer, J. Lim, S. F. Huelga, M. B. Plenio, C. Rozzi, G. Cerullo, E. Molinari and C. Lienau, *Nat. Commun.*, 2016, 7, 13742.

24 A. D. Sio, F. Camargo, K. Winte, E. Sommer, F. Branchi, G. Cerullo and C. Lienau, *Eur. Phys. J. B*, 2018, **91**, 236.

25 A. G. Dijkstra, T. la Cour Jansen and J. Knöster, *J. Chem. Phys.*, 2008, **128**, 164511.

26 W. Barford, J. Gardner and J. R. Mannouch, *Faraday Discuss.*, 2020, **221**, 281.

27 Y. Lee, S. Das, R. M. Malamakal, S. Meloni, D. M. Chenoweth and J. M. Anna, *J. Am. Chem. Soc.*, 2017, **139**, 14733.

28 J. D. Schultz, T. Kim, J. P. O'Connor, R. M. Young and M. R. Wasielewski, *J. Phys. Chem. C*, 2022, **126**, 120–131.

29 P. Malý, J. Lüttig, A. Turkin, J. Dostál, C. Lambert and T. Brixner, *Chem. Sci.*, 2020, **11**, 456.

30 R. Binder, D. Lauvergnat and I. Burghardt, *Phys. Rev. Lett.*, 2018, **120**, 227401.

31 M. F. Gelin, D. Egorova and W. Domcke, *Acc. Chem. Res.*, 2009, **42**, 1290.

32 J. Krčmář, M. F. Gelin and W. Domcke, *Chem. Phys.*, 2013, **422**, 53.

33 J. Krčmář, M. F. Gelin, D. Egorova and W. Domcke, *J. Phys. B: At., Mol. Opt. Phys.*, 2014, **47**, 124019.
34 R. Hegger, R. Binder and I. Burghardt, *J. Chem. Theory Comput.*, 2020, **16**, 5441.
35 M. Feit, J. A. Fleck and J. Steiger, *J. Comput. Phys.*, 1982, **47**, 412.
36 M. Feit and J. A. Fleck, *J. Chem. Phys.*, 1984, **80**, 2578.
37 J. B. Delos, W. R. Thorson and S. K. Knudson, *Phys. Rev. A: At., Mol., Opt. Phys.*, 1972, **6**, 709.
38 G. Billing, *Chem. Phys. Lett.*, 1983, **100**, 535.
39 R. Binder and I. Burghardt, *J. Chem. Phys.*, 2020, **152**, 204120.
40 D. V. Makhov and W. Barford, *Phys. Rev. B: Condens. Matter Mater. Phys.*, 2010, **81**, 165201.
41 O. R. Tozer and W. Barford, *J. Chem. Phys.*, 2015, **143**, 084102.
42 R. Binder, M. Bonfanti, D. Lauvergnat and I. Burghardt, *J. Chem. Phys.*, 2020, **152**, 204119.
43 R. Binder, S. Römer, J. Wahl and I. Burghardt, *J. Chem. Phys.*, 2014, **141**, 014101.
44 M. H. Beck, A. Jäckle, G. A. Worth and H. D. Meyer, *Phys. Rep.*, 2000, **324**, 1.
45 H. Wang, *J. Phys. Chem. A*, 2015, **119**, 7951.
46 See ref. 14, ESI.
47 W. Popp, M. Polkehn, R. Binder and I. Burghardt, *J. Phys. Chem. Lett.*, 2019, **10**, 3326.
48 A. Scott, *Phys. Rep.*, 1992, **217**, 1.
49 F. C. Spano and C. Silva, *Annu. Rev. Phys. Chem.*, 2014, **65**, 477.
50 A. B. Trofimov and J. Schirmer, *J. Phys. B: At., Mol. Opt. Phys.*, 1995, **28**, 2299.
51 H. Li, R. Nieman, A. J. A. A. H. Lischka and S. Tretiak, *J. Chem. Theory Comput.*, 2014, **10**, 3280–3289.
52 Y. Song, C. Hellmann, N. Stingelin and G. D. Scholes, *J. Chem. Phys.*, 2015, **142**, 212410.
53 G. Strang, *SIAM J. Numer. Anal.*, 1968, **5**, 506.
54 L. Chen, R. Borrelli, D. V. Shalashilin, Y. Zhao and M. Gelin, *J. Chem. Theory Comput.*, 2021, **17**, 4316.
55 N. Makri, *J. Chem. Phys.*, 2018, **149**, 214108.
56 L. Chen, R. Borelli and Y. Zhao, *J. Phys. Chem. A*, 2017, **121**, 8757.
57 L. Chen, M. F. Gelin and D. V. Shalashilin, *J. Chem. Phys.*, 2019, **151**, 244116.
58 T. Ma, M. Bonfanti, P. Eisenbrandt, R. Martinazzo and I. Burghardt, *J. Chem. Phys.*, 2018, **149**, 244107.
59 Y. Tanimura, *J. Phys. Soc. Jpn.*, 2006, **75**, 082001.
60 L. Chen, Y. Zhao and Y. Tanimura, *J. Phys. Chem. Lett.*, 2015, **6**, 3110.
61 W. J. D. Beenken, *Phys. Status Solidi A*, 2009, **206**, 2750–2756.
62 A. G. Dijkstra and Y. Tanimura, *New J. Phys.*, 2010, **12**, 055005.
63 N. Zhou, L. Chen, Z. Huang, K. Sun, Y. Tanimura and Y. Zhao, *J. Phys. Chem. A*, 2016, **120**, 1562–1576.
64 A. J. Kiessling and J. A. Cina, *J. Chem. Phys.*, 2020, **152**, 244311.

Faraday Discussions

PAPER

THz non-linear optical response in cuprates: predominance of the BCS response over the Higgs mode

Mattia Udina, [*a] Jacopo Fiore, [a] Tommaso Cea,[b] Claudio Castellani,[a] Goetz Seibold [c] and Lara Benfatto [a]

Received 21st January 2022, Accepted 10th February 2022
DOI: 10.1039/d2fd00016d

Recent experiments with strong THz fields in unconventional cuprate superconductors have clearly evidenced an increase of the non-linear optical response below the superconducting critical temperature T_c. As in the case of conventional superconductors, a theoretical estimate of the various effects contributing to the non-linear response is needed in order to interpret the experimental findings. Here, we report a detailed quantitative analysis of the non-linear THz optical kernel in cuprates within a realistic model, accounting for the band structure and disorder level appropriate for these systems. We show that the BCS quasiparticle response is the dominant contribution for cuprates, and its polarization dependence accounts very well for the third-harmonic generation measurements. On the other hand, the polarization dependence of the THz Kerr effect is only partly captured by our calculations, suggesting the presence of additional effects when the system is probed using light pulses with different central frequencies.

1 Introduction

The recent technological advances in the generation of strong terahertz (THz) pulses has triggered intense activity aimed at using light to selectively excite fundamental modes in condensed-matter systems.[1–3] In particular, the THz range is the relevant frequency window for phononic excitations and collective modes in broken-symmetry states, *e.g.* magnons in the magnetically-ordered phase and complex (amplitude and phase) fluctuations of the superconducting (SC) order parameter. In the case of phonons, it has been rapidly understood that experiments with THz pulses[4,5] closely mirror the experiments done with intense near-infrared (NIR) and visible (VIS) light fields.[6] This is, for example, the case for

[a]*Department of Physics and ISC-CNR, "Sapienza" University of Rome, P.le Aldo Moro 5, Rome, 00185, Italy. E-mail: mattia.udina@uniroma1.it; lara.benfatto@roma1.infn.it*
[b]*Imdea Nanoscience, Faraday 9, 28015 Madrid, Spain*
[c]*Institut für Physik, BTU Cottbus-Senftenberg, PBox 101344, 03013 Cottbus, Germany*

pump–probe protocols, where a weak probe pulse with a variable time delay t_{pp} with respect to the pump detects the relative changes in the reflectivity or transmission through the sample, which can be periodically modulated due to the excitation of Raman-active phonons. In the case of electronvolt (eV) light, the interaction between ultrashort optical pulses and lattice degrees of freedom in non-absorbing materials has been unambiguously interpreted[6,7] as an impulsive-stimulated Raman scattering (ISRS) process, where the phonon generation occurs in a difference frequency of two high-energy photons taken from the pump field. In a close analogy, experiments done with THz pulses achieve the same goal by a sum-frequency process,[4,5] leading to a sharp response whenever the pump frequency Ω matches half of the phonon energy $\omega_{\rm ph}$, *i.e.* $\Omega = \omega_{\rm ph}/2$.

The case of Raman-active phonons provides a benchmark example of a more general mechanism allowing for the impulsive excitation of many different modes with a Raman-like symmetry. Indeed, as we will discuss in detail in this manuscript, the experimental findings can be understood in a rather general way by considering that a Raman-active phonon leads to a strong resonance in the non-linear optical kernel $K(\omega)$ at the phonon frequency $\omega_{\rm ph}$. As a consequence, one would expect the same reasoning to hold also for SC collective excitations, for example,[8,9] once $\omega_{\rm ph}$ is replaced by the characteristic energy scale $\omega_{\rm res}$, where the corresponding non-linear kernel is resonant. However, even within this scheme, the interpretation of experiments in this class of materials has been much more controversial.[10] The reason is that even though all measurements in conventional superconductors[11–17] can be reconciled with a non-linear kernel peaked at $\omega_{\rm res} = 2\Delta$, where Δ is the SC gap, the identification of the relevant excitations responsible for such resonance has been debated. Indeed, in a superconductor, both the response described by the Bardeen–Cooper–Schrieffer (BCS) theory, probing the quasiparticle continuum, and the amplitude fluctuations of the SC order parameter, also named the Higgs mode, are resonant at 2Δ. As a consequence, only a precise theoretical estimate of the relative intensity of the two contributions, or the analysis of their dependence on the polarization of the pump field with respect to the main crystallographic axes, can be used to disentangle the origin of the 2Δ resonance.

Trying to understand which one, among the BCS and the Higgs contributions, is the main source for this nonlinear effect has been the subject of much experimental[11–17] and theoretical work[8,13–15,18–25] in the last few years. In the attempt to resolve such a controversy, it turned out that despite both VIS and THz pulses triggering collective excitations *via* an ISRS excitation process, one finds that in a superconductor, the non-linear optical kernel K controlling the response is *different* in the two cases, *i.e.*

$$K^{\rm THz}_{\rm Raman} \neq K^{\rm eV}_{\rm Raman}. \tag{1}$$

In other words, even though in both cases only Raman-like excitations are involved, the microscopic fermionic processes mediating the light-mode coupling are different in the case of a difference-frequency (in the visible) or of a sum-frequency (in the THz) excitation. More specifically, while $K^{\rm eV}_{\rm Raman}$ in the widely used effective-mass approximation[26] essentially probes the BCS continuum *via* lattice-modulated charge fluctuations, as given by diamagnetic-like coupling of electrons to light, $K^{\rm THz}_{\rm Raman}$ also probes the BCS continuum *via* paramagnetic-like coupling of the electron current to light. Such a difference arises once impurity effects are taken into account,[22–25] since current is no more conserved in the presence of disorder. Such a difference in turns leads to distinct selection rules for THz-driven or VIS-driven excitations, eventually making the Higgs-mode excitation irrelevant in the clean limit[8,26] and sizeable in the strong-disorder limit.[22–25]

The above result is particularly relevant in the context of recent experiments on unconventional cuprate superconductors. So far, the only observation of marked 2Δ oscillations in cuprates has been done using visible light,[27] and they have been ascribed to the BCS response, in agreement with the usual interpretation of conventional Raman measurements in cuprates.[8,26] Nonetheless, the specular experiments performed with THz light have been almost exclusively interpreted so far to be a response of the Higgs mode,[28–31] despite the lack of a theoretical calculation of $K^{\mathrm{THZ}}_{\mathrm{Raman}}$ for these systems. The aim of the present manuscript is to provide such a quantitative analysis, by means of the direct computation of the non-linear THz Raman tensor within a realistic microscopic model for cuprates. In particular, we will take advantage of the numerically-exact calculation in the presence of disorder recently discussed in ref. 25 to provide a quantitative estimate of $K^{\mathrm{THZ}}_{\mathrm{Raman}}$ as a function of the band parameters and/or the doping level. As we shall see, for the band structure and disorder level of cuprates, the contribution of the Higgs mode is still fairly subdominant with respect to the BCS response, whose polarization dependence can be understood well by taking into account both diamagnetic and paramagnetic processes at an appropriate disorder level. While this result completely accounts for the polarization dependence observed for third-harmonic generation experiments in transmission,[29,31] where the signal is pretty much isotropic, it leaves nonetheless some open questions for the polarization observed *via* THz pump–optical probe experiments,[28,30] that report instead a sizeable anisotropy of the signal for overdoped samples. We will argue below that such a difference can be ascribed to the peculiar role of paramagnetic-like processes when THz and VIS light pulses act simultaneously, leading in principle to an intermediate effect between the two extreme cases encoded in eqn (1).

2 Non-linear response from resonant modes

Before discussing the specific case of cuprates, let us provide a general interpretative scheme to understand how the measured quantity in different unconventional THz spectroscopic techniques can be related to the same non-linear optical response, setting the basis for a more rigorous understanding of the experimental results. Since a detailed discussion has been already provided in ref. 9, here we will recast the main results only. In general, one can distinguish between two classes of experiments (see Fig. 1): (a) measurements of the third-harmonic generation (THG) in transmission and (b) pump–probe protocols, both in transmission and reflection configurations. In the former case, one excites the sample with an intense THz laser pulse $E_{\mathrm{pump}}(t)$ and records the transmitted electric field $E_{\mathrm{out}}(t)$. These experiments are usually performed with multi-cycle laser pulses, such that the spectrum of the incoming radiation $E_{\mathrm{pump}}(\omega)$ is strongly peaked around a central frequency $\omega = \Omega$. In this situation, the THG process manifests as a significant spectral component of $E_{\mathrm{out}}(\omega)$ at 3Ω, with different amplitudes according to the incoming Ω value or to the specific temperature of the sample. In the case of (b), instead, the system can be excited either with a single-cycle or with a multi-cycle THz laser pulse. Single-cycle pulses

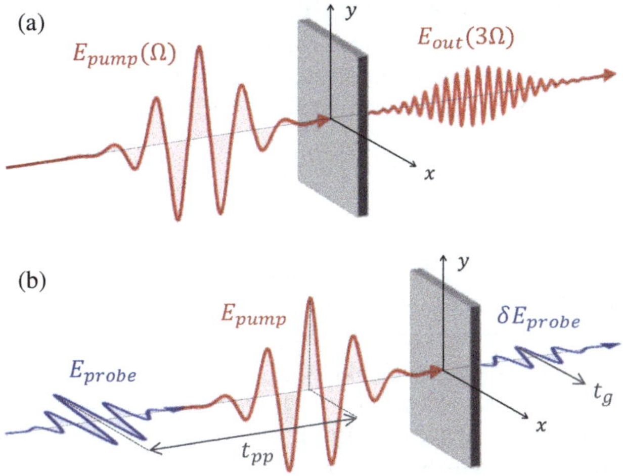

Fig. 1 Schematics of unconventional THz spectroscopic techniques in transmission. In a typical THG experiment (a), the sample is perturbed using a multi-cycle (narrowband) THz pump pulse at Ω and one collects the induced transmitted component of the field at 3Ω. In pump–probe measurements (b), instead, one first perturbs the sample with an intense (multi-cycle or single-cycle) pump field and then records the differential transmitted component, with and without the pump, of a weak probe pulse, as a function of the time delay t_{pp} between the two pulses at a fixed observation time t_g.

last for less than 1 ps, and are thus associated with a relatively broad spectrum around the central frequency Ω. The subsequent detection process occurs using a weak THz or NIR/VIS probe pulse. In both cases, the probing field is delayed by t_{pp} with respect to the pump laser pulse and the recorded signal is usually a differential change δE_{probe} in the transmitted or reflected probe field measured with and without the pump, recorded as a function of t_{pp}.

Following the field-theory approach developed *e.g.* in ref. 8, 9 and 15, in order to reproduce the experimental findings, one has to compute the third-order current flowing inside the sample, given in full generality by the partial derivative with respect to the external e.m. field of the fourth-order action $S^{(4)}$, written in terms of the e.m. vector potential **A**, where $\mathbf{E} \equiv -\partial_t \mathbf{A}$, *i.e.*

$$S^{(4)}[\mathbf{A}] = e^4 \int \mathrm{d}\Omega_1 \mathrm{d}\Omega_2 \mathrm{d}\Omega_3 \sum_{\alpha\beta\gamma\delta} A_\alpha(\Omega_1) A_\beta(\Omega_2) \times K_{\alpha\beta\gamma\delta}(\Omega_1, \Omega_2, \Omega_3) A_\gamma(\Omega_3) A_\delta(-\Omega_1 - \Omega_2 - \Omega_3), \quad (2)$$

with e is the electron charge and $K_{\alpha\beta\gamma\delta}$ is a third-order tensor, which depends in the most general case on four spatial indexes and three incoming frequencies. While eqn (2) accounts for all the possible third-order processes contributing to the nonlinear current, we will focus for the moment on a sub-set of processes, allowing us to rewrite the effective action as

$$S^{(4)}[\mathbf{A}] = \int \mathrm{d}\Omega \sum_{\alpha\beta\gamma\delta} \overline{A}_{\alpha\beta}{}^2(\Omega) K_{\alpha\beta;\gamma\delta}(\Omega) \overline{A}_{\gamma\delta}{}^2(-\Omega) = \int \mathrm{d}t \mathrm{d}t' \sum_{\alpha\beta\gamma\delta} A_{\alpha\beta}{}^2(t) K_{\alpha\beta;\gamma\delta}(t - t') A_{\gamma\delta}{}^2(t'), \quad (3)$$

where we define $\overline{A}_{\alpha\beta}^{2}(\Omega)\equiv\int \mathrm{d}\Omega A_\alpha(\Omega)A_\beta(\Omega-\omega)$ as the Fourier transform of $A_\alpha(t)$ $A_\beta(t)$ and we put the semicolon between the spacial indices in K to underline that the nonlinear kernel has a Kubo-like structure, with two vertices carrying two field components each. The nonlinear current along the generic direction α then reads

$$J_\alpha^{\mathrm{NL}}(t)=-2e^4\sum_{\beta\gamma\delta}A_\beta(t)\int \mathrm{d}t' K_{\alpha\beta;\gamma\delta}(t-t')A_\gamma(t')A_\delta(t'). \tag{4}$$

When dealing with THG measurements, only the $A_{\mathrm{pump}}(t)$ field is present, meaning that, if light is shed along *e.g.* the *y* crystallographic axis and one collects the transmitted field component at 3Ω along the same direction (see Fig. 1a), the resulting THG intensity is proportional to $|J_y^{\mathrm{NL}}(3\Omega)|^2$, since the linear response to the incoming field does not contain additional harmonics. In particular, by approximating the incident multi-cycle incoming field with a monochromatic one $A_y(t)=A_0\cos(\Omega t)$, the intensity of the THG signal can be written as

$$I^{\mathrm{THG}}(\Omega)\sim\left|\int J_y^{\mathrm{NL}}(t)e^{i3\Omega t}\right|^2\propto\left|K_{yy;yy}(2\Omega)A_0{}^3\right|^2, \tag{5}$$

showing that I^{THG} scales as the squared modulus of the nonlinear kernel evaluated at *twice* the frequency of the incoming pump field.

For realistic multi-cycle band pulses, the pump field can be described by periodic oscillations convoluted with a Gaussian decay,[9] *i.e.* $A_y(t)=A_0e^{-(2t\sqrt{\ln 2}/\tau_p)^2}\cos(\Omega t)$. In this case, the non-linear current J^{NL} is given by the more general convolution[8,9]

$$J_y^{\mathrm{NL}}(\omega)=-\int \mathrm{d}\omega' A_y(\omega-\omega')K_{yy;yy}(\omega')\overline{A}_y{}^2(\omega'). \tag{6}$$

A typical spectrum of $J^{\mathrm{NL}}(\omega)$ is shown in Fig. 2b, with the blue and red lines corresponding to two possible values of the central frequency Ω of the pump. Here the non-linear kernel $K(\omega)$ is taken with a marked resonance at $\omega_{\mathrm{res}}\simeq 1.4$ THz, and its form is the one expected for a superconductor (see below). As one can see, since $\overline{A}^2(\omega)$ has spectral components both around $\omega\simeq 0$ and $\omega\simeq 2\Omega$, see Fig. 2a,

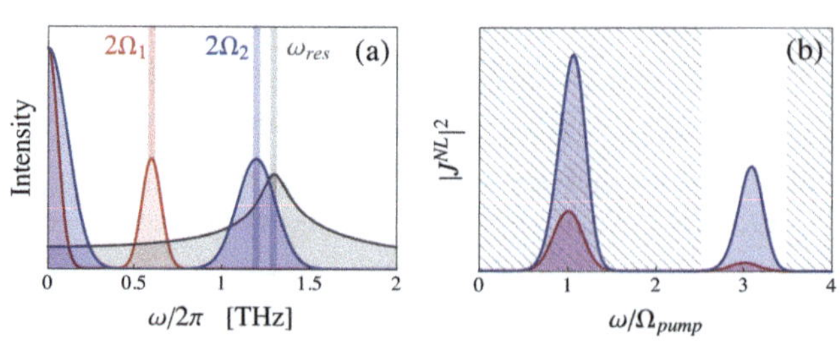

Fig. 2 THG with a narrow-band pulse. (a) Spectral component of the non-linear kernel $K(\omega)$ resonant at $\omega_{\mathrm{res}}/2\pi\simeq 1.4$ THz (grey line) along with the squared pump pulse $\overline{A}^2(\omega)$ for two values of the central frequency, $\Omega_1/2\pi=0.3$ THz ($\tau=9.5$ ps, red line) and $\Omega_2/2\pi=0.6$ THz ($\tau=4.8$ ps, blue line). (b) Corresponding spectra of the non-linear current $J^{\mathrm{NL}}(\omega)$ computed using eqn (6). The dashed area represents the spectral region not relevant for the THG.

the non-linear current $J^{NL}(\omega)$ has components around $\omega \simeq \Omega$, weighted approximately[8] with $2K(0) + K(2\Omega)$, and around $\omega \simeq 3\Omega$, weighted approximately with $K(2\Omega)$, see also eqn (5) above. As a consequence, the maximum value of the intensity of $J^{NL}(\omega)$ around $\omega \approx 3\Omega$ is obtained when $\Omega \approx \omega_{res}/2$, so that the overlap of $\bar{A}^2(\omega \approx 2\Omega)$ with $K(\omega \approx \omega_{res})$ is the largest, see Fig. 2b. In other words, THG is maximized when two photons of the pump field resonantly excite the collective mode responsible for the divergence of $K(\omega)$.

The connection with THz sum-frequency two-photon processes is also evident in the typical experimental set-up behind pump–probe protocols. For what concerns this kind of experiment, indeed, the differential transmitted field δE_{probe} can be related once more to the nonlinear current (4), where now both the pump and probe pulses, as well as their relative time delay t_{pp}, must be taken into account. In particular, by explicitly rescaling the pump field as $A_{pump}(t) = \tilde{A}_{pump}(t + t_{pp})$ in such a way that both $\tilde{A}_{pump}$ and A_{probe} are centered around $t = 0$, and by fixing the observation time at t_g, it can be easily shown[9,15] that in the cross-polarized configuration depicted in Fig. 1b, the measured quantity reads

$$\delta E_{probe}(t_{pp}) \propto A_{probe}(t_g) \int dt' K_{xx;yy}(t_g + t_{pp} - t') \times \left[\tilde{A}_{pump}(t')\right]^2. \qquad (7)$$

Since the acquisition time t_g is fixed, the previous expression shows that the probe field simply acts as a multiplying factor, setting the overall amplitude and phase of the oscillations, while the time-evolution of the signal is controlled by the convolution between the kernel and the squared pump field. More interestingly, if we Fourier transform eqn (7) with respect to t_{pp}, we find a very compact expression for the power spectrum of the differential transmitted field:

$$\delta E_{probe}(\omega) \propto K_{xx;yy}(\omega) \overline{A}_{pump}{}^2(\omega). \qquad (8)$$

Eqn (8) allows one to predict the presence of oscillations in $\delta E_{probe}(t_{pp})$ at the ω_{osc} frequency, which dominates the convolution between the kernel and the squared pump pulse. In addition, eqn (8) provides a simple way to understand what marks the difference between pump–probe experiments performed using a pump pulse with a central frequency Ω in the NIR/VIS or in the THz range. For the sake of clarity, let us focus again on a non-linear kernel $K(\omega)$, which displays a marked maximum at a given frequency ω_{res} in the THz range, see Fig. 3. When dealing with NIR/VIS pulses, $2\Omega \gg \omega_{res}$, so the only relevant overlap in eqn (8) occurs between $K(\omega)$ and the $\omega \simeq 0$ peak of $\overline{A}_{pump}{}^2(\omega)$. This means that the mode is excited *via* a difference-frequency process, in which two photons taken from the pump have energies $\omega_1 \simeq \Omega$ and $\omega_2 \simeq \Omega - \omega_{res}$ (see Fig. 3a and b), in full analogy with stimulated Raman scattering. On the contrary, for THz pump pulses designed to have $2\Omega \sim \omega_{res}$, what matters in eqn (8) is the overlap of the non-linear kernel with the 2Ω peak in $\overline{A}_{pump}{}^2(\omega)$ and one is dealing with a sum-frequency excitation, where $\omega_{1,2} \simeq \omega_{res}/2$ (see Fig. 3c and d). The basic mechanism is then the same already highlighted before when discussing THG processes with multi-cycle THz pulses.

Once the general mechanism is established, it is useful to further distinguish between three different cases which determine the final time evolution of the

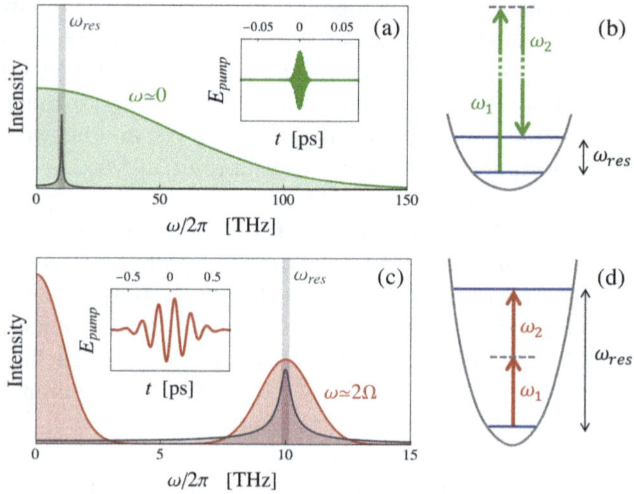

Fig. 3 Difference-frequency *vs.* sum-frequency excitations in pump–probe protocols, for a generic non-linear kernel $K(\omega)$ resonant at ω_{res} (grey line). (a) When VIS pump pulses are applied (inset), the kernel overlaps with the $\omega \simeq 0$ peak in $\bar{A}_{pump}^2(\omega)$ (green line). (b) Two photons at energies ω_1 and ω_2, taken from the (relatively broad) pump spectrum, lead to the difference-frequency excitation of the mode, *i.e.* $\omega_{res} = \omega_1 - \omega_2$. (c) When THz pump pulses are applied (inset), the kernel overlaps with the $\omega \simeq 2\Omega$ peak in $\bar{A}_{pump}^2(\omega)$ (red line). (d) The two ω_1, ω_2 photons lead to the sum-frequency excitation of the mode at $\omega_{res} = \omega_1 + \omega_2$.

$\delta E_{probe}(t_{pp})$ signal. Let us first consider the case of short single-cycle THz light pulses whose duration τ_p is much shorter than the time scale τ_{res} of the ω_{res} resonance, as set by the inverse width of the ω_{res} peak in the non-linear kernel $K(\omega)$, *i.e.* $\tau_p < \tau_{res}$, and having a central frequency $2\Omega \simeq \omega_{res}$. In this case, which is usually named as "quench", one finds that $\bar{A}_{pump}^2(\omega)$ is rather flat around ω_{res} (Fig. 4a) and the spectrum of the differential transmitted field (8) essentially follows the optical kernel, *i.e.* $\delta E_{probe}(\omega) \sim K(\omega)$. Notice that this condition is also fulfilled for eV light pulses, since their typical duration $\tau_p \sim 10$ fs makes $\tau_p \ll \tau_{res}$. As a consequence, one still obtains that $\bar{A}^2(\omega \approx 0)$ is very flat around the resonance, see Fig. 3a, and consequently $\delta E_{probe}(\omega) \sim K(\omega)$. In both cases, the main outcome is that in the time domain, the differential electric field is expected to oscillate at the resonance frequency of the kernel, *i.e.* $\omega_{osc} \approx \omega_{res}$, see Fig. 4b.

Notice also that for THz light pulses, the possibility to observe ω_{res} oscillations strongly relies not only on the antiadiabatic condition $\tau_p < \tau_{res}$ but also on the matching condition $2\Omega \simeq \omega_{res}$ of the pulse central frequency. Indeed, when the pump field moves out of resonance, as happens *e.g.* for $\Omega \ll \omega_{res}$, the only relevant overlap in eqn (8) is between $\bar{A}^2(\omega \approx 0)$ and $K(\omega \simeq 0)$, leading to a nearly instantaneous contribution in $\delta E_{probe}(t_{pp})$ following the squared pump field in time, also referred to in the literature as the THz Kerr effect (Fig. 4d). An analogous effect is found when the system is driven by a narrow multi-cycle pump pulse, such that $\tau_p > \tau_{res}$. Such a condition, which is usually named a "drive" pulse, translates in a $\bar{A}_{pump}^2(\omega)$ narrower than the ω_{res} peak in $K(\omega)$, see Fig. 4e. As a consequence, the convolution (eqn (8)) is dominated by the spectral components of $\bar{A}_{pump}^2(\omega)$ and $\delta E_{probe}(t_{pp})$ is once more proportional to the squared field in time,

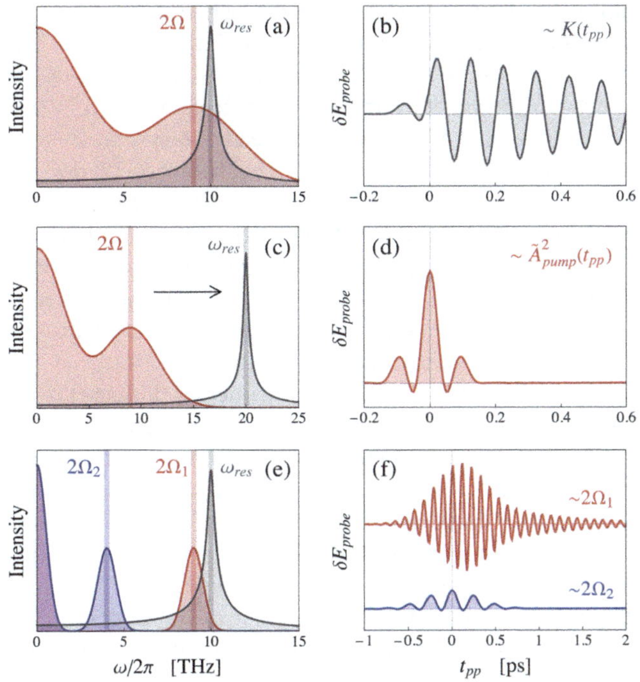

Fig. 4 Pump–probe experiments with THz fields. When using single-cycle pump pulses, if the resonance in the nonlinear kernel (grey line) is close to $\omega = 2\Omega$ (a), the squared pump field ($\tau = 0.2$ ps, red line) is approximated well by a constant in eqn (8), and the differential transmitted probe field $\delta E_{\text{probe}}(t_{\text{pp}})$ oscillates at ω_{res}, following $K(t_{\text{pp}})$ in time (b). When, instead, $\omega_{\text{res}} \gg 2\Omega$ (c), the convolution is dominated by $\bar{A}_{\text{pump}}{}^2(\omega)$ and the differential probe field δE_{probe} follows the squared pump field in the t_{pp} time domain, leading to the so-called THz Kerr effect (d). Analogously, for multi-cycle pump fields, only the response at 2Ω dominates (e), and due to the stronger duration of the pulse ($\tau_{1,2} = 1$ ps) one can clearly identify in $\tilde{A}^2(t_{\text{pp}})$ marked oscillations at 2Ω (f), with an amplitude that is strongly enhanced when $\Omega \simeq \omega_{\text{res}}/2$ (red line) as compared to the case when $\Omega \ll \omega_{\text{res}}/2$ (blue line).

oscillating at $\omega_{\text{osc}} \approx 2\Omega$, *i.e.* at *twice* the central frequency of the narrow-band light pulse, see Fig. 4f.

3 Superconducting modes and the role of disorder

The results discussed in the previous section are rather general, since they only require the existence of a marked resonance in the non-linear kernel $K(\omega)$ at a given frequency ω_{res}, whatever its origin. Thus, the same scheme can be used to understand the non-linear excitation of phonons, as well as of electronic collective modes,[9] as they emerge across a phase transition to a SC or a charge-density-wave (CDW) state. Let us then see how one can interpret the experiments in conventional superconductors, *e.g.* NbN,[11–14] Nb_3Sn[16] and MgB_2,[15,17] in light of such a general paradigm. THG experiments have been performed in disordered NbN[13] by fixing the central frequency Ω of a multi-cycle pulse while changing the temperature of the sample. The general result is an enhancement of the THG below T_c, with a maximum at the temperature where $\Omega = \Delta(T)$. In light of the previous discussion, see eqn (5) and Fig. 2, this implies that the SC non-linear optical kernel $K_{SC}(\omega)$ has a maximum at $\omega_{res} = 2\Delta(T)$. Analogously, pump–probe protocols with the same multi-cycle pulse have reported[13] 2Ω oscillations, as for the case shown in Fig. 4f. When instead the system is quenched with a single-cycle THz pulse,[11,12] the $\delta E_{probe}(t_{pp})$ signal shows oscillations at $\omega_{osc} = 2\Delta$, leading again to a kernel $K_{SC}(\omega)$ resonant at 2Δ, as explained while discussing Fig. 4b above. When other collective modes are present, as is the case for the Leggett mode in MgB_2, for example, one can selectively see the resonance at twice the gap[17] or at the Leggett-mode frequency[15] by tuning the pump frequency. In the latter case, the strong increase of 2Ω oscillations in the pump–probe signal have been also tested[15] when $\Omega \simeq \omega_{res}/2$, as shown in Fig. 4f.

Focusing now on the case of NbN, the experimental findings are all consistent with a SC non-linear optical kernel $K_{SC}(\omega)$ resonant at $\omega_{res} = 2\Delta(T)$. However, as mentioned in the Introduction, there has been for a while a debate in the literature about the identification of the relevant excitations responsible for such a resonance. Since this issue has been discussed at length in several manuscripts,[8,10,14,19–25] here we just summarize the main points in a schematic way. On very general grounds, to derive the non-linear kernel appearing in eqn (2) one needs to integrate out all the electronic degrees of freedom of a model system where interacting electrons are coupled to the external gauge field **A**. Let us start from the case of a clean system. Assuming that interaction terms are gauge-invariant, the gauge field only couples to the kinetic term of the Hamiltonian, so one can expand the Hamiltonian $H(\mathbf{A})$ in the presence of the e.m. field as:

$$H(\mathbf{A}) \simeq H(\mathbf{A} = 0) + \mathbf{j}\cdot\mathbf{A} + \rho_{\alpha\beta}A_\alpha A_\beta + \mathcal{O}(A^3), \tag{9}$$

where **j** is the electronic paramagnetic current and $\rho_{\alpha\beta}$ simply reduces to $\delta_{\alpha\beta}n/m$ for free electrons, where n is the electron density and m the mass. Notice that in a lattice model, the simple expression in eqn (9) should be extended to include all the bands describing electrons moving in a periodic potential. To derive $S^{(4)}[\mathbf{A},]$ one needs to sum over all possible electronic processes, leading to a response of order A^4. In terms of the usual Feynman diagrams expansion, this leads to all possible electronic loops with four external e.m. legs, as exemplified in Fig. 5. In general, terms of order A^4 can be obtained, *e.g.* with correlation functions

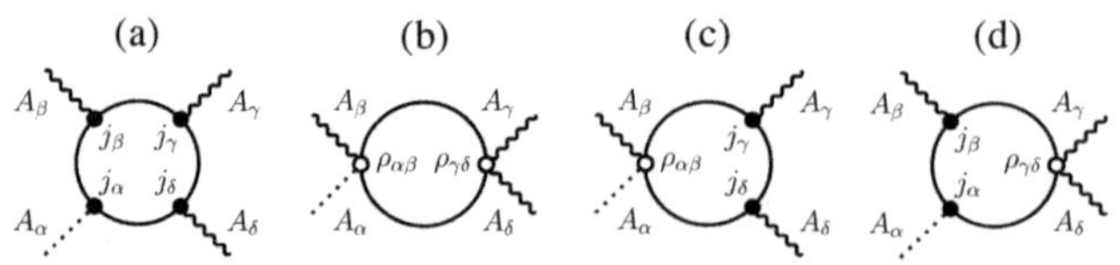

Fig. 5 Schematic of some of the Feynman diagrams contributing to the non-linear optical response in electronic systems. Here, solid lines denote the electronic Green's functions and wavy lines the e.m. gauge field, with α, β, γ, δ being spatial indexes. The dotted line denotes the derivative with respect to one field component, as needed to compute the current according to eqn (4). (a) Paramagnetic-like processes are built with current-like e.m. field insertions, denoted by a full circle. (b) Diamagnetic-like processes are built with density-like e.m. field insertions, denoted by an empty circle. Diagrams with two diamagnetic vertexes have the form of Kubo-like response functions. (c and d) Mixed terms.

including four paramagnetic-like terms $\mathbf{j}\cdot\mathbf{A}$ or two diamagnetic-like terms $\rho_{\alpha\beta}A_\alpha A_\beta$ from eqn (9), plus all possible mixed combinations. When dealing with high-energy NIR/VIS light pulses, interband transitions should be also considered, as for the usual Raman response. In the case of non-resonant excitations, a very popular description for the overall non-linear response of metals relies on the so-called effective-mass approximation[26] that reduces the sum of *all* possible interband processes to the computation of lattice-modulated density fluctuations for the band at the Fermi level:

$$K_{\text{Raman}}^{\text{eV}} = \langle \rho_{\text{R}}^{\text{i,s}} \rho_{\text{R}}^{\text{i,s}} \rangle, \tag{10}$$

where $\rho_{\text{R}}^{\text{i,s}} = \sum_{\mathbf{k},\sigma} \gamma^{\text{i,s}}(\mathbf{k}) c_{\mathbf{k},\sigma}^{\dagger} c_{\mathbf{k},\sigma}$ is a lattice-modulated density-like electronic operator, with $c_{\mathbf{k},\sigma}^{(\dagger)}$ being the annihilation (creation) operator for an electron with momentum $\mathbf{k}$ and spin σ. The form factor $\gamma^{\text{i,s}}(\mathbf{k})$ is proportional to the momentum derivatives of the band dispersion $\varepsilon(\mathbf{k})$ along the crystallographic axes, in a combination dictated by the polarization of the incident (i) and scattered (s) light. Within this scheme, which has been widely used in the past to interpret Raman experiments in cuprates,[26] the enhancement of the non-linear response at 2Δ in the SC state is simply a consequence of the fact that the density-like response evaluated at the BCS level probes the quasiparticle continuum at $\mathbf{q} = 0$, that is pushed at 2Δ below T_c. On the other hand, the Higgs response appears as a vertex correction of such a density-like response in the amplitude channel, and it turns out to be quantitatively negligible,[8,21] since amplitude fluctuations are weakly coupled to density fluctuations in the particle–hole symmetric BCS case.[8,26] In short, whenever the non-linear kernel $K_{\text{SC}}(\omega)$ reduces to a density-like response, as in the case of eqn (10), the Higgs contribution is quantitatively subdominant with respect to the BCS one.

In the case of experiments with a THz driving pump field, a similar approach has been followed so far. Since in this case the energy of the field is comparable to intraband transitions, the non-linear response has been derived by coupling the gauge field directly to the electronic band at the Fermi level, *via* a Peierls-like substitution in the kinetic term of the Hamiltonian.[8,13,14,18,19,21,22,24,25] The expansion of the Hamiltonian then leads again to a structure similar to eqn (9), provided that the diamagnetic term is replaced directly with a density-modulated Raman operator $\rho_{\text{R},\alpha\beta} = \sum_{\mathbf{k},\sigma} \Gamma_{\alpha\beta}(\mathbf{k}) c_{\mathbf{k},\sigma}^{\dagger} c_{\mathbf{k},\sigma}$, where the modulation prefactor $\Gamma_{\alpha\beta}(\mathbf{k}) \sim \partial^2\varepsilon_{\mathbf{k}}/\partial k_\alpha k_\beta$ depends on the direction of the applied field with respect to the crystallographic axes. In the clean case, only diamagnetic-like processes are relevant,[8,21] so the non-linear kernel $K_{\text{Raman}}^{\text{THz,clean}}$ has the same behavior of $K_{\text{Raman}}^{\text{ev}}$ and it is dominated by the BCS response:

$$K_{\text{Raman}}^{\text{THz,clean}} = \langle \rho_{\text{R}} \rho_{\text{R}} \rangle. \tag{11}$$

However, as shown by several authors in the last few years,[22–25] once *disorder* is taken into account, the paramagnetic current is no longer conserved at the BCS level and, also, contributions mediated by the paramagnetic-like coupling term $\mathbf{j}\cdot\mathbf{A}$ in eqn (9) are finite, becoming quantitatively larger than diamagnetic ones already at an intermediate disorder level. As a consequence, for a realistic SC system, the THz response is controlled by intraband excitations mediated by both diamagnetic and paramagnetic correlation functions, *i.e.*

$$K^{\mathrm{THz,dirty}}_{\mathrm{Raman}} \sim \langle \rho_{\mathrm{R}}\rho_{\mathrm{R}}\rangle + \langle j_{\mathrm{P}}j_{\mathrm{P}}j_{\mathrm{P}}j_{\mathrm{P}}\rangle, \tag{12}$$

where the precise dependence on frequencies and spatial indexes has been omitted for simplicity. As far as the Higgs contribution is concerned, it turns out that paramagnetic-like processes can mediate a sizeable coupling of the e.m. field to Higgs fluctuations, such that at strong disorder, the Higgs contribution can even dominate over the BCS one. For thin films of conventional NbN, it seems now plausible to ascribe the 2Δ resonance of the non-linear kernel to an excitation of the Higgs mode.[22–25] However, cuprates are a much cleaner system, and preliminary studies in ref. 25 suggest that the BCS response is still dominant for this level of disorder, and that collective SC phase-density fluctuations can give a bigger contribution to the non-linear response, as compared to the Higgs one. In the next section, we will see how detailed quantitative calculations with a realistic band structure support a predominant role of the BCS response.

4 Non-linear response in cuprates

So far, the THz-induced non-linear response in cuprates has been investigated *via* THG experiments in transmission,[29,31] focusing on several classes of materials, and by means of pump–probe protocols aimed at measuring the THz Kerr effect in Bi-based compounds as a function of doping.[28,30] In an analogy with previous work on conventional superconductors, the experiments are performed by varying the temperature at a fixed pump frequency. While all experiments clearly show a strong enhancement of the non-linear response below T_c, that is then naturally ascribed to the SC phase transition, the present data do not allow one to clearly identify a resonance of $K_{\mathrm{SC}}(\omega)$ at twice the gap maximum. This is in part understood by noting that in cuprates, the pump pulse has a central frequency much smaller than twice the gap. More specifically, a multi-cycle pulse with central frequency $\Omega = 0.7$ THz has been used for THG measurements,[29,31] and a single-cycle THz pump field with central frequency $\Omega = 0.6$ THz and a probe in the VIS has been used for the THz Kerr effect.[28,30] In both cases, the pump frequency ($\simeq 3$ meV) is much smaller than the gap value at $T = 0$, which lies at around 10–20 meV in cuprates. As a consequence, the resonance condition $\Omega = \Delta(T)$ only occurs very near to T_c, making it difficult in general to observe the resonant enhancement of the non-linear response. In addition, *d*-wave symmetry of the order parameter can in part smear out the divergence of the response at 2Δ, with a maximum Δ gap, in analogy with what is observed in Raman.[26] On the other hand, the polarization results are rather robust, and the analysis of their behavior deep in the SC phase can help in disentangling the nature of the relevant modes involved in the non-linear response. Concerning the THG experiments, one usually measures the third-harmonic emission in the field direction by changing the angle θ that the pump field forms with the in-plane x crystallographic axis, see Fig. 6, where the xy plane represents the CuO_2 plane of cuprates. Since $I^{\mathrm{THG}} \propto |J^{\mathrm{NL}}(3\Omega)|^2$, see eqn (5), the relevant quantity to be computed is the non-linear current $j^{\mathrm{NL}}_{\parallel}$ in the field direction. The experimental results of ref. 29 and 31 are pretty much isotropic, at least within the error bars, see *e.g.* the data in Fig. 6b,

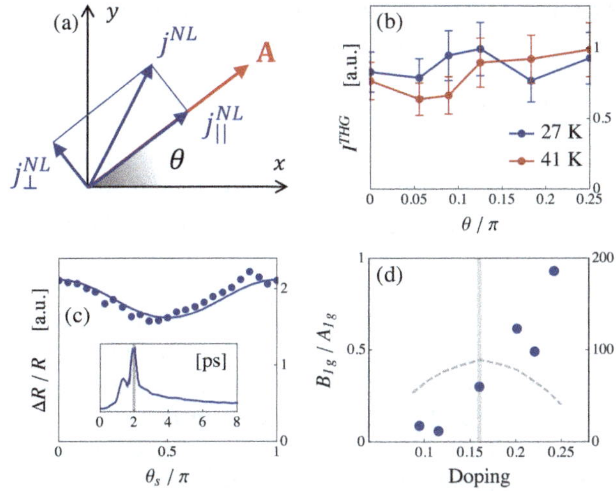

Fig. 6 (a) Definition of the polarization dependence for THG experiments in cuprates, with the square lattice formed by Cu atoms in the CuO_2 planes. For a generic angle θ between the applied field **A** and the *x* direction, the measured THG response is in the field direction, so it depends on the component $j_{\parallel}^{NL}$ of the non-linear current in the θ direction. (b) THG results for an optimally doped $La_{1.84}Sr_{0.16}CuO_4$ ($T_c = 45$ K) sample at two temperatures below T_c, taken from ref. 29. (c) Polarization dependence of the THz Kerr effect in an optimally-doped Bi-2212 compound ($T_c = 90$ K) from ref. 28. The inset shows the measured instantaneous response at a function of t_{pp}, while the main panel shows the recorded intensity at $t_{pp} = 2$ ps (marked by a vertical grey bar in the inset) as a function of θ_S for $\theta_P = 0°$. According to eqn (13) a K_{B1g} component appears as a minimum at $\theta_s = \pi/2$. (d) Doping dependence of the ratio K_{B1g}/K_{A1g} in Bi-2212 as a function of doping, taken from ref. 28. The dashed gray line denotes the corresponding T_c values, referring to the right vertical axis. The gray vertical line indicates the optimal doping.

taken from ref. 29. For THz pump–optical probe experiments, one can vary both the pump (P) and probe (S) angles $\theta_{P,S}$ with respect to the crystallographic axes, and study the angular-dependence of the time-resolved modulation at a fixed t_{pp}. As we explained above, at low T one is always in the condition $\Omega \ll \omega_{res}$, so the time-dependent response is expected to scale in the t_{pp} domain with the square of the pump field, see Fig. 4d, as indeed observed experimentally, see the inset in Fig. 6c. However, in contrast to the case of THG experiments, THz Kerr effect measurements reveal a modulation of the signal (see Fig. 6c) that increases as the doping increases, see Fig. 6d. In ref. 28, such a modulation has been analyzed in terms of a decomposition of $j_{\parallel}^{NL}$ in the symmetry projections for the D_{4h} point group relevant for cuprates:

$$j_{\parallel}^{NL}(\theta_S,\theta_P) = K_{A1g} + K_{B1g}\cos 2\theta_S \cos 2\theta_P + K_{B2g}\sin 2\theta_S \sin 2\theta_P. \qquad (13)$$

As has been discussed in ref. 25, such a decomposition is meaningful when the non-linear kernel $K_{\alpha\beta\gamma\delta}$ only admits diamagnetic Kubo-like diagrams, which is not necessarily the case in the presence of disorder. Nonetheless, by preserving the notation of ref. 28 based on the decomposition in eqn (13), the analysis of the experimental data gives $K_{B2g} = 0$ and a ratio K_{B1g}/K_{A1g} increasing with doping from almost 0 in underdoped samples to almost 1 in overdoped ones, see Fig. 6d.

It is worth noting that by using the same decomposition for THG measurements in transmission as well, one would obtain that

$$j_{\parallel}^{\mathrm{NL}}(\theta) = K_{\mathrm{A1g}} + K_{\mathrm{B1g}}\cos^2(2\theta) + K_{\mathrm{B2g}}\sin^2(2\theta). \tag{14}$$

As a consequence, the isotropy of the THG signal in cuprates, see Fig. 6b, points to the existence only of the K_{A1g} component, as seen at low doping in pump–probe experiments, see Fig. 6d, but this makes the lack of a strong B_{1g} component around optimal levels and overdoping puzzling, as we will comment on further below.

All the experimental findings summarized so far help in disentangling the contribution to the non-linear response coming from the BCS response or from the other collective modes. Indeed, these results can be understood from the general scheme outlined in Section 2 by assuming that a sizeable non-linear kernel $K_{\mathrm{SC}}(\omega)$ emerges below T_c, with a peculiar polarization dependence. To clarify the nature of the non-linear response in cuprates, we then study an extended version of the attractive Hubbard model that we used in ref. 25, accounting for the effects of disorder and doping on a realistic band structure. More specifically, we start from the electronic Hamiltonian

$$H = -\sum_{ij\sigma} t_{\mathrm{ij}} c_{\mathrm{i}\sigma}^{\dagger} c_{\mathrm{j}\sigma} - |U| \sum_{i} n_{\mathrm{i}\uparrow} n_{\mathrm{i}\downarrow} + \sum_{i\sigma} V_{\mathrm{i}} n_{\mathrm{i}\sigma}, \tag{15}$$

where the local potential V_{i} is taken from a flat distribution $-V \leq V_{\mathrm{i}} \leq +V$. We set $t_{\mathrm{ij}} = t$ for the nearest neighbors hopping and a different $t_{\mathrm{ij}} = -t'$ for the next-nearest neighbors hopping to better reproduce the Fermi surface of cuprates. For numerical reasons, calculations are performed on a 32×32 lattice, and we set $U/t = 1.4$. This is the lowest value for which finite size effects can be neglected. For a nearest-neighbor hopping $t \sim 200$–300 meV and a next-nearest-neighbors hopping of $t'/t \sim -0.2$, one obtains gap values $\Delta \sim 25$–55 meV compatible with ARPES measurements in overdoped Bi-2212 close to the nodal region, where pseudogap effects should be small.[28] We consider the doping levels $p = 0.1$ for the underdoped regime and $p = 0.2$ for the overdoped. Notice that the polarization dependence, which is the main focus here, depends specifically on the band structure and disorder level, so even though the model given in eqn (15) does not include the *d*-wave symmetry of the order parameter, it can nonetheless provide a good quantitative account of the realistic situation, especially at pump frequencies far from the resonance.

In order to establish a close connection with the experiments, we compute the transport scattering rate τ for each doping and disorder level, following the procedure outlined in ref. 25. Experimental data in ref. 28 indicate a disorder level $\gamma/2\Delta \sim 0.85$, which is intermediate between our $V/t = 0.1$, corresponding to $\gamma/2\Delta \sim 0.03$ (0.04) for $p = 0.1$ ($p = 0.2$), and $V/t = 0.5$, corresponding to $\gamma/2\Delta \sim 1.12$ (1.79) for $p = 0.1$ ($p = 0.2$). By following the time evolution of the mean-field density matrix stemming from eqn (15), we calculate the third-harmonic current by selectively including the charge, phase and amplitude fluctuations, distinguishing between the paramagnetic and diamagnetic processes. We then focus on the zero-frequency value of the non-linear current as a good approximation in the out-of-resonance condition. Further details on the numerical procedure have been reported in ref. 25.

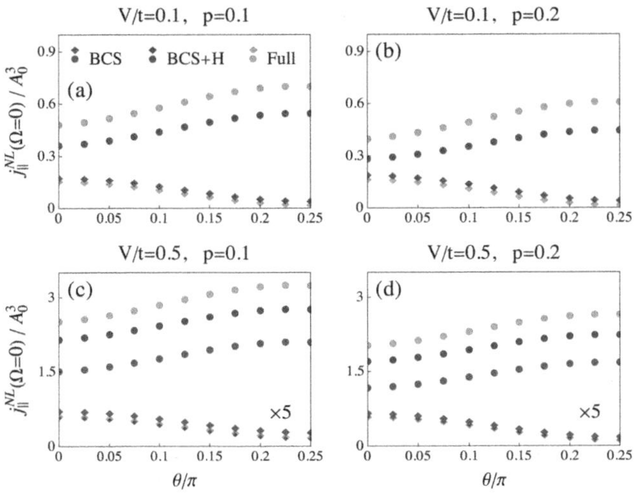

Fig. 7 Numerical results for the non-linear current $j_{\parallel}^{\mathrm{NL}}(\theta)$ at $\Omega = 0$ as a function of the angle θ for $V/t = 0.1$ (upper row) and $V/t = 0.5$ (lower row) at two doping levels. Diamonds refer to the diamagnetic contribution and circles to the paramagnetic one. "BCS" labels the pure BCS response, probing the quasiparticle continuum, "BCS + Higgs" the response including vertex correction in the SC amplitude (Higgs) channel, and "Full" the results obtained by including vertex corrections in all channels (SC amplitude, SC phase and charge). For more details on the notation, see ref. 25.

Fig. 7 shows the $j_{\parallel}^{\mathrm{NL}}(\theta)$ component of the nonlinear current as a function of the angle θ, where we keep separate the BCS contribution, the full response including all SC fluctuations, and the contribution of the BCS + Higgs fluctuations only. Diamonds denote the diamagnetic contribution, see eqn (11), which is also present in the perfectly clean case ($V = 0$), while circles denote the paramagnetic contribution, which only arises in the presence of disorder, see eqn (12). At the lowest disorder level $V/t = 0.1$, the Higgs contribution (blue diamonds/circles) is quantitatively negligible, and for $V/t = 0.5$ it only adds a correction at most in the order of 30% of the BCS one, while phase modes (green diamonds/circles) give a sizeable contribution already at $V/t = 0.1$. As a consequence, one can safely conclude that THG measurements in cuprates should be ascribed to the BCS response, further enhanced by the contribution of phase fluctuations, while the Higgs response is largely subleading. Concerning the polarization dependence, one can see that the paramagnetic part becomes rapidly predominant at $\theta = 0$, but the overall modulation of the two contributions has a similar strength, so that the overall response is pretty much isotropic at $V/t = 0.1$, see Fig. 8, and only slightly modulated at $V/t = 0.5$, in excellent agreement with the THG results, see Fig. 6b.

Finally, we would like to comment on the polarization dependence expected in THz pump–optical probe measurements. The relevant processes for this kind of experiments are still depicted by the diagrams in Fig. 5, but the frequencies running in the fermionic loops are different with respect to the case of simple THG experiments. Indeed, for an optical probe, two external lines carry a large frequency in the visible range, leading to a large frequency running in the internal loop. As observed in ref. 22, in this situation disorder effects are expected to be

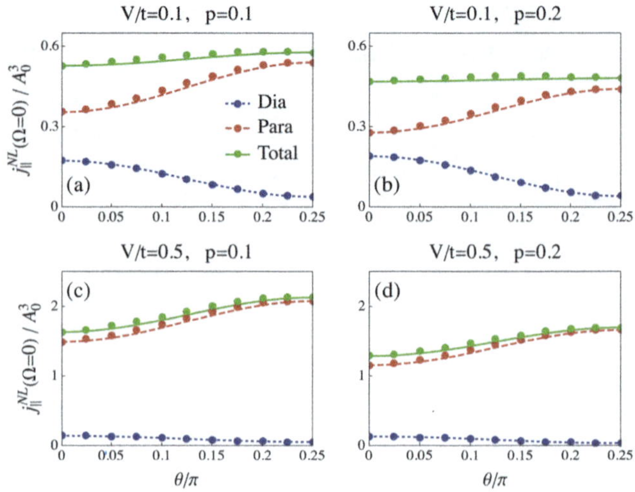

Fig. 8 Polarization dependence of the BCS contribution to the non-linear current at $\Omega = 0$, as given by diamagnetic (left) or paramagnetic (right) diagrams, for two different levels of doping and disorder. Solid lines represent a fit of the diamagnetic term with K_{A1g} + K_{B1g} contributions, and of the paramagnetic one with K_{A1g} + K_{B2g} contributions. The parameters of the fit are reported in Tables 2–3.

less efficient in triggering a finite paramagnetic response, in contrast to what happens when only a frequency in the order of the THz pump field is involved. As a consequence, one can speculate that for THz Kerr effect measurements, the relevant non-linear kernel can be a combination of the diamagnetic response plus only part of the paramagnetic one. If this is the case, it can be worth analyzing separately the angular dependence of the two contributions, as done in Fig. 8 for the BCS part only. Notice that while for pump–probe experiments one can identify separately the K_{B1g} and K_{B2g} contributions by tuning the pump θ_P and the probe θ_S angles independently, see eqn (13), in the case of THG experiments, $j_{\parallel}^{NL}$ depends only on the pump angle θ, see eqn (14), leaving some ambiguity in the identification of the various angle-dependent terms. In the specific case of our calculations, we fit the diamagnetic term with the sum of the K_{A1g} and a K_{B1g} contributions, using the fact that for these Kubo-like diagrams, the K_{B2g} term in the clean limit can only scale with the sub-leading t' next-nearest-neighbors hopping term. On the other hand, for paramagnetic-like diagrams, one cannot establish *a priori* a prevalence of one asymmetric channel over the other, so one can only fix in principle the relative weight of the combination of two terms, for example by rewriting eqn (14) as $j_{\parallel}^{NL}(\theta) = K_{A1g} + K_{B2g} + (K_{B1g} - K_{B2g})\cos^2(2\theta)$, or the analogous expression where only a $\sin^2(2\theta)$ is left. Once such an ambiguity was clarified, and in order to simplify the analysis, we decided to conventionally fit the paramagnetic term as the sum of the K_{A1g} and K_{B2g} terms only by simply observing that the signal increases as θ increases. The relative weights of the various channels are reported in Tables 1–3. As one can see, the diamagnetic term has a sizeable B_{1g} component with an increasing ratio K_{B1g}/K_{A1g} as the doping increases, especially for larger disorder. Even though these ratios are larger than the experimental findings of ref. 28, one could expect that a partial compensation

Table 1 Results of the fitting procedure for the diamagnetic BCS-only contribution to $j_{\parallel}^{NL}(\theta)$ through eqn (14), obtained by conventionally setting $K_{B2g} = 0$

	$V/t = 0.1$			$V/t = 0.5$		
	K_{A1g}	K_{B1g}	K_{B1g}/K_{A1g}	K_{A1g}	K_{B1g}	K_{B1g}/K_{A1g}
$p = 0.1$	0.038	0.135	3.545	0.054	0.086	1.576
$p = 0.2$	0.040	0.150	3.732	0.035	0.096	2.756

Table 2 Results of the fitting procedure for the paramagnetic BCS-only contribution to $j_{\parallel}^{NL}(\theta)$ through eqn (14), obtained by conventionally setting $K_{B1g} = 0$

	$V/t = 0.1$			$V/t = 0.5$		
	K_{A1g}	K_{B2g}	K_{B2g}/K_{A1g}	K_{A1g}	K_{B2g}	K_{B2g}/K_{A1g}
$p = 0.1$	0.354	0.185	0.521	1.487	0.587	0.395
$p = 0.2$	0.278	0.162	0.581	1.152	0.503	0.437

Table 3 Results of the fitting procedure for the sum of diamagnetic and paramagnetic BCS-only contributions to $j_{\parallel}^{3}(\theta)$. The A_{1g} component is taken as the sum of the separate results from the diamagnetic- and paramagnetic-only contributions

	$V/t = 0.1$		$V/t = 0.5$	
	K_{B1g}/K_{A1g}	K_{B2g}/K_{A1g}	K_{B1g}/K_{A1g}	K_{B2g}/K_{A1g}
$p = 0.1$	0.344	0.470	0.055	0.381
$p = 0.2$	0.470	0.508	0.081	0.424

from the paramagnetic channel can explain the difference with THG measurements, and account for the observed doping dependence of the K_{B1g}/K_{A1g} ratio. Finally, it is worth mentioning that a third possible mechanism has been recently proposed in ref. 32, based on two-plasmon excitation processes that are beyond the approximation studied here. By accounting for this additional channel within an *XY* model description of plasma modes, one finds an additional contribution to the non-linear kernel having $K_{B1g}/K_{A1g} = 0.5$. So far, the quantitative relevance of this effect with respect to the BCS response has not been estimated, making a direct comparison with experiments difficult. On the other hand, since the energy scale setting the strength of two-plasmon excitations is the superfluid stiffness, one would expect a larger contribution in overdoped samples, where it becomes quantitatively larger. A closer analysis of this problem, along with a direct estimate of the effective relevance of paramagnetic processes for the THz Kerr effect, will certainly help elucidating the nature of the THz non-linear response in cuprate superconductors, and will thus deserve future work.

Conflicts of interest

There are no conflicts to declare.

Acknowledgements

This work has been supported by Sapienza University *via* Ateneo 2019 RM11916B56802AFE and Ateneo 2020 RM120172A8CC7CC7, and by the Italian MIUR project PRIN 2017 No. 2017Z8TS5B. G. S. acknowledges financial support from the Deutsche Forschungsgemeinschaft under SE 806/19-1.

Notes and references

1 T. Kampfrath, K. Tanaka and K. A. Nelson, *Nat. Photonics*, 2013, **7**, 680–690.
2 X. C. Zhang, A. Shkurinov and Y. Zhang, *Nat. Photonics*, 2017, **11**, 16–18.
3 Y. Laplace and A. Cavalleri, *Adv. Phys.: X*, 2016, **1**, 387–411.
4 S. Maehrlein, A. Paarmann, M. Wolf and T. Kampfrath, *Phys. Rev. Lett.*, 2017, **119**, 127402.
5 C. L. Johnson, B. E. Knighton and J. A. Johnson, *Phys. Rev. Lett.*, 2019, **122**, 073901.
6 C. Giannetti, M. Capone, D. Fausti, M. Fabrizio, F. Parmigiani and D. Mihailovic, *Adv. Phys.*, 2016, **65**, 58–238.
7 R. Merlin, *Solid State Commun.*, 1997, **102**, 207–220.
8 T. Cea, C. Castellani and L. Benfatto, *Phys. Rev. B*, 2016, **93**, 180507.
9 M. Udina, T. Cea and L. Benfatto, *Phys. Rev. B*, 2019, **100**, 165131.
10 R. Shimano and N. Tsuji, *Annu. Rev. Condens. Matter Phys.*, 2020, **11**, 103–124.
11 R. Matsunaga and R. Shimano, *Phys. Rev. Lett.*, 2012, **109**, 187002.
12 R. Matsunaga, Y. I. Hamada, K. Makise, Y. Uzawa, H. Terai, Z. Wang and R. Shimano, *Phys. Rev. Lett.*, 2013, **111**, 057002.
13 R. Matsunaga, N. Tsuji, H. Fujita, A. Sugioka, K. Makise, Y. Uzawa, H. Terai, Z. Wang, H. Aoki and R. Shimano, *Science*, 2014, **345**, 1145–1149.
14 R. Matsunaga, N. Tsuji, K. Makise, H. Terai, H. Aoki and R. Shimano, *Phys. Rev. B*, 2017, **96**, 020505.
15 F. Giorgianni, T. Cea, C. Vicario, C. P. Hauri, W. K. Withanage, X. Xi and L. Benfatto, *Nat. Phys.*, 2019, **15**, 341–346.
16 X. Yang, C. Vaswani, C. Sundahl, M. Mootz, L. Luo, J. H. Kang, I. E. Perakis, C. B. Eom and J. Wang, *Nat. Photonics*, 2019, **13**, 707–713.
17 S. Kovalev, T. Dong, L.-Y. Shi, C. Reinhoffer, T.-Q. Xu, H.-Z. Wang, Y. Wang, Z.-Z. Gan, S. Germanskiy, J.-C. Deinert, I. Ilyakov, P. H. M. van Loosdrecht, D. Wu, N.-L. Wang, J. Demsar and Z. Wang, *Phys. Rev. B*, 2021, **104**, L140505.
18 N. Tsuji and H. Aoki, *Phys. Rev. B: Condens. Matter Mater. Phys.*, 2015, **92**, 064508.
19 N. Tsuji, Y. Murakami and H. Aoki, *Phys. Rev. B*, 2016, **94**, 224519.
20 T. Cea and L. Benfatto, *Phys. Rev. B*, 2016, **94**, 064512.
21 T. Cea, P. Barone, C. Castellani and L. Benfatto, *Phys. Rev. B*, 2018, **97**, 094516.
22 M. Silaev, *Phys. Rev. B*, 2019, **99**, 224511.
23 Y. Murotani and R. Shimano, *Phys. Rev. B*, 2019, **99**, 224510.
24 N. Tsuji and Y. Nomura, *Phys. Rev. Res.*, 2020, **2**, 043029.
25 G. Seibold, M. Udina, C. Castellani and L. Benfatto, *Phys. Rev. B*, 2021, **103**, 014512.
26 T. P. Devereaux and R. Hackl, *Rev. Mod. Phys.*, 2007, **79**, 175–233.
27 B. Mansart, J. Lorenzana, A. Mann, A. Odeh, M. Scarongella, M. Chergui and F. Carbone, *Proc. Natl. Acad. Sci. U. S. A.*, 2013, **110**, 4539–4544.

28 K. Katsumi, N. Tsuji, Y. I. Hamada, R. Matsunaga, J. Schneeloch, R. D. Zhong, G. D. Gu, H. Aoki, Y. Gallais and R. Shimano, *Phys. Rev. Lett.*, 2018, **120**, 117001.

29 H. Chu, M.-J. Kim, K. Katsumi, S. Kovalev, R. D. Dawson, L. Schwarz, N. Yoshikawa, G. Kim, D. Putzky, Z. Z. Li, H. Raffy, S. Germanskiy, J.-C. Deinert, N. Awari, I. Ilyakov, B. Green, M. Chen, M. Bawatna, G. Cristiani, G. Logvenov, Y. Gallais, A. V. Boris, B. Keimer, A. P. Schnyder, D. Manske, M. Gensch, Z. Wang, R. Shimano and S. Kaiser, *Nat. Commun.*, 2020, **11**, 1793.

30 K. Katsumi, Z. Z. Li, H. Raffy, Y. Gallais and R. Shimano, *Phys. Rev. B*, 2020, **102**, 054510.

31 H. Chu, *et al.*, *Fano interference of the Higgs mode in cuprate high-Tc superconductors*, arXiv:2109.09971, 2021.

32 F. Gabriele, M. Udina and L. Benfatto, *Nat. Commun.*, 2021, **12**, 752.

Faraday Discussions

ROYAL SOCIETY OF CHEMISTRY

PAPER

Signatures of collective modes in fifth harmonic generation of BCS superconductor

Pascal Derendorf, Marvin A. Müller and Ilya M. Eremin *

Received 22nd March 2022, Accepted 11th April 2022
DOI: 10.1039/d2fd00067a

Recent advances in the field of THz spectroscopy allow for controlled experiments to measure signatures of collective excitations in the conventional s-wave superconductor in the fifth harmonic generation current (FHG). Here, we analyze this process theoretically within the Anderson pseudospin formalism and use a periodic multicycle pulse setup, where the driving electromagnetic field points in the direction of a lattice vector. We investigate the interplay of the Higgs mode contribution to the fifth harmonic generation current and compare it to other contributing mechanisms, such as charge density fluctuations (CDF). Similar to the third harmonic generating current we show that the signal in the FHG is also dominated by the CDF. Most importantly, we predict a double peak signature in the frequency dependence of the intensity amplitude of the FHG current with one peak located at $\Omega = \Delta_0$ ($4\Omega = 4\Delta_0$) and another one at $\Omega = \Delta_0/2$ ($4\Omega = 2\Delta_0$). The resonant enhancement in the latter case is indicative of the higher order coupling to the Higgs mode or CDF, while the former is reminiscent of the THG describing the coupling of CDF and the Higgs mode with 2 single photons.

1 Introduction

An experimental observation of collective modes of a superconducting ground state such as amplitude (Higgs) mode is experimentally a challenging task as this mode does not couple linearly to light. This can be easily seen by considering the effective action of a conventional superconductor in the presence of an electromagnetic vector potential, **A** (ref. 1),

$$S_{\mathrm{sc}}(\Delta_0 + H) = \int \mathrm{d}^3 r \mathrm{d}t \left[-\alpha|\Delta_0 + H|^2 + \beta|\Delta_0 + H|^4 + \gamma\left|\left(\frac{\partial}{\partial r} + 2ieA\right)(\Delta_0 + H)\right|^2 \right] \tag{1}$$

where H is the amplitude fluctuation around the ground state Δ_0 and we omit for simplicity the phase fluctuations, as they can be gauged away in a superconductor

Institut für Theoretische Physik III, Ruhr-Universität Bochum, 44801 Bochum, Germany. E-mail: Ilya.Eremin@rub.de

and become plasmons once the long-range Coulomb interaction is taken into account.[2] Note, α, β and γ are parameters, which could be determined from a microscopic theory. The last term in eqn (1) leads to a quadratic coupling of the amplitude fluctuations to the electromagnetic vector potential. The linear coupling, necessary to make the Higgs mode visible in a linear response, can only be achieved in a system where superconductivity coexists with charge density wave (CDW).[3–7]

In addition to the nonlinear coupling to light, the energy of the Higgs mode frequency $\Omega_{\rm H}$ is typically a few meV in conventional superconductors, implying that real-time oscillations occur on a picosecond timescale and require ultrafast THz spectroscopy. Recent advances in its technical development enabled the construction of THz-pump THz-probe spectroscopy in the conventional s-wave superconductor, $Nb_{1-x}Ti_xN$.[8,9] The idea of this experimental setup is that a single-cycle (pump) pulse with an energy on the order of Δ_0 excites the collective Higgs mode in the superconducting condensate non-adiabatically. This leads to an amplitude oscillation of the order parameter Δ_0 with the Higgs mode frequency $\Omega_{\rm H} = 2\Delta_0$. These oscillations can be detected with a second, weaker pulse that comes with a delay δt and probes these dynamics. The transmitted electric field can be measured using electro-optic probing.[9] The probe pulse is applied with an orthogonal polarization to the pump pulse. Thus, the pump pulse can be filtered out from the probe pulse using a wire grid polarizer. A detailed description of the experimental setup can be found in ref. 10.

While a non-adiabatic quench with a single-cycle pulse on the picosecond scale is an experimental challenge, one can consider to drive the system periodically with a multi-cycle pulse of frequency Ω. Since the Higgs mode couples to light in the second order, it is driven at twice the driving frequency 2Ω. The driven oscillation can then be measured as an induced current. Varying the action S with respect to the electromagnetic vector potential $\mathbf{A}$ yields the current $\mathbf{j}$. The latter is the leading order in the amplitude fluctuation H, as given by,[1]

$$\mathbf{j}(t) = -\frac{\delta S}{\delta \mathbf{A}} = -\left(8e^2\gamma\Delta_0\right)H(t)\mathbf{A}(t). \tag{2}$$

Since the electromagnetic vector potential oscillates with the driving frequency $\mathbf{A}(t) \sim e^{i\Omega t}$ and the Higgs mode oscillates with twice the driving frequency $H(t) \sim e^{2i\Omega t}$, one obtains a total current, which oscillates with three times the driving frequency $\mathbf{j}(t) \sim e^{3i\Omega t}$.[11,12] This generation of a third-harmonic is a process, which only occurs inside the superconducting state at $T < T_c$ and should be resonantly enhanced once 2Ω activates the Higgs mode $2\Omega = \Omega_{\rm H} = 2\Delta_0$.

In ref. 9 the first successful measurement of the third-harmonic generation in the conventional s-wave superconductor $Nb_{1-x}Ti_xN$ was reported. However, soon after it was realized that the activation of the Higgs mode is not the only process measured in the current produced by the third-harmonic generation. Since the frequency of the Higgs mode is given at $\Omega_{\rm H} = 2\Delta_0$, it coincides with the onset of the particle-hole continuum and the energy needed to break the Cooper pairs. In particular, ref. 13 has shown that in the clean BCS superconductor the charge density fluctuation contribution to the third harmonic generation current is three orders of magnitude larger than that from the Higgs mode. Subsequent studies analyzed how impurities affect this ratio.[14–18] Obviously, this shows that research on the detection of the collective modes in conventional superconductors due to a driving field is still ongoing.

Very recently, studying NbN thin films with a strong terahertz (THz) wave, the authors of ref. 19 found a higher order nonlinear optics effect, *i.e.* fifth harmonic generation, which they assign to the higher order coupling between the Higgs mode and the electromagnetic field. Here we investigate the interplay of the Higgs mode contribution to the fifth harmonic generation current and compare it to other contributing mechanisms, such as charge density fluctuations (CDF) of the BCS superconductor. Similar to the third harmonic generating current, we show that the signal in the FHG is also dominated by the CDF in the clean case. Most importantly, we predict a double peak signature in the frequency dependence of the intensity amplitude of the FHG current with one peak located at $4\Omega = 4\Delta_0$ and another one at $4\Omega = 2\Delta_0$. The resonant enhancement in the latter case is indicative of the higher order coupling to the Higgs mode or CDF, while the former is reminiscent of the THG describing the coupling of CDF and the Higgs mode with 2 single photons.

2 Theory

In order to derive the generated fifth-harmonic current we start with a BCS Hamiltonian, written in terms of Anderson pseudospins,[20] defined as $\mathbf{s_k} = \frac{1}{2}\Psi^\dagger_{\mathbf{k}}\sigma\Psi_{\mathbf{k}}$, with $\boldsymbol{\sigma} = (\sigma_x,\sigma_y,\sigma_z)^{\mathrm{T}}$. In particular, the components of the pseudospins are $\sigma^x_{\mathbf{k}} = \frac{1}{2}(c_{-\mathbf{k}\downarrow}c_{\mathbf{k}\uparrow} + c^\dagger_{\mathbf{k}\uparrow}c^\dagger_{-\mathbf{k}\downarrow})$, $\sigma^y_{\mathbf{k}} = \frac{i}{2}(c_{-\mathbf{k}\downarrow}c_{\mathbf{k}\uparrow} - c^\dagger_{\mathbf{k}\uparrow}c^\dagger_{-\mathbf{k}\downarrow})$ and $\sigma^z_{\mathbf{k}} = \frac{1}{2}(c^\dagger_{\mathbf{k}\uparrow}c_{\mathbf{k}\uparrow} - c_{-\mathbf{k}\downarrow}c^\dagger_{-\mathbf{k}\downarrow})$. Then the BCS Hamiltonian acquires the form

$$H = \sum_{\mathbf{k}} B_{\mathbf{k}} \cdot \vec{\sigma}_{\mathbf{k}}, \tag{3}$$

with $\mathbf{B_k} = (2\Delta',2\Delta'',\xi_{\mathbf{k}-e\mathbf{A}(t)} + \xi_{\mathbf{k}+e\mathbf{A}(t)})$ and we set $\hbar = c = 1$. These pseudospins inherit the Lie algebra from the Pauli matrices and the equations of motion follow immediately from the spin-algebra

$$\frac{\mathrm{d}}{\mathrm{d}t}\vec{\sigma}_{\mathbf{k}} = i[H, \vec{\sigma}_{\mathbf{k}}] = \mathbf{B_k} \times \vec{\sigma}_{\mathbf{k}}. \tag{4}$$

The superconducting gap equation acquires the form

$$\Delta = V\sum_{\mathbf{k}}(\langle\sigma^x_{\mathbf{k}}\rangle - i\langle\sigma^y_{\mathbf{k}}\rangle). \tag{5}$$

where V is the attractive interaction in the isotropic s-wave channel. The driving electromagnetic field can be described with the following vector potential

$$\mathbf{A}(t) = A_0(t)\begin{pmatrix}\cos\theta \\ \sin\theta\end{pmatrix} = A_0 \sin(\Omega t)\begin{pmatrix}\cos\theta \\ \sin\theta\end{pmatrix} \tag{6}$$

where θ is the polarization angle in the x–y plane.

The time evolution of the gap Δ can be derived *via* the time evolution of the Anderson-pseudospins. The pseudospins and the superconducting gap can be conveniently separated into

$$\begin{aligned} \sigma_k^a(t) &= \langle \sigma_k^a \rangle(0) + \delta\sigma_k^a(t) \\ \Delta(t) &= \Delta_0 + \delta\Delta(t) \end{aligned} \tag{7}$$

where $\langle\sigma_k^a\rangle(0)$ and Δ_0 are the values at equilibrium, $\delta\sigma_k^a(t)$ and $\delta\Delta(t)$ are the deviations from the equilibrium state and $a = \{x, y, z\}$. For $t = 0$ the pseudospins should be in an equilibrium state meaning that $\delta\sigma_k^a(t = 0) = 0$ and $\langle\sigma_k^a\rangle(0)$ are the averages without an applied em-field. At finite temperatures we have $\langle\sigma_k^x\rangle(0) = \frac{\Delta_0}{2E_k}\tanh\left(\frac{E_k}{2k_BT}\right)$, $\langle\sigma_k^y\rangle(0) = 0$, and $\langle\sigma_k^z\rangle(0) = \frac{-\xi_k}{2E_k}\tanh\left(\frac{E_k}{2k_BT}\right)$ where $E_k = \sqrt{\xi_k{}^2 + \Delta_0{}^2}$ is the Bogoliubov-quasiparticle energy. In order to calculate the time evolution of the pseudospins up to the 4th-order in $A_0(t)$, necessary for the fifth harmonic generation, we expand the z-component of the pseudomagnetic field:

$$\xi_{\mathbf{k}-e\mathbf{A}(t)} + \xi_{\mathbf{k}+e\mathbf{A}(t)} \approx 2\xi_{\mathbf{k}} + \xi_{\mathbf{k}}^{A^2}(t) + \frac{1}{12}\xi_{\mathbf{k}}^{A^4}(t) \tag{8}$$

where the linear and the cubic terms in $A_0(t)$ vanish under the assumption of the inversion symmetry. The term $\xi_{\mathbf{k}}^{A^2}(t)$ is given by:

$$\xi_{\mathbf{k}}^{A^2}(t) = e^2 A_0{}^2(t) D_{\xi_{\mathbf{k}}}^{\parallel}(\theta) \tag{9}$$

where $D_{\xi_{\mathbf{k}}}^{\parallel}(\theta)$ denotes

$$D_{\xi_{\mathbf{k}}}^{\parallel}(\theta) = \cos^2\theta\partial_{k_xk_x}{}^2\xi_{\mathbf{k}} + \sin^2\theta\partial_{k_yk_y}{}^2\xi_{\mathbf{k}} + 2\cos\theta\sin\theta\partial_{k_xk_y}{}^2\xi_{\mathbf{k}} \tag{10}$$

For $\xi_{\mathbf{k}}^{A^4}(t)$ we get:

$$\xi_{\mathbf{k}}^{A^4}(t) = e^4 A_0{}^4(t) \Lambda_{\xi_{\mathbf{k}}}^{\parallel}(\theta) \tag{11}$$

with

$$\begin{aligned} \Lambda_{\xi_{\mathbf{k}}}^{\parallel}(\theta) &= \cos^4\theta\partial_{k_xk_xk_xk_x}{}^4\xi_{\mathbf{k}} + \sin^4\theta\partial_{k_yk_yk_yk_y}{}^4\xi_{\mathbf{k}} \\ &\quad + 4\cos^3\theta\sin\theta\partial_{k_xk_xk_xk_y}{}^4\xi_{\mathbf{k}} + 6\cos^2\theta\sin^2\theta\partial_{k_xk_xk_yk_y}{}^4\xi_{\mathbf{k}} \\ &\quad + 4\cos\theta\sin^3\theta\partial_{k_xk_yk_yk_y}{}^4\xi_{\mathbf{k}} \end{aligned} \tag{12}$$

Using these expressions the equations of motion become

$$\partial_t \begin{pmatrix} \delta\sigma_{\mathbf{k}}^x(t) \\ \delta\sigma_{\mathbf{k}}^y(t) \\ \delta\sigma_{\mathbf{k}}^z(t) \end{pmatrix} = \begin{pmatrix} -2(\Delta_0 + \delta\Delta'(t)) \\ 2\delta\Delta''(t) \\ 2\varepsilon_{\mathbf{k}} + \xi_{\mathbf{k}}^{A^2}(t) + \frac{1}{12}\xi_{\mathbf{k}}^{A^4}(t) \end{pmatrix} \times \begin{pmatrix} \langle\sigma_{\mathbf{k}}^x\rangle(0) + \delta\sigma_{\mathbf{k}}^x(t) \\ \langle\sigma_{\mathbf{k}}^y\rangle(0) + \delta\sigma_{\mathbf{k}}^y(t) \\ \langle\sigma_{\mathbf{k}}^z\rangle(0) + \delta\sigma_{\mathbf{k}}^z(t) \end{pmatrix}. \tag{13}$$

The obtained equations can be now linearized assuming $eA_0 \ll 1$:

$$\partial_t\delta\sigma_{\mathbf{k}}^x(t) = -2\xi_{\mathbf{k}}\delta\sigma_{\mathbf{k}}^y(t) - \frac{\xi_{\mathbf{k}}}{E_{\mathbf{k}}}\delta\Delta''(t) \tag{14}$$

$$\partial_t \sigma_{\mathbf{k}}^{y}(t) = 2\xi_{\mathbf{k}}\delta\sigma_{\mathbf{k}}^{x}(t) + 2\Delta_0\delta\sigma_{\mathbf{k}}^{z}(t) + \frac{1}{2E_{\mathbf{k}}}\left[\Delta_0\left(\xi_{\mathbf{k}}^{A^2}(t) + \frac{1}{12}\xi_{\mathbf{k}}^{A^4}(t)\right) - 2\xi_{\mathbf{k}}\delta\Delta'(t)\right] \tag{15}$$

$$\partial_t \sigma_{\mathbf{k}}^{z}(t) = -2\Delta_0\delta\sigma_{\mathbf{k}}^{y}(t) - \frac{\Delta_0}{E_{\mathbf{k}}}\delta\Delta''(t). \tag{16}$$

The obtained equations can be solved by a transformation in the Laplacian space $f(t)\rightarrow f(s) = \int_0^\infty \mathrm{e}^{-st}f(t)\mathrm{d}t$, which yields

$$\delta\sigma_{\mathbf{k}}^{x}(s) = \frac{\xi_{\mathbf{k}}\left(2\xi_{\mathbf{k}}\delta\Delta'(s) - \Delta_0\left(\xi_{\mathbf{k}}^{A^2}(s) + \frac{1}{12}\xi_{\mathbf{k}}^{A^4}(s)\right) - s\delta\Delta''(s)\right)}{E_{\mathbf{k}}\left(4E_{\mathbf{k}}^{\ 2} + s^2\right)} \tag{17}$$

$$\delta\sigma_{\mathbf{k}}^{y}(s) = \frac{-\left(2\xi_{\mathbf{k}}s\delta\Delta'(s) - s\Delta_0\left(\xi_{\mathbf{k}}^{A^2}(s) + \frac{1}{12}\xi_{\mathbf{k}}^{A^4}(s)\right) + 4E_{\mathbf{k}}^{\ 2}\delta\Delta''(s)\right)}{2E_{\mathbf{k}}\left(4E_{\mathbf{k}}^{\ 2} + s^2\right)} \tag{18}$$

$$\delta\sigma_{\mathbf{k}}^{z}(s) = \frac{\Delta_0\left(2\xi_{\mathbf{k}}\delta\Delta'(s) - \Delta_0\left(\xi_{\mathbf{k}}^{A^2}(s) + \frac{1}{12}\xi_{\mathbf{k}}^{A^4}(s)\right) - s\delta\Delta''(s)\right)}{E_{\mathbf{k}}\left(4E_{\mathbf{k}}^{\ 2} + s^2\right)} \tag{19}$$

Assuming the simplest tight-binding dispersion on a square lattice and a constant density of states at the Fermi level we further obtain

$$\delta\Delta'(s) = \frac{1}{2}\left(\alpha_1\Delta_0 e^2 A_0^{\ 2}(s) + \frac{1 - \frac{1}{2}\sin^2 2\theta}{12}\tilde{\alpha}_1\Delta_0 e^4 A_0^{\ 4}(s)\right) \times \left(1 - \frac{1}{\lambda\left(4\Delta_0^{\ 2} + s^2\right)F(s)}\right) \tag{20}$$

where the function, $A_0^{\ 2}(s) = A_0^{\ 2}\frac{2\Omega^2}{s(s^2 + 4\Omega^2)}$, and $A_0^{\ 4}(s) = A_0^{\ 4}\left(\frac{24\Omega^4}{s(s^4 + 20s^2\Omega^2 + 64\Omega^4)}\right)$. We also employ $\alpha_1 = -\frac{a^2}{2}$ and $\tilde{\alpha}_1 = \frac{a^4}{2}$, a is a lattice constant.

Similarly, the imaginary part of the gap oscillation can also be calculated to be:

$$\delta\Delta''(s) = -\frac{\alpha_0}{s}\Delta e^2 A_0^{\ 2}(s) - \frac{1 - \frac{1}{2}\sin^2(2\theta)}{12s}\tilde{\alpha}_0\Delta e^4 A_0^4(s) \tag{21}$$

where $\alpha_0 = -\frac{\varepsilon_{\mathrm{F}}a^2}{2}$ and $\tilde{\alpha}_0 = \frac{\varepsilon_{\mathrm{F}}a^4}{2}$ and ε_{F} is the Fermi energy. Obviously we can split the real and imaginary parts of the gap oscillation into terms $\propto A_0^{\ 2}(s)$ and $\propto A_0^{\ 4}(s)$.

The fifth harmonic generation (FHG) current intensity is proportional to the absolute square of the electric current $I \propto |\vec{j}|^2$ and can be written:

$$I_{\mathrm{FHG}} \propto |\vec{j}^{(5)}(5\Omega)|^2 \tag{22}$$

Incorporating electromagnetic fields into the electric current *via* the Peierls-substitution, we arrive at

$$\vec{j}(t) = e\sum_{\mathbf{k}} \vec{v}_{\mathbf{k}-e\mathbf{A}(t)} \langle \hat{n}_{\mathbf{k}}(t) \rangle. \tag{23}$$

To calculate the response in the FHG we need to expand the velocity in terms of $\mathbf{A}(t)$ up to the third order

$$\begin{aligned}\vec{v}_{-e\mathbf{A}(t)} = \vec{v}_{\mathbf{k}} - e\sum_{i} A_j(t)\partial_{k_i}\vec{v}_{\mathbf{k}} + \frac{1}{2}e^2 \sum_{i,j=\{x,y\}} A_i(t)A_j(t)\partial_{k_i}\partial_{k_j}\vec{v}_{\mathbf{k}} \\ - \frac{1}{6}e^3 \sum_{i,j,l=\{x,y\}} A_i(t)A_j(t)A_l(t)\partial_{k_i}\partial_{k_j}\partial_{k_l}\vec{v}_{\mathbf{k}} + \mathcal{O}\left({A_0}^4\right)\end{aligned} \tag{24}$$

Expressing $\langle \hat{n}k(t) \rangle$ through $\delta\sigma_k^z(t)$ and splitting the current into the following terms:

$$\vec{j}(t) = \vec{j}^{\,(0)}(t) + \vec{j}^{\,(1)}(t) + \vec{j}^{\,(3)}(t) + \vec{j}^{\,(4)}(t) + \vec{j}^{\,(5)}(t) \tag{25}$$

In particular, the fifth harmonic generated current, $\vec{j}^{\,(5)}(t)$, can be split into parallel and perpendicular components with respect to the driving vector potential

$$\vec{j}^{\,(5)}_{\|\perp}(t) = \vec{j}^{\,(5)}(t)\cdot\hat{e}_{\|\perp} = -2e^2 A_0(t)\sum_{\mathbf{k}} D^{\|\perp}_{\varepsilon_{\mathbf{k}}}(\theta)\delta\sigma^z_{\mathbf{k}_{A^4}}(t) - \frac{1}{3}e^3 {A_0}^3(t)\sum_{\mathbf{k}} \varLambda^{\|\perp}_{\xi_{\mathbf{k}}}(\theta)\delta\sigma^z_{\mathbf{k}_{A^2}}(t) \tag{26}$$

with

$$D^{\perp}_{\xi_{\mathbf{k}}}(\theta) = \sin\theta\cos\theta\left(\partial_{k_xk_x}{}^2\xi_{\mathbf{k}} - \partial_{k_yk_y}{}^2\xi_{\mathbf{k}}\right) + \partial_{k_xk_y}{}^2\xi_{\mathbf{k}}\left(\sin^2\theta - \cos^2\theta\right) \tag{27}$$

$$\varLambda^{\perp}_{\xi_{\mathbf{k}}}(\theta) \approx \cos^3\theta\sin\theta\partial_{k_xk_xk_xk_x}{}^4\xi_{\mathbf{k}} - \sin^3\theta\cos\theta\partial_{k_yk_yk_yk_y}{}^4\xi_{\mathbf{k}}. \tag{28}$$

Performing the Fourier transform one obtains the frequency dependent fifth harmonic generated current

$$\begin{aligned}\vec{j}^{\,(5)}_{\|\perp}(\omega = 5\Omega) = -ie^2 A_0\sum_{k} D^{\|\perp}_{\varepsilon_k}(\theta)\delta\sigma^z_{k_{A^4}}(s = 4i\Omega) + \frac{1}{6}e^2 \varLambda^{\|\perp}_{\varepsilon_k}(\theta) \\ \times {A_0}^2\left[\frac{3}{4}\delta\sigma^z_{k_{A^2}}(s = 4i\Omega) - \frac{1}{4}\delta\sigma^z_{k_{A^2}}(s = 2i\Omega)\right]\end{aligned} \tag{29}$$

Similar to the third harmonic generated currents it is convenient to divide the total FHG into contributions stemming from the amplitude (Higgs), phase, and charge density (CDF) fluctuations. In particular, for the parallel component of the current Higgs contribution, proportional to the amplitude of the order parameter, $\delta\varDelta'$ has the form

$$\vec{j}_{\|}^{(5)\mathrm{H}}(5\Omega) = -ie^2 A_0 \sum_k \frac{\Delta_0 D^{\|}_{\xi_k}(\theta)}{E_k\left(4E_k^{\,2} - 16\Omega^2\right)} 2\xi_k \delta\Delta'_{A^4}(s = 4i\Omega)$$

$$+\frac{ie^4 A_0^{\,3}}{24} \sum_k \frac{\Delta_0 \Lambda^{\|}_{\xi_k}(\theta)}{E_k\left(4E_k^{\,2} - 4\Omega^2\right)} 2\xi_k \delta\Delta'_{A^2}(s = 2i\Omega) \qquad (30)$$

The term proportional to $\delta\Delta''$ can be dubbed the phase fluctuation contribution

$$\vec{j}_{\|}^{(5)\mathrm{P}}(5\Omega) = +ie^2 A_0 \sum_k \frac{\Delta_0 D^{\|}_{\xi_k}(\theta)}{E_k\left(4E_k^{\,2} - 16\Omega^2\right)} 4i\Omega \delta\Delta''_{A^4}(s = 4i\Omega)$$

$$-\frac{ie^4 A_0^{\,3}}{24} \sum_k \frac{\Delta_0 \Lambda^{\|}_{\xi_k}(\theta)}{E_k\left(4E_k^{\,2} - 4\Omega^2\right)} 2i\Omega \delta\Delta''_{A^2}(s = 2i\Omega) \qquad (31)$$

One has to bear in mind that to compute the full contribution of the phase fluctuations for the charged superconductors one would have to go beyond the Anderson pseudospin formalism, which we leave to a future study. Finally the contributions proportional to $\xi_k^{A^4}$ and $\xi_k^{A^2}$ are frequently denoted charge density fluctuations (CDF)

$$\vec{j}_{\|}^{(5)\mathrm{CDF}}(5\Omega) = ie^2 A_0 \sum_k \frac{\Delta_0^{\,2} D^{\|}_{\xi_k}(\theta)}{E_k\left(4E_k^{\,2} - 16\Omega^2\right)} \frac{1}{12} \xi_k^{A^4}(s = 4i\Omega)$$

$$-\frac{ie^4 A_0^{\,3}}{24} \sum_k \frac{\Delta_0^{\,2} \Lambda^{\|}_{\xi_k}(\theta)}{E_k\left(4E_k^{\,2} - 4\Omega^2\right)} \xi_k^{A^2}(s = 2i\Omega) \qquad (32)$$

We omitted the contributions for the terms of $\propto \delta\sigma_{k_{A^2}}(s = 4i\Omega)$ as they do not fulfil a resonance condition and would only clutter the expressions for the current contributions further.

We note that the contribution to the FHG from the coupling of the Higgs (CDF) mode with the photons can be visualized diagrammatically, see Fig. 1. The diagram shown in (a) is a generalization of the THG, while the other two, (b) and (c), are the new processes involving the coupling of the collective excitations to two and four photons.

3 Numerical results

In what follows we present our numerical results for the fifth harmonic generation current, $\vec{j}^{(5)}$, and compare them with $\vec{j}^{(3)}$. The expression $\vec{j}^{(3)}$ can be obtained straightforwardly and can be found for example in ref. 21.

We employ the tight-binding nearest neighbor energy dispersion and choose the Fermi energy $\varepsilon_F = -400$ meV, $T_D = 600$ K, $\Delta_0 \approx 20.25$ meV, $t = 10\Delta_0$, $\theta = 0$, $\lambda = 0.6$, and the lattice constant $a = 1$. We also introduce the damping factor $\gamma = 0.035\Delta_0 \ll \Delta_0$.

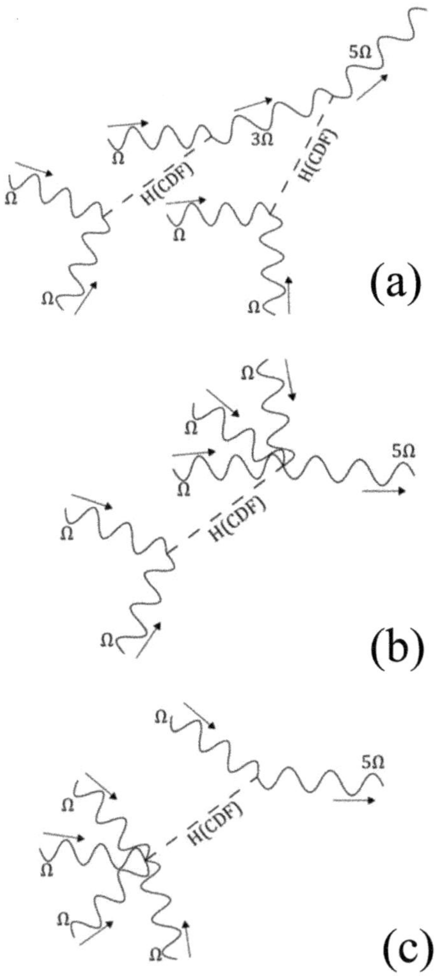

Fig. 1 Diagrammatic representation of the fifth harmonic generation mediated either by the Higgs mode (*H*) or charge density fluctuations (CDF). While (a) refers to the general process, which evolves from the third-harmonic generation, (b) and (c) are the new processes, which involve initial coupling of two (b) and four (c) photons with the Higgs (CDF) mode.

In Fig. 2 we show the amplitudes of the FHG response in the emitted electromagnetic field through Higgs, phase, and CDF contributions and the total current $j_{\parallel}^{(5)}(5\Omega)$. We observe two sharp peaks, one at $\Omega = 0.5\Delta_0$ $(4\Omega = 2\Delta_0)$ and another at $\Omega = \Delta_0$ $(4\Omega = 4\Delta_0)$. The first peak is indicative of four photons exciting a collective mode, whereas the second peak refers to the excitement of the collective mode by two individual photons. The latter process is similar to the third harmonic generation current and yields a signature at the same frequency, $\Omega = \Delta_0$. We also show these processes diagrammatically in Fig. 1. Similar also to the third harmonic generation current, we observe that the Higgs mode contribution is around two and three orders of magnitude smaller than that of the CDF. Interestingly enough the phase fluctuations contribuiton is of the same

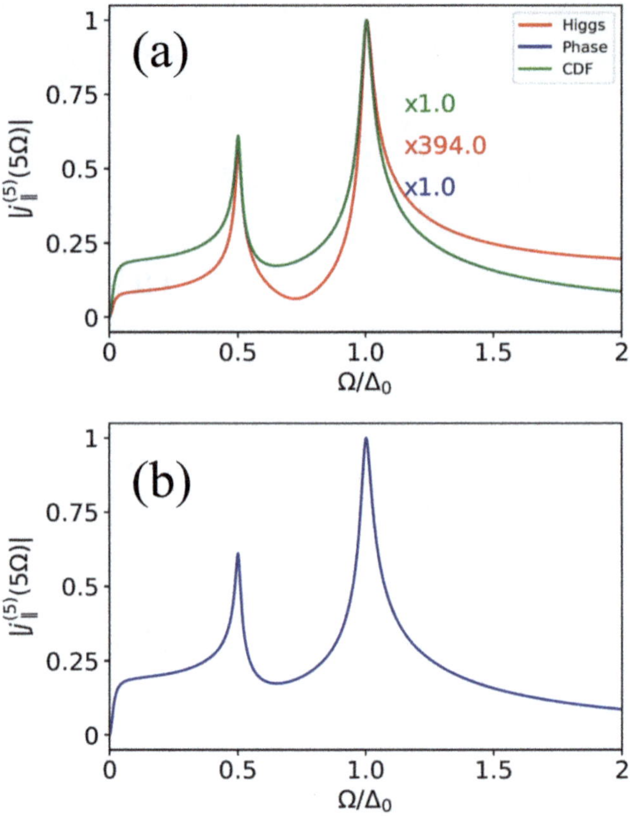

Fig. 2 Frequency dependence of the normalized amplitudes of the individual contributions (Higgs, phase and CDF) to the parallel component of the fifth harmonic generation current at $T = 0$ K (a). Note the phase fluctuation contribution and the CDF contribution lie on top of each other. The total current is shown in (b).

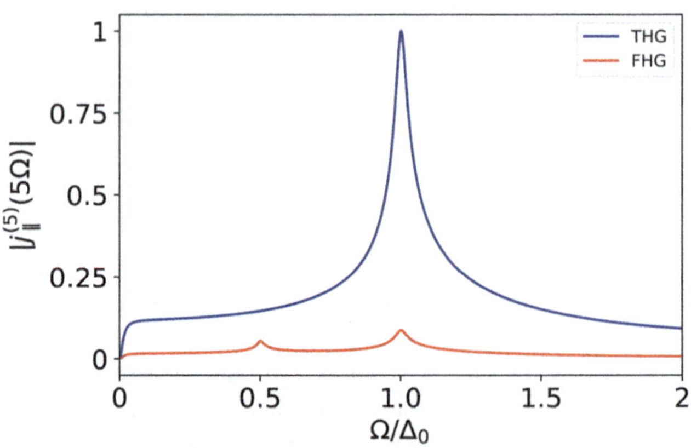

Fig. 3 Comparison between the fifth harmonic generation and the third harmonic generation currents assuming $eA_0 = 1$. The values are normalized to the maximum value of the THG current.

magnitude as the CDF, which is not accidental. To understand this, recall that the CDF term is the term, which is directly proportional to the vector potential ${A_0}^2$ and ${A_0}^4$, which allows us to identify it as also related to the fluctuations of the vector potential. As is well known, the phase fluctuations are related to the vector potential fluctuations upon the gauge transformation $\mathbf{A} \rightarrow \mathbf{A} - \nabla \rightarrow \phi$ where ϕ is a phase of the superconducting order parameter. So the fact that we see similar magnitude dynamics in CDF and in the phase fluctuations term is not a coincidence because these terms are related by the gauge transformations. Note, in a continuum model both contributions would be identical. Overall we find that the non-equilibrium response of a BCS superconductor is dominated by the charge density fluctuations rather than the direct contribution of the Higgs modes at least in the clean case. In Fig. 3 we show a comparison between the FHG and THG frequency dependencies assuming $eA_0 = 1$ for the sake of comparison. We observe that even in this case, the difference in magnitude is of the order of 20 which makes any experimental measurements of the FHG quite demanding. Finally in Fig. 4 we present the temperature dependence of the FHG current assuming BCS temperature dependence of the superconducting gap and $2\Delta_0/k_BT_c = 3.53$ for two characteristic frequencies. For $\Omega = 0.35\Delta_0$ the temperature

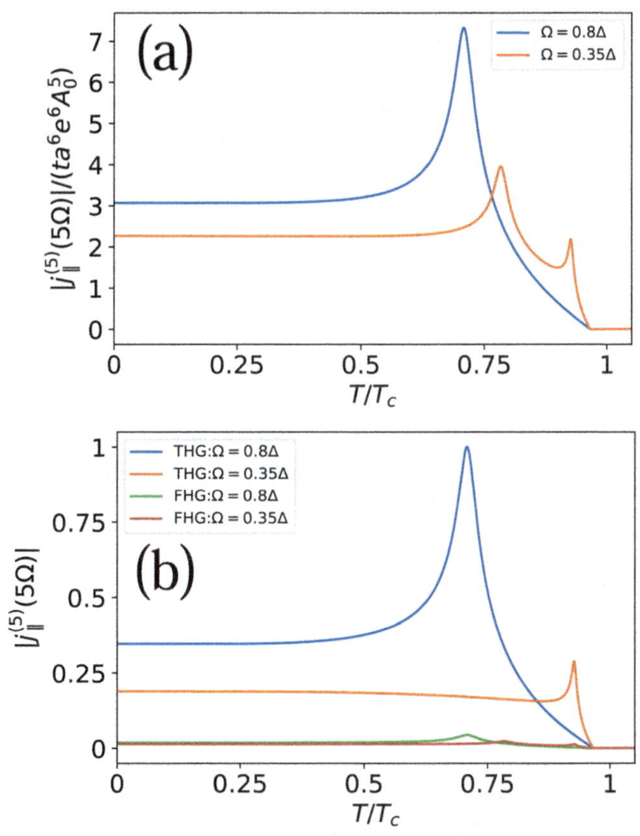

Fig. 4 Temperature dependence of the FHG current for two characteristic frequencies (a) and their comparison to the THG is shown in (b).

dependence shows two characteristic features associated with $\Omega = \Delta(T)/2$, *i.e.* four photons exciting a predominantly CDF mode, and $\Omega = \Delta(T)$, which refers to the excitement of the collective mode by the two individual photons. For larger frequencies the FHG looks similar to the THG except of much smaller intensity as can be seen from Fig. 4(b).

4 Conclusions

To conclude, motivated by the recent observation of the fifth harmonic generation in conventional superconductors, we analyze this process within the Anderson pseudospin formalism in a BCS superconductor. Similar to the third harmonic generating current we found that the signal in the FHG is also dominated by the charge density fluctuations, and the amplitude mode contribution is three orders of magnitude smaller. Most importantly, FHG intensity amplitude can show a double peak signature in the frequency and temperature dependence of the intensity amplitude of the FHG current with one peak located at $\Omega = \Delta_0$ ($4\Omega = 4\Delta_0$) and another one at $\Omega = \Delta_0/2$ ($4\Omega = 2\Delta_0$). The resonant enhancement in the latter case is indicative of the higher order coupling to the CDF, while the former is reminiscent of the THG describing the coupling of CDF and the Higgs mode with 2 single photons.

Author contributions

PD and MAM performed the analytical and numerical calculations under the supervision of IME. PD and IME wrote the manuscript and prepared the figures. All authors discussed the numerical and analytical results.

Conflicts of interest

There are no conflicts to declare.

Acknowledgements

We are thankful to Zhe Wang for sharing with us the unpublished experimental results of his group and for fruitful discussions. This work was supported by a joint NSFC-DFG grant (ER 463/14-1).

Notes and references

1 R. Shimano and N. Tsuji, *Annu. Rev. Condens. Matter Phys.*, 2020, **11**, 103–124.
2 P. W. Anderson, *Phys. Rev.*, 1963, **130**, 439–442.
3 P. B. Littlewood and C. M. Varma, *Phys. Rev. Lett.*, 1981, **47**, 811–814.
4 P. B. Littlewood and C. M. Varma, *Phys. Rev. B: Condens. Matter Mater. Phys.*, 1982, **26**, 4883–4893.
5 A. Moor, P. A. Volkov, A. F. Volkov and K. B. Efetov, *Phys. Rev. B: Condens. Matter Mater. Phys.*, 2014, **90**, 024511.
6 T. Cea and L. Benfatto, *Phys. Rev. B: Condens. Matter Mater. Phys.*, 2014, **90**, 224515.

7 M. A. Sentef, A. Tokuno, A. Georges and C. Kollath, *Phys. Rev. Lett.*, 2017, **118**, 087002.
8 R. Matsunaga, Y. I. Hamada, K. Makise, Y. Uzawa, H. Terai, Z. Wang and R. Shimano, *Phys. Rev. Lett.*, 2013, **111**, 057002.
9 R. Matsunaga, N. Tsuji, H. Fujita, A. Sugioka, K. Makise, Y. Uzawa, H. Terai, Z. Wang, H. Aoki and R. Shimano, *Science*, 2014, **345**, 1145–1149.
10 R. Matsunaga and R. Shimano, *Phys. Rev. Lett.*, 2012, **109**, 187002.
11 N. Tsuji and H. Aoki, *Phys. Rev. B: Condens. Matter Mater. Phys.*, 2015, **92**, 064508.
12 Y. Murotani, N. Tsuji and H. Aoki, *Phys. Rev. B*, 2017, **95**, 104503.
13 T. Cea, C. Castellani and L. Benfatto, *Phys. Rev. B*, 2016, **93**, 180507.
14 T. Jujo, *J. Phys. Soc. Jpn.*, 2018, **87**, 024704.
15 Y. Murotani and R. Shimano, *Phys. Rev. B*, 2019, **99**, 224510.
16 M. Silaev, *Phys. Rev. B*, 2019, **99**, 224511.
17 R. Haenel, P. Froese, D. Manske and L. Schwarz, *Phys. Rev. B*, 2021, **104**, 134504.
18 G. Seibold, M. Udina, C. Castellani and L. Benfatto, *Phys. Rev. B*, 2021, **103**, 014512.
19 Z. Wang, J. Xue, H. Shi, X. Jia, T. Lin, L. Shi, T. Dong, F. Wang and N.-L. Wang, *Phys. Rev. B*, 2022, **105**, L100508.
20 P. W. Anderson, *Phys. Rev.*, 1958, **112**, 1900–1916.
21 L. Schwarz and D. Manske, *Phys. Rev. B*, 2020, **101**, 184519.

Faraday Discussions

DISCUSSIONS

Theory of out of equilibrium light-induced phenomena: general discussion

Marc Alías-Rodríguez, Martina Basini, Lara Benfatto, Yorrick Boeije, Irene Burghardt, Andrew Burnett, Lin Chen, Eric Collet, Rory Cowin, Ilya Eremin, Graham Fleming, Aswathy V. Girija, Kunio Ishida, Shinichiro Iwai, J. Olof Johansson, Steven L. Johnson, Kota Katsumi, James McCusker, Christophe Odin, Matteo Puviani, Jan M. Rost, Habib Rostami, Mattia Udina and Julia Weinstein

DOI: 10.1039/d2fd90036j

James McCusker opened a discussion of the paper by Marc Alías-Rodríguez: The spin–orbit coupling constant between the $^1A_1$ and $^5T_2$ states is roughly an order of magnitude larger than what has been determined experimentally from variable-temperature ultrafast time-resolved absorption measurements.[1] If you were to replace your calculated value with the experimental one, how do you think this would change your mechanistic picture?

1 M. C. Carey, S. L. Adelman and J. K. McCusker, *Chem. Sci.*, 2019, **10**, 134–144.

Marc Alías-Rodríguez replied: In the thermal process the HS state lifetime is determined through the Fermi golden rule, within the Condon approximation. This assumes that the transition rate is the direct product between the vibrational overlap and the (effective) spin–orbit coupling, where the latter is determined through the effective Hamiltonian theory and the direct spin–orbit matrix elements using the Douglas–Kroll–Hess (DKH) effective one-electron operator. The spin–orbit could be replaced by the experimental one, then the transition rate will be, roughly speaking, one order of magnitude slower and the HS lifetime one order of magnitude larger.

Regarding the photophysical process, the model Hamiltonian has been constructed using the DKH direct spin–orbit matrix elements for the spin–orbit and the two-electron terms are neglected. Therefore, the coupling between states with $\Delta S \geq 2$, such as $^1A_{1g}$ and $^5T_{2g}$, is strictly 0. However, the effective coupling is present *via* the intermediate $^3T_{1g}$ state. Actually, it is not clear to me whether the $^5T_{2g}$ to $^1A_{1g}$ transition occurs directly or this always happens *via* the intermediate $^3T_{1g}$; in any of the cases, I would expect similar HS lifetime and the same mechanism (with the obvious difference of the participation of the $^3T_{1g}$ or not).

Aswathy V. Girija remarked: In the model, you have not included vibronic spin orbit coupling and direct coupling between the modes, and you have mentioned that including those would increase the computation cost – how do you go about improving this model? How else could you approach this without it being computationally too intense?

Marc Alías-Rodríguez answered: The model could be improved with inclusion of new modes or by increasing the number of parameters in the Hamiltonian. The model includes the main vibrational modes and seems to be enough to qualitatively describe the photophysics of this system. However, this model Hamiltonian could be enhanced to enhance the quantitative results obtained. The way to do it would be the following: (a) introduce the potentials along other relevant vibrational modes such as N–Fe–N bending modes, and ligand torsions; (b) include the intra-mode coupling parameters to construct the model Hamiltonian; (c) include the singlet–quintet direct coupling by the use of the explicit two-electron spin–orbit operator for the SOC; (d) explore the influence of the vibronic spin–orbit coupling.

J. Olof Johansson commented: You mention in your paper (https://doi.org/10.1039/d2fd00027j) that you do not observe involvement of the triplet state in the decay from the quintet to singlet state. K. F. Purcell has identified that the triplet state is important in this decay but that it becomes activated by torsional twisting modes.[1] Given that you do not take into account twisting modes, could this affect your conclusion?

1 K. F. Purcell, *J. Am. Chem. Soc.*, 1979, **101**, 5147–5152.

Marc Alías-Rodríguez responded: In the photophysical process, we have observed that the transition from the vibrationally hot HS state, $^5T_2$*, to the $^1A_1$ LS state happens *via* the intermediate state $^3T_1$. However, the singlet–quintet coupling was strictly 0 in our model because we employed the effective one-electron operator. To determine whether the direct path or the path *via* the intermediate triplet is the dominant path and the importance of twisting torsional modes to activate this state, an enhanced model Hamiltonian would be required. This Hamiltonian should be constructed using the explicit two-electron spin–orbit operator and the multi-dimensional surface should be extended along the most relevant torsional twisting modes.

Eric Collet remarked: In your paper (https://doi.org/10.1039/d2fd00027j) you refer to our experimental data, where there is a signature of a transient state, intermediate between the excited ^5E state and the final LS state. We have interpreted this as the optical signature of the ^3T state. In your paper you propose a different pathway, where the intermediate state is an excited ^5T* state, which is highly excited. Can you estimate the phonon population in this highly excited state or the relative energy level compared to the ground state? Can you estimate the variation of optical absorption in the ^5T* state compared to the ground ^5T state to explain what is reported in the ultrafast experimental data?

Marc Alías-Rodríguez replied: Our hypothesis, apart from the results obtained with the wave packet dynamics, is based on the fact that the ^3T–^5E spin–orbit coupling is very small because of the two-electron difference between their main configurations. We have tried to calculate the time-resolved transition absorption spectra from the data we obtained from our simulations. However, the excited state absorption requires double excitations which are not obtained using the standard linear-response TD-DFT and the use of multi-configurational methods such as CASSCF/CASPT2 is totally unpractical for obtaining the MLCT states at large Fe–N distances. Unfortunately, there is no one-to-one correspondence between our simulations and the spectroscopic data you obtained. I hope new theoretical tools will be developed in the near future to beat this limitation.

James McCusker commented: As a follow-up, the Arrhenius barrier you determined for conversion from the $^5T_2$ state to the $^1A_1$ ground state is roughly 3–4 times larger than the one determined experimentally.[1] Would it be possible to use the experimental value as a fixed term in your simulations, and how do you think that increase in the barrier would affect the dynamics (including any tunneling contributions to ground-state recovery)?

1 M. C. Carey, S. L. Adelman and J. K. McCusker, *Chem. Sci.*, 2019, **10**, 134–144.

Marc Alías-Rodríguez answered: In the thermal process, the adiabatic energy difference between the states can be modified, therefore, the experimental value could be used. The larger the HS–LS energy difference, the shorter the HS state lifetime, as can be observed by comparing $Fe(bpy)_3$ and $Fe(mtz)_6$.

Rory Cowin remarked: If you were able to include the vibronic spin–orbit coupling in your calculations on the photoexcitation pathway, how would you expect this to affect the dynamics? More specifically, how would this stabilise the ^5E_g state, and would it be significant enough to make up the order of magnitude difference with the experimental lifetime?

Marc Alías-Rodríguez responded: To include the vibronic spin–orbit coupling is not trivial because this is computed within the adiabatic basis and the model Hamiltonian is expressed in the diabatic one. However, it can be observed that the spin–orbit coupling almost remains constant along the main modes. For this reason, I think that the vibronic spin–orbit coupling will not dramatically affect the dynamics of this system.

Matteo Puviani opened a discussion of the paper by Kunio Ishida: In the first model shown, a linear electron–phonon coupling was used. What would change using a non-linear electron–phonon coupling? Would the phase diagram change qualitatively?

Kunio Ishida answered: If the second-order term of the electron–phonon coupling is taken into account, the curvature of potential energy surfaces becomes different between the electronic ground state and the excited state.

Since the phase boundary in the phase diagram is determined by the energy gain by lattice deformation, such interaction will change the shape of the boundary. However, we do not think it is a qualitative change.

Graham Fleming asked: Did you explore how properties of the quantum light such as entanglement time or the bandwidth of the pump source for down conversion influence the interchain entanglement spins?

Kunio Ishida replied: Both the coherence time of light and the bandwidth of the incident light will disturb the entanglement generation between remote systems, since photon states act as a reservoir which also causes the decoherence of photons. Hence, they are particularly important for discussing the possibility of entanglement sources (quantum memory) for long-distance quantum communications. Decoherence of entangled photons has been studied by Antonelli *et al.*[1] and Brodsky *et al.*,[2] for example. Entanglement generation by such partially entangled photons has not been studied to our knowledge, but we expect that we will be provided with knowledge of the role of the quantum nature of light in entanglement generation in materials by studying this problem.

When light is well-approximated by a coherent state of photons $|\alpha\rangle$, it is an eigenstate of the annihilation operator of the photon c, *i.e.*, $c|\alpha\rangle = \alpha|\alpha\rangle$. Although $|\alpha\rangle$ is not an eigenstate of the creation operator $c^\dagger$, we obtain $c^\dagger|\alpha\rangle = \alpha^*|\alpha\rangle + |r\rangle$ after a simple algebraic calculation, where $\langle r|\alpha\rangle = 0$ and $\langle r|r\rangle = 1$.

The classical correspondence of α is the complex amplitude of the light's electric field, and the quantum mechanical effect is mainly caused by $|r\rangle$, which is often regarded as negligible for large values of $|\alpha|$. However, our calculated results show that $|r\rangle$ induces entanglement and/or correlation between electrons/phonons in irradiated materials. Thus, the time evolution of $|r\rangle$ is a key to understanding the efficiency of entanglement generation. We have not found a similar property for entangled photons, which will help us clarify the effect of photon decoherence by traveling far from the light source.

1 C. Antonelli, M. Shtaif and M. Brodsky, *Phys. Rev. Lett.*, 2011, **106**, 080404.
2 M. Brodsky, E. C. George, C. Antonelli and M. Shtaif, *Opt. Lett.*, 2011, **36**, 43–45.

Lara Benfatto enquired: How can you measure the quantum entanglement that you define theoretically in the experiments? In particular, how can you measure it using (optical and/or THz) light?

Kunio Ishida answered: Since electrons, phonons, and photons are entangled with each other, the scattered light carries information on the entanglement in the electron–phonon systems. In the case of remote electron–phonon systems ($J = 0$), for example, we showed that the phonon–phonon entanglement is detected by the heterodyne detection, in which the entanglement will be observed as small deviations in the coherent phonon signals.[1] Since, however, such quantum correlation would vanish immediately after photoexcitation, both high time resolution and low detection limit are required to detect those signals by optical pump–probe experiments.

Detection by THz light is an interesting problem considering that the vibration frequency in some materials lies in the THz range. In this case, phonon entanglement will affect the signals of time-domain THz spectroscopy, though the intensity of the signal is weak.

On the other hand, entanglement in a material is measured by quantum state tomography, which determines the density matrix of the system experimentally. We are interested in the detection of phonon entanglement by this method. However, the electron–phonon system is much more complicated than the system with several qubits, and we have not obtained signals expected for phonon entanglement yet.

1 K. Ishida and H. Matsueda, *J. Phys. Soc. Jpn.*, 2021, **90**, 104714.

Steven L. Johnson commented: I note that in your framework you are currently treating the phonon interaction using Einstein model optical phonon modes, and so there is no interaction between the sites *via* these modes. Have you considered other models of phonon modes (*e.g.* Debye model modes) where you might introduce correlations between sites?

Kunio Ishida responded: Phonon dispersion is introduced by the intersite phonon interaction. Thus, the entanglement between phonons on different sites is modulated, and the entanglement dynamics is quantitatively different from that in dispersionless phonon systems. When we consider the interaction with an acoustic phonon mode, electronic interaction with that mode is relevant to the bond length change between sites. Hence, we expect that the electronic exchange interaction J is modulated as well. If the quantum-mechanical nature of the vibration mode is neglected, the modulation of J does not affect the spin state except in the vicinity of the phase boundary. If the quantization of the vibration mode is taken into account, spin-triplet states may appear in the Schmidt decomposition of the wavefunction (*cf.* eqn (24) of the paper, https://doi.org/10.1039/d2fd00007e) depending on the form of the electron–phonon interaction. Thus, intersite correlation would change by introducing coupling with acoustic phonons. However, so far, we consider that the coupling with optical phonons gives us a more remarkable change in the spin-phonon states, *e.g.*, dynamical Jahn–Teller states, and that various types of quantum state control by light are possible in the system with optical phonons.

Lara Benfatto remarked: You mentioned one can think about extending calculations to multisite systems. So far, even the solution of the two-site problem looks challenging, due to the need to account for electrons, phonons and photons, all at a full quantum level. How far can one go beyond the two-site problem?

Kunio Ishida replied: The dimension of the Hilbert space is 2^{30} even for the two-site system with photons, and we require a high computational cost to solve the time-dependent Schrodinger equation. Thus, honestly, we think that it is hopeless to extend our calculation to larger systems in a straightforward manner. Instead, we consider using the singular value decomposition of the wavefunction of the entire electron–phonon–photon system. Since only the singular vectors corresponding to large singular values contain important information on the transient dynamics of the system, we expect to choose a small number of bases that accurately reproduce the results at a full quantum level. To treat a larger system, we are trying to construct a basis set from these singular vectors, which makes it possible to reduce the number of bases for a larger system. Our preliminary results show that this method will work, and we are preparing another paper regarding this method. If such a "reconstructed" basis set is still too large to treat larger systems, we would require further approximation on the wavefunction. We are interested in the notion of the tensor network in quantum many-body physics to solve this problem, since it works well in the case of the one-dimensional antiferromagnetic Heisenberg model, *i.e.*, a long spin chain without phonon degrees of freedom.

However, the tensor network has mainly been applied to obtain the ground state wavefunction, and thus it is an open and challenging problem to apply this method to the physics of transient phenomena.

In particular, we need further analysis of the wavefunction to resolve the spin-phonon configurations that mainly contribute to the transient states, in order to understand how many sites we can treat in our model.

Eric Collet asked: Is there any possibility of taking advantage of quantum entanglement control of electron–phonon systems by light to drive symmetry change or symmetry breaking and therefore to control some functions of materials?

Kunio Ishida responded: We consider some possibilities regarding the relationship between entanglement generation and photoinduced cooperativity. First, photoinduced phase transition is triggered by the injection of small fluctuations to materials' states by light. We are studying the effect of fluctuation induced by quantum entanglement. In other words, if the entanglement between separated sites (molecules or unit cells) enhances fluctuations that bring about cooperative effects, our results show that the quantum effect of light is important and plays a certain role in transient dynamics. Secondly, we are also aware of the role of the entanglement as an order parameter of quantum phase transitions (*cf.* ref. 1). In this case, even when a highly nonequilibrium state of the system does not allow us to determine an order parameter by physical insights, we expect that some entanglement properties, *e.g.*, quantum mutual information, would give us a clue to understanding the transient cooperativity. In this context, the control of entanglement will lead us to control some functions of materials.

Another possibility is related to the measurement induced phase transitions. Obviously, entanglement between various configurations is relevant to the result of the measurement, and thus the control of entanglement also affects the phase transition.

In any case, we are still seeking evidence that clarifies the role of quantum entanglement in the transient regime.

1 R. Singh, J. H. Bardarson and F. Pollmann, *New J. Phys.*, 2016, **18**, 023046.

Aswathy V. Girija commented: I have a general question – it is mentioned in the paper (https://doi.org/10.1039/d2fd00007e) that for a system to be used for entanglement storage, we might not need single photon sources. How can you realise this experimentally? This is with reference to the second model.

Kunio Ishida answered: As shown in ref. 1, quantum entanglement was generated between remote diamond crystals by an optical pulse, *i.e.*, without single-photon sources. By using the Raman scattering process, the entangled states involve the phonon degrees of freedom, which is advantageous for obtaining entangled states owing to the bosonic nature of phonons. We expect that spin chains play a similar role to phonons as an additional degree of freedom for each qubit.

1 K. C. Lee, M. R. Sprague, B. J. Sussman, J. Nunn, N. K. Langford, X.-M. Jin, T. Champion, P. Michelberger, K. F. Reim, D. England, D. Jaksch and I. A. Walmsley, *Science*, 2011, **334**, 1253–1256.

Aswathy V. Girija remarked: More generally – the realisation was that we do not necessarily need single photon sources to store entanglement. Will this make it easier to have those systems in practice?

Kunio Ishida replied: Yes. Single-photon sources still have several difficulties in their use as practical devices. Thus, we aim to reveal the conditions and/or apparatus to obtain quantum states (not limited to qubits) as strongly entangled as possible. This is why we treated "many" (<40) photons to excite the material systems. Although 40 photons are not enough to discuss the quantum effects of "real" optical pulses, we expect to obtain a clue to reveal the entanglement properties with multiple photons, at least.

Jan M. Rost commented: With cavities one can experimentally generate in a controlled way a few photons and entangle them with the target – is that a possibility for realizing the systems you presented the theory for?

Kunio Ishida responded: Cavities are suitable for suppression of dissipation/decoherence due to electromagnetic fields, and I agree that using cavities is a standard way to observe/realize entanglement between electrons and/or phonons. However, our target is to realize the calculated situations without cavities. In this case, the dissipation due to photon modes would become a serious problem, but the quantum state will still remain entangled at ultrashort time scale. Furthermore, we aim to show the possibility of photoinduced quantum phase transition in electron–phonon systems, where the weak entanglement generated by photons would trigger the macroscopic symmetry breaking. If this hypothesis holds true, the initially generated entanglement will turn to or generate other types of quantum correlation and thus we should pursue their quantum dynamics. We are searching for the indication of such a process by changing the values of parameters and the system size.

Shinichiro Iwai enquired: Is it possible to introduce electron correlation effects to the study? May I ask whether you can predict the influence of the Coulomb repulsion interaction?

Kunio Ishida answered: As the antiferromagnetic interaction between electrons is introduced in our model, the electronic correlation is taken into account to some extent. Furthermore, if we regard the "quasi-spin" states as Frenkel excitons, the exchange interaction corresponds to the hopping of an exciton which lets it move in the system. However, it would be a model of hard-core bosons with multiple excitons, and thus discussion on the itinerancy of electrons is still limited.

When the itinerancy of electrons is more explicitly taken into account, we need to consider the Hilbert space with much higher dimension than that in our study. In this case, the quantum state of the material travels in a wider space, which will cause another type of entanglement generation. Although we have not studied this case so far, our results show that the low-lying spin excitations play an important role in the photoinduced entanglement generation. They may be a clue to understanding the similar effect in strongly correlated electron systems. Competition of the electron itinerancy, the Coulomb repulsion, and the electron–phonon interaction has always been an interesting problem in understanding the various phases of materials, as shown in the traditional "t–U–s triangle". We consider that, in addition to them, the photon degrees of freedom will play a certain role in the transient regime, and that further study on the material–photon systems will be more important in spite of huge computational costs.

Ilya Eremin opened the discussion of the paper by Matteo Puviani: Does the signal to the higher harmonics in the incoherent regimes come from the imaginary part of Delta? If the main contribution for the incoherent part comes from phase fluctuations (*i.e.* Im Delta) there is a danger because this approach is not gauge invariant.

Matteo Puviani replied: The signal to the higher harmonics comes from the third component of the pseudospin, which contains both the real and imaginary part of the order parameter. However, the main contribution still comes from the fluctuations of the real part.

Habib Rostami remarked: As mentioned in your talk, you dynamically break inversion symmetry due to a quenched pump pulse leading to an SHG signal. I can see a strong intensity peak at zero frequency in Fig. 7 of your paper (https://doi.org/10.1039/d2fd00010e). Do you also observe a second-order rectification current? If yes, how does it scale with the relaxation rate? How does it depend on the pairing symmetry?

Matteo Puviani answered: We haven't checked for the presence of a second-order rectification current, and we thank you for the suggestion.

In our model we don't have any relaxation process for the quasiparticles.

Kota Katsumi asked: Why do you have an increase in superconducting order parameter? Could you explain the mechanism of the enhancement of the order parameter?

Matteo Puviani responded: The increase of the order parameter is the result of the increased coherence of incoherent pairs when a low-frequency pulse is used with specific polarization, as shown in ref. 1.

In our model we treat the phase of superconducting fluctuations as formed by incoherent Cooper pairs, *i.e.* with an artificial equilibrium state obtained by adding random noise to the phase of the original state pairs (therefore the global order parameter is lower than the original one). As a result, when the right pulse excites the quasiparticles, it increases the phase coherence (*i.e.* reduces the phase incoherence) towards the original condition, resulting in an increased global order parameter (even though the pairing strength is not increased).

1 F. Giusti, A. Marciniak, F. Randi, G. Sparapassi, F. Boschini, H. Eisaki, M. Greven, A. Damascelli, A. Avella and D. Fausti, *Phys. Rev. Lett.*, 2019, **122**, 067002.

Mattia Udina commented: Both the experimental technique and the theoretical description detailed in your paper (https://doi.org/10.1039/d2fd00010e) look very general. Which features are specific for superconducting cuprates and which, instead, can be generalized to other systems in a straightforward way?

Matteo Puviani answered: It's true that the approach is rather general. The specific features of superconducting cuprates are the enhancement of the order parameter in the presence of incoherent Cooper pairs, and the broadening of the peaks in the 2D Fourier spectra at a frequency which is twice the superconducting gap. Both these results in fact rely on the d-wave pairing symmetry. The presence of a sizable dynamical inversion symmetry breaking depends mostly on the parameters which are typical of high-temperature superconductors and the pulses used, however in principle it should not be physically limited to cuprates (maybe less sizable in other materials).

The other features, such as the higher harmonics and their dynamical modulation, can also be obtained in other superconductors.

Lara Benfatto remarked: The way you treat the incoherent superconducting (SC) state is different from what one usually has in mind when talking about phase fluctuations. Indeed, one usually has in mind thermal fluctuations of the global U(1) phase of the SC order. Here instead you introduce fluctuations on a variable connected to the momentum dependence of the Cooper pair, *i.e.* to the relative motion of the two electrons of the Cooper pair. What is the physical meaning of this phase and how does it relate to the real global phase?

Matteo Puviani replied: Experiments in cuprates have revealed the presence of superconducting fluctuations up to tens of Kelvin above the superconducting critical temperature: this feature is commonly taken to imply the presence of Cooper pairs losing phase coherence. This means that the global phase coherence vanishes above the critical temperature, while the pairing strength does not, together with phase correlations which are local in space and time.

This was the main motivation for employing a generalized BCS model with a new equilibrium state created artificially by adding random noise to the original phase in momentum space. Physically, this is similar to many Josephson couplings between Cooper pairs (in momentum space), each of which carries its own phase. The resulting global order parameter will be smaller than the equilibrium one (with fully coherent pairs) and it will have a global phase related to the sum of the phases of each Cooper pair.

Lara Benfatto asked: Related to my previous question, by using this "fluctuating" internal phase can you reproduce the typical pseudogap feature seen experimentally, *e.g.* a Fermi surface with antinodal states remaining gapped above T_c?

Matteo Puviani answered: We didn't check for this feature in our work since the calculations were performed at $T = 0$, imitating the behavior above the critical temperature by inserting phase incoherence.

However, previous studies[1,2] have shown that a local superconducting gap distribution was able to explain the presence of an effective average gap above T_c in photoemission experiments and the exponential tail exhibited by several observables.

1 D. Pelc, M. Vučković, M. S. Grbić, M. Požek, G. Yu, T. Sasagawa, M. Greven and N. Barišić, *Nat. Commun.*, 2018, **9**, 4327.
2 F. Giusti, A. Marciniak, F. Randi, G. Sparapassi, F. Boschini, H. Eisaki, M. Greven, A. Damascelli, A. Avella and D. Fausti, *Phys. Rev. Lett.*, 2019, **122**, 067002.

Christophe Odin commented: Thank you for your nice work and talk.

I was wondering why you use a two-dimensional Fourier transform to interpret your results. Usually it is used to see correlations from diagonal peaks, or disentangle interactions, when the arrow of time is not important. In my opinion, in your case, using a Fourier transform in the other dimension, the delay time dimension, somehow loses the temporal sequence or causality of the process of the two excitations by mixing all delay-times.

A 1D Fourier transform of real time as a function of the quench-drive delay shows how the "frequency content" of the signal evolves as a function of the quench-drive delay time. For instance, we already see that frequencies at 1 and 3 (in reduced units) are present for all delay times. These two contributions appear as strong peaks at $\omega_{\Delta t} = 0$ (time average of the signal over all delay times) in the 2D Fourier transform as expected. For delay times larger than roughly -2 ps, other broad contributions (in the real frequency space) appear as a function of delay time, and a modulation of the amplitudes of the 1D spectra with a period roughly equal to 1/5 of the unit delay time seems to be introduced. Are these effects a consequence of the stripes in the 2D spectrum?

Matteo Puviani responded: The 2D Fourier transform has been used because it allows the response coming from the four-wave mixing processes to be analyzed: in particular, the 2Δ resonances due to the mixing of quench and drive pulses. These would not be straightforwardly visible in a simple 1D Fourier transform as a function of the quench-drive delay time.

It is true, however, that the Fourier transform is applied to the entire time evolution, and therefore the information about the excitation process is somehow diluted: namely, the effect of the quenching is different if the pulse is before or after the driving. This point could be overcome by selectively and separately Fourier transforming some regions of the time-dependent response, in order to distinguish the processes occurring, for example, before and after the quench pulse.

We thank you for the comment and we will try to improve this analysis in future work.

It is also true that this framework allows interactions to be disentangled: therefore it can be generalized to other superconductors (*e.g.*: multi-band, superconductors with CDW, *etc.*) in order to study and characterize the coupling between different modes (Higgs, Leggett, CDW amplitudon, *etc.*).

Finally, the broad contributions which modulate the amplitude of the single-Fourier transformed spectrum as a function of the delay time (Fig. 3(b) of the paper, https://doi.org/10.1039/d2fd00010e) are related to the diagonal responses in the 2D spectrum (Fig. 3(c) of the paper). Indeed, the diagonal features in the 2D plot collapse in the 1D spectrum and spread out in the quench-drive delay time.

Shinichiro Iwai asked: Why did you choose the driving frequency of 5 THz?

Matteo Puviani replied: For the results of the first part of our paper (https://doi.org/10.1039/d2fd00010e) (cuprates with full coherence) we used a driving frequency $\omega \approx 5$ THz. This value ($\omega \ll 2\Delta$, Δ being the superconducting gap) allowed us not to break too many quasiparticles (and therefore allowed our results to be relevant for comparison with future experiments), as well as to satisfy other resonance conditions (such as $3\omega = 2\Delta$) in order to distinctly see the features in the 2D Fourier transform plot.

Shinichiro Iwai remarked: The possible mechanism in cuprates for superconductivity is an antiferromagnetic interaction. The energy scale is 0.1–0.2 eV in cuprates (maybe) which is larger than that discussed here. What happens for such higher frequency driving?

Matteo Puviani answered: For higher frequency driving other mechanisms and effects (difficult to treat simultaneously microscopically in theory) can play a role: such as, for example, heating, breaking of Cooper pairs (leading to suppression of the order parameter), excitation of other modes, *etc.*

Ilya Eremin commented: Breaking the symmetry with light should activate the subdominant channels (like extended s-waves) which will be a competing process.

Matteo Puviani replied: Thanks for the comment. In our work we modelled the cuprates with an (extended) BCS model (to include phase incoherence), in a range of doping and temperature for which we could safely neglect the effect of subdominant pairing channels.

It would be interesting for future work to study the effect of subdominant channels, and in particular the other resulting modes (*e.g.* Bardasis–Schrieffer), even for different parameter regimes.

Andrew Burnett enquired: In order to be able to practically implement quench-drive spectroscopy, how narrow does the drive pulse need to be? Have you explored how sensitive your 2D plots are to the drive pulse properties in terms of the spectral signals that could be determined?

Matteo Puviani responded: We haven't systematically checked the full dependence of the spectral signal on the driving pulse properties. However, in a previous study we have compared the use of a Gaussian-shaped driving pulse to the asymmetric one used in this work. In particular, while the main features remained the same, the intensity of the 2Δ peaks along the diagonal lines as well as the $2\omega \pm \Delta$ peaks along the vertical line in the 2D Fourier plots was smaller with a Gaussian driving pulse.

In order to implement quench-drive spectroscopy on cuprates the requirement for the time duration/frequency width of the driving pulse is of the order of $1/\Delta t \leq$ 1 THz.

Kota Katsumi remarked: There are some studies about using THz 2D spectroscopy. In the case of superconductors, can you extract any information on collective modes using this 2D frequency?

Matteo Puviani replied: With the approach that we used in this work for cuprates, it is not possible to disentangle the contribution of collective modes, such as amplitude (Higgs) oscillations, from the quasiparticles' response, since they have the same energy. However, performing symmetry/angle-resolved 2D THz spectroscopy could also offer new hints for such a problem.

Moreover, we do believe that – even in different superconductors – other modes with different energies (Leggett mode, Josephson plasma mode, *etc.*), and which may couple to either the Higgs or the quasiparticles' excitations, could be analyzed and disentangled with such 2D THz spectroscopy, leading to off-diagonal terms in the 2D Fourier transform plot.

Lara Benfatto said: I have a comment on the possibility of using light to disentangle the Higgs mode from quasiparticles. The point is that, at least in the weak-coupling regime, the Higgs mode is nothing other than a "projection" of the quasiparticle continuum starting at 2Δ. As a consequence, it is very hard to imagine how one can disentangle the two by just looking for excitations occurring at 2Δ.

Graham Fleming opened a discussion of the paper by Irene Burghardt: Could you please expand on your proposals to obtain spatial resolution from the measurements? Is this based on the 5th order component of the full nonlinear response?

Irene Burghardt answered: Indeed, I was referring to recent experiments by the Brixner group where fifth-order 2D spectroscopy is performed to measure exciton dynamics and exciton–exciton annihilation. Specifically, in the work by P. Malý *et al.*,[1] which is cited in our paper (https://doi.org/10.1039/d2fd00014h), exciton–exciton interaction two-dimensional (EEI2D) spectroscopy is used to elucidate exciton transport, highlighting a transition from coherent to diffusive motion. Information about the exciton delocalization length was inferred indirectly, by matching the spectroscopic observations with a Frenkel-exciton model whose predictions depend on the exciton overlap and, hence, the spatial extension of the excitonic wavefunction.

1 P. Malý, J. Lüttig, A. Turkin, J. Dostál, C. Lambert and T. Brixner, *Chem. Sci.*, 2020, **11**, 456–466.

Graham Fleming said: It will be important to isolate specific orders of the nonlinear response to obtain the spatial information you describe. Tobias Brixner has recently done some very interesting work along these lines which suggests that such a decomposition to isolate, for example, the 5th order response may be practically realizable.

Irene Burghardt responded: I entirely agree. Besides their development of the fifth-order EEI2D spectroscopy, the Brixner group further explored various types of four-wave mixing and six-wave mixing experiments which reveal a plethora of information about excitonic systems, including transfer dynamics and delocalization.[1]

1 P. Malý, S. Mueller, J. Lüttig, C. Lambert and T. Brixner, *J. Chem. Phys.*, 2020, **153**, 144204.

J. Olof Johansson commented: During the introductory lecture (https://doi.org/10.1039/d2fd00098a), the possibility of controlling light-induced processes was discussed. Do you think it is possible to spatially direct the migration of excitons *via* chemical or mechanical (for example *via* strain) "engineering" of the nature of the low-frequency modes you mention in your paper (https://doi.org/10.1039/d2fd00014h)?

Irene Burghardt replied: Yes, absolutely, this is a very interesting perspective. In a less specific scenario, the engineering of excitons both by chemical composition and by molecular packing is already an active area. For example, the general understanding is that more crystalline materials sustain longer exciton diffusion lengths. Torsional (dis)order has been engineered to some extent, by modifying ring substituents, side chains, and chain lengths. Mechanical strain would be highly interesting as an additional parameter. Some effects are subtle, though – *e.g.*, dynamical disorder induced by ring torsions seems detrimental to some extent because the desirable π-conjugation is weakened; but at the same time, torsional fluctuations induce exciton transport. While a global effect is the expected increase of exciton diffusion coefficients with temperature, this mechanism also provides a handle for directed exciton motion, or energy flow, at the single-molecule level. One could therefore envisage a molecular-level control scheme where the evolution of excitonic wavefunctions is guided by soft modes, notably torsional modes or low-frequency intermolecular modes.

Aswathy V. Girija remarked: My question is on exciton transport. Your calculations assume a 'single molecule' perspective. In experiments, which are often in ensemble, it is quite difficult to quantitatively estimate the exciton diffusion constant, and it's reported only in a few systems, such as P3HT. My question is, can we infer from the theoretical perspective how quantitatively exciton diffusion constants can be calculated in real systems? Would that help in correctly estimating exciton diffusion in polymers?

Irene Burghardt responded: Yes, it is correct that the present calculations of 2DES spectra assume a "single molecule" perspective, and they are actually restricted to a zero-temperature system. However, we previously used a more complete set-up to simulate exciton transport at finite temperature, in a full quantum dynamical setting, and obtained diffusion coefficients from mean-squared displacements.[1] These calculations employ ensembles of stochastic quantum realizations of exciton dynamics coupled to multiple torsional modes that in turn interact with a thermalized environment. From such simulations, a detailed understanding of ensemble averaging can be deduced, which could in principle be made to match experimental conditions. As you mentioned, experimentally determined diffusion coefficients (typically relying on photoluminescence quenching or exciton–exciton annihilation experiments) are not necessarily highly accurate. However, new approaches like fifth-order exciton–exciton-interaction 2D spectroscopy (EEI2D) developed by the Brixner group promise substantial progress. Using this approach, Brixner and collaborators highlighted the presence of a subdiffusive exciton transport mechanism in certain systems.[2] Quantum simulations can provide complementary information regarding the diffusion regime (normal, subdiffusive, superdiffusive) and, importantly, regarding the coherent or hopping type mechanism of exciton diffusion. By combining these novel spectroscopic techniques with both quantum dynamical and kinetic modeling – including a realistic description of both static and dynamic disorder – a more quantitative picture of exciton diffusion from the single-molecule *vs.* ensemble perspectives should result.

1 R. Binder and I. Burghardt, *Faraday Discuss.*, 2020, **221**, 406–427.
2 P. Malý, J. Lüttig, A. Turkin, J. Dostál, C. Lambert and T. Brixner, *Chem. Sci.*, 2020, **11**, 456–466.

Yorrick Boeije said: According to your calculations low-frequency torsional modes are involved in intra-chain exciton transport. Are there also low-frequency modes involved in inter-chain transport?

Irene Burghardt replied: Yes, there is indeed an analogous mechanism for inter-chain transport, which is triggered, *e.g.*, by inter-chain distance modes, along with slip-stacking or tilt type modes. In a very recent analysis,[1] we carried out simulations of exciton transport in a regioregular poly(3-hexylthiophene) (rr-P3HT) model system, including inter-chain modes whose thermally induced fluctuations are found to play a similar role to fluctuations of the torsional modes in the intra-chain process that we analysed earlier.[2] In the relevant molecular materials, the inter-chain morphology is known to play a crucial role in determining whether inter-chain exciton migration makes an appreciable contribution.

1 M. Mondelo-Martell, D. Brey and I. Burghardt, *J. Chem. Phys.*, 2022, DOI: 10.1063/5.0104729.
2 R. Binder and I. Burghardt, *Faraday Discuss.*, 2020, **221**, 406–427.

Yorrick Boeije remarked: The adiabatic potential energy surface computations demonstrate the presence of curve crossings between higher-lying excited states and S_1, resulting in internal conversion to S_1 driven by high-frequency polaronic modes. Have you also investigated the relevance of curve crossings for the S_1 to S_0 internal conversion?

Irene Burghardt answered: In fact, our present study is restricted to the excitonic manifold and does not include the S_0 electronic ground state, under the assumption that the decay to S_0 is comparatively slow. It is known that the fluorescence lifetime is of the order of one nanosecond in oligothiophenes, such that nonradiative decay to S_0 by internal conversion and intersystem crossing exhibits the same, slow time scale. By contrast, our simulation time scale relates to the femtosecond to picosecond range. However, this may be different in other systems – especially chromophores exhibiting accessible S_1/S_0 conical intersections – such that one would include the S_0 state in these cases.

Julia Weinstein enquired: If quasi-stationary dynamics of a collective mode is a generic feature of ultrafast photoinduced dynamics in molecular materials, would it be possible to perturb the dynamics in some way by a picosecond narrowband IR pulse? Thank you very much for your thoughts on this matter.

Irene Burghardt responded: Thank you very much for this interesting question. It is indeed conceivable that one would perturb the dynamics of a collective phonon mode and effectively destabilize the exciton-polaron quasi-particle. This would modify the spatial extension of the excitonic wavefunction and could eventually lead to "detrapping" of the exciton by vibronic excitation, entailing a modification of the transport properties.

Besides this type of control scenario, novel two-dimensional electronic-vibrational spectroscopies (2DEVS) as developed by Professor Fleming and collaborators could possibly inform about the dynamics of quasi-stationary exciton-polarons in molecular materials. More generally, vibrational and mixed vibrational-electronic spectroscopies are increasingly gaining ground in the detailed understanding of the structural dynamics of these multi-chromophoric systems, including time-resolved IR and Raman spectroscopies such as Femtosecond Stimulated Raman Spectroscopy (FSRS).[1,2]

1 K. Bera, S. Y. Kwang and R. R. Frontiera, *J. Phys. Chem. C*, 2020, **124**, 25163–25174.
2 A. E. Bragg, W. Yu, J. Zhou and T. Magnanelli, *J. Phys. Chem. Lett.*, 2016, 7, 3990–4000.

Lin Chen asked: Would you please clarify the definition of collective polaron modes?

Irene Burghardt answered: Thank you for this question. Indeed, in our paper (https://doi.org/10.1039/d2fd00014h) we introduced the term "polaronic mode" in quotation marks to signal that the expression may be non-standard in a molecular-level context. A physical definition, as given in the manuscript, is that of a collective high-frequency phonon mode that travels with the time-evolving exciton. Due to the strong vibronic coupling of molecular high-frequency modes – notably C–C stretch type modes – a hybridized quasi-particle species is generated within tens of femtoseconds, *i.e.*, an exciton-polaron. This quasi-particle represents the exciton dressed with a lattice deformation induced by a collective phonon mode that reacts immediately to any displacement of the exciton. This mode consists physically of individual molecular modes (*e.g.*, of C–C stretch character) belonging to neighboring molecular building blocks of the material. As the exciton migrates, it is accompanied ("trapped") by the collective polaronic mode. As shown in experiments[1,2] and also in our previous simulations,[3] one should think of quasi-stationary trapping in the sense that high-frequency oscillations of this collective mode persist into the picosecond range. As far as the formal description of this collective mode is concerned, a fully microscopic Hamiltonian can be used as a starting point to derive a time-dependent effective mode that is coupled to the electronic subsystem.

1 Y. Song, C. Hellmann, N. Stingelin and G. D. Scholes, *J. Chem. Phys.*, 2015, **142**, 212410.
2 A. De Sio, F. Troiani, M. Maiuri, J. Réhault, E. Sommer, J. Lim, S. F. Huelga, M. B. Plenio, C. A. Rozzi, G. Cerullo, E. Molinari and C. Lienau, *Nat. Commun.*, 2016, **7**, 13742.
3 R. Binder and I. Burghardt, *Faraday Discuss.*, 2020, **221**, 406–427.

Lin Chen commented: More recent developments in conjugate polymers for solar cell materials adapt the intrachain charge transfer character with alternating "donor"–"acceptor" building blocks in a repeating unit, resulting in so called "charge transfer" polymers. How would you expect the vibrational mode to affect exciton diffusion in such conjugated polymers?

Irene Burghardt responded: Indeed, new materials with alternating donor–acceptor building blocks could feature more complex exciton transport properties than homopolymer species. In particular, intra-oligomer excitation energy transfer and charge transfer could compete and interact. (We studied such effects in collaboration with experiments in ref. 1 for conjugated donor–acceptor block co-oligomers.) As far as I recall, your group actually carried out a highly informative comparative spectroscopic study of exciton diffusion in charge-transfer polymers (notably PTB7) as compared with homopolymers (P3HT) a few years ago;[2] these are very interesting results highlighting the complexity of the elementary transfer processes. As for the role of soft vibrations, many charge-transfer polymers exhibit ring torsional modes and thienyl type substituents, such that similar basic mechanisms of thermally promoted exciton transport to those in homopolymer systems should be operative.

1 L. Liu, P. Eisenbrandt, T. Roland, M. Polkehn, P.-O. Schwartz, K. Bruchlos, B. Omiecienski, S. Ludwigs, N. Leclerc, E. Zaborova, J. Léonard, S. Méry, I. Burghardt and S. Haacke, *Phys. Chem. Chem. Phys.*, 2016, **18**, 18536–18548.
2 S. Cho, B. S. Rolczynski, T. Xu, L. Yu and L. X. Chen, *J. Phys. Chem. B*, 2015, **119**, 7447–7456.

Eric Collet remarked: In your paper (https://doi.org/10.1039/d2fd00014h) you discuss the role of structural change induced by low-frequency torsional modes, along with quasi-stationary trapping effects induced by high-frequency polaronic modes. Your results focus on dynamics located at the molecular level. Is there any possibility of extending your work to molecular crystals, for which polaronic dynamics may be localized at the molecular or inter-molecular level?

Irene Burghardt replied: Yes, our analysis can indeed be extended to molecular crystals, where exciton delocalization and phonon-induced trapping phenomena occur across molecular units. On the theory side, one would follow a general approach relying on vibronic lattice Hamiltonians and quantum dynamical simulations, similar to our recent reports on regioregular polymer materials[1] and self-assembled mesomorphic nanostructures.[2] On the experimental side, various

2D electronic spectroscopy experiments have been carried out, *e.g.*, on thin films of regioregular poly(3-hexylthiophene) (P3HT)[3] or oligoacene molecular crystals to investigate singlet fission.[4] In these systems, the details of intermolecular packing and the interplay between local trapping and π-stacking interactions modulated by intermolecular vibrational modes are of key importance.

1 W. Popp, M. Polkehn, R. Binder and I. Burghardt, *J. Phys. Chem. Lett.*, 2019, **10**, 3326–3332.
2 D. Brey, W. Popp, P. Budakoti, G. D'Avino and I. Burghardt, *J. Phys. Chem. C*, 2021, **125**, 25030–25043.
3 A. De Sio, F. Troiani, M. Maiuri, J. Réhault, E. Sommer, J. Lim, S. F. Huelga, M. B. Plenio, C. A. Rozzi, G. Cerullo, E. Molinari and C. Lienau, *Nat. Commun.*, 2016, **7**, 13742.
4 A. A. Bakulin, S. E. Morgan, T. B. Kehoe, M. W. B. Wilson, A. W. Chin, D. Zigmantas, D. Egorova and A. Rao, *Nat. Chem.*, 2016, **8**, 16–23.

James McCusker commented: One of the great challenges in studying these sorts of systems is the large degree of vibronic coupling that exists, perhaps to the point where a question arises whether the Born–Oppenheimer approximation can really be invoked. I understand that this is a challenging problem, but where does theory stand in terms of tackling systems for which the B–O approximation simply cannot be used?

Irene Burghardt responded: Thank you for this very relevant question. In fact, the breakdown of the Born–Oppenheimer approximation is the rule rather than the exception in photoinduced processes in functional molecular materials, due to strong vibronic interactions. Since non-Born–Oppenheimer (or non-adiabatic) effects can affect the elementary transfer steps, they are the main reason for using quantum wavepacket methods, rather than trajectory-based propagation. Non-adiabatic effects may have a tangible impact both on the early-time dynamics following photoexcitation, and on exciton or carrier transport. A prominent example of the first case are H-aggregates such as regioregular poly(3-hexylthiophene) (rr-P3HT) where the upper exciton band edge is photoexcited and rapid de-excitation occurs by internal conversion (see, *e.g.*, our study in ref. 1). As for transport properties, there is an ongoing discussion about the role of non-Born–Oppenheimer effects in promoting exceptionally efficient exciton transport, *e.g.*, in highly ordered nanofiber materials.

In order to account for non-adiabatic effects, full quantum-dynamical methods are the strategy of choice, but these are numerically demanding for large systems, and need to be combined with pre-computed electronic structure-based potentials. Even so, much progress has been achieved in this direction using multi-configurational methods that open an avenue for accurate quantum propagation for molecular aggregate systems comprising tens to hundreds of electronic states and vibrational modes.[2] At the other end of the spectrum are semiclassical trajectory-based methods, which are less accurate but more flexible since they are compatible with an on-the-fly protocol (similar to *ab initio* molecular dynamics (AIMD)). In our work, we generally side with accurate quantum wavepacket propagation, or with mixed quantum-classical approaches as exemplified in the present paper (https://doi.org/10.1039/d2fd00014h). Current efforts on the theory side move towards systematic method comparison and benchmarking, with particular focus on non-Born-Oppenheimer effects.[3]

1 W. Popp, M. Polkehn, R. Binder and I. Burghardt, *J. Phys. Chem. Lett.*, 2019, **10**, 3326–3332.

2 W. Popp, D. Brey, R. Binder and I. Burghardt, *Annu. Rev. Phys. Chem.*, 2021, **72**, 591–616.
3 W. T. Peng, D. Brey, S. Giannini, D. Dell'Angelo, I. Burghardt and J. Blumberger, *J. Phys. Chem. Lett.*, 2022, **13**, 7105–7112.

James McCusker asked: Generally speaking, how extensive are the interactions between the theory and experimental communities when it comes to work on these sorts of topics? It seems to me that both would benefit tremendously from more bilateral collaborations, but I'm not really sure how common this is.

Irene Burghardt answered: I entirely agree that interaction between experiment and theory is highly beneficial in this area. Further, in the context of nonlinear spectroscopies, it is increasingly a necessity to analyze spectra based on theoretical models and simulations. The same holds for coherent control schemes. However, a challenge for collaborations between experiment and theory is that many theoretical models are still restricted to a few electronic states and vibrational modes, while the relevant systems are high-dimensional and truly complex. In favorable cases, electronic structure analyses of the molecular building blocks together with (quasi-)crystalline structures (*e.g.*, obtained by electron diffraction) are available, such as to permit quantitative dynamical modeling. In recent years, significant progress has been made in these directions, and the number of collaborations between experiment and theory is steadily growing.

Another aspect is that theory can provide important guidance to experiment. For example, the theory of vibronic absorption and emission spectra of molecular aggregates[1] has been widely adopted by experimentalists, and transient localization theory[2] has created a new paradigm for carrier transport in molecular aggregates. Finally, theoretical simulations have the advantage that temporal, spatial and frequency resolution are available, which can bring significant insight for the interpretation of experiments.

1 F. C. Spano and C. Silva, *Annu. Rev. Phys. Chem.*, 2014, **65**, 477–500.
2 S. Fratini, D. Mayou and S. Ciuchi, *Adv. Func. Mater.*, 2016, **26**, 2292–2315.

Kota Katsumi opened a discussion of the paper by Mattia Udina: The final slide showed the intensity of BCS and Higgs responses. You said the BCS response is dominant even in the case of a paramagnetic response, but for me the difference from the Higgs response is at most a factor of 2 or 3. What determines the relative ratio between the two in the case of a paramagnetic response?

Mattia Udina replied: The ratio between the Higgs and the quasiparticle contribution to the paramagnetic response is set by the $\gamma/2\Delta$ value, with γ the effective quasiparticle scattering rate and Δ the superconducting gap value. In the dirty limit ($\gamma/2\Delta \gg 1$), as valid *e.g.* for strongly disordered NbN samples, the Higgs mode can overcome the quasiparticle contribution. In the clean limit ($\gamma/2\Delta \simeq 0$), instead, the quasiparticle contribution is orders of magnitude larger than the Higgs mode. Since cuprates are relatively clean systems ($\gamma/2\Delta \sim 0.85$), the ratio between the quasiparticle and Higgs contributions is always larger than 1. In particular, in the present work we have considered two limiting cases: $\gamma/2\Delta \sim 0.01$ and $\gamma/2\Delta \sim 1.12$. In the lower limit, one finds that the Higgs is completely negligible while it adds a correction at most of the order of 30% of the quasiparticle contribution in the upper limit.

Kota Katsumi asked: Which do you think is the most important factor that determines the intensity of the Higgs and BCS "paramagnetic" responses?

Mattia Udina answered: As we detailed in ref. 1, the intensity of the paramagnetic response strongly depends on the $\gamma/2\Delta$ ratio, with γ the effective quasiparticle scattering rate and Δ the superconducting gap value. By assuming that Δ is kept fixed for a given superconducting system, the strongest effect on the amplitude of the paramagnetic response is then given by the disorder level, which is intrinsically related to the scattering rate γ.

1 G. Seibold, M. Udina, C. Castellani and L. Benfatto, *Phys. Rev. B*, 2021, **103**, 014512.

Steven L. Johnson commented: In the existing published literature (cited in your paper, https://doi.org/10.1039/d2fd00016d and in the introductory lecture, https://doi.org/10.1039/d2fd00098a) there are several reports of effects that have been attributed to Higgs mode oscillations in superconductors, which are subsequently often disputed as being attributable to quasiparticle dynamics (a.k.a. charge density fluctuations). My understanding is that it is difficult to produce "smoking gun" evidence for the Higgs mode since it is not possible to uniquely identify *via* experimental features like frequency, and in the end it seems to require detailed modelling to claim that a given oscillation is partially attributable to the Higgs mode (or not). Based on experimental reports to date that you are aware of, is there strong evidence for the Higgs mode having already been observed?

Mattia Udina responded: As far as I know, the only unambiguous experimental signature of the superconducting Higgs mode has been obtained in $NbSe_2$, where superconductivity emerges at low temperature ($T_c = 7$ K) after a charge-density-wave (CDW) transition at higher temperatures. In this material, a strong resonance in the Raman response below T_c was measured long ago and related to the Higgs mode, coupled to the Raman-active CDW soft phonon mode. In particular, the CDW gap formation pushes the quasiparticle continuum away from the Higgs pole and allows for the appearance of a well-defined divergence in the Raman response below $2\Delta_{SC}$, which can be assigned to the Higgs mode (see *e.g.* ref. 1).

1 R. Grasset, T. Cea, Y. Gallais, M. Cazayous, A. Sacuto, L. Cario, L. Benfatto and M.-A. Méasson, *Phys. Rev. B*, 2018, **97**, 094502.

Ilya Eremin answered: There were initially proposals on the different polarizations of the Higgs mode as compared to the contribution of the quasiparticles. It turned out, however, that both contributions show a very similar behavior in the disordered case, so the situation is still ambiguous. Nevertheless the ability of THz pump probe techniques to see signatures of superconductivity is important because it opens a perspective to see further interesting effects related to superconductivity (like Bardasis–Schrieffer modes *etc.*). Furthermore, there have been further successful proposals and realization of the Higgs mode activation by sending a current through the sample.

Steven L. Johnson addressed Mattia Udina and Ilya Eremin: Can you conceive of any way to design an experiment to observe the excitations of the Higgs mode unambiguously (*i.e.* in a manner that cannot see contributions from other excitations)?

Mattia Udina replied: Unfortunately, neither the time evolution nor the polarization dependence of the signal in pump probe experiments provides a definitive answer to this problem. New insights could be (maybe) obtained by considering different experimental techniques, such as 2D pump–probe spectroscopy or time-resolved ARPES experiments.

Ilya Eremin responded: There was a proposal to drive current and see the Higgs mode, and this was recently realized.[1,2]

1 A. Moor, A. F. Volkov and K. B. Efetov, *Phys. Rev. Lett.*, 2017, **118**, 047001.
2 S. Nakamura, Y. Iida, Y. Murotani, R. Matsunaga, H. Terai and R. Shimano, *Phys. Rev. Lett.*, 2019, **122**, 257001.

Steven L. Johnson enquired: What is the overall impact of an unambiguous demonstration of an excitation of the Higgs mode? Is there a potential application connected with driving the Higgs mode? What do we learn from trying to understand the distinction between Higgs mode contributions *versus* quasiparticle contributions?

Ilya Eremin replied: The initial hope was that third harmonic generation and pump–probe techniques would allow the Higgs mode to be probed directly. The truth is that these experiments are able to see signatures of superconductivity which is good news. Whether or not these signatures found so far are Higgs is another question and the answer depends on the details of the modeling. However, what is more interesting is the possibility of seeing further effects connected to superconductivity like signatures of the subdominant modes (Bardasis–Schrieffer, Leggett). This is even more important than Higgs because it allows seeing the fine structure of the superconducting pairing interaction.

Lara Benfatto responded: A comment on why it matters if the THG in superconductors comes from Higgs or quasiparticles. The discussion started in the effort to understand what happens in conventional superconductors. Now, conventional superconductors by themselves are boring, so the real goal was more to use them as a benchmark to understand what this kind of spectroscopy can do. After a few years of experimental and theoretical work we can now say that in some regimes time-resolved THz spectroscopy can see something different than ordinary Raman. This now gives a different perspective on the possibility of using these protocols for nonconventional superconductors, where you don't understand many things. The hope is that these techniques see something you don't normally see.

Mattia Udina answered: The answer to this question is given by Lara Benfatto's comment.

Ilya Eremin commented: In continuation from Steven L. Johnson's question, does the mode have the same energy and can we talk about a well defined mode? Is higher harmonic generation a signature of the Higgs mode? A second question is what is the effect of disorder? Does the Higgs contribution become undamped? Does this mean it suppresses the contribution of quasiparticles? Experimental groups will keep calling this the Higgs mode – instead of fighting, should we find a better way of naming it? Are we in a regime where the onset of quasiparticles coincides with the onset of the mode? Would a true Higgs mode solve the problem/make experimental groups happy? Does the Higgs mode becoming dominant mean that quasiparticles are suppressed?

Mattia Udina responded: Disorder effects allow for sizable paramagnetic coupling between the electrons and light. While the diamagnetic coupling always leads to a negligible contribution of the Higgs as compared to the quasiparticle contribution, the ratio between the two can be reversed at strong disorder when accounting for the paramagnetic coupling. Interestingly, both the Higgs and the quasiparticle contributions increase when disorder effects are taken into account. As a general rule, the Higgs mode can be seen as a projection of the quasiparticle continuum in a specific channel, so it cannot be defined without quasiparticles.

Lara Benfatto remarked: The initial interpretation of superconducting signatures as due to Higgs was natural: one sends light, this quenches the superconducting phase, and it then recovers with oscillations of the order parameter. However, this is not what happens, since light does not simply quench the SC coupling. Light couples to electron–hole pairs, that can then excite other degrees of freedom, as happens *e.g.* for phonons. In the case of superconductors one then discovers that this quasiparticle signal is stronger than the Higgs one. So the reason why it is difficult to see the Higgs at small disorder is not because it is overdamped (*i.e.* its spectral function deviates from the delta-like behavior expected for real propagating modes). The reason why the Higgs is not visible is simply because it is not strongly Raman active, and then it weakly couples to light – in the very same way as non-Raman active phonons are not seen in pump–probe spectroscopy. With disorder one activates new electronic processes that increase the effective Raman visibility of the Higgs. Why then does the "Higgs" interpretation remain popular? In my view because there is strong pressure from part of the community to push theoretical studies which are very interesting by themselves but not necessarily related to experiments. I refer to many reports studying the response of the system after a quench of the SC coupling. This is a legitimate theoretical problem, but experiments are more complicated as I mentioned above. The real experiment is carried out by sending light into the sample, and driving some excitation. The probe then detects this non-equilibrium state, and its response is different with respect to the case without the pump. The assumption that the only effect of the first, strong pump pulse is just a quenching of the SC order parameter is not really theoretically justified. Describing the full process leading pump-excited states to interfer with the probe is more complicated, but it is out of the strong feedback between theory and experiments that something new will come about.

Matteo Puviani asked: Why did you only consider the zero-frequency limit in the results? Up to which frequency limit is this approximation valid?

In addition, in the zero-frequency limit, with also vanishing light momentum, wouldn't the nonlinear paramagnetic contribution already be nonzero even in the clean limit?

Mattia Udina answered: The zero-frequency limit is expected to be a good approximation as soon as the energy mismatch between the pump frequency and the superconducting gap value is large enough. This is guaranteed in THz pump–probe and THG experiments exploring the nonlinear response of cuprate superconductors, while the full frequency dependence of the signal may be relevant when considering *e.g.* superconducting NbN samples. At zero disorder, the paramagnetic contribution (also the one associated with the linear optical response) is different from zero in the static limit ($\omega = 0, q \rightarrow 0$), while it vanishes in the opposite dynamical limit ($q = 0, \omega \rightarrow 0$) considered here.

Matteo Puviani remarked: In the paper (https://doi.org/10.1039/d2fd00016d) an attractive Hubbard model has been used, with an s-wave order parameter. Would it be possible to use a d-wave order parameter? Would the use of a d-wave order parameter change the results qualitatively and/or quantitatively (and why would you expect so)?

Mattia Udina replied: The model can in principle be generalized to account for the d-wave symmetry of the superconducting order parameter. Nonetheless, we expect that this will not significantly affect our results concerning the effect of disorder on the polarization dependence of the nonlinear current, which is the focus of the present work. Indeed, as shown by previous work (see *e.g.* ref. 1), this should be mainly controlled by band-structure effects.

1 R. Matsunaga, N. Tsuji, K. Makise, H. Terai, H. Aoki and R. Shimano, *Phys. Rev. B*, 2017, **96**, 020505.

Matteo Puviani commented: In the paper (https://doi.org/10.1039/d2fd00016d), a Hubbard model with short-range attractive interaction has been used. What about a long-range Coulomb repulsion? Could it be implemented in that model? How would you expect the long-range Coulomb repulsion to affect the results in the zero-frequency limit?

Mattia Udina answered: Long-range Coulomb forces have not been explicitly included in our model, meaning that the phase mode preserves in principle a sound-like dispersion in the homogeneous case. As is well known, Coulomb interactions play a crucial role in the spectrum of the phase/charge mode at long wavelengths ($q \rightarrow 0$), pushing it towards the plasma frequency. On the other hand, the effect of Coulomb interactions on phase fluctuations at large q could be expected to be small. In the presence of disorder, one mixes to some extent the fluctuations at different momenta, so it is not evident *a priori* how the results obtained in our model will be quantitatively and qualitatively modified in the presence of long-range forces.[1] This is an interesting issue that will deserve future investigation.

1 G. Seibold, M. Udina, C. Castellani and L. Benfatto, *Phys. Rev. B*, 2021, **103**, 014512.

Matteo Puviani said: It is known that the A_{1g} symmetry quasiparticles' response in Raman spectroscopy is strongly suppressed in cuprates (by Coulomb interactions/phase due to gauge invariance). Is this also true in your model? Could you comment on that?

Mattia Udina replied: It is true that the full density response is suppressed by charge/phase fluctuations in superconducting systems, as expected due to charge conservation (see *e.g.* ref. 1). This effect is taken into account in our model, since we are including both amplitude and phase oscillations. Nonetheless, notice that this cancellation is exact only in continuum models, where the quasiparticle response coincides with the (full) density response. On the contrary, lattice-modulated density fluctuations do not follow the same conservation law and, therefore, they always lead to a sizable quasiparticle response in the superconducting state, as indeed demonstrated by Raman measurements.

1 T. Cea and L. Benfatto, *Phys. Rev. B*, 2016, **94**, 064512.

Matteo Puviani opened a discussion of the paper by Ilya M. Eremin: In the calculation of the fifth-harmonic generation two peaks appear in the frequency spectrum. Would it be possible to calculate the phase shift between the generated higher harmonic and the fundamental? Would you expect two "phase jumps" in correspondence with the two amplitude peaks?

Ilya Eremin responded: Many thanks for this suggestion, we will certainly do it.

Matteo Puviani commented: In the paper (https://doi.org/10.1039/d2fd00067a), diagrams representing the processes responsible for the fifth harmonic generation are shown. Why isn't a triangular diagram with diamagnetic vertices present? Would it also be relevant?

Ilya Eremin replied: These diagrams were drawn for illustrative purposes, *i.e.* we just illustrate the expression we found in the pseudospin approach. Certainly further diagrams might be necessary but this would go beyond the scope of the present study.

Lara Benfatto remarked: You mentioned that you aren't doing a gauge-invariant calculation, but you are doing it as you are adding phase fluctuation. What did you mean?

Ilya Eremin responded: Yes, the phase fluctuations are included within the pseudospin formalism and we see their magnitude is similar to the charge density fluctuation contributions (term directly proportional to A^3 and A^5) yet we do not include the long-range Coulomb interaction, which would allow the transformation of the phase fluctuations to the plasmons to be treated.

Lara Benfatto commented: An easy way to check if you are including phase fluctuations in a gauge-invariant way is to compute the third-harmonic response in the continuum limit, *i.e.* at very small doping. In this regime the diamagnetic response maps into the density–density response and as such it must vanish at finite frequency (and zero momentum) because of gauge invariance. In our diagrammatic approach we can easily check this result.

Lara Benfatto said: You showed two sets of experiments reporting fifth-harmonic generation. One is in a paper already published. The other is about MgB_2, and it is unpublished. Can you comment more on the reason why this should be a signature of the Higgs mode?

Ilya Eremin answered: In the experiments of Zhe Wang's group from TU Dortmund, which I showed (he kindly allowed me to show these unpublished data) they saw a signal in fifth harmonic generation coming from the lower gap, Δ_π. It is believed that the π-band is in the dirty limit, thus based on the current theoretical understanding they concluded that it is likely the signal comes from the Higgs mode of the gap in the π-band. But of course one would need to do the actual analysis to understand this better.

Lara Benfatto remarked: Related to my previous question, actually we have a calculation on MgB_2 which shows that it is true that the smaller gap is in the dirty limit, but for reasonable disorder values estimated from available experiments on MgB_2 in this regime, quasiparticles from the π band still dominate the non-linear response. Jacopo Fiore is presenting a poster (P07) showing these results.

Ilya Eremin responded: This is great news, I will let my colleague know.

Jan M. Rost asked: What are the prospects for investigating superconductivity in the context of topological effects? Have there been first attempts at this?

Ilya Eremin replied: Investigations on topological semimetals in non-linear optics are known. At the same time topological superconductors are a bit more elusive. We either do not have clear experimental evidence that these superconductors exist or the gaps in these systems are tiny. Of course ideally one should look/search for the magnetoelectric (axionic) effects. There are quite a number of theoretical studies but much fewer experimental ones.

Habib Rostami addressed Mattia Udina and Ilya Eremin: What is the effect of laser-induced heating of electrons on your response function? Can the temperature dependence of the nonlinear response be used as another channel to distinguish Higgs from the quasiparticle contribution? Consider that with the ultrashort pulses, we can have very high electronic temperature ($\sim$1000 K) which is not in equilibrium with the lattice (phonon) temperature.

Mattia Udina answered: Electron heating effects are not accounted for in our model, which is based on a quasi-equilibrium description. Nonetheless, experimental data show no significant variations in the time evolution of the oscillating signal when the pump fluence is increased, except for a gradual decrease in the amplitude of the 2Δ oscillations (see ref. 1), which is consistent with the expected smearing of both the quasiparticle and Higgs response.

1 R. Matsunaga, Y. I. Hamada, K. Makise, Y. Uzawa, H. Terai, Z. Wang and R. Shimano, *Phys. Rev. Lett.*, 2013, **111**, 057002.

Ilya Eremin replied: Usually within terahertz pulses in superconductors, the effect of the heating is reduced. The reason is that the frequency of the THz pulse is usually well below the $2\Delta_0$ where the electronic spectrum is gapped, and the quasiparticles cannot be heated. This is also confirmed experimentally where the gap signatures observed experimentally coincide with values known from equilibrium.

Martina Basini addressed Mattia Udina and Ilya Eremin: How do you usually evaluate the THz-induced temperature in the (electronic) system?

Mattia Udina responded: The effect of field-induced temperature variations is not considered in our calculations, which still rely on a quasi-equilibrium microscopic description. Nonetheless, absorption effects on the electron population, and the subsequent electronic temperature variations, could in principle be associated *e.g.* with the step-like background observed in some THz pump–probe experiments in metallic or superconducting compounds, where the amplitude of the step-like background increases with increasing pump intensity (see *e.g.* ref. 1).

1 R. Matsunaga, Y. I. Hamada, K. Makise, Y. Uzawa, H. Terai, Z. Wang and R. Shimano, *Phys. Rev. Lett.*, 2013, **111**, 057002.

Ilya Eremin replied: Well usually this is done by looking at the quasiparticle distribution function (in trARPES) allowing you to estimate the effective electronic temperature.

Lara Benfatto addressed Irene Burghardt and Ilya Eremin: What is the next goal of the theory – what is the big issue we should try to solve? Can you comment on this considering each of our separate sub-fields?

Irene Burghardt answered: From a general vantage point, the central big issue remains harnessing the multitude of molecular excited-state pathways in these multi-chromophoric functional materials, such as to gain molecular-level control of the desired energy and charge transfer processes. This goal obviously concerns experiment and theory alike.

From the viewpoint of theory and computation, an immense incentive is provided by the rapid progress in spectroscopic techniques. Besides many variants of nonlinear electronic and electronic-vibrational spectroscopies, novel ultrafast transient microscopy experiments provide a direct imaging of exciton and carrier transport in nanostructured materials. A big goal on the computational side is therefore to keep up with and contribute to these developments. Molecular-level quantum simulations provide both spatial and temporal resolution that directly connects to these experimental observations. Moreover, simulations can be extended to systematically explore quantum control schemes from the spatio-temporal perspective. From these directions, one would expect to not only achieve a better molecular-level understanding, but also to obtain general insight into the role of quasi-particles, collective effects, and coherent dynamics in photoexcited molecular materials.

In the specific context of our work presented in this Discussion, our immediate next goal is to perform computations of various nonlinear electronic and electronic-vibrational spectroscopies based upon accurate quantum propagation in many dimensions. The computational set-up for 2DES that was presented in this Discussion is accurate but minimalist and can be combined with accurate propagation for tens to hundreds of vibrational modes.[1] Much progress has been made in recent years in calculating nonlinear spectra from classical trajectories, and we plan to expand work in this direction towards accurate quantum propagation for molecular materials, including the use of multiconfigurational Gaussian wavepackets.

Finally, on the computational side, a major goal remains to efficiently combine accurate quantum dynamics with accurate excited-state electronic structure calculations. This is far from having been achieved. Ideally, one would merge two existing approaches, *i.e.*, on-the-fly trajectory based calculations *vs.* quantum wavepacket dynamics based on pre-computed, parametrized potentials. The prevailing excited-state electronic-structure methods employed for molecular nanoscale systems, especially Time-Dependent Density Functional Theory (TD-DFT), Density Functional based Tight Binding (DFTB) and related methods, are not necessarily accurate, while high-level methods tend to be prohibitively expensive when it comes to computing potential energy surface scans. Here, machine learning (ML) approaches could play an important role in the potential construction, such that the combination of quantum wavepacket propagation with ML potentials is a promising direction.

1 W. Popp, D. Brey, R. Binder and I. Burghardt, *Annu. Rev. Phys. Chem.*, 2021, **72**, 591–616.

Ilya Eremin responded: I guess the main question is to understand whether THG (as well as FHG) could be used to see more than just signatures of the superconducting gaps. For example the pair density wave state or hidden order in exotic superconductors, *i.e.* relatively clean systems, could be an interesting direction.

Conflicts of interest

There are no conflicts to declare.

Faraday Discussions

PAPER

Out-of-equilibrium dynamics driven by photoinduced charge transfer in CsCoFe Prussian blue analogue nanocrystals

S. Zerdane, *[ab] M. Hervé, [a] S. Mazerat,[c] L. Catala, [c] R. Alonso-Mori,[d] J. M. Glownia,[d] S. Song,[d] M. Levantino, [e] T. Mallah, *[c] M. Cammarata *[ae] and E. Collet *[a]

Received 21st January 2022, Accepted 16th March 2022

DOI: 10.1039/d2fd00015f

In this paper we study the out-of-equilibrium dynamics associated with photoinduced charge-transfer (CT) in cyanide-bridged Co–Fe Prussian blue analogue nanocrystals. In these coordination networks, the structural trapping of the photoinduced CT polaron involves local electronic and structural reorganizations. Femtosecond X-ray and optical absorption spectroscopies show that the local structural trapping process occurs on similar timescale for particles with 11 nm and 70 nm sizes. The local photoinduced spin transition, elongating the Co–N bonds and driving the $Co^{III}Fe^{II} \rightarrow Co^{II}Fe^{III}$ CT, activates coherent lattice torsion modes. The elastic deformation waves, launched by these bond elongations, drive macroscopic volume expansion and breathing of the particles. The timescale of this macroscopic deformation depends strongly on the size of the particle, which is more evidence of the multiscale nature of photoinduced phenomena in molecular materials.

Introduction

One of the most fascinating properties of cyanide-bridged bimetallic coordination networks (M–N–C–M′) is their ability to exhibit photoinduced charge transfer (CT) between the constitutive transition metal ions M′ and M.[1–14] This process is responsible for the emergence of various exotic functionalities such as magnetic order at low temperature. In the case of Co–Fe systems forming 3D cubic structures (Fig. 1), the cyanide groups form octahedral ligand fields, splitting the 3d orbitals of Co and Fe into t_{2g} and e_g sets. It was shown that photoexcitation of the

[a]Univ. Rennes, CNRS, IPR (Institut de Physique de Rennes) – UMR 6251, F-35000 Rennes, France. E-mail: eric.collet@univ-rennes1.fr

[b]SwissFEL, Paul Scherrer Institut, Villigen, PSI, Switzerland. E-mail: serhane.zerdane@psi.ch

[c]Institut de Chimie Moléculaire et des Matériaux d'Orsay, CNRS, Université Paris-Saclay, Orsay, France. E-mail: talal.mallah@u-psud.fr

[d]Linac Coherent Light Source, SLAC National Accelerator Laboratory, Menlo Park, CA, USA

[e]ESRF—The European Synchrotron, Grenoble, France. E-mail: marco.cammarata@esrf.fr

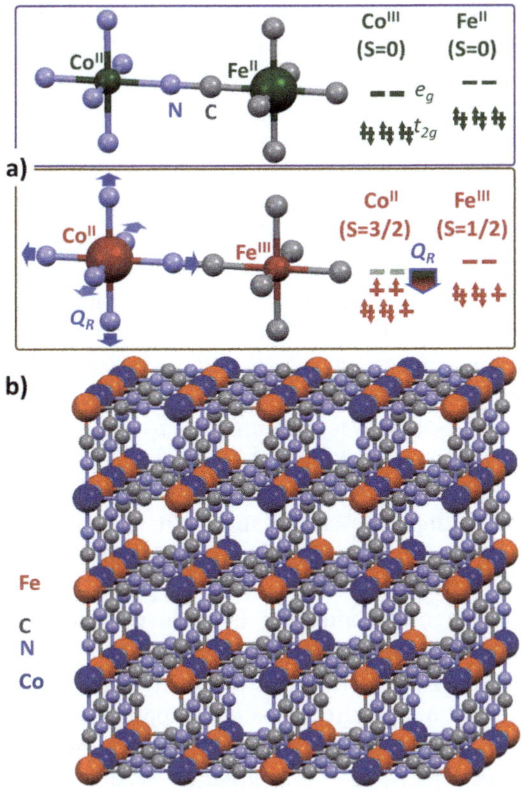

Fig. 1 (a) Representation of the $Co^{III}Fe^{II}$ and $Co^{II}Fe^{III}$ states. The charge transfer is accompanied by a spin transition on the Co, elongating the Co–N bonds, as e_g orbitals are populated. (b) Schematic view of the cyanide bridged CoFe cubic network.

low temperature diamagnetic $Co^{III}(S = 0)Fe^{II}(S = 0)$ state generates a CT $Co^{II}(S = 3/2)Fe^{III}(S = 1/2)$ state.[1] This seminal piece of work opened new ways to control various CT-based functions by light in M–N–C–M′ systems.[3,12,14–16] Chemical engineering made it possible to synthesize molecular analogues made of the Co–N–C–Fe photo-switchable unit and forming chains, $[Co_4Fe_4]$ cubes, $[Co_2Fe_2]$ squares or [CoFe] pairs.[17–26] The photoinduced intermetallic CT was also observed in various M–N–C–M′ systems made of different metal ions, including Fe–Fe,[27,28] Mn–Fe[29–31] and V–Cr[32] systems or Co–W[11,33] systems, which underlined the generalization of the photoinduced intermetallic CT process.

Understanding and mastering the ultra-fast photoinduced intermetallic CT process is a real challenge because it requires probing the system on the sub-100 femtosecond (fs) timescale of electronic dynamics and molecular motions, as underlined in a recent review by Johansson[34] about ultrafast studies of Fe–Fe, Co–Fe, Mn–Fe, and V–Cr PBAs.[27,28,35–37]

We recently used femtosecond X-ray and optical absorption spectroscopies to study the CT process in a photo-excited $Co^{III}(S = 0)Fe^{II}(S = 0)$ material (Fig. 1).[38] We studied nanocrystals with ≃11 nm size of the coordination network Cs{Co-[Fe(CN)$_6$]}, which were characterized by various complementary techniques including transmission electron microscopy, XPS, magnetism, optical spectroscopy, X-ray diffraction as well as EPR spectroscopy.[14] The photoinduced spin transition on the Co is responsible for the large lattice expansion, with an increase of the lattice parameter from 10.07 to 10.15 Å. The photoactivity of the nanocrystals[14] compared to bulk crystals is ascribed to their nanometric size, lowering the elastic energy cost for macroscopic photoinduced CT.

Ultra-fast pump–probe measurements allowed understanding the photoinduced CT process.[38] The element-specific XANES technique allowed for disentangling electronic and structural dynamics. XANES studies at the LCLS X-ray free-electron laser (X-FEL) revealed that it is the spin-transition, which is photoinduced first on the Co site within ∼50 fs, that drives the subsequent Fe → Co charge-transfer within ∼200 fs. Femtosecond optical spectroscopy confirmed that the process occurs within ∼200 fs. In addition, the structural trapping of the $Co^{II}(S = 3/2)Fe^{III}(S = 1/2)$ state is accompanied by coherent lattice vibrations, which corresponds to a torsion of the M–N–C–M′ bridge.[38] A similar study of the photoinduced charge transfer was performed in the RbMnFe PBA, which has a similar structure to the CoFe crystal shown in Fig. 1. Femtosecond optical pump–probe spectroscopy experiments highlighted the local and ultrafast nature of the photoinduced $Mn^{III}Fe^{II} \rightarrow Mn^{II}Fe^{III}$ CT process.[36] Photoexcitation of α spins around 540 nm, corresponding to the Mn(d–d) band, drives reverse Jahn–Teller distortion through the population of antibonding Mn–N orbitals, and induces CT within ∼190 fs. The process also launches the coherent lattice torsion mode of the M–N–C–M′ bridge during the self-trapping of the CT small-polaron. Time-resolved X-ray diffraction on the picosecond timescale evidenced the macroscopic volume expansion of the RbMnFe crystal due to the structural trapping process, as the small CT polaron associated with the elongation of the Mn–N bonds expands the polymeric M–N–C–M′ lattice.[37]

Here we study out-of-equilibrium dynamics in CoFe nanocrystals on the femtosecond and picosecond timescales. In particular, we are interested in the effect of the size of the crystalline particles on the femtosecond photoinduced CT process and the subsequent macroscopic changes. We focus on two nanocrystal sizes, 11 nm and 70 nm.

Results and discussion

Characterization of the 11 nm and 70 nm CoFe nanocrystals

As explained in previous papers,[14,38] colloidal solutions of CsCoFe Prussian blue analogue nanoparticles were prepared by adding an aqueous solution of $K_3[Fe^{III}(CN)_6]$ to an aqueous solution containing $[Co^{II}(H_2O)_6]Cl_2$ and CsCl in order to obtain, first, a solution containing the 11 nm nanocrystals. Then, the 70 nm nanocrystals were obtained by performing a seed-mediated growth of the 11 nm nanocrystals up to 70 nm using a procedure similar to one already reported.[39,40] Dynamic light scattering measurements (Fig. 2) of the colloidal solutions show objects with a hydrodynamic diameter centred at around 11 nm (**1**) and 70 nm (**2**), confirmed by transmission electron microscopy (TEM) imaging. For sample (**1**), there is a majority (∼70%) of low-spin and photoactive diamagnetic $Co^{III}Fe^{II}$ pairs, located mainly in the core of the ≃11 nm nanocrystals. There is also a mixture of $Co^{II}Fe^{II}$ and $Co^{II}Fe^{III}$ species present mainly within the shell of the objects estimated to have a thickness around 1 nm. For sample (**2**) made of

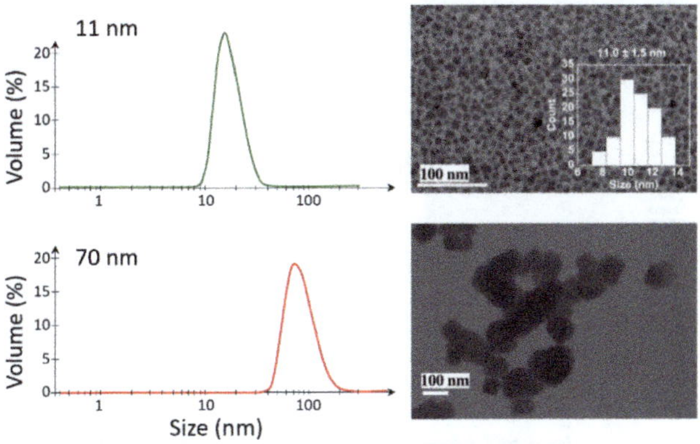

Fig. 2 Dynamic light scattering and transmission electron microscopy images of the CsCoFe nanocrystals with characteristic sizes: (**1**) (11 nm, top) and (**2**) (70 nm, bottom).

$\simeq$70 nm nanocrystals, the smaller shell–core ratio favours a higher fraction of photoactive diamagnetic $Co^{III}Fe^{II}$ pairs.

Hereafter we use ultrafast time-resolved XANES and optical spectroscopy to compare the effect of the size of the particles on the out-of-equilibrium dynamics following fs laser excitation.

Ultrafast and local self-trapping dynamics

Bleuzen *et al.* have demonstrated that X-ray Absorption Near-Edge Structure (XANES) is a relevant technique for studying photoinduced CT phenomena in Co–Fe PBAs.[16,41] At the Fe edge the formal $Fe^{II} \rightarrow Fe^{III}$ change of oxidation is characterized by a rigid spectral shift ($\approx$0.6 eV) towards higher energy, as there is no important structural reorganization around Fe. At the Co edge, the $Co^{III} \rightarrow Co^{II}$ oxidation change is associated with an opposite spectral shift towards lower energy. In addition, the spin transition on the Co, increasing Co–N distances as antibonding e_g orbitals get populated, shifts the first EXAFS feature towards lower energy ($\simeq$7770 eV) in the HS Co^{II} state.

Fig. 3 shows XANES spectra of (**1**) in the ground state, measured at the LCLS around Co and Fe edges. The XANES white line, with maxima at 7725 eV and 7728 eV, is characteristic of the presence of both Co^{II} and Co^{III} ions in (**1**).[38,41] As already explained in our previous study of the fs dynamics of photoexcited 11 nm CoFe particles,[38] we used the optical pump/XANES probe technique with a time resolution $\sigma_X \approx 25$ fs RMS ($IRF_{FWHM} = 58$ fs) at the XPP instrument of the LCLS X-FEL.[42–44] We acquired measurements from an optimized free-flowing jet of (**1**) or (**2**) CoFe nanocrystals in aqueous solutions, excited at 540 nm by fs optical pulses and probed by fs X-ray pulses.

Weak XANES changes are observed 3 ps after photoexcitation as only a fraction of the active sites are excited (Fig. 3). Fig. 4 shows the time-resolved changes of X-ray absorption intensity (ΔI) around the Co and Fe K-edges (left and right) both for 11 nm (**1**) and 70 nm (**2**) samples (top and bottom respectively).

In our previous study,[38] we have shown that the photoinduced $Co^{III}Fe^{II} \rightarrow Co^{II}Fe^{III}$ CT is characterised by the ultrafast spectral shifts in opposite directions at the Fe and Co edges. The simultaneous decrease of X-ray absorption below the

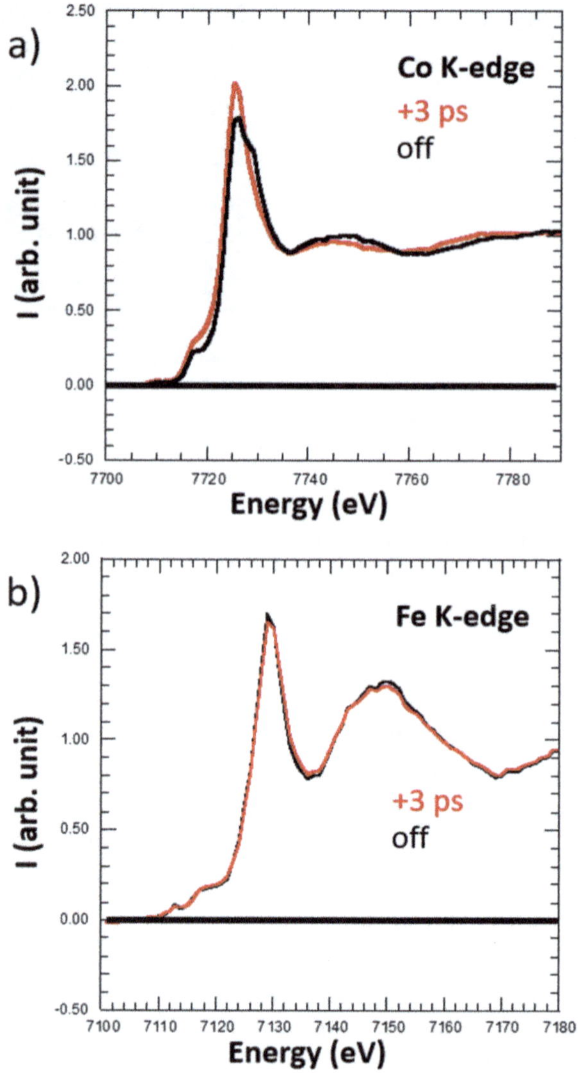

Fig. 3 XANES spectra of the CoFe nanoparticles at the Co (a) and Fe (b) edges without laser excitation (black) and 3 ps after photoexcitation (red).

Fe K-edge (7131 eV) and increase above this value is due to the change from Fe^{II} to Fe^{III}, while the opposite simultaneous increase of X-ray absorption below the Co K-edge (7727 eV) and decrease above this value is due to the change from Co^{III} to Co^{II}. Fig. 4 shows that the dynamics are quite similar for the 11 nm and 70 nm particles. On the one hand, the photoinduced spin transition occurring first within $\simeq$50 fs on the Co is responsible for the ultrafast increase of X-ray absorption at 7770 eV. This is the direct signature of the fast elongation of the Co–N bonds, as discussed previously for the case of the 11 nm nanocrystals, and observed here also for the 70 nm ones.[38] On the other hand, changes are slower at the Fe edge (7121 and 7129 eV), which is the direct signature of the slower CT,

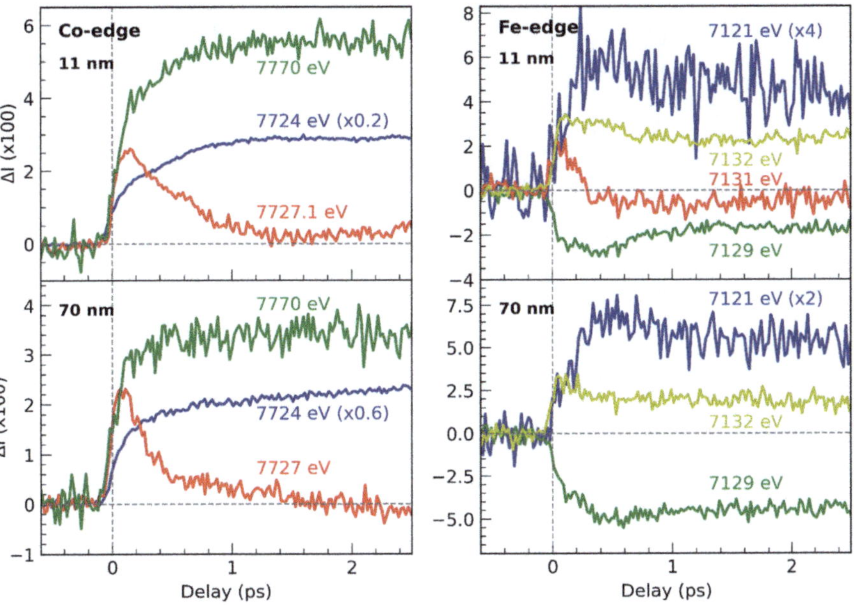

Fig. 4 Femtosecond optical pump/X-ray probe time scans, showing the changes of X-ray absorption intensity ΔI at selected energies around the Co (left) and Fe (right) K-edges after photoexcitation at 540 nm, measured for 11 nm (top) and 70 nm (bottom) particles. Red curves are measured close to isosbestic points.

which occurs in a second stage within $\simeq$200 fs. The transient absorption components around 7727 eV and 7131 eV, corresponding to isosbestic points at Co and Fe edges between $Co^{III}Fe^{II}$ and photoinduced $Co^{II}Fe^{III}$ states, are associated with the appearance of an intermediate state (or several), decaying within $\simeq$200 fs. This decaying intermediate also contributes to the fast component at 7132 eV. Therefore, for both the 11 nm and the 70 nm particles, the main XANES signatures confirm that the photoinduced transformation process is a spin-transition-induced charge-transfer process, as discussed in more detail previously.[38]

Fig. 5 shows time-resolved optical density changes (ΔOD), following excitation with fs pulses at 540 nm, of the same samples (**1**) or (**2**). We recorded ΔOD changes for different probe wavelengths with high signal to noise and a time resolution of $\sigma_{OD} \approx 80$ fs. The spectral changes, with bleaching of the 540 nm band and the appearance of new bands above and below, is also characteristic of the photoinduced $Co^{III}Fe^{II} \rightarrow Co^{II}Fe^{III}$ CT.[38] For the 11 nm particles, the isosbestic point around 627 nm provides a clear measurement of the decay of the intermediate state, in good agreement with XANES data. The ΔOD signals monitored for 11 and 70 nm particles show different relative changes in amplitude and the isosbestic point is shifted around 670 nm in the 70 nm particles. This is probably due to the slightly different chemical compositions and the different ratio of photoactive diamagnetic $Co^{III}Fe^{II}$ pairs. However, the dynamics of the ΔOD signals show similar timescales for the 11 nm and 70 nm particles, with clear signatures of the intermediate electronic state at the isosbestic points

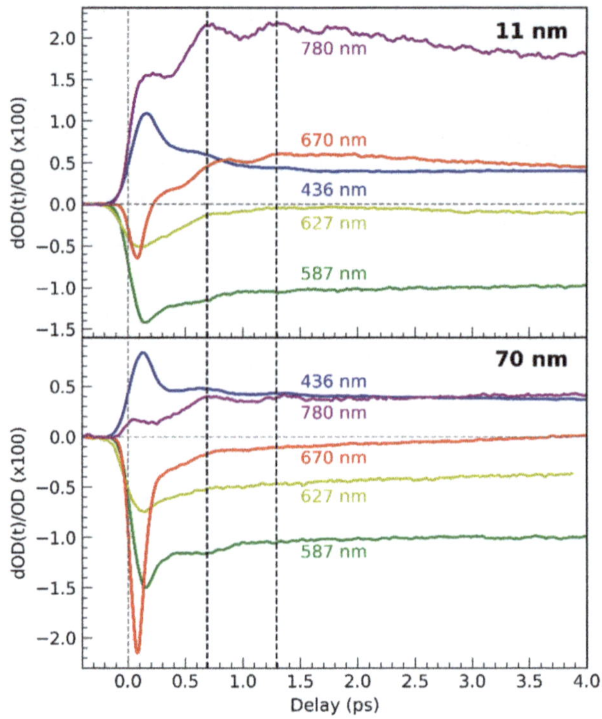

Fig. 5 Time-resolved relative variation of OD at different probe wavelengths after fs laser excitation at 540 nm for 11 nm (top) and 70 nm (bottom) particles.

in agreement with time-resolved XANES data. Another key feature of the dynamics is the appearance of damped coherent oscillation at 1.65 THz (~620 fs period, Fig. 5), harder to distinguish in XANES data, due to limited signal to noise ratio. However, these oscillations are clearly in phase for the 11 nm and 70 nm particles, as indicated by the dashed lines in Fig. 5. All these XANES and optical data show that the early response of photoexcited CoFe particles does not depend on the size of the nanocrystals, which agrees with the local nature of the self-trapping process of the photoinduced CT polaron already highlighted in both CoFe and MnFe materials.[36–38] As previously discussed in detail for the CoFe PBA, the very first step of the photoinduced dynamics corresponds to the light-induced excited spin-state trapping process (LIESST) on the Co. The initial photoexcited $Co^{III*}(S=0)Fe^{II}(S=0)$ state populates antibonding $Co(e_g)$ states, which drives Co–N bond expansion and the spin transition towards the $Co^{III}(S=2)Fe^{II}(S=0)$.[38] This bond elongation occurs within $\simeq 50$ fs, as characterized by the fast increase in absorption at 7770 eV for the 11 nm particles. Fig. 4 shows that the timescales are very similar in the case of 70 nm particles. The formation of the short-lived intermediate states, like the $Co^{III*}(S=0)Fe^{II}(S=0)$ or $Co^{III*}(S=2)Fe^{II}(S=0)$, contributes to the fast signal increase within the time resolution and its decay within $\simeq 200$ fs around the isosbestic points at the Fe edge, Co edge or OD data.

These are mediators towards the $Co^{II}(S=3/2)Fe^{III}(S=1/2)$ state, and the 200 fs timescale corresponds therefore to the one of the Fe-to-Co CT. Indeed, there is no fast component around the Fe K-edge, except around the isosbestic point, where the

INT states contributes. The spectral shift at the Fe edge, within 200 fs, is therefore only attributed to CT since there is no structural signature of the process around the Fe edge. This 200 fs CT timescale is similar for the 11 nm and 70 nm particles.

The coherent oscillations with 620 fs period (1.62 THz), are attributed to a global torsion mode of the Fe–C–N–Co bridged network, as shown by DFT calculations.[38] In a similar Mn–Fe system, the activation of such a coherent mode after photo-induced CT was also reported and attributed to the coherent activation of the torsion mode of the M–N–C–M′ bridge.[45] Indeed, the crystalline network has no time to expand on this timescale and a lattice torsion is necessary to accommodate the Co–N bond elongation trapping the photoinduced CT. The in-phase oscillation of these coherent oscillations is another indication that the sub-ps dynamics associated with the structural trapping of the CT process, forming a small polaron stabilized by the spin transition on the Co, is similar for 11 nm and 70 nm particles, confirming the local nature of the process.

Breathing of the crystalline particles

Fig. 6 shows ΔOD, probed around the isosbestic points, following excitation with fs pulses at 540 nm, of the same samples made of 11 nm or 70 nm nanocrystals,

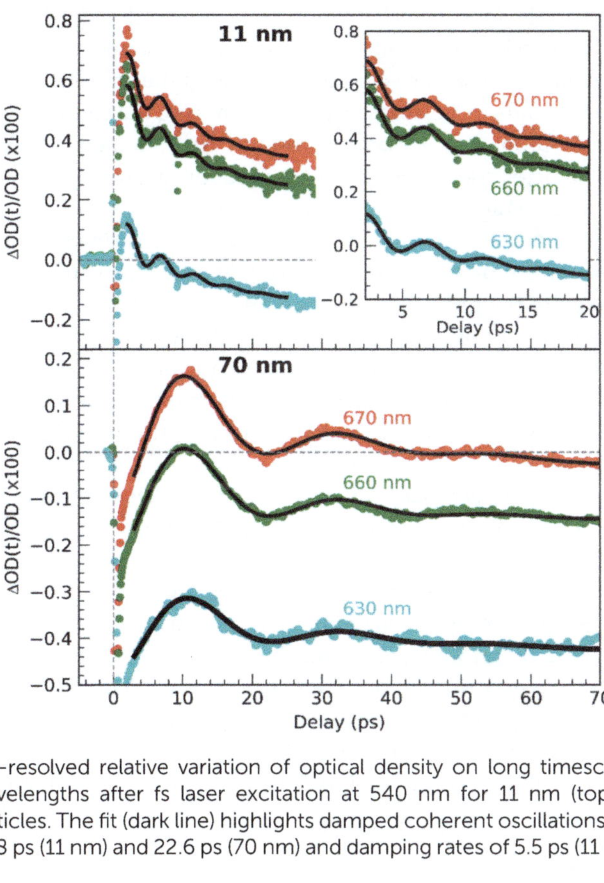

Fig. 6 Time-resolved relative variation of optical density on long timescale probed at different wavelengths after fs laser excitation at 540 nm for 11 nm (top) and 70 nm (bottom) particles. The fit (dark line) highlights damped coherent oscillations with different periods of 4.8 ps (11 nm) and 22.6 ps (70 nm) and damping rates of 5.5 ps (11 nm) and 13.23 ps (70 nm).

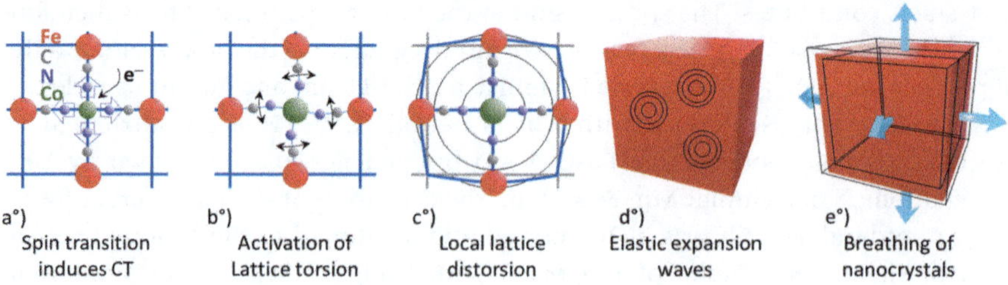

Fig. 7 Multiscale dynamics in PBA nanocrystals, from local trapping of CT state to macroscopic crystal breathing.

on the 70 ps timescale. In addition to the ΔOD signals within ≃1 ps shown in Fig. 5, Fig. 6 shows that the dynamics on longer timescales differ for the 11 nm and the 70 nm nanocrystals. For **1** (11 nm particles), the fit of the data reveals damped coherent oscillations, with a period of 4.8 (0.1) ps and a damping rate of 5.5 (0.3) ps. This dynamic is clearly different in the case of **2** (70 nm particles): the fit of the data gives an oscillating period of 22.6 (0.2) ps and a damping rate of 13.2 (0.3) ps. In the following we discuss the origin of this process and its dependence on the size of the nanocrystals.

In the case of the similar RbMnFe system, time-resolved X-ray diffraction of photoexcited crystals, with typical 300 nm size, revealed that the formation of the small CT polarons, with M–N bond elongation, is the source of elastic deformation waves driving macroscopic volume expansion of the crystals.[14,37] This volume expansion corresponds to a displacive process, shifting the equilibrium volume.[45] As elastic waves propagate within the crystals with sound velocity (c = 4300 ms^{-1}),[46] the characteristic acoustic timescale τ_a for macroscopic volume expansion scales to the average crystal size (r).[47]

For **1** (r = 11 nm), we estimate $\tau_a = r/c = 2.6$ ps. This timescale corresponds well to the first maximum of the OD oscillation around 2.5 ps in Fig. 6. Due to the displacive nature of the process, the crystal breathes around its new equilibrium volume with corresponding period $T_b = 2\tau_a$, as the acoustic waves have to propagate back and forth during the contraction/expansion of the crystal. Therefore, the measured breathing period T_b = 4.8(1) ps corresponds well to $2\tau_a$, even though the size distribution of the particles, in the 9–13 nm range, can broaden the breathing period $T_b = 2\tau_a$ in the 4–6 ps range.

For **2** (r = 70 nm), we estimate $\tau_a = r/c = 16.3$ ps, which corresponds well to the first maximum of the OD oscillation around 12 ps in Fig. 6, given the size dispersion of the particles (50–90 nm). The breathing period T_b = 32 ps is also quite consistent with the measured 22.6(1) ps breathing period.

The macroscopic structural dynamics associated with the breathing of the particles depends therefore strongly on their size. The damping rate is much faster (5.5 ps) for the 11 nm nanocrystals than for the 70 nm ones. This may be due to the fact that 11 nm particles are made of a ≃7 nm core of active $Co^{III}Fe^{II}$ species surrounded by a shell of passive $Co^{II}Fe^{III}$ and $Co^{II}Fe^{II}$ species, which may damp the oscillation compared to the 70 nm particle with larger core/shell ratio.

Conclusions

Fig. 7 shows a summary of the out-of-equilibrium dynamics induced by a fs laser pulse in photoactive CT PBA, for which the CT couples to a local bond elongation (spin transition). The optical excitation of the d–d transition centred on the Co(M) drives the spin transition on the Co (Fig. 7a), as the optical population of Co(e_g) antibonding orbitals elongates Co–N bonds within 50 fs, as revealed by XANES. This process reduces the Co(t_{2g})–Co(e_g) gap, which favours CT and stabilizes the $Co^{II}Fe^{III}$ state and the CT process occurs within 200 fs. The local bond elongation requires local lattice torsion (Fig. 7b) as the lattice has to accommodate to the local volume change. This process activates coherently lattice torsion modes. The local lattice distortions due to the local trapping of small CT polarons (Fig. 7c) are the source of elastic deformation waves propagating through the sample (Fig. 7d), responsible for macroscopic volume expansion and breathing of the nanocrystals around the new equilibrium volume. The formation of the polaron, associated with the spin transition, the consecutive CT and the local distortion, is a local process and the dynamics do not depend on the size of the crystalline particles. On the contrary, the propagation of the elastic deformation waves, responsible for macroscopic volume expansion, is size dependent.

Author contributions

E. C., M. C. and T. M. conceived the project. S. M., L. C. and T. M. synthesized and characterized particles. M. C., S. Z., R. A.-M., J. M. G., S. S. and E. C. performed the fs XANES experiment. S. Z. and M. H. performed the optical study. S. Z., M. C. and E. C. analysed the data. E. C., S. Z. and M. C. wrote the paper. All authors contributed to discussions and gave comments on the manuscript.

Conflicts of interest

There are no conflicts to declare.

Acknowledgements

E. C. and M. C. acknowledge the support of Rennes Métropole, ANR (ANR-13-BS04-0002 FEMTOMAT, ANR-19-CE30-0004 ELECTROPHONE and ANR-19-CE29-0018 MULTICROSS) Centre National de la Recherche Scientifique (CNRS, PEPS SASLELX), Fonds Européen de Développement Régional (FEDER) and Région Bretagne (ARED 8925/XFELMAT). T. M. and L. C. thanks the IUF (Institut Universitaire de France) and the ANR (ANR-13-BS04-0002 FEMTOMAT) for financial support. Use of the Linac Coherent Light Source, SLAC National Accelerator Laboratory, is supported by the US Department of Energy, Office of Science, Office of Basic Energy Sciences under contract no. DE-AC02-76SF00515.

References

1 O. Sato, T. Iyoda, A. Fujishima and K. Hashimoto, *Science*, 1996, **272**, 704–705.
2 S. Ohkoshi and H. Tokoro, *Acc. Chem. Res.*, 2012, **45**, 1749–1758.

3 C. Cartier dit Moulin, F. Villain, A. Bleuzen, M.-A. Arrio, P. Sainctavit, C. Lomenech, V. Escax, F. Baudelet, E. Dartyge, J.-J. Gallet and M. Verdaguer, *J. Am. Chem. Soc.*, 2000, **122**, 6653–6658.

4 O. N. Risset, P. A. Quintero, T. V. Brinzari, M. J. Andrus, M. W. Lufaso, M. W. Meisel and D. R. Talham, *J. Am. Chem. Soc.*, 2014, **136**, 15660–15669.

5 D. Aguila, Y. Prado, E. S. Koumousi, C. Mathoniere and R. Clerac, *Chem. Soc. Rev.*, 2016, **45**, 203–224.

6 S. Ferlay, T. Mallah, R. Ouahes, P. Veillet and M. Verdaguer, *Nature*, 1995, **378**, 701–703.

7 D. M. Pajerowski, B. Ravel, C. H. Li, M. F. Dumont and D. R. Talham, *Chem. Mater.*, 2014, **26**, 2586–2594.

8 K. Zhang, S. Kang, Z. S. Yao, K. Nakamura, T. Yamamoto, Y. Einaga, N. Azuma, Y. Miyazaki, M. Nakano, S. Kanegawa and O. Sato, *Angew. Chem., Int. Ed.*, 2016, **55**, 6047–6050.

9 O. N. Risset, T. V. Brinzari, M. W. Meisel and D. R. Talham, *Chem. Mater.*, 2015, **27**, 6185–6188.

10 T. Yoshida, K. Nakabayashi, H. Tokoro, M. Yoshikiyo, A. Namai, K. Imoto, K. Chiba and S.-i. Ohkoshi, *Chem. Sci.*, 2020, **11**, 8989–8998.

11 N. Ozaki, H. Tokoro, Y. Hamada, A. Namai, T. Matsuda, S. Kaneko and S. Ohkoshi, *Adv. Funct. Mater.*, 2012, **22**, 2089–2093.

12 M. Verdaguer, A. Bleuzen, V. Marvaud, J. Vaissermann, M. Seuleiman, C. Desplanches, A. Scuiller, C. Train, R. Garde, G. Gelly, C. Lomenech, I. Rosenman, P. Veillet, C. Cartier and F. Villain, *Coord. Chem. Rev.*, 1999, **190–192**, 1023–1047.

13 J. M. Cain, A. C. Felts, M. W. Meisel and D. R. Talham, *Chem. Mater.*, 2021, **33**, 246–255.

14 L. Trinh, S. Zerdane, S. Mazérat, N. Dia, D. Dragoe, C. Herrero, E. Rivière, L. Catala, M. Cammarata, E. Collet and T. Mallah, *Inorg. Chem.*, 2020, **59**, 13153–13161.

15 N. Shimamoto, S. Ohkoshi, O. Sato and K. Hashimoto, *Inorg. Chem.*, 2002, **41**, 678–684.

16 J. D. Cafun, J. Lejeune, F. Baudelet, P. Dumas, J. P. Itie and A. Bleuzen, *Angew. Chem., Int. Ed.*, 2012, **51**, 9146–9148.

17 D. Li, R. Clérac, O. Roubeau, E. Harté, C. Mathonière, R. Le Bris and S. M. Holmes, *J. Am. Chem. Soc.*, 2008, **130**, 252–258.

18 Y. Zhang, D. Li, R. Clérac, M. Kalisz, C. Mathonière and S. M. Holmes, *Angew. Chem., Int. Ed.*, 2010, **49**, 3752–3756.

19 E. S. Koumousi, I.-R. Jeon, Q. Gao, P. Dechambenoit, D. N. Woodruff, P. Merzeau, L. Buisson, X. Jia, D. Li, F. Volatron, C. Mathonière and R. Clérac, *J. Am. Chem. Soc.*, 2014, **136**, 15461–15464.

20 R. Lescouezec, J. Vaissermann, C. Ruiz-Perez, F. Lloret, R. Carrasco, M. Julve, M. Verdaguer, Y. Dromzee, D. Gatteschi and W. Wernsdorfer, *Angew. Chem., Int. Ed.*, 2003, **42**, 1483–1486.

21 J. Mercurol, Y. Li, E. Pardo, O. Risset, M. Seuleiman, H. Rousseliere, R. Lescouezec and M. Julve, *Chem. Commun.*, 2010, **46**, 8995–8997.

22 M. Nihei, Y. Sekine, N. Suganami, K. Nakazawa, A. Nakao, H. Nakao, Y. Murakami and H. Oshio, *J. Am. Chem. Soc.*, 2011, **133**, 3592–3600.

23 Y.-Z. Zhang, P. Ferko, D. Siretanu, R. Ababei, N. P. Rath, M. J. Shaw, R. Clérac, C. Mathonière and S. M. Holmes, *J. Am. Chem. Soc.*, 2014, **136**, 16854–16864.

24 Q. P. Xuan, J. Glatz, A. Benchohra, J.-R. Jiménez, R. Plamont, L.-M. Chamoreau, A. Flambard, Y. Li, L. Lisnard, D. Dambournet, O. J. Borkiewicz, M.-L. Boillot, L. Catala, A. Tissot and R. Lescouëzec, *J. Mater. Chem. C*, 2021, **9**, 8882–8890.

25 S. Kamilya, S. Ghosh, Y. Li, P. Dechambenoit, M. Rouzières, R. Lescouëzec, S. Mehta and A. Mondal, *Inorg. Chem.*, 2020, **59**, 11879–11888.

26 S. F. Jafri, E. S. Koumousi, M.-A. Arrio, A. Juhin, D. Mitcov, M. Rouzières, P. Dechambenoit, D. Li, E. Otero, F. Wilhelm, A. Rogalev, L. Joly, J.-P. Kappler, C. Cartier dit Moulin, C. Mathonière, R. Clérac and P. Sainctavit, *J. Am. Chem. Soc.*, 2019, **141**, 3470–3479.

27 D. C. Arnett, P. Voehringer and N. F. Scherer, *J. Am. Chem. Soc.*, 1995, **117**, 12262–12272.

28 D. Weidinger, D. J. Brown and J. C. Owrutsky, *J. Chem. Phys.*, 2011, **134**, 124510.

29 S. Ohkoshi, H. Tokoro, M. Utsunomiya, M. Mizuno, M. Abe and K. Hashimoto, *J. Phys. Chem. B*, 2002, **106**, 2423–2425.

30 H. Tokoro and S. Ohkoshi, *Bull. Chem. Soc. Jpn.*, 2015, **88**, 227–239.

31 K. Kato, Y. Moritomo, M. Takata, M. Sakata, M. Umekawa, N. Hamada, S. Ohkoshi, H. Tokoro and K. Hashimoto, *Phys. Rev. Lett.*, 2003, **91**, 255502.

32 J. O. Johansson, J. W. Kim, E. Allwright, D. M. Rogers, N. Robertson and J. Y. Bigot, *Chem. Sci.*, 2016, **7**, 7061–7067.

33 S. Chorazy, J. Wang and S. Ohkoshi, *Chem. Commun.*, 2016, **52**, 10795–10798.

34 K. Barlow and J. O. Johansson, *Phys. Chem. Chem. Phys.*, 2021, **23**, 8118–8131.

35 J. Zimara, H. Stevens, R. Oswald, S. Demeshko, S. Dechert, R. A. Mata, F. Meyer and D. Schwarzer, *Inorg. Chem.*, 2021, **60**, 449–459.

36 G. Azzolina, H. Tokoro, K. Imoto, M. Yoshikiyo, S.-i. Ohkoshi and E. Collet, *Angew. Chem., Int. Ed.*, 2021, **60**, 23267–23273.

37 G. Azzolina, R. Bertoni, C. Mariette, M. Cammarata, E. Trzop, C. Ecolivet, M. Sander, M. Levantino, H. Tokoro, K. Imoto, M. Yoshikiyo, S. Ohkoshi and E. Collet, *J. Mater. Chem. C*, 2021, **9**, 6773–6780.

38 M. Cammarata, S. Zerdane, L. Balducci, G. Azzolina, S. Mazerat, C. Exertier, M. Trabuco, M. Levantino, R. Alonso-Mori, J. M. Glownia, S. Song, L. Catala, T. Mallah, S. F. Matar and E. Collet, *Nat. Chem.*, 2021, **13**, 10–14.

39 L. Catala, D. Brinzei, Y. Prado, A. Gloter, O. Stéphan, G. Rogez and T. Mallah, *Angew. Chem., Int. Ed.*, 2009, **48**, 183–187.

40 Y. Prado, S. Mazerat, E. Rivière, G. Rogez, A. Gloter, O. Stéphan, L. Catala and T. Mallah, *Adv. Funct. Mater.*, 2014, **24**, 5402–5411.

41 A. Bordage and A. Bleuzen, *Radiat. Phys. Chem.*, 2020, **175**, 108143.

42 H. T. Lemke, K. S. Kjær, R. Hartsock, T. Brandt van Driel, M. Chollet, J. M. Glownia, S. Song, D. Zhu, E. Pace, S. F. Matar, M. N. Nielsen, M. Benfatto, K. J. Gaffney, E. Collet and M. Cammarata, *Nat. Commun.*, 2017, **8**, 15342.

43 M. Chollet, R. Alonso-Mori, M. Cammarata, D. Damiani, J. Defever, J. T. Delor, Y. Feng, J. M. Glownia, J. B. Langton, S. Nelson, K. Ramsey, A. Robert, M. Sikorski, S. Song, D. Stefanescu, V. Srinivasan, D. Zhu, H. T. Lemke and D. M. Fritz, *J. Synchrotron Radiat.*, 2015, **22**, 503–507.

44 M. Harmand, R. Coffee, M. R. Bionta, M. Chollet, D. French, D. Zhu, D. M. Fritz, H. T. Lemke, N. Medvedev, B. Ziaja, S. Toleikis and M. Cammarata, *Nat. Photonics*, 2013, **7**, 215–218.

45 S. I. Ohkoshi, M. Yoshikiyo, A. Namai, K. Nakagawa, K. Chiba, R. Fujiwara and H. Tokoro, *Sci. Rep.*, 2017, **7**, 8088.

46 K. Boukheddaden, E. D. Loutete-Dangui, E. Codjovi, M. Castro, J. A. Rodriguéz-Velamazán, S. Ohkoshi, H. Tokoro, M. Koubaa, Y. Abid and F. Varret, *J. Appl. Phys.*, 2011, **109**, 013520.

47 R. Bertoni, M. Lorenc, H. Cailleau, A. Tissot, J. Laisney, M. L. Boillot, L. Stoleriu, A. Stancu, C. Enachescu and E. Collet, *Nat. Mater.*, 2016, **15**, 606–610.

Faraday Discussions

ROYAL SOCIETY OF CHEMISTRY

PAPER

Nonequilibrium dynamics of α-$RuCl_3$ – a time-resolved magneto-optical spectroscopy study

Julian Wagner, [a] Anuja Sahasrabudhe,[a] Rolf Versteeg, [a] Zhe Wang, [b] Vladimir Tsurkan,[cd] Alois Loidl, [c] Hamoon Hedayat *[a] and Paul H. M. van Loosdrecht *[a]

Received 10th January 2022, Accepted 24th February 2022
DOI: 10.1039/d2fd00006g

We present time-resolved magneto-optical spectroscopy on the magnetic Mott–Hubbard-insulating Kitaev spin liquid candidate α-$RuCl_3$ to investigate the nonequilibrium dynamics of its antiferromagnetically ordered zigzag groundstate after photoexcitation. A systematic study of the transient magnetic linear dichroism under different experimental conditions (temperature, external magnetic field, photoexcitation density) gives direct access to the dynamical interplay of charge excitations with the zigzag ordered state on ultrashort time scales. We observe a rather slow initial demagnetization (few to 10s of ps) followed by a long-lived non-thermal antiferromagnetic spin-disordered state (100–1000s of ps), which can be understood in terms of holons and doublons disordering the antiferromagnetic background after photoexcitation. Varying temperature and fluence in the presence of an external magnetic field reveals two distinct photoinduced dynamics associated with the zigzag and quantum paramagnetic disordered phases. The photo-induced non-thermal spin-disordered state shows universal compressed-exponential recovery dynamics related to the growth and propagation of zigzag domains on nanosecond time scales, which is interpreted within the framework of the Fatuzzo–Labrune model for magnetization reversal. The study of nonequilibrium states in strongly correlated materials is a relatively unexplored topic, but our results are expected to be extendable to a large class of Mott–Hubbard insulator materials with strong spin–orbit coupling.

1. Introduction

To probe material properties using ultrafast laser pulses is a fascinating aspect of modern experimental condensed matter physics, which has been shown to be

[a]*Universität zu Köln, II. Physikalisches Institut, Zülpicher Straße 77, Köln D-50937, Germany. E-mail: hedayat@ph2.uni-koeln.de; pvl@ph2.uni-koeln.de*
[b]*Department of Physics, TU Dortmund University, Otto-Hahn-Str. 4, 44227 Dortmund, Germany*
[c]*Experimental Physics V, Center for Electronic Correlations and Magnetism, University of Augsburg, 86135 Augsburg, Germany*
[d]*Institute of Applied Physics, Chisinau, MD 2028, Republic of Moldova*

very powerful in accessing new transient phases, which are nonexistent under thermal equilibrium conditions.[1,2] A prime example is the ultrafast manipulation of magnetic order.[3,4] Especially, the understanding of the underlying dynamical processes, associated time scales and how magnetic order can be controlled is at the forefront of non-equilibrium condensed matter research. Of particular interest are Mott-insulators, which exhibit strong electron–electron interactions and significant correlations between electronic and magnetic degrees of freedom giving rise to exotic properties and groundstates.[5] An interesting case is found in the magnetic Mott–Hubbard-insulator α-$RuCl_3$, which is believed to be a prime candidate to realize the so-far elusive Kitaev quantum spin liquid ground state, resulting from the presence of strong bond-dependent anisotropic coupling among spins accompanied by quantum fluctuations.[6,7] The trihalide α-$RuCl_3$ is a layered material, in which edge-sharing $RuCl_6$ octahedra with Ru^{3+} ions in the center form a two-dimensional honeycomb structure (*cf.* Fig. 1(a)). The partially filled t_{2g} orbitals are coupled *via* nearly 90° superexchange paths.[8] The $4d^5$ electronic configuration of Ru is in a low-spin state with $S = 1/2$, giving rise to $J_{\text{eff}} = 1/2$ pseudospins due to the presence of the octahedral crystal-field splitting and large spin–orbit coupling.[9,10] The on-site interaction U opens a gap inside the $J_{\text{eff}} = 1/2$ band, creating the upper Hubbard band (UHB) and lower Hubbard band (LHB), respectively. These properties are essential prerequisites to realize the Kitaev model in condensed matter systems. Thus, significant efforts have focused on the rich equilibrium phase diagram to find experimental fingerprints of this highly-entangled topological state of matter, which is characterized by spin-flip

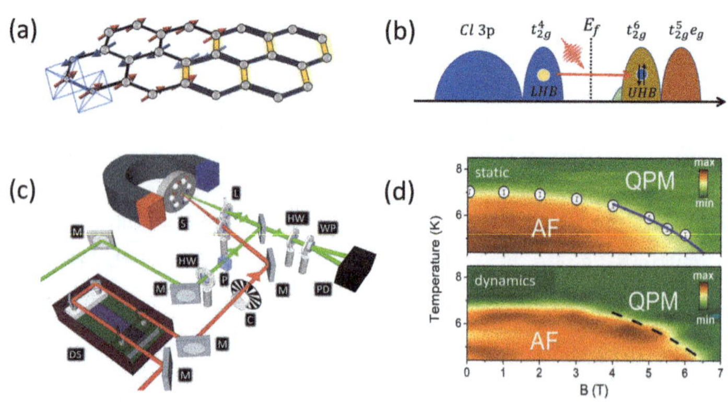

Fig. 1 Time-resolved magneto-optical experiment. (a) Crystal structure and zigzag spin orientation. The color contrast of yellow and dark blue bonds indicates small inequivalence in one of the Ru–Ru bond lengths.[13] (b) Schematical electronic structure and photoexcitation process across the Mott–Hubbard gap. (c) Experimental setup (HW: half-wave plate, P: polarizer, DS: delay stage, C: chopper, M: (beam-splitting) mirror, L: lens, S: sample, WP: Wollaston prism, PD: photo-diode). (d) In-plane magnetic field and temperature phase diagram of α-$RuCl_3$ constructed from the static change in rotation θ_{MLD} obtained by iso-magnetic MLD measurements with a fluence of 0.6 W cm^{-2} (upper panel) and extracted from the time-integrated transient MLD response $\Delta\theta_{\text{MLD}}(t)$ for an applied pump fluence of 3 μJ cm^{-2} (lower panel). White circles in the upper panel indicate the Néel temperature derived from the derivative of the rotation w.r.t. the temperature.[14] The color code displays the amplitude of the rotation from 0 to 70 mdeg (upper panel) and on a relative scale in the lower panel.

excitations fractionalizing into itinerant Majorana fermions and emergent gauge fields.[11,12]

However, experimentally it has been found that α-$RuCl_3$ establishes long-range antiferromagnetic zigzag (AFM) order below the Néel temperature of ≈7 K and in-plane applied magnetic fields ≲7 T, indicating the presence of significant non-Kitaev interactions.[15,16] Above the Néel temperature the zigzag magnetic order is destroyed and α-$RuCl_3$ exhibits a quantum paramagnetic phase (QPM) characterized by strong quantum fluctuations. We have investigated the zigzag phase in one of our recent studies, where we report on the anisotropic magneto-optical response under equilibrium conditions and show that magneto-optical spectroscopy is a versatile tool to study the antiferromagnetically zigzag ordered ground state by purely optical means.[14] Currently, the discussion on α-$RuCl_3$ equilibrium physics is still highly controversial, while there are experimental and theoretical arguments supporting that the magnetically ordered ground state of α-$RuCl_3$ is unstable under external perturbations and metastable states are likely to occur under nonequilibrium conditions.[17–19]

Further, besides being a spin liquid candidate material, the presence of Mott–Hubbard physics accompanied by strong spin–orbit coupling and the role of excitons in photoexcited α-$RuCl_3$ makes it enormously complex with unique photoinduced dynamics providing an opportunity to investigate many-body interactions. Understanding dynamical excitonic correlations and their interaction with a magnetically ordered background is a fundamental scientific question.[20] Optical spectroscopy and ellipsometry revealed a narrow line shape and asymmetry of the peak in the optical conductivity at around 1.1 eV associated with the optical gap, which suggest strong electron–hole interaction, and a small shoulder extending into the Mott gap indicating the presence of an excitonic bound state.[21,22] Further, the excitonic fingerprint shows remarkable temperature dependence in equilibrium, indicating a coupling between high-energy excitations and the magnetic structure.[22]

Here, open questions regarding the interaction of Mott–Hubbard excitons with other degrees of freedom and associated time scales arise. In our previous time-resolved magneto-optical spectroscopy study on the spin-liquid candidate α-$RuCl_3$, we showed that the transient magnetic linear dichroism can be used to study the optically induced quench of antiferromagnetic order followed by a transient spin disordered state.[17] It has been known for decades that ultrafast laser pulses can be used to induce ultrafast demagnetization.[23,24] The demagnetization dynamics has been studied extensively and described successfully within different frameworks, *e.g.* the well-established phenomenological multi-temperature model,[25] a time-dependent Ginzburg–Landau theory approach,[17] a combination of both models[26] or another recently developed general theory of out-of-equilibrium order parameter correlations.[27] Still, the microscopic interactions leading to magnetization dynamics in highly correlated materials has to be better understood. Especially the ultrafast melting of antiferromagnetic order in Mott-insulators has been argued to be fundamentally different from metals and common band insulators due to the presence of enhanced electron–electron interactions. The (de)magnetization dynamics are believed to be mainly governed by a redistribution of charge carriers among neighboring lattice sites, leading to a fast generation of doubly occupied and empty sites, called doublons and holons. The coupling between charge excitations and the spin order has been argued to be the dominant mechanism leading to ultrafast spin dynamics.[28] Specifically, the motion of holons and doublons can create trails of defects in the antiferromagnetic background, transferring energy from the hot photocarriers to the antiferromagnetic order when these quasiparticles hop around. Local spin perturbations are then dispersed through magnons into the whole system. The recombination of holon–doublon pairs has been argued to be mediated by multimagnon emission, which provides an efficient electronic demagnetization mechanism to transfer the excess energy to the system.[17,29] This raises several questions, *e.g.*, of how an applied magnetic field affects the dynamics of photo-excited holons and doublons. Further, it is important to understand whether the origin of the photo-induced dynamics is thermal or non-thermal and what are the potential mechanisms driving the recovery dynamics. In this regard, a more comprehensive study is needed to unravel the underlying physics.

To investigate the dynamics of the coupling between the charge and magnetic sector in α-$RuCl_3$, we performed time-resolved optical pump–probe experiments, in which the system is excited by a pump laser pulse over the Mott–Hubbard gap $\Delta_{MH} \approx 1.1$ eV (ref. 22) and the transient magnetization dynamics are probed with time-delayed weaker probe laser pulses. Pumping α-$RuCl_3$ with 1.55 eV photons, which have an excess energy over the MH gap, leads to multiple hopping processes *via* the impact ionization relaxation mechanism[30] (*cf.* Fig. 1(b)). The pump photons induce intersite d–d charge transfer transitions resulting in doubly occupied and non-occupied sites creating multiple spinless quasiparticles, the doublons and holons. A recent time-resolved two-photon photoemission spectroscopy (2PPES) study reported the condensation of these quasiparticles forming Mott–Hubbard excitons within a few picoseconds after photoexcitation,[31] which are expected to interfere with the zigzag ordered antiferromagnetic background.

In this report, we systematically investigate the magnetization dynamics and the subsequent recovery processes from fs to ns time scales by transient magnetic linear dichroism measurements, which have been performed for different applied pump fluences varying by more than one order of magnitude, at different temperatures below and above the Néel temperature $T_N \approx 7$ K and in external magnetic field strengths from 0 to ± 7 T applied along the *ab* honeycomb-planes. Our experiments suggest that the presence of photo-induced holons and doublons gives rise to a transient long-lived spin disordered state, which cannot be reached by just tuning the temperature or magnetic field.

2. Methods

For the study of the dynamics of the antiferromagnetically ordered zigzag ground state in α-$RuCl_3$, we apply a table-top pump–probe spectroscopy technique making use of the magnetic linear dichroism (MLD) effect in reflection geometry. Magnetic linear dichroism is a quadratic magneto-optical effect, in which a rotation of the light polarization (or a change of its ellipticity) occurs due to a difference in the complex index of refraction for normal incidence light with a polarization plane oriented parallel and perpendicular to the antiferromagnetic Néel vector, respectively.[32,33] The magnetic linear dichroism in antiferromagnets is sensitive to both the size and the orientation of the Néel vector, *i.e.* the order parameter. It has been used previously to determine the orientation of the Néel vector,[34] to study zigzag antiferromagnetically ordered phases in equilibrium[14,35] and to track ultrafast dynamics of critical phenomena in antiferromagnetically ordered systems near a magnetic phase transition.[17,36]

For the study of the magneto-optical response the α-$RuCl_3$ samples grown by vacuum sublimation[37] were placed in a helium-cooled cryostat (Oxford Spectromag) with temperatures down to 2.2 K inside the coils of a superconducting magnet with magnetic field strengths up to $\pm$7 T. Characterization of the sample was done applying a static magnetic linear dichroism experiment,[14] which revealed a Néel temperature of $\approx$7 K indicating good sample quality.[13,38] Fig. 1(c) illustrates the experimental setup. The magnetic field was applied within the crystallographic *ab*-plane, *i.e.* within the honeycomb layers. The polarization of the incident probe light was rotated by a half waveplate and set to the polarization orientation setting, which gave the maximum signal in zero magnetic field. The pump beam passes through a computer-controlled delay line and is mechanically chopped at a frequency of 680 Hz. The pump beam polarization is set cross-polarized to the probe beam in order to minimize potential pump beam scattering to reach the detector. The measurements of the magneto-optical response were carried out in the so-called Voigt geometry[39] at near-normal incidence, such that the light wave propagation vector $\mathbf{k}$ of the probe laser pulses was perpendicular to the honeycomb layer *ab* planes ($\mathbf{k} \perp \mathbf{ab}$) and magnetic field vector $\mathbf{B_{ab}}$ (see Fig. 1(a)). The time-resolved magneto-optical experiment was performed using 800 nm ($\approx$1.55 eV) pump pulses with a temporal width of 40 fs, and probe pulses of 512 nm ($\approx$2.42 eV) with a temporal width of 250 fs, generated with a laser system based on a LightConversion Pharos equipped with LightConversion optical parametric amplifiers as core components. The penetration depth of the pump is larger than for the probe, which helps to probe a homogeneously excited area.[17] The pump and probe beams were focused down to a radius of 40 μm and 25 μm, respectively. The repetition rate of the amplified laser system was set to f = 30 kHz in order to ensure that the system can relax back to the ground state between consecutive pulses and photoinduced thermal heating is reduced. Detection of the transient polarization rotation $\Delta\theta(t)$ was done using a balanced-detection configuration[40] consisting of a $\lambda/2$ wave plate, a Wollaston prism, and a balanced photodiode. The signal from the photodiodes was sent through a pre-amplifier before reading out of the balanced signal by a Zurich Instruments HF2LI lock-in amplifier.

3. Results and discussion

First, we show that changes in the magnetic linear dichroism $\Delta\theta_{MLD}$ are related to the magnetic order and probe the zigzag antiferromagnetic phase. The top panel of Fig. 1(d) shows the phase diagram as a function of temperature T and in-plane applied magnetic field B obtained under equilibrium conditions (same data as in ref. 14). The variation of θ_{MLD} for different temperatures and fields reconstructs the already observed B–T phase diagram of α-$RuCl_3$.[41] We note that the absolute value of θ_{MLD} varies by the formation of three equal zigzag domains depending on their angle with respect to the polarization orientation of the incident probe beam, such that it is non-trivial to extract the absolute change in rotation.[14,34] To check if the pump induced transient $\Delta\theta_{MLD}(t)$ is also directly related to the zigzag order, we measured $\Delta\theta_{MLD}(t)$ after photoexcitation and plotted the time-integrated signals as a function of equilibrium temperatures and fields in a false-color plot (see Fig. 1(b) lower panel). The fact that both from the transient $\Delta\theta_{MLD}(t)$ and the static θ_{MLD} the B–T phase diagram can be reconstructed demonstrates the sensitivity of our approach to the zigzag antiferromagnetic order.

The fluence-dependent measurements of the induced change in the magnetic linear dichroism, $\Delta\theta_{MLD}(t)$, at a temperature of 3 K, *i.e.*, deep in the zigzag ordered phase, are shown in Fig. 2(a). For small pump fluences, up to 7 μJ cm^{-2}, the transient magneto-optical response $\Delta\theta_{MLD}(t)$ reaches its maximum on a timescale of ~400 ps and then relaxes back to its pre-pump value within the accessible time-window of 4 ns. However, for increasing pump fluences the changes in $\Delta\theta_{MLD}(t)$ get more pronounced and a ns long-lived (metastable) state with plateauish, non-decaying behavior is observed. No further increase of $|\Delta\theta_{MLD}|$ is observed above a pump fluence of 70 μJ cm^{-2}. We point out that while reducing the fluence again, all measurements are reproducible indicating no photo-induced sample damage. Further, the system can relax back to the equilibrium state between consecutive pulses as we do not observe any residual background, not even with the highest applied fluence, at negative time delays.

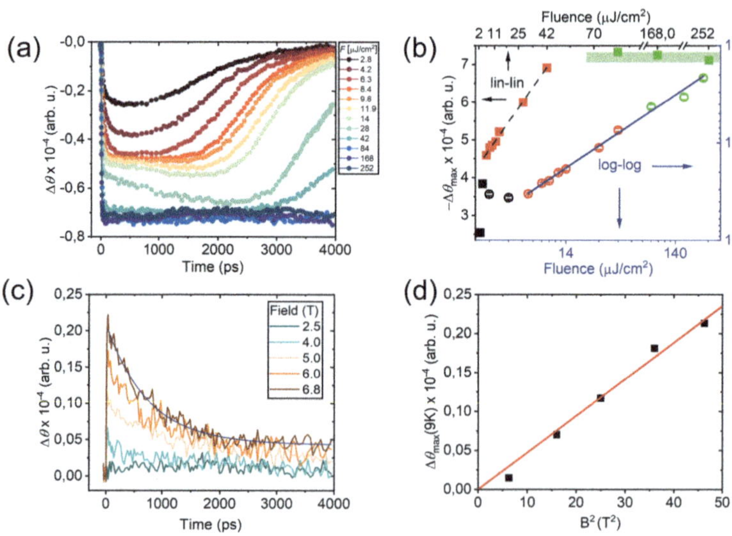

Fig. 2 Comparison of the magneto-optical response in the zigzag AFM and QPM phase. (a) Fluence dependence of $\Delta\theta_{MLD}(t)$ in zero magnetic field at 3 K for applied pump fluences varying between 3 to 250 μJ cm^{-2}. For applied fluences above 70 μJ cm^{-2}, the maximum reached $\Delta\theta_{MLD}$ is constant. (b) Maximum photo-induced polarization rotation $\Delta\theta_{MLD}$ (squares, black axes) on a lin–lin scale and extracted demagnetization time constants τ_D (open circles, blue axes) on a log–log scale. A fit to extract power law scaling of τ_D is shown by the blue solid line. The color code of the symbols indicates the distinct fluence regimes: black, red and green represent the low, intermediate and high fluence regimes, respectively. Maximum demagnetization is indicated by the greenish background. (c) Field dependence of $\theta_{MLD}(t)$ at an applied fluence of 80 μJ cm^{-2} in the quantum paramagnetic phase at 9 K. The blue solid line is a single exponential fit with an offset. (d) Maximum amplitude of $\Delta\theta_{MLD}$ at 9 K for an applied pump fluence of 80 μJ cm^{-2} as a function of applied magnetic field, which scales linearly with B^2 as expected for second-order magnetic linear dichroism in the field-polarized QPM state.

Fig. 2(b) shows the fluence dependence of the amplitude $\Delta\theta_{MLD}(t)$ at a time delay of 500 ps as well as the demagnetization time constants τ_D associated with the initial changes in the magneto-optical response after pump excitation. The fluence dependence of the amplitude $\Delta\theta_{MLD}(t)$ indicates three distinct fluence regimes. In the low fluence regime (LFR) there is an initial sharp increase in $\Delta\theta_{MLD}(t)$ occurring for applied pump fluences up to 7 μJ cm^{-2} (black symbols), followed by an intermediate rise between 10 and 50 μJ cm^{-2} (red symbols) called the intermediate fluence regime (IFR). We observe saturation in the high fluence regime (HFR) above 70 μJ cm^{-2} (green symbols), *i.e.*, no further demagnetization occurs. In contrast, the demagnetization time constant τ_D extracted from single-exponential fits to the initial dynamics shows only two regimes. First, it is almost constant ~30 ps for fluences up to 7 μJ cm^{-2}. Above this threshold fluence τ_D decreases non-linearly for increasing pump fluences F and follows a clear power-law scaling $\propto F^{-0.7}$ (blue line in Fig. 2(b)). The fluence dependent behavior of $\Delta\theta_{MLD}(t)$ (amplitude of the demagnetization) and τ_D (risetime of the demagnetization) provide essential information on the photoexcited state. Together with other experimental observations presented later in this paper and other recently reported results,[17] we will show that all three regimes are non-thermal. In the LFR, the ultrafast created holons and doublons have already formed a bound pair. These excitons show an excitation density independent recombination process, which is indicated by the constant demagnetization time τ_D associated with the recombination *via* magnon emission.[29] If the holons and doublons would not be already bound, the recombination process would depend on the probability to encounter the oppositely charged particle that causes a non-exponential and excitation density dependent recombination process.[42,43] In the IFR, the excess amount of excited carriers gives rise to the screening effect, which causes a faster breaking of initially formed excitons into holons and doublons, which experimentally scales with the power-law $\propto F^{-0.7}$. The antiferromagnetic system is completely disordered in the highest excitation regime, such that no further change in $\Delta\theta_{MLD}(t)$) is observed, but the screening effect is continuously enhanced (decrease in τ_D), as expected. The mechanism within the LFR is conceivable, as already observed in similar Mott–Hubbard systems. At low excitation densities, the holons and doublons form bound pairs, a Mott–Hubbard exciton, (also reported for α-$RuCl_3$ in ref. 31), which then recombine within 10s of ps, while their excess energy is dispersed through magnons and potentially other bosonic excitations into the whole system. The clearly different non-linear fluence dependence of the initial demagnetization dynamics within the IFR thus can be understood in terms of an increasing unbound holon–doublon density perturbing the zigzag order. The experimental observations indicate that the dominating demagnetization mechanisms in the LFR and IFR seem to be different. For sufficiently small excitation densities holon–doublon pair recombination *via* magnon emission appears to dominate, the hopping of unbound holons and doublons in the IFR effectively perturbs the long-range zigzag order. Theory predicts that in the strong excitation regime indeed a larger fraction of the excitons separates into unbound doublons and holons.[20] The most intriguing observation in the IFR is the analogous power-law scaling behavior ($\propto F^{-0.7}$) observed for the time scale of the gap collapse after photoexcitation in the excitonic insulator materials, *e.g.* Ta_2NiSe_5 (ref. 44) and 1T-$TiSe_2$ (ref. 45 and 46), associated with a free carrier screening effect which is responsible for excitons breaking into electron and holes. However, the reported time scales are in the order of 100s of fs and different to the 10s to few ps timescale observed in our experiment. This could be related to the small bandgap in Ta_2NiSe_5 and 1T-$TiSe_2$, which is an order of magnitude smaller than for α-$RuCl_3$. The characteristic build-up time for carrier screening in response to an ultrashort laser excitation is the plasma oscillation period, which scales with $n^{-0.5}$, where n is the carrier density.[47] Since α-$RuCl_3$ is a Mott-insulator and the carrier density in equilibrium expected to be small, the photo-excited carrier density should scale linearly with the applied fluence and hence a similar scaling behavior is expected. Indeed, the pump-induced changes in $\Delta\theta_{MLD}(t)$ in the IFR scales linearly with the applied fluence. Therefore, the excitation density dependent screening effect plays an important role in the dynamics in the IFR. In the HER, the photo-induced magnitude of the demagnetization is constant (in contrast with a saturating behavior of the risetime) suggesting a completely spin-disordered system. In the following, we elaborate in more detail on the interaction between holons, doublons and Mott–Hubbard excitons and the underlying magnetically ordered background.

To prove experimentally that the presence of long-range zigzag antiferromagnetic correlations is crucial to observe the $\Delta\theta_{MLD}(t)$ dynamics in Fig. 2(a), we conducted measurements at 9 K in the disordered QPM phase of α-$RuCl_3$ (see Fig. 2(c)). We point out that even at a high pump fluence of 80 μJ cm^{-2} (*cf.* saturation of $\Delta\theta_{MLD}(t)$ at 3 K) transient changes in the polarization rotation $\Delta\theta_{MLD}(t)$ are very small in zero magnetic field in the QPM phase. This is the behavior that is naturally expected, since in the QPM no long-range magnetic order exists and hence no magnetic linear dichroism is expected to be observed. However, by applying an external field, a transient change in $\Delta\theta_{MLD}(t)$ can be observed experimentally. The sign of $\Delta\theta_{MLD}(t)$ at 9 K is opposite to the one at 3 K. This is more evidence that the probed demagnetization dynamics originates from two distinct magnetic phases. The maximum photo-induced changes in $\Delta\theta_{MLD}(t)$ scale quadratically $\propto (\mu_0 H)^2$ in the externally applied field, which indicates that we are probing the magnetic linear dichroism in the field-aligned QPM phase with a finite net magnetization $\propto \chi H$ (*cf.* Fig. 2(d)). Moreover, the dynamics of $\Delta\theta_{MLD}(t)$ has a clearly different character than the dynamics in the AFM phase governed by a field-independent quick rise followed by a single exponential decay with a time constant of ~800 ps for all applied field strengths showing relaxation on ns time scales. As expected, there is no signature of a long-lived plateauish state.

After comparing $\Delta\theta_{MLD}(t)$ in the zigzag AFM ordered and QPM phases, we now turn to temperature dependent high fluence measurements at 6.2 and 6.8 T (F = 80 μJ cm^{-2}) (see Fig. 3(a)). Higher field measurements allow us to understand the nature of the long-lived plateauish state under higher pump laser excitation and to detect the QPM contribution in $\Delta\theta_{MLD}(t)$ more evidently. It is crucial to understand whether there is a significant effect of laser induced heating on the transient magneto-optical response. If the heating effect is negligible, a temperature induced crossover from the AFM to the QPM upon increasing the bath temperature should be visible in the MLD transients. As the bath temperature is raised the sign of $\Delta\theta_{MLD}(t)$ changes from negative to positive. We interpret this behavior as being indicative of the thermal crossover from the zigzag AFM towards the QPM phase (*cf.* Fig. 2(a and c)). Further, in the early dynamics, one can immediately recognize a mixture of two dynamical components, which

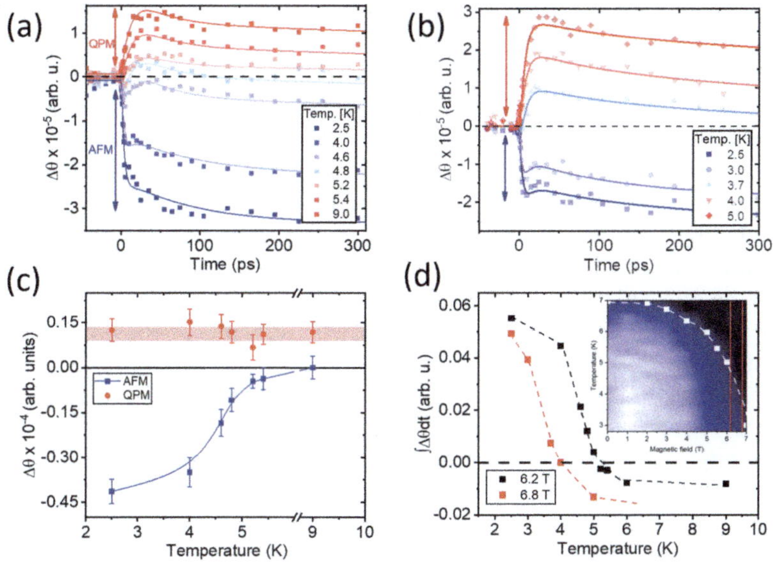

Fig. 3 Coexistence of the QPM and AFM phase dynamics. Temperature dependence of $\theta_{MLD}(t)$ at an applied magnetic field of (a) 6.2 T and (b) 6.8 T for an applied pump fluence of $F = 80\ \mu J\ cm^{-2}$. The solid lines display the fits described in the main text. (c) Extracted amplitudes of the two component exponential fit related to the zigzag AFM and the QPM phase for the fits derived from (a). The zigzag AFM component vanishes towards crossing the phase boundary, while the QPM dynamical component survives. (d) Changes of the integrated magneto-optical response as a function of temperature. The inset shows the $B-T$ phase diagram and red lines indicate the applied field strengths of 6.2 and 6.8 T, respectively.

motivated us to apply a multi-exponential fit with two distinct dynamical components related to the AFM and QPM state.

$$\Delta\theta_{MLD}(t) = A_{AFM}(1 - \exp(-t/\tau_{AFM})) \times \exp(-t/\tau_1) + A_{QPM}(1 - \exp(-t/\tau_{QPM})) \times \exp(-t/\tau_2) \qquad (1)$$

Here, A_{AFM}, A_{QPM} correspond to the contribution of the AFM and QPM phases to the amplitude of the transient MLD; τ_{AFM}, τ_{QPM} display the demagnetization time-constants for the two distinct phases and τ_1, τ_2 are added to capture the recovery dynamics on the 100s of ps time scales. An additional Gaussian response function capturing the laser profile was added, but does not influence the fit due to the different time scales. Interestingly, the time constants τ_{AFM} and τ_{QPM} of the two distinct states differ significantly, while being temperature-insensitive. We obtained ~5 ps and ~10 ps time constants for AFM and QPM demagnetization, respectively. This observation indicates that the system is optically driven into a non-thermal state. In addition, the extracted amplitudes of the fit (shown in Fig. 3(c)) show an interesting behavior. While the component related to the zigzag AFM dynamics vanishes approaching a temperature of ≈5 K for an applied field strength of 6.2 T, the QPM component is temperature independent. The vanishing of the AFM component at around 5 K is perfectly in line with static MLD measurements (*cf.* Fig. 3(d)) indicating negligible laser heating effects even at the highest pump fluences. These findings indicate that the overall pump induced heating is negligible on macroscopic scales and that we observe the emergence of a photo-induced bi-partite phase, which is non-accessible by just tuning the bath temperature. These observations support the interpretation that the long-lived state at high fluences is not appearing due to a change in the temperature of the system but has its origin in the transient photo-induced change of the electronic and magnetic structure. Similar phase-coexistence in the vicinity of magnetic phase transitions has been observed previously. The mixture of two dynamical components far from equilibrium has been understood in terms of transiently quenched magnetic interactions by laser irradiation, which triggers the local formation and nucleation of magnetic domains of the other thermodynamic phase.[48]

Having established the non-thermal nature of the fluence dependent $\Delta\theta_{\mathrm{MLD}}(t)$ shown in Fig. 2(a), we now turn to discuss the possible underlying mechanisms and the origin of the long-lived state. The initial demagnetization time is on the order of 10s of ps and no faster component on sub-ps time scales has been observed in our measurements. Previous magneto-optical studies on magnetically ordered systems reported similar time scales for an initial photo-induced demagnetization process,[49,50] although the relevance of the involved microscopic degrees of freedom in different systems lead to a wide spreading of time scales in the literature.[51] Recent 2PPES[31] directly revealed a sub-ps lifetime of doublons in the UHB after photo-excitation across the Mott gap, which is initially unexpected for such a large Mott-gapped insulator. In the same study some signatures of sub-ps creation of Mott–Hubbard excitons have been reported, which has been interpreted as a reason for the fast decay of the doublons in the UHB in line with ref. 29. This fact underlines that the transient magnetic linear dichroism signal monitors the magnetic dynamics driven by the photo-excitations rather than the purely electronic response of the photo-excited charge carriers in the UHB. To reveal the underlying processes, which lead to the demagnetization dynamics displayed by the transient magnetic linear dichroism, the here presented comprehensive study under different experimental conditions, *i.e.*, varying the applied pump fluence, bath temperature and external applied magnetic field, is crucial. At this point, the formation of holon–doublon pairs and Mott–Hubbard excitons is important, which have been observed in different Mott-insulating systems with an AFM ordered groundstate.[26,28,52] A recent study on the spin–orbit coupled Mott insulator $Sr_3Ir_2O_7$ (ref. 52) revealed that the photo-induced density of spin-less doublons is indeed crucial in understanding the dynamics after demagnetization. As mentioned above, the photodoping of a Mott insulator creates holons and doublons in the LHB and UHB, respectively. By this charge excitation multiple spinless quasiparticles are created, such that magnetic moments are effectively removed from the lattice. When these charge excitations hop through the lattice, they continuously perturb and destroy the magnetic order.[53] This can be interpreted as creating spin defects, which naturally quench the zigzag order locally and hence decrease the zigzag antiferromagnetic order parameter. The strong charge–spin coupling has been theoretically proposed to display an efficient nonthermal demagnetization process.[54] In this regard, the fluence dependent MLD transients reveal the influence of the photo-excited holon–doublon density on the zigzag ordered state. At low excitation densities the changes in the orbital occupancies given by the present holon and doublons induce only minor spin defects and hence perturbations to the zigzag antiferromagnetic background. Any local frustration in the zigzag order given by misaligned pseudospins decreases the effective hopping between the sites. Hence, the surrounding intact zigzag order prevents the holons and doublons moving around, since this is energetically unfavorable in terms of the magnetic correlations. This is displayed by the experimental data, where no or short-lived plateauish state can be observed at low excitation fluences, but the photo-induced changes in the transient MLD recover (*cf.* Fig. 2(a)), *i.e.*, the zigzag order is not perturbed beyond the initially created spin defects. Moreover, the amount of emitted magnons due to the excess energy during the recombination of holon–doublon pairs does not suffice to fully spin disorder the system. Beyond the creation of holons and doublons it is theoretically predicted that the photodoping results in enhanced initial excitonic correlations, the behavior opposite to that caused by simple heating.[20] It has been argued that long-lived excitations are generic for Mott–Hubbard insulators with large U giving rise to the excitonic Mott–Hubbard insulator rather than a transient metallic state after photodoping,[55] which has been commonly observed for narrow gap excitonic- and Mott-insulating materials.[44,56,57] The Hubbard U in α-$RuCl_3$ has been reported to take a value of $U \approx 4.35$ eV.[58] In α-$RuCl_3$ the holons and doublons are found to form indeed an excitonic state in the Mott gap due to direct Coulomb interactions with a long-lived character, which is in accord with theory predicting in-gap states to be longer lived compared to short-lived excitonic states within the doublon–holon continuum.[20] Different to the hopping of single holons and doublons the movement of holons and doublons paired up to a Mott–Hubbard exciton does not lead to a further disordering of the zigzag ordered groundstate, hence leaving zero net disorder behind, and they only contribute to the magnetization through magnon emission in the recombination process. At the time, the presence of Mott–Hubbard excitons can be regarded as a bottleneck for the re-establishment of long-range antiferromagnetic zigzag order, as in α-$RuCl_3$ the holon–doublon bound together to an exciton are long-lived with lifetimes exceeding 100s of ps.[31] Thinking of an antiferromagnetically disordered zigzag groundstate by the presence of Mott–Hubbard excitons with long lifetime and breaking to single holons and doublons is in good agreement with the start of the recovery process at long times for low pump fluences. For increasing fluence, *i.e.*, a higher excitation density the screening effect leads to a faster and larger amount of spin defects, the dynamics becomes different. As discussed above, the power law scaling of the initial demagnetization time τ_D with the fluence might indicate a breakdown of the Mott–Hubbard excitons, such that the number of free holons and doublons increases. Theory confirms that in the strong excitation regime a larger fraction of the excitons separates into unbound doublons and holons.[20] For a larger amount of subsequent spin defects the zigzag order is already strongly perturbed, such that the holons and doublons then move easier through the lattice. In this regard they can separate further leaving a trace of disordered spins behind. In this way, the spin disordered state can be understood as a metastable state driven by the photo-induced holon–doublon density. In analogy, it has been reported for the magnetoresistive pyrochlore $Tl_2Mn_2O_7$ that the temporal persistence of the photoexcited carrier density can be strongly influenced by spin disorder.[59] For a certain amount of spin defects the long-range antiferromagnetic order cannot be re-established by the magnetic exchange interactions. Hence, the fluence dependent demagnetization time fits to the picture of initially created spin defects causing a faster disordering of the zigzag order accompanied by a dramatic increase in the lifetime of the disordered state, which is in line with our observations. Monte Carlo simulations of a simple 3D spin-1/2 system showed that the amount of spin-flips, treated as the creation of double occupied sites, has significant influence on the recovery dynamics. For weak perturbations a quick recovery is seen, whereas the recovery dramatically slows down upon strong perturbations and a long-lived spin-disordered state appears.[52]

Next, we turn to the field-dependence of $\Delta\theta_{MLD}(t)$ at a temperature of 3 K and an applied pump fluence of $F = 15\ \mu J\ cm^{-2}$, which is shown in Fig. 4(a). One can clearly see that in general the total change in $\Delta\theta_{MLD}(t)$ decreases with increasing applied magnetic field, which can be naturally assigned to the field-driven decrease of the order parameter. In small magnetic fields the dynamics is very similar, while for fields larger than 2 T the signal amplitude $\Delta\theta_{MLD}(t)$ changes and the plateau observed in the fluence dependence data in 0 T starts to become longer lived. Recent measurements found that at zero magnetic field the zigzag ground state is degenerate, *i.e.*, there can be three differently oriented zigzag domains with the zigzag chain directions deviating by 120° from each other. Applying a small in-plane magnetic field induces a change in the domain population that lifts the degeneracy and fits the observations of an initial small

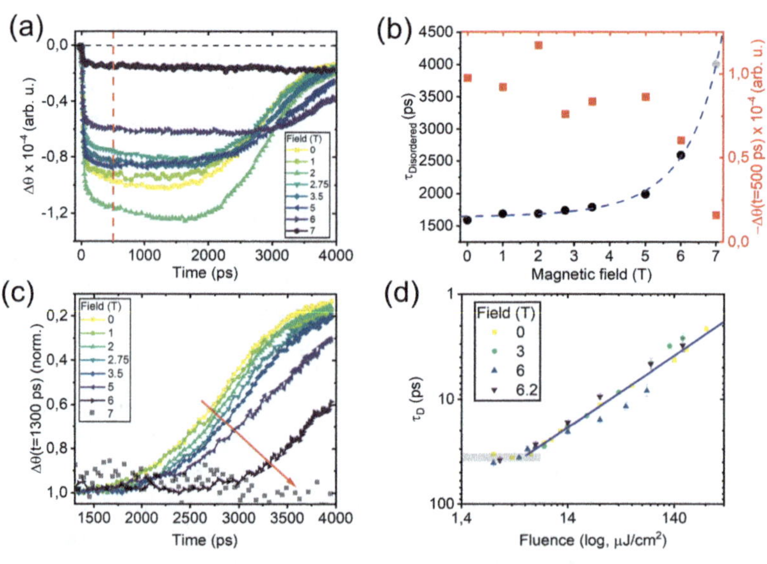

Fig. 4 Field-dependent magneto-optical response in the zigzag AFM phase. (a) Field dependence of $\theta_{MLD}(t)$ at 3 K for an applied pump fluence of 15 $\mu J\ cm^{-2}$. (b) Extracted changes in θ_{MLD} (t = 500 ps) (red squares), which follows the static magneto-optical response. The black circles indicate the lifetime of the plateau, which diverges towards approaching the equilibrium critical field strength above which α-$RuCl_3$ enters the quantum disordered paramagnetic phase. (c) Normalized transients of $\Delta\theta$ (t = 1300 ps) shown in (a) for the long time delays. The red arrow indicates a slowing down of the recovery dynamics. (d) Fluence dependent demagnetization constants τ_D at different applied field strength on a log–log scale. The power law scaling discussed in the main text is indicated by the blue solid line.

change in $|\Delta\theta(t)|$ between 0 and 1 T. In addition, under an in-plane magnetic field oriented perpendicular to one of the symmetry equivalent {1, 1, 0} axes a field-driven metamagnetic transition has been identified at around 1.5 to 2 T.[14] Interestingly, the amplitude changes in $\Delta\theta_{\mathrm{MLD}}$ ($t = 500$ ps) displayed in Fig. 4(b) show an initial rise and sharp drop at around 2 T, which is related to the domain reorientation and clearly matches the reported hysteresis in the zigzag phase,[14] in line with expectations. In addition, the lifetime of the metastable plateauish regime tends to diverge for increasing field strength approaching the field-driven phase transition boundary becoming longer lived than the accessible time delay, such that a lower bound of 4 ns at a field of 7 T has been assigned (see Fig. 4(b)). As the external applied field approaches the critical field value, magnetic correlations get suppressed and fluctuations in the order parameter occur in the vicinity of the phase transition.[27] In this picture holons and doublons then move on a many body fluctuating background with decreased magnetic correlations, which hinders the re-establishment of magnetic long-range order. Besides the longer lived spin-disordered phase also the recovery dynamics under an in-plane applied magnetic field slow down as shown in the normalized transient $\Delta\theta$ ($t = 1300$ ps) in Fig. 4(c). In contrast, the initial fluence dependent demagnetization dynamics show the power law scaling $\propto F^{-0.7}$ for all applied magnetic field strengths (see Fig. 4(d)). This provides more proof for our initial discussion about the purely electronic mechanism behind the demagnetization. The holon–doublon density is indeed crucial in understanding the photo-induced quench of the zigzag antiferromagnetic state. Up to a threshold fluence τ_{D} is constant, which points towards the initial formation of Mott–Hubbard excitons by the photo-excited holons and doublons referring to ref. 31. Above this threshold fluence the amount of excess free holons and doublons increases and causes a screening effect, such that the formed excitons break up into free holons and doublons. This corresponds to a larger amount of spin-defects disturbing the AFM background, which leads to the emergence of the long-lived plateauish regime in $\Delta\theta_{\mathrm{MLD}}(t)$.

From the experimental data, we can conclude that the field dependent transient MLD responses for the zigzag AFM and the QPM phase are intrinsically different. Considering the results for the field-aligned QPM phase (*cf.* Fig. 2(b)) we can make the following statements. Although the excitation mechanism above and below T_{N} is the same, the difference in the demagnetization dynamics indicates clearly that the presence of antiferromagnetic correlations influences the dynamics of photo-excited holons and doublons significantly and is crucial in understanding the formation of a long-lived spin-disordered state. From inhomogeneous nonequilibrium dynamical mean-field theory, it is predicted that antiferromagnetic correlations indeed influence charge excitations.[60] Based on our experimental data we conclude that the presence of AFM long range interactions restricts the movement of holons and doublons for sufficiently low excitation densities. Further, a certain amount of holons and doublons needs to be created to break up the already formed excitons and to decrease the AFM correlations *via* the creation of a high density of local spin defects. In addition, the magnetic field also reduces AFM correlations and leads to a slowing down of the recovery dynamics in the zigzag AFM phase as shown in Fig. 4(c). In contrast, the recovery of $\Delta\theta_{\mathrm{MLD}}(t)$ at 9 K is similar for all applied field strengths and does not show a field dependence. This is another indicator that the recovery dynamics are influenced by the interaction between the holons and doublons and the underlying long range magnetic order.

Up to now we have mainly discussed the demagnetization and the induced metastable phase. We now turn to the recovery, we discuss the recovery dynamics of the transient magnetic linear dichroism, which displays how the photo-induced magnetically disordered state recovers back to its pre-excitation equilibrium state. Differently from the observed non-linear fluence dependence of the initial demagnetization dynamics, the fluence dependence of the recovery dynamics indicates a universal scaling behavior. Although the starting point of the recovery process is non-linear in the applied fluence and shows a field dependent behavior due to the reduced antiferromagnetic background (Fig. 3(c)), in the following, we argue that the mechanism of recovery is similar for all measured conditions. In Fig. 5(a) we show the rescaled long-time recovery of $\Delta\theta_{MLD}(t)$ at 2.5 K and zero applied field. The MLD data has been normalized to its amplitude value at which the recovery process starts and the time axis has been rescaled by $t_{1/2}$, the time at which $\Delta\theta_{MLD}(t)$ has already recovered by 50%. As can be seen clearly, the rescaled transients fall on top of each other pointing towards an underlying universal decay process. The recovery dynamics can be best fit using a compressed exponential decay function $\propto \exp(-(t/\tau)^{\beta})$, with an extracted fluence-independent exponent $\beta = (2.3 \pm 0.2)$. The recovery time constant ~1 ns of the unscaled data transients matches the time scale of slow diffusion driven processes, which are often observed in pump–probe experiments, when the heat in the excited probe area is transferred along a temperature gradient to unpumped, colder regions. Although we cannot exclude the presence of diffusion driven recovery, the fluence-independent compressed exponential behavior points to a minor heating effect, which should otherwise influence the dynamics significantly. A compressed exponential behavior has been observed in the dynamics of antiferromagnetic domain wall dynamics,[61] for a variety of soft matter systems undergoing 'jamming' transitions, disordered glassy states and has also been interpreted in terms of ballistic motion of elastic deformation in response to heterogeneous local stress.[62] Further, the phenomenon of compressed

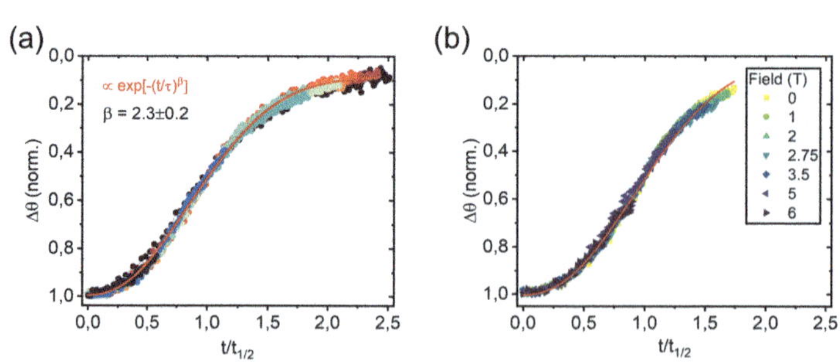

Fig. 5 Universal recovery dynamics in the zigzag AFM phase. (a) Rescaled fluence dependence of the long-time recovery $\theta_{MLD}(t)$ in zero magnetic field at 3 K for applied pump fluences varying between 3 to 250 μJ cm^{-2}. (b) Rescaled field dependence of the long-time recovery $\theta_{MLD}(t)$ at 3 K for applied pump fluences of 15 μJ cm^{-2}. The data in (a) and (b) is normalized to its value at which the recovery process starts and the time axis is rescaled to $t_{1/2}$ as discussed in the main text. The red solid line indicates a compressed exponential fit associated with domain propagation dynamics.

exponential relaxation has been discussed in terms of magnetization reversal in magnetic thin films.[63] Here, the Fatuzzo–Labrune model describes magnetization reversal in a simple and intuitive way in terms of an initial and probabilistic nucleation process, which is followed by either further nucleation governed magnetization reversal or by a domain wall propagation driven process. Although, the Fatuzzo–Labrune model makes an oversimplification of the present magnetic domains treating them as circularly shaped, it can give useful qualitative information. The time evolution of the dynamics is generally captured by $\exp(-(t/\tau)^{\beta})$, while the value of β gives information about the underlying recovery processes[63,64] and allows to distinguish between domain wall propagation and nucleation driven relaxation. It was established that the shape of the relaxation curves depends on the relative importance of domain wall motion and nucleation processes. The relaxation curves show an S-shape (compressed exponential) for relaxation dynamics dominated by domain wall propagation ($\beta > 1$), while for nucleation dominated reversal the relaxation curve takes a (stretched) exponential shape ($\beta \leq 1$).[63–65]

Therefore, examination of the shape of the scaled recovery transient curves suggests that the reversal is mainly governed by domain wall propagation (see Fig. 5). Normally, magnetization reversal is a slow process on time scales extending up to several seconds or even minutes.[66,67] However, magnetization reversal in these studies is usually triggered by an external applied magnetic field under purely equilibrium conditions. Here, the situation is drastically different since we report on the dynamics of the magnetic sector under nonequilibrium conditions. This might pave the way to understand the clearly different time scales we observe here. The pump laser excitation quenches the zigzag antiferromagnetic order locally within the area of the laser spot. A metastable state is created, which is, as discussed above, mainly described by the presence of holons and doublons interpreted as local spin defects in the zigzag phase. The compressed exponential clearly indicates a relaxation process driven by domain wall propagation within the here proposed model. The picture of domain wall propagation also holds under applied magnetic fields as can be deduced from Fig. 5(b). The recovery dynamics in field can also be rescaled and follow a compressed exponential recovery with an exponent of $\beta = (2.21 \pm 0.04)$ in line with the fluence dependence recovery in 0 applied magnetic field. Hence, it seems that after a certain time the zigzag AFM order builds up locally again, *i.e.* a first zigzag domain is formed within a spin-disordered state. This domain then expands in time, *i.e.* its domain wall propagates throughout the probed area in contrast to a multiple domain nucleation scenario. Which process helps to form this initial starting domain is not evident. The presence of local defects and domain wall pinning are considerable,[67] but experimental probes, which visualize the domain structure directly,[68] are necessary to address this question in more detail. It is interesting that the temporal behavior governed by a compressed exponential decay matches the observations for magnetic reordering processes in thermodynamic equilibrium pretty well. This interpretation supports our argument about the strong effect of magnetic background on the fast recovery in the LFR and the long-lived plateau in the IFR. Here the domain wall propagation is indicative of locally strong antiferromagnetic order which assists holons and doublons to relax and becomes dominant after initial random nucleation in the beginning of the recovery process.

4. Summary and conclusions

In this report, we presented the time-resolved dynamics of the spin liquid candidate α-$RuCl_3$ throughout the whole low-temperature zigzag antiferromagnetically ordered phase upon photo-excitation above the Mott–Hubbard gap. The photo-excitation creates holons and doublons in the lower and upper Hubbard bands, respectively. Due to strong charge–spin coupling in the spin–orbit assisted Mott insulator, the effect of charge excitations on the antiferromagnetically ordered state can be efficiently probed *via* transient magnetic linear dichroism. Our measurements reveal that the density of these spinless quasiparticles opens a nonequilibrium way to destabilize the zigzag antiferromagnetic order and can be treated as an additional dimension to the α-$RuCl_3$ phase diagram to create a non-thermal photo-induced long-lived and spin-disordered state. The presence of holons and doublons and their effect on the antiferromagnetically ordered background can be understood within a picture of spin defects that move around and disturb the zigzag order. Further, our measurements indirectly point towards the presence of holon–doublon pairs forming Mott–Hubbard excitons with 30 ps lifetime at low excitation densities. The results show the demagnetization is a consequence of the relaxation of excited holons and doublons and it occurs through competing mechanisms, either magnon emission or hopping of unbound carriers. The dominating demagnetization process seems to be dependent on the excitation density. Whereas, for small excitation densities the exciton recombination *via* magnon emission is dominant, the hopping of unbound holons and doublons effectively perturbs the zigzag order upon high excitation densities. Further, the magnetic field dependent transient magnetic linear dichroism shows pseudocritical behavior, especially a longer lifetime of the photoinduced metastable state reminiscent of the field driven decrease of zigzag antiferromagnetic correlations. Besides the creation and nature of the long-lived spin-disordered state, our study also indicates that the recovery dynamics can be understood in terms of a magnetic reordering process rather than simple heat diffusion driving thermalization of the magnetic, lattice and charge degrees of freedom. The recovery dynamics follow a universal compressed exponential form with a compressing exponent $\beta = (2.3 \pm 0.2)$ indicative for a domain wall propagation dominated reordering process.

Our study revealed that photoexcitation of charge carriers can be used to destabilize magnetic order in systems with strongly intertwined degrees of freedom like α-$RuCl_3$. It is interesting to study other spin liquid candidates (*e.g.* Na_2IrO_3 (ref. 69)) and other magnetic Mott insulators to clarify that the presence of charge excitations does play a critical role in stabilizing hidden phases, helps to quench antiferromagnetic order and to access metastable non-thermal states. In this regard, there remain still some important open questions that need to be addressed in future experiments. First, a broad spectral probing of the magneto-optical response would give a deeper understanding of the interactions between the charge, lattice and magnetic sectors. Further, it would be interesting to study the electronic and lattice dynamics at low temperatures and long delays more directly by complementary time-resolved techniques,[70] like time-resolved ARPES,[71] XRD,[72] Raman,[73] *etc.* These studies will reveal the intrinsic time scales and interactions of holons and doublons and other quasiparticles and could answer whether a transient lattice deformation can quench magnetic exchange interactions, while simultaneously enhancing the anisotropic Kitaev interactions to drive α-$RuCl_3$ into the Kitaev spin liquid state. Moreover, it would be important to understand theoretically the qualitative correspondence between the here observed compressed exponential decay under nonequilibrium conditions and magnetization reversal in ferromagnets in equilibrium.

Conflicts of interest

There are no conflicts to declare.

Acknowledgements

We are indebted to Fulvio Parmigiani, Daniel I. Khomskii, Sebastian Diehl and Alessio Chiocchetta for fruitful discussions. The authors acknowledge financial support funded by the Deutsche Forschungsgemeinschaft (DFG) through project no. 277146847-CRC1238, Control and Dynamics of Quantum Materials (subproject no. B05). AL and VT acknowledge support by the Deutsche Forschungsgemeinschaft (DFG) through the Transregional Research Collaboration TRR 80: From Electronic Correlations to Functionality (Augsburg, Munich, and Stuttgart). VT acknowledges the support *via* the project ANCD 20.80009.5007.19 (Moldova).

References

1 A. de la Torre, D. M. Kennes, C. Martin, S. Gerber, J. W. McIver and M. A. Sentef, Nonthermal Pathways to Ultrafast Control in Quantum Materials, *arXiv*, 2021, preprint, arXiv:2103.14888.

2 J. Li, H. U. R. Strand, P. Werner and M. Eckstein, Theory of photoinduced ultrafast switching to a spin-orbital ordered hidden phase, *Nat. Commun.*, 2018, **9**, 1–7.

3 A. M. Kalashnikova, A. Vol'demarovich Kimel and R. V. Pisarev, Ultrafast optomagnetism, *Phys.-Usp.*, 2015, **58**, 969.

4 A. Kirilyuk, A. V. Kimel and T. Rasing, Ultrafast optical manipulation of magnetic order, *Rev. Mod. Phys.*, 2010, **82**, 2731.

5 J. George and G. Khaliullin, Mott insulators in the strong spin-orbit coupling limit: from Heisenberg to a quantum compass and Kitaev models, *Phys. Rev. Lett.*, 2009, **102**, 017205.

6 L. Savary and B. Leon, Quantum spin liquids: a review, *Rep. Prog. Phys.*, 2017, **80**, 016502.

7 A. Kitaev, Anyons in an exactly solved model and beyond, *Ann. Phys.*, 2006, **321**, 2–111.

8 A. Glamazda, P. Lemmens, S.-H. Do, Y. S. Kwon and K.-Y. Choi, Relation between Kitaev magnetism and structure in α-$RuCl_3$, *Phys. Rev. B*, 2017, **95**, 174429.

9 A. Loidl, P. Lunkenheimer and V. Tsurkan, On the proximate Kitaev quantum-spin liquid α-$RuCl_3$: thermodynamics, excitations and continua, *J. Phys.: Condens. Matter*, 2021, **33**, 443004.

10 T. Simon, Kitaev Materials, *arXiv*, 2017, preprint, arXiv:1701.07056 [cond-mat.str-el].

11 Y. Kasahara, T. Ohnishi, Y. Mizukami, O. Tanaka, S. Ma, K. Sugii, N. Kurita, H. Tanaka, J. Nasu, Y. Motome, *et al.*, Majorana quantization and half-integer thermal quantum Hall effect in a Kitaev spin liquid, *Nature*, 2018, **559**, 227–231.

12 J. Wen, S.-L. Yu, S. Li, W. Yu and J.-X. Li, Experimental identification of quantum spin liquids, *npj Quantum Mater.*, 2019, **4**, 1–9.

13 H. B. Cao, A. Banerjee, J.-Q. Yan, C. A. Bridges, M. D. Lumsden, D. G. Mandrus, D. A. Tennant, B. C. Chakoumakos and S. E. Nagler, Low-temperature crystal and magnetic structure of α-$RuCl_3$, *Phys. Rev. B*, 2016, **93**, 134423.

14 J. Wagner, A. Sahasrabudhe, R. Versteeg, L. Wysocki, Z. Wang, V. Tsurkan, A. Loidl, D. Khomskii, H. Hedayat and P. van Loosdrecht, Magneto-optical Study of Metamagnetic Transitions in the Antiferromagnetic Phase of α-$RuCl_3$, *npj Quantum Mater.*, 2022, **7**, 28.

15 J. M. Fletcher, W. E. Gardner, A. C. Fox and G. Topping, X-ray, infrared, and magnetic studies of α- and β-ruthenium trichloride, *J. Chem. Soc. A*, 1967, 1038–1045.

16 J. A. Sears, L. E. Chern, S. Kim, P. J. Bereciartua, S. Francoual, Y. B. Kim and Y.-J. Kim, Ferromagnetic Kitaev interaction and the origin of large magnetic anisotropy in α-$RuCl_3$, *Nat. Phys.*, 2020, **16**, 837–840.

17 R. B. Versteeg, A. Chiocchetta, F. Sekiguchi, A. I. R. Aldea, A. Sahasrabudhe, K. Budzinauskas, Z. Wang, V. Tsurkan, A. Loidl, D. I. Khomskii, *et al.*, Nonequilibrium Quasistationary Spin Disordered State in the Kitaev-Heisenberg Magnet α-$RuCl_3$, *arXiv*, 2020, preprint, arXiv:2005.14189 [cond-mat.str-el].

18 H. Suzuki, H. Liu, J. Bertinshaw, K. Ueda, H. Kim, S. Laha, D. Weber, Z. Yang, L. Wang, H. Takahashi, *et al.*, Proximate ferromagnetic state in the Kitaev model material α-$RuCl_3$, *Nat. Commun.*, 2021, **12**, 1–9.

19 L. Janssen and M. Vojta, Heisenberg-Kitaev physics in magnetic fields, *J. Phys.: Condens. Matter*, 2019, **31**, 423002.

20 N. Bittner, D. Golež, M. Eckstein and P. Werner, Photoenhanced excitonic correlations in a Mott insulator with nonlocal interactions, *Phys. Rev. B*, 2020, **101**, 085127.

21 J. S. Luke, T. Yao, A. A. Reijnders, H.-S. Kim, K. W. Plumb, Y.-J. Kim, H.-Y. Kee and K. S. Burch, Spin-orbit excitations and electronic structure of the putative Kitaev magnet α-$RuCl_3$, *Phys. Rev. B*, 2016, **93**, 075144.

22 J. S. Luke, C. H. Sohn, H. J. Park, S. Y. Kim, K. W. Kim, J. A. Sears, Y.-J. Kim and T. W. Noh, Optical probe of Heisenberg-Kitaev magnetism in α-$RuCl_3$, *Phys. Rev. B*, 2016, **94**, 195156.

23 E. Beaurepaire, J.-C. Merle, A. Daunois and J.-Y. Bigot, Ultrafast spin dynamics in ferromagnetic nickel, *Phys. Rev. Lett.*, 1996, **76**, 4250.

24 E. Carpene, H. Hedayat, F. Boschini and C. Dallera, Ultrafast demagnetization of metals: Collapsed exchange *versus* collective excitations, *Phys. Rev. B: Condens. Matter Mater. Phys.*, 2015, **91**, 174414.

25 N. Sirica, H. Hedayat, D. Bugini, M. R. Koehler, L. Li, D. S. Parker, D. G. Mandrus, C. Dallera, E. Carpene and N. Mannella, Disentangling electronic, lattice, and spin dynamics in the chiral helimagnet $Cr_{1/3}NbS_2$, *Phys. Rev. B*, 2021, **104**, 174426.

26 D. J. Lovinger, M. Brahlek, P. Kissin, D. M. Kennes, A. J. Millis, R. Engel-Herbert and R. D. Averitt, Influence of spin and orbital fluctuations on Mott–Hubbard exciton dynamics in $LaVO_3$ thin films, *Phys. Rev. B*, 2020, **102**, 115143.

27 P. E. Dolgirev, H. M. Marios, A. Zong, N. Gedik and E. Demler, Self-similar dynamics of order parameter fluctuations in pump-probe experiments, *Phys. Rev. B*, 2020, **101**, 174306.

28 D. Afanasiev, A. Gatilova, D. J. Groenendijk, B. A. Ivanov, M. Gibert, S. Gariglio, J. Mentink, J. Li, N. Dasari, M. Eckstein, *et al.*, Ultrafast spin dynamics in photodoped spin-orbit Mott insulator Sr_2IrO_4, *Phys. Rev. X*, 2019, **9**, 021020.

29 Z. Lenarčič and P. Prelovšek, Ultrafast charge recombination in a photoexcited Mott–Hubbard insulator, *Phys. Rev. Lett.*, 2013, **111**, 016401.

30 H. Yamakawa, T. Miyamoto, T. Morimoto, T. Terashige, H. Yada, N. Kida, M. Suda, H. M. Yamamoto, R. Kato, K. Miyagawa, *et al.*, Mott transition by an impulsive dielectric breakdown, *Nat. Mater.*, 2017, **16**, 1100–1105.

31 D. Nevola, B. Alexander, A. Kumar, S. Sridhar, J. Frick, S. O'Donnell, H. Ade, P. A. Maggard, A. F. Kemper, K. Gundogdu, *et al.*, Timescales of excited state relaxation in α-$RuCl_3$ observed by time-resolved two-photon photoemission spectroscopy, *Phys. Rev. B*, 2021, **103**, 245105.

32 J. Ferré and G. A. Gehring, Linear optical birefringence of magnetic crystals, *Rep. Prog. Phys.*, 1984, **47**, 513.

33 K. F. Mak, J. Shan and D. C. Ralph, Probing and controlling magnetic states in 2D layered magnetic materials, *Nat. Rev. Phys.*, 2019, **1**, 646–661.

34 V. Saidl, P. Němec, P. Wadley, V. Hills, R. P. Campion, V. Novák, K. W. Edmonds, F. Maccherozzi, S. S. Dhesi, B. L. Gallagher, *et al.*, Optical determination of the Néel vector in a CuMnAs thin-film antiferromagnet, *Nat. Photonics*, 2017, **11**, 91–96.

35 Q. Zhang, K. Hwangbo, C. Wang, Q. Jiang, J.-H. Chu, H. Wen, D. Xiao and X. Xu, Observation of Giant Optical Linear Dichroism in a Zigzag Antiferromagnet $FePS_3$, *Nano Lett.*, 2021, **21**, 6938–6945.

36 X.-X. Zhang, S. Jiang, J. Lee, C. Lee, K. F. Mak and J. Shan, Spin dynamics slowdown near the antiferromagnetic critical point in atomically thin $FePS_3$, *Nano Lett.*, 2021, **21**, 5045–5052.

37 A. Sahasrabudhe, K. Das, S. Reschke, R. German, T. C. Koethe, J. Buhot, D. Kamenskyi, C. Hickey, P. Becker, V. Tsurkan, *et al.*, High-field quantum disordered state in α-$RuCl_3$: spin flips, bound states, and multiparticle continuum, *Phys. Rev. B*, 2020, **101**, 140410.

38 X. Mi, X. Wang, H. Gui, M. Pi, T. Zheng, K. Yang, Y. Gan, P. Wang, A. Li, A. Wang, *et al.*, Stacking faults in α-$RuCl_3$ revealed by local electric polarization, *Phys. Rev. B*, 2021, **103**, 174413.

39 N. Tesařová, T. Ostatnick, V. Novák, K. Olejník, J. Šubrt, H. Reichlová, C. T. Ellis, A. Mukherjee, J. Lee, G. Matos Sipahi, *et al.*, Systematic study of magnetic linear dichroism and birefringence in (Ga, Mn) As, *Phys. Rev. B: Condens. Matter Mater. Phys.*, 2014, **89**, 085203.

40 R. P. Prasankumar and A. J. Taylor, *Optical Techniques for Solid-State Materials Characterization*, CRC Press, 2016, ch. 13.

41 Z. Ma, K. Ran, J. Wang, B. Song, Z. Cai, S. Li and J. Wen, Recent progress on magnetic-field studies on quantum-spin-liquid candidates, *Chin. Phys. B*, 2018, **27**, 106101.

42 Z. Lenarčič and P. Peter, Charge recombination in undoped cuprates, *Phys. Rev. B: Condens. Matter Mater. Phys.*, 2014, **90**, 235136.

43 Z. Lenarčič, Nonequilibrium properties of Mott insulators: doctoral thesis, PhD thesis, Univerza v Ljubljani, 2015.

44 K. Okazaki, Y. Ogawa, T. Suzuki, T. Yamamoto, T. Someya, S. Michimae, M. Watanabe, Y. Lu, M. Nohara, H. Takagi, *et al.*, Photo-induced semimetallic states realised in electron–hole coupled insulators, *Nat. Commun.*, 2018, **9**, 1–6.

45 T. Rohwer, S. Hellmann, M. Wiesenmayer, C. Sohrt, A. Stange, B. Slomski, A. Carr, Y. Liu, L. M. Avila, M. Kalläne, *et al.*, Collapse of long-range charge order tracked by time-resolved photoemission at high momenta, *Nature*, 2011, **471**, 490–493.

46 H. Hedayat, C. J. Sayers, D. Bugini, C. Dallera, D. Wolverson, T. Batten, S. Karbassi, S. Friedemann, G. Cerullo, J. van Wezel, *et al.*, Excitonic and lattice contributions to the charge density wave in 1 T-$TiSe_2$ revealed by a phonon bottleneck, *Phys. Rev. Res.*, 2019, **1**, 023029.

47 M. Wagner, A. S. McLeod, S. J. Maddox, Z. Fei, M. Liu, R. D. Averitt, M. M. Fogler, R. B. Seth, K. Fritz and D. N. Basov, Ultrafast dynamics of surface plasmons in InAs by time-resolved infrared nanospectroscopy, *Nano Lett.*, 2014, **14**, 4529–4534.

48 I. Radu, C. Stamm, N. Pontius, T. Kachel, P. Ramm, J.-U. Thiele, H. A. Dürr and C. H. Back, Laser-induced generation and quenching of magnetization on FeRh studied with time-resolved X-ray magnetic circular dichroism, *Phys. Rev. B: Condens. Matter Mater. Phys.*, 2010, **81**, 104415.

49 D. Afanasiev, B. A. Ivanov, A. Kirilyuk, T. H. Rasing, R. V. Pisarev and A. V. Kimel, Control of the ultrafast photoinduced magnetization across the Morin transition in $DyFeO_3$, *Phys. Rev. Lett.*, 2016, **116**, 097401.

50 B. Bergman, G. Ju, J. Hohlfeld, R. J. M. van de Veerdonk, J.-Y. Kim, X. Wu, D. Weller and B. Koopmans, Identifying growth mechanisms for laser-induced magnetization in FeRh, *Phys. Rev. B: Condens. Matter Mater. Phys.*, 2006, **73**, 060407.

51 B. Koopmans, G. Malinowski, F. Dalla Longa, D. Steiauf, M. Fähnle, T. Roth, M. Cinchetti and M. Aeschlimann, Explaining the paradoxical diversity of ultrafast laser-induced demagnetization, *Nat. Mater.*, 2010, **9**, 259–265.

52 E. Pastor, D. Moreno-Mencía, M. Monti, A. S. Johnson, N. Fleischmann, C. Wang, Y. Shi, X. Liu, D. G. Mazzone, M. P. M. Dean, *et al.*, Non-thermal Breaking of Magnetic Order via Photo-Generated Spin Defects, *arXiv*, 2021, preprint, arXiv:2104.04294.

53 J. Li, N. Dasari and M. Eckstein, Ultrafast Dynamics in Relativistic Mott Insulators, *arXiv*, 2020, preprint, arXiv:2010.09253.

54 K. Balzer, F. A. Wolf, I. P. McCulloch, P. Werner and M. Eckstein, Nonthermal melting of Néel order in the Hubbard model, *Phys. Rev. X*, 2015, **5**, 031039.

55 M. Eckstein and P. Werner, Photoinduced states in a Mott insulator, *Phys. Rev. Lett.*, 2013, **110**, 126401.

56 L. Perfetti, P. A. Loukakos, M. Lisowski, U. Bovensiepen, M. Wolf, H. Berger, S. Biermann and G. Antoine, Femtosecond dynamics of electronic states in the Mott insulator 1T-TaS_2 by time resolved photoelectron spectroscopy, *New J. Phys.*, 2008, **10**, 053019.

57 V. R. Morrison, R. P. Chatelain, K. L. Tiwari, H. Ali, A. Bruhács, M. Chaker and B. J. Siwick, A photoinduced metal-like phase of monoclinic VO_2 revealed by ultrafast electron diffraction, *Science*, 2014, **346**, 445–448.
58 S. Sinn, C. H. Kim, B. H. Kim, K. D. Lee, C. J. Won, J. S. Oh, M. Han, Y. J. Chang, N. Hur, H. Sato, *et al.*, Electronic structure of the Kitaev material α-$RuCl_3$ probed by photoemission and inverse photoemission spectroscopies, *Sci. Rep.*, 2016, **6**, 1–7.
59 R. P. Prasankumar, H. Okamura, H. Imai, Y. Shimakawa, Y. Kubo, S. A. Trugman, A. J. Taylor and R. D. Averitt, Coupled charge-spin dynamics of the magnetoresistive pyrochlore $Tl_2Mn_2O_7$ probed using ultrafast midinfrared spectroscopy, *Phys. Rev. Lett.*, 2005, **95**, 267404.
60 M. Eckstein and P. Werner, Ultrafast separation of photodoped carriers in Mott antiferromagnets, *Phys. Rev. Lett.*, 2014, **113**, 076405.
61 O. G. Shpyrko, E. D. Isaacs, J. M. Logan, Y. Feng, G. Aeppli, R. Jaramillo, H. C. Kim, T. F. Rosenbaum, Z. Paul, M. Sprung, *et al.*, Direct measurement of antiferromagnetic domain fluctuations, *Nature*, 2007, **447**, 68–71.
62 L. Cipelletti, L. Ramos, S. Manley, E. Pitard, D. A. Weitz, E. E. Pashkovski and M. Johansson, Universal non-diffusive slow dynamics in aging soft matter, *Faraday Discuss.*, 2003, **123**, 237–251.
63 H. Xi, K.-Z. Gao, J. Ouyang, Y. Shi and Y. Yang, Slow magnetization relaxation and reversal in magnetic thin films, *J. Phys.: Condens. Matter*, 2008, **20**, 295220.
64 M. Labrune, S. Andrieu, F. Rio and P. Bernstein, Time dependence of the magnetization process of RE-TM alloys, *J. Magn. Magn. Mater.*, 1989, **80**, 211–218.
65 F. Romanens, S. Pizzini, F. Yokaichiya, M. Bonfim, P. Yan, J. Camarero, J. Vogel, J. Sort, F. Garcia, B. Rodmacq, *et al.*, Magnetic relaxation of exchange biased Pt/Co multilayers studied by time-resolved Kerr microscopy, *Phys. Rev. B: Condens. Matter Mater. Phys.*, 2005, **72**, 134410.
66 A. A. Adjanoh, R. Belhi, J. Vogel, M. Ayadi and K. Abdelmoula, Compressed exponential form for disordered domain wall motion in ultra-thin Au/Co/Au ferromagnetic films, *J. Magn. Magn. Mater.*, 2011, **323**, 504–508.
67 J. Pommier, P. Meyer, G. Pénissard, J. Ferré, P. Bruno and D. Renard, Magnetization reversal in ultrathin ferromagnetic films with perpendicular anisotropy: Domain observations, *Phys. Rev. Lett.*, 1990, **65**, 2054.
68 S.-W. Cheong, M. Fiebig, W. Wu, L. Chapon and V. Kiryukhin, Seeing is believing: visualization of antiferromagnetic domains, *npj Quantum Mater.*, 2020, **5**, 1–10.
69 J. Chaloupka and G. Khaliullin, Magnetic anisotropy in the Kitaev model systems Na_2IrO_3 and α-$RuCl_3$, *Phys. Rev. B*, 2016, **94**, 064435.
70 H. Hedayat, C. J. Sayers, A. Ceraso, J. v. Wezel, S. R. Clark, C. Dallera, G. Cerullo, E. Da Como and E. Carpene, Investigation of the non-equilibrium state of strongly correlated materials by complementary ultrafast spectroscopy techniques, *New J. Phys.*, 2021, **23**, 033025.
71 H. Hedayat, D. Bugini, H. Yi, C. Chen, X. Zhou, G. Cerullo, C. Dallera and E. Carpene, Ultrafast evolution of bulk, surface and surface resonance states in photoexcited Bi_2Te_3, *Sci. Rep.*, 2021, **11**, 1–8.
72 S. Gerber, S.-L. Yang, D. Zhu, H. Soifer, J. A. Sobota, S. Rebec, J. J. Lee, T. Jia, B. Moritz, C. Jia, *et al.*, Femtosecond electron-phonon lock-in by photoemission and x-ray free-electron laser, *Science*, 2017, **357**, 71–75.

73 R. B. Versteeg, J. Zhu, P. Padmanabhan, C. Boguschewski, R. German, M. Goedecke, P. Becker and P. H. M. van Loosdrecht, A tunable time-resolved spontaneous Raman spectroscopy setup for probing ultrafast collective excitation and quasiparticle dynamics in quantum materials, *Struct. Dyn.*, 2018, **5**, 044301.

Faraday Discussions

ROYAL SOCIETY OF CHEMISTRY

PAPER

Ultrafast branching in intersystem crossing dynamics revealed by coherent vibrational wavepacket motions in a bimetallic Pt(II) complex†

Pyosang Kim, *[ab] Andrew J. S. Valentine,[c] Subhangi Roy,[d] Alexis W. Mills,[c] Felix N. Castellano, *[d] Xiaosong Li *[c] and Lin X. Chen *[ab]

Received 17th January 2022, Accepted 7th February 2022

DOI: 10.1039/d2fd00009a

Ultrafast excited state processes of transition metal complexes (TMCs) are governed by complicated interplays between electronic and nuclear dynamics, which demand a detailed understanding to achieve optimal functionalities of photoactive TMC-based materials for many applications. In this work, we investigated a cyclometalated platinum(II) dimer known to undergo a Pt–Pt bond contraction in the metal–metal-to-ligand-charge-transfer (MMLCT) excited state using femtosecond broadband transient absorption (fs-BBTA) spectroscopy in combination with geometry optimization and normal mode calculations. Using a sub-20 fs pump and broadband probe pulses in fs-BBTA spectroscopy, we were able to correlate the coherent vibrational wavepacket (CVWP) evolution with the stimulated emission (SE) dynamics of the ^{1}MMLCT state. The results demonstrated that the 145 cm^{-1} CVWP motions with the damping times of ~0.9 ps and ~2 ps originate from coherent Pt–Pt stretching vibrations in the singlet and triplet MMLCT states, respectively. On the basis of excited state potential energy surface calculations in our previous work, we rationalized that the CVWP transfer from the Franck–Condon (FC) state to the ^{3}MMLCT state was mediated by a triplet ligand-centered (^{3}LC) intermediate state through two step intersystem crossing (ISC) on a time scale shorter than a period of the Pt–Pt stretching wavepacket motions. Moreover, it was found that the CVWP motion had 110 cm^{-1} frequency decays with the damping time of ~0.2 ps, matching the time constant of 0.253 ps, corresponding to a redshift in the SE feature at early times. This observation indicates that the Pt–Pt bond contraction changes the stretching frequency from 110 to 145 cm^{-1} and stabilizes the ^{1}MMLCT state relative to the ^{3}LC state with a ~0.2 ps time scale. Thus, the ultrafast ISC from the

[a]Department of Chemistry, Northwestern University, Evanston, Illinois 60208, USA. E-mail: pyosang.kim@northwestern.edu

[b]Chemical Science and Engineering Division, Argonne National Laboratory, Lemont, Illinois 60349, USA

[c]Department of Chemistry, University of Washington, Seattle, Washington 98195, USA

[d]Department of Chemistry, North Carolina State University, Raleigh, North Carolina 27695-8204, USA

† Electronic supplementary information (ESI) available. See DOI: 10.1039/d2fd00009a

^{1}MMLCT to the ^{3}LC states occurs before the Pt–Pt bond shortening. The findings herein provide insight into understanding the impact of Pt–Pt bond contraction on the ultrafast branching of the ^{1}MMLCT population into the direct (^{1}MMLCT → ^{3}MMLCT) and indirect ISC pathways (^{1}MMLCT → ^{3}LC → ^{3}MMLCT) in the Pt(II) dimer. These results revealed intricate excited state electronic and nuclear motions that could steer the reaction pathways with a level of detail that has not been achieved before.

Introduction

Transition metal complexes (TMCs) have received tremendous attention in solar energy conversion, photocatalysis and optoelectronics due to their versatile photochemical properties, enabling a variety of functions as electron donors/acceptors, photoluminescence emitters, and redox/catalytic centers.[1–8] Recent advances in femtosecond optical and X-ray spectroscopy/scattering have brought new insights into the electronic and nuclear dynamics of excited state TMCs in an ultrafast time regime.[9–16] These studies have shown that electronic transitions induced by light and nuclear motions may occur on similar time scales along the excited state trajectories. The observations indicate that the nuclear and electronic coordinates are coupled, and the excited state trajectories could go across multiple potential energy surfaces (PESs) before relaxation to vibrational equilibrium. Therefore, a precise understanding of the interplay between nuclear and electronic degrees of freedom in the ultrafast regime is particularly crucial in the development of advanced TMC-based materials for the aforementioned applications.

Cyclometalated platinum(II) dimer complexes offer an ideal platform to investigate the interplay of electronic and nuclear dynamics in TMCs, because they display distinct structural changes in the excited state upon the electronic transitions, and their structural dynamics have been linked to their emission properties as phosphorescence emitters in organic light-emitting diodes (OLEDs).[8,17] Previous experimental and theoretical studies on these systems have revealed strong structural dependent photophysical properties, which enable tunability in their absorption/emission properties *via* chemical synthesis, such as by changing the Pt–Pt bond distance. As the Pt–Pt distance decreases, the interaction between the $5d_{z^2}$ orbitals of the two Pt(II) atoms becomes stronger, and the energy splitting between the Pt–Pt σ bonding and σ* antibonding molecular orbitals (MOs) increases, as shown in Fig. 1a. Since the close Pt–Pt distance can raise the energy of the σ* antibonding MO higher than those of the other d orbitals, the lowest electronic transition transforms from a localized ligand-centered (LC) and/or a metal-to-ligand-charge transfer (MLCT) in one half of the molecule with a pseudo two-fold symmetry to a delocalized metal–metal-to-ligand-charge transfer (MMLCT).[18–23] This transformation from the LC/MLCT to the MMLCT transitions results in a concomitant redshift in the absorption and phosphorescence spectra. Moreover, the MMLCT excitation moves an electron from the σ* antibonding MO to the π* MO of cyclometalating ligands, thereby increasing the Pt–Pt bond order by ~0.5 due to the removal of the antibonding electron, and resulting in a significant Pt–Pt bond contraction in the excited state. Indeed, the photoinduced Pt–Pt bond shrinkage has been observed in the triplet MMLCT state through X-ray transient absorption and scattering experiments.[24,25]

Although the Pt–Pt bond contraction has been known as the primary structural change in the MMLCT excited state of the Pt(II) dimers, a question remains of how the Pt–Pt bond shortening directly affects the excited state processes, including intersystem crossing (ISC) to the ^{3}MMLCT state. To address this question, our previous works utilized optical transient absorption spectroscopy equipped with a ~35 fs pump pulse to create coherent vibrational wavepacket (CVWP) motions and track their evolution along the excited state trajectories. Upon the photoexcitation of the ^{1}MMLCT state, we have observed a strong beating signal corresponding to Pt–Pt stretching CVWP motions in pyrazolate-, 2-thiolpyridyl- and 2-hydroxypyridyl-bridged Pt(II) dimers.[13,26,27] Furthermore, a comparative study on the Pt–Pt stretching CVWP dynamics has been conducted in a series of structurally related 2-hydroxypyridyl-bridged Pt(II) dimers. We found that the Pt–Pt stretching CVWP generated in the Franck–Condon (FC) region can be sustained during the ISC by adding phenyl substituents on the bridging ligands. With the help of quantum mechanical calculations of excited state PESs, it was shown that the interaction between the phenyl rings and Pt(II) atoms stabilizes the energy of the ^{1}MMLCT state relative to the ^{3}LC state and, thus, locates the ^{3}LC states in the vicinity of the FC state. Thus, ultrafast ISC can take place from the ^{1}MMLCT to the ^{3}LC intermediate state, followed by subsequent internal conversion (IC) to the ^{3}MMLCT state. However, determining the population decay of the ^{1}MMCLT state and corresponding CVWP dynamics was difficult in the previous study, largely due to the lack of stimulated emission (SE) features in the probe spectrum (520–700 nm) used in the TA measurements. Hence, it is still unclear how the photoinduced Pt–Pt contraction modulates the energy of the ^{1}MMLCT state relative to the ^{3}LC state beyond the FC region and then affects the ultrafast ISC pathway.

Using femtosecond broadband transient absorption spectroscopy (fs-BBTA), the work presented herein investigated the CVWP dynamics during the ISC processes in one of the 2-hydroxypyridyl-bridged Pt(II) dimers that showed long-lived vibrational coherence in our previous study (inset in Fig. 1b). The broadband probe spectrum covering the range of 520–850 nm enabled us to clearly identify the SE dynamics and relevant CVWP motions appearing in the range of 600–850 nm. Also, sub-20 fs pump pulses enhance the amplitude of CVWP motions associated with the photoinduced structural changes including the Pt–Pt contraction as compared to the previous measurements as the synchronization of the CVWP motion improves. By correlating the temporal evolution of SE and CVWP signals, we demonstrate two intriguing facts of ISC processes in the Pt(II) dimer: (1) the initially populated MMLCT state branches into the direct (^{1}MMLCT → ^{3}MMLCT) and indirect ISC path (^{1}MMLCT → ^{3}LC → ^{3}MMLCT) and (2) the ultrafast branching occurs before completing the Pt–Pt bond contraction on the ~0.2 ps time scale. Therefore, the photoinduced Pt–Pt contraction is a key structural factor to determine the ultrafast ISC and the CVWP transfer in the vicinity of the FC region.

Experimental section

Sample preparation

The Pt(II) dimer was synthesized and structurally characterized as described in our previous publication.[28]

Steady-state measurements

Steady-state absorption and emission spectra were measured by a Shimadzu UV-3600 UV-Vis-NIR spectrophotometer and a Horiba Scientific model Fluorolog-3 spectrometer, respectively. Both measurements were performed in 1 cm path-length quartz cuvettes in THF at room temperature.

Femtosecond broadband transient absorption spectroscopy

For the fs-BBTA measurements, a Yb:KGW regenerative amplifier (Light Conversion, Pharos-10 W) was used to generate 1030 nm fundamental pulses at a 10 kHz repetition rate with a pulse width of 160 fs. The fundamental beam was split into two pulses using a 90/10 beam splitter for the generation of a visible pump and a broadband probe pulse. The former was accomplished by a home-built noncollinear optical parametric amplifier (NOPA) for the pump pulses centered at 540 nm with a 60 nm bandwidth.[29] The pump pulses were compressed with a chirped mirror pair (Laser Quantum, DCM-12) and a fused silica wedge pair (Newport) to a 17.5 fs full width at half maximum (FWHM) at the sample position, which was characterized by a second-harmonic generation frequency-resolved optical gating (SHG-FROG) with a 5 μm thick beta barium borate (BBO) crystal (United Crystals Inc.) (Fig. S1†). For the broadband probe, the low-power beams were focused on a 3 mm thick YAG window (EKSMA) by a 10 cm plano-convex lens to generate a white light continuum (WLC) covering the 520–850 nm range. The WLC was subsequently compressed by a chirped mirror pair (Laser Quantum, DCM-9). The pump and probe beams were focused on the sample contained in a 1 mm path-length quartz cuvette using a 25 cm concave mirror. The optical time delay between the pump and probe pulses was controlled by a motorized linear stage (Newport, XMS160-S). To measure the TA spectra, the "pump-on" and "pump-off" probe spectra were obtained by a chopper operating at 500 Hz and detected with a spectrometer (Andor, Kymera 328i) and a sCMOS camera (Andor, Zyla-5.5). The TA measurements were carried out with the pump polarization at a magic angle (54.7°) with respect to the probe polarization to prevent polarization-dependent signals.

Quantum mechanical calculations

All calculations were performed at University of Washington using GAUSSIAN16.[30] The 6-31g(d) basis set was employed for all light atoms, while the LanL2DZ effective core potential and associated basis set[31] was used for Pt atoms. Density functional calculations were performed with the ωB97X-D functional.[32] This functional was chosen because it is range-separated, to better describe charge-transfer excited states,[33] and it contains dispersion corrections, necessary to capture π–π interactions.[32] Excited states were evaluated using time-dependent density functional theory (TD-DFT). Energetic minima found by ground and excited state geometry optimizations were confirmed by normal mode calculations at the same level of theory.

Results and discussion

Excited state dynamics

The steady-state absorption and emission spectra of the Pt(II) dimer are depicted in Fig. 1b. In the previous report by Chakraborty *et al.*, three electronic transition

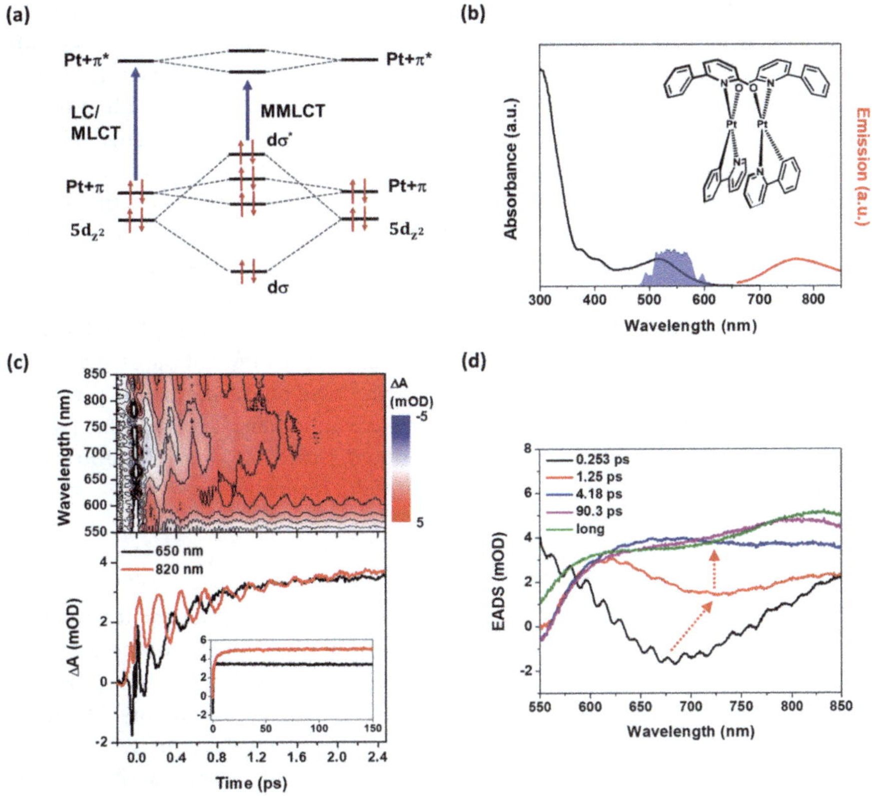

Fig. 1 (a) Molecular orbital energy scheme for the Pt(II) dimer, illustrating the interaction between the 5d$_{z^2}$ orbitals of Pt(II) atoms and its impact on the transformation of the lowest electronic transition from LC/MLCT to MMLCT. (b) Steady-state absorption (black) and emission spectra (red) for the Pt(II) dimer in THF and the pump pulse spectrum (blue) used in the fs-BBTA. The inset shows a schematic of the molecular structure of the Pt(II) dimer. (c) The fs-BBTA data map (top) and the kinetic traces (bottom) at 650 and 820 nm probe wavelengths. The inset exhibits the TA kinetic traces up to 150 ps delay at the same probe wavelengths. (d) Evolution-associated difference spectra (EADS) obtained by the global kinetic analysis of fs-BBTA data scanned with a time delay of 150 ps. The red dotted arrows highlight the decay dynamics of the SE signal.

bands were confirmed by TD-DFT calculations: (1) the ^{1}LC band below 350 nm, (2) a mixture of ^{1}LC and ^{1}MLCT bands in the 350–450 nm range and (3) ^{1}MMLCT bands above 450 nm.[28] This work also assigned the broad emission spectrum around 780 nm to the ^{3}MMLCT band.

We performed the fs-BBTA measurements to identify the excited state dynamics associated with the ISC from the singlet to the triplet MMLCT state upon photoexcitation in the 500–590 nm pump spectral range (Fig. 1c). To get details of the excited state dynamics for the Pt(II) dimer, global kinetic analysis was carried out on the BBTA data scanned up to 150 ps using the Glotaran program,[34] as shown in Fig. 1d. The best fits to the TA kinetics have four time constants of 0.253, 1.25, 4.18 and 90.3 ps and a constant offset originating from the 55.9 ns decay of the ^{3}MMLCT state.[28] The evolution-associated difference spectra (EADS) for the time component of 0.253 ps clearly shows the stimulated emission (SE) feature emerging between the edge of the ^{1}MMLCT absorption (~600 nm) and the peak of the ^{3}MMLCT emission (770 nm). When converted to the 1.25 ps EADS, the SE signal decays with a redshift to ~725 nm and a concomitant growth of the ESA signal around 625 nm. Then, the SE feature is completely quenched in the 4.18 ps EADS, leading to the further growth of the excited state absorption (ESA) signals in the range of 625–850 nm. Finally, the 4.18 and 90.3 ps EADSs display the rise of ESA features at around 825 nm. Although the ^{1}MMLCT fluorescence was not detected in the steady-state measurement, the SE kinetics can be reasonably connected to the quenching dynamics of the initially populated ^{1}MMLCT state. In particular, since the SE decay dynamics do not accompany the recovery of the ground state absorption (GSB) signal at around 550 nm and the decay of ESA signal, we determined the ISC rate from the singlet to triplet MMLCT state to be 1.25 ps. The origin of the SE dynamics on the 0.253 ps time scale will be discussed later. Based on the previous TA study of pyrazolate-bridged Pt(II) dimers,[23] the slower two components, 4.18 and 90.3 ps, are herein assigned to the intramolecular vibrational redistributions (IVR) and vibrational cooling occurring in the ^{3}MMLCT manifold, respectively.

Coherent vibrational wavepacket dynamics

The ultrashort pump pulse (17.5 fs) used in the fs-BBTA measurement launched CVWPs modulating the BBTA signals as a function of the pump–probe delay, as shown in Fig. 1c. To identify the frequencies associated with the CVWP motions, we performed fast Fourier transformation (FFT) on the oscillatory residuals extracted from the global kinetic analysis of the BBTA data. In Fig. 2a, the wavelength-resolved FFT power map, obtained by the FFT of the oscillations in the delay time window of 0.1–2.5 ps, shows a dominant peak at 145 cm^{-1}, centered around 580, 700 and 840 nm. The broadband probe enabled us to newly measure the 145 cm^{-1} CVWP motions around 840 nm, in comparison to our previous study in which this probe wavelength was inaccessible.[27] Three minor peaks were also observed at 110, 228 and 292 cm^{-1}, except the 914 cm^{-1} Raman mode of the THF solvent.[35] The FFT map displays a decrease of the amplitude at ~650 and ~750 nm, indicating a π phase change, the so-called "node", in the 145 cm^{-1} CVWP motions. As shown in Fig. 2b, the wavelength-resolved residual map clearly shows the spectral positions of the nodes and their temporal evolution along the probe wavelength. For the node at ~650 nm in the FFT map, it appears at 600 nm from ~0.3 ps and then moves to 650 nm. Similarly, the other node transforms from 675 nm to 770 nm. The CVWP motions created in the PES show the phase shift of the oscillations at the minimum of the PES, resulting in the zero-amplitude node, as shown in Fig. 2.[36–38] Therefore, the spectral position of the node can be a strong evidence for the origin of the CVWP; the ground state wavepacket produces the node around the absorption (GSB) maximum, while the node of the excited state wavepacket arises near the fluorescence (SE) or ESA maximum.[39] The two nodes observed in the 145 cm^{-1} CVWP motions emerge beyond the ground state absorption region (500–600 nm, Fig. 1b). Thus, the 145 cm^{-1} CVWP motions are expected to be the excited state wavepackets. In particular, the node shift from 675 to 770 nm shown in the residual map occurs near the SE maxima of the 0.253 and 1.25 ps EADS (Fig. 1d), indicating that the 145 cm^{-1} CVWP motions generating the node at ~750 nm in the FFT map originate from the ^{1}MMLCT state.

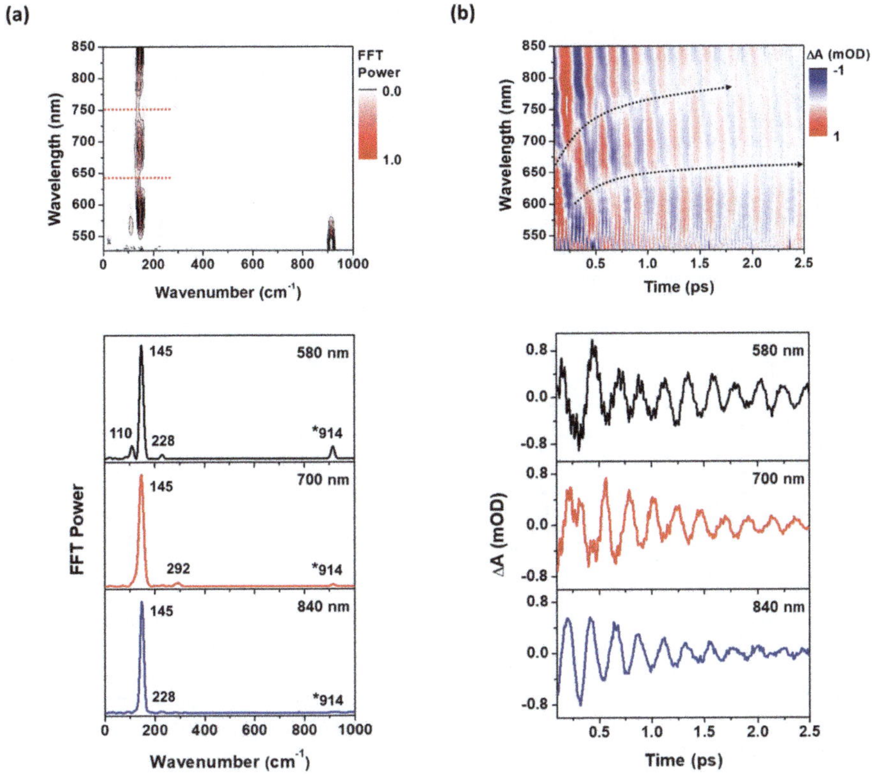

Fig. 2 (a) Wavelength-resolved FFT power map (top) and FFT spectra (bottom) at 580, 700 and 840 nm. The FFT was performed on the oscillations in 0.1–2.5 ps window along the probe wavelength. The oscillatory residuals in 0–0.1 ps window were truncated to avoid the contribution of cross-phase modulation to the FFT. The red dotted lines in the FFT map indicate the node positions that the π phase jump occur. (b) Wavelength-resolved residual map (top) and oscillatory signals at 580, 700 and 840 nm. The dotted arrows highlight the node shift along the time delay.

The oscillatory residuals in the time window of 2.5–5 ps were FFT-analyzed separately in order to investigate the spectral positions of the two nodes beyond the 2.5 ps delay (Fig. 3a). In comparison with the wavelength-resolved FFT map in the 0.1–2.5 ps window (Fig. 2b), the FFT map in the 2.5–5 ps window clearly shows an absence of the node at around 750 nm, while the node at around 650 nm is slightly redshifted to ~675 nm and preserved over 2.5 ps. The damping times of the oscillations near 580, 700, 750 and 840 nm were fitted to a combination of sines multiplied by exponential functions (Fig. 3b). The results are summarized in Table 1. The 145 cm^{-1} CVWPs detected at around 580 and 750 nm show a very long dephasing time of ~2 ps, while the ones at around 700 and 840 nm exhibit the damping time of ~0.9 ps. This fit explains that the beating node near 750 nm vanishes twice as fast as the other observed at around 650 nm. Therefore, the temporal change of the FFT map for the 145 cm^{-1} CVWPs suggests that the two nodes near 650 nm and 750 nm correspond to the different excited states. As has already been mentioned, the node shift from 675 to 770 nm matches to the redshift of the SE signals well (Fig. 1d). Also, the damping time of

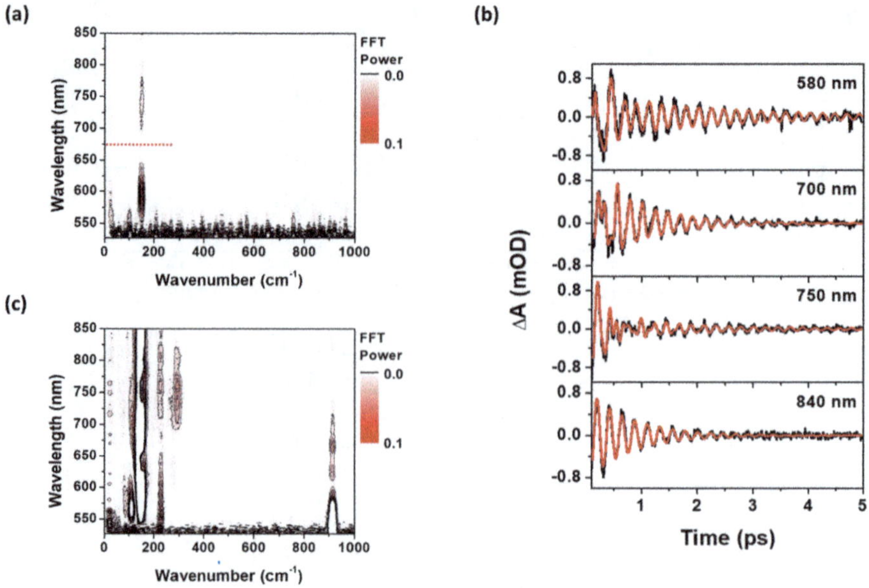

Fig. 3 (a) Wavelength-resolved FFT power map obtained by the FFT on the oscillations in the time window of 2.5–5 ps. The red dotted line indicates the spectral position of the node. (b) The oscillatory residuals (black) with the fit to exponentially damped sine functions (red) at 580, 700, 750 and 840 nm. (c) Zoom in of the FFT power map of the oscillations in the 0.1–2.5 ps window to clearly show the amplitude distribution of the 110, 228 and 292 cm^{-1} frequencies.

Table 1 Dephasing times of the oscillations at 580, 700, 750 and 840 nm

Parameter	580 nm	700 nm	750 nm	840 nm
τ_1 (145 cm^{-1})	2.35 ± 0.18 ps	0.95 ± 0.04 ps	0.23 ± 0.01 ps 2.11 ± 0.28 ps	0.78 ± 0.02 ps
τ_2 (228 cm^{-1})	0.58 ± 0.30 ps	0.63 ± 0.14 ps	0.75 ± 0.10 ps	—
τ_3 (292 cm^{-1})	—	0.44 ± 0.05 ps	0.68 ± 0.08 ps	—
τ_4 (110 cm^{-1})	0.24 ± 0.02 ps	0.16 ± 0.01 ps	—	—

~0.9 ps indicates that the 145 cm^{-1} CVWP motions at 700 and 840 nm decay within the ISC rate of 1.25 ps. Both findings justify that the node at ~750 nm stems from the ^{1}MMLCT state. In contrast, the node at ~650 nm persists over the ISC rate. Furthermore, the node shift from 650 to 675 nm seems to be largely due to the growth of the ESA feature induced by the quenching of SE signals, as shown in the transition from 1.25 ps to 4.18 ps EADSs (Fig. 1d). Thus, the 145 cm^{-1} CVWP motions generating the node at around 650 nm can be reasonably assigned to the nuclear motions populated in the ^{3}MMLCT state.

The lower frequency mode, 110 cm^{-1} CVWP, was observed for the same Pt(II) dimer in our previous work.[27] The other ultrafast TA studies also showed a similar frequency in the pyrazolate- and 2-thiolpyridyl-bridged Pt(II) dimers equipped with the same cyclometalating ligands.[26,40] Since the lower frequency oscillation had a strong amplitude mainly near the ground state absorption region, the previous studies reported that its origin was the ground state wavepacket. Although the current fs-BBTA measurement also showed a similar trend, the damping time of this oscillation suggests a possibility of the excited state wavepacket. In general, the ground state CVWP has a long-lived coherence lifetime compared to that of the excited state CVWP, because the ground state wavepacket does not undergo electronic transitions, vibrational relaxations, solvent-related reorganization and photoinduced structural changes, strongly affecting the vibrational dephasing. Very surprisingly, the 110 cm^{-1} oscillation at 580 nm for the Pt(II) dimer decays 5–10 times faster than the 145 cm^{-1} excited state CVWP. The damping time of 0.24 ps is in good agreement with the 0.253 ps time scale, corresponding to the redshift of the SE features. Moreover, the current fs-BBTA clearly unraveled that the CVWP motions near 700 nm contain the 110 cm^{-1} vibration, as shown in Fig. 3a, and the oscillatory fit showed the damping time of 0.16 ps, supporting that the 110 cm^{-1} CVWP is highly possibly the excited state wavepacket.

The 228 and 292 cm^{-1} frequency vibrations are observed beyond the ground state absorption region (Fig. 3c). Also, the oscillatory fit results show that their damping times are 0.6–0.8 ps. Although it is hard to determine the spectral position of the node for these vibrations in the FFT map due to the weak amplitudes, their vibrational dephasings exhibit similar time scales to the damping time of 145 cm^{-1} CVWP at 700 and 840 nm, suggesting that these nuclear motions originate from the ^{1}MMLCT state and dephase during the ISC with the rate of 1.25 ps.

Nuclear motions associated with photoinduced structural changes

To assess the correlation between the photoinduced nuclear reorganization and the observed CVWP dynamics, the structure optimization and the normal mode calculations were performed as pictured in Fig. 4. The optimized molecular geometries of the ^{1}MMLCT and ^{3}MMLCT states show significant changes from that of the ground state in terms of Pt–Pt bond distance and dihedral angle of the two planes of the cyclometalating ligands: (1) the Pt–Pt bond is shortened from 2.94 Å in the ground state to 2.70 Å and 2.69 Å in the singlet and triplet MMLCT states, respectively, (2) the distance between the ligand planes is decreased from 3.47 Å in the ground state to 3.20 Å and 3.22 Å in the singlet and triplet MMLCT states, respectively, and (3) the twisting angle is reduced from 15.06° in the ground state to 11.01° and 11.31° in the singlet and triplet MMLCT states, respectively. In particular, the bond contraction in the MMLCT states has been directly observed by the X-ray transient absorption and scattering measurements of the structurally related pyrazole-bridged Pt(II) dimers.[24,25,41]

In Fig. 4b, the normal mode calculations of the ground state showed the Pt–Pt stretching vibration at a frequency of 114.7 cm^{-1}. For the ^{1}MMLCT and ^{3}MMLCT states, the Pt–Pt stretching vibrations were also found at frequencies of 149.7 and 145.5 cm^{-1} in the computations. The Pt–Pt stretching frequency is upshifted in the excited MMLCT states because the Pt–Pt bond order increases by 0.5 due to the lack of one electron in the antibonding dσ^* orbital. The calculated Pt–Pt stretching vibrational frequencies match the observed 110 and 145 cm^{-1} CVWPs well. The other normal modes at around 228 and 292 cm^{-1} in the excited MMLCT states involved the interligand vibrations oscillating parallelly and perpendicularly to the Pt–Pt bond axis. Therefore, as the above nuclear motions modulate the

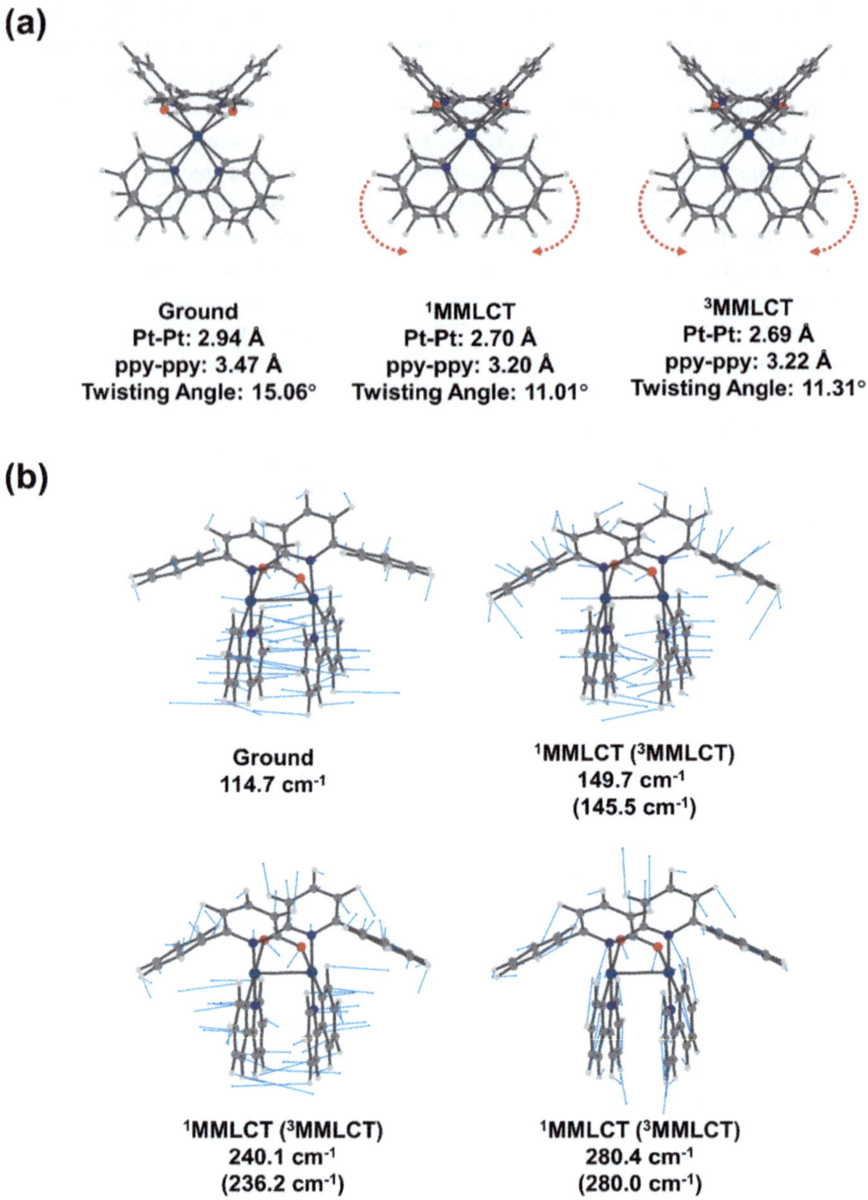

Fig. 4 (a) Optimized molecular structures for the ground, ^{1}MMCLT and ^{3}MMLCT states. The red dotted arrows highlight the decrease in the twisting angle between the ppy ligand planes. (b) Calculated normal modes for the ground, ^{1}MMCLT and ^{3}MMLCT states.

Pt–Pt distance and the distance and twisting angle between the two ligands, they can be assigned to the MMLCT active modes.

Intersystem crossing pathways

As depicted in Fig. 5, the excited state PES calculations along the axis of the Pt–Pt bond distance in our previous work proposed that the coherent Pt–Pt stretching motions can be populated in the ^{3}MMLCT state through an ultrafast ISC pathway mediated by an intermediate ^{3}LC state lying close to the ^{1}MMLCT state.[27] The PES calculations further showed a curve crossing between the ^{1}MMLCT and the ^{3}LC states occurring in the vicinity of the FC region, implying the presence of a conical intersection. In the present work, the analyses of the excited state and CVWP dynamics and the normal mode calculations confirm that the short- and long-lived 145 cm^{-1} CVWPs are the coherent Pt–Pt stretching nuclear motions generated in the ^{1}MMLCT and ^{3}MMLCT states, respectively. This is consistent with our previous TA results for the same compound. However, the present BBTA measurement clearly unveiled SE kinetics corresponding to the population dynamics of the ^{1}MMLCT state. This observation demonstrates that the Pt–Pt stretching vibration in the ^{1}MMLCT state decays largely due to the ISC to the ^{3}MMLCT state. Furthermore, since the decay (1.25 ps) and damping time (~0.9 ps) of the SE and CVWP signals are much longer than the period of 145 cm^{-1} oscillation (230 fs), the direct ISC pathway (^{1}MMLCT → ^{3}MMLCT) is impossible to transfer the coherent Pt–Pt stretching motion to the ^{3}MMLCT state. Therefore, the Pt–Pt stretching CVWP motions in the ^{3}MMLCT state is a strong indication that a fraction of the ^{1}MMLCT population branches into the ^{3}MMLCT state as fast as the period of coherent Pt–Pt stretching motion with the 145 cm^{-1} frequency.

Along this line, the 110 cm^{-1} CVWP with the damping time of ~0.2 ps captured in the present study can be strong evidence for the Pt–Pt stretching CVWP transfer occurring near the FC state. Considering the molecular orbital energy level for the Pt(II) dimer (Fig. 1a), the Pt–Pt bond shrinking can regulate the dσ* orbital energy (Fig. 1a) and, thus, alter the MMLCT transition energy through the changes in this nuclear coordinate. As was already mentioned, if the 110 cm^{-1}

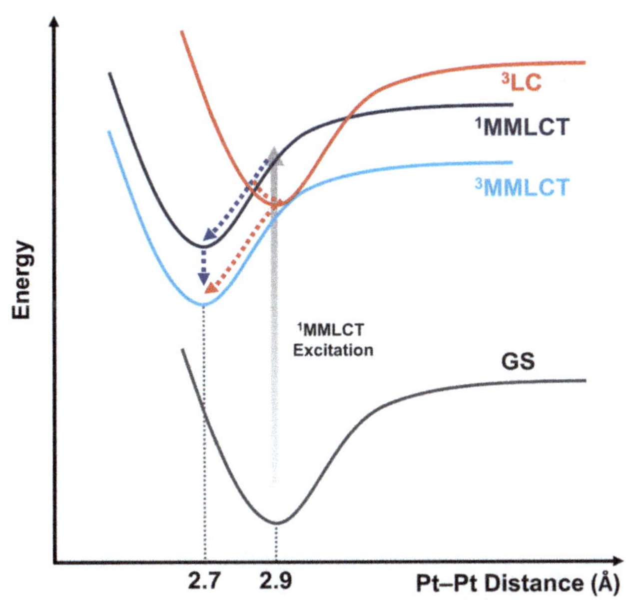

Fig. 5 Schematic illustration of the excited state PESs. The initial population of the ^{1}MMLCT state branches into two ISC pathways: the direct (^{1}MMLCT → ^{3}MMLCT, blue dotted arrow) and indirect ISC paths (^{1}MMLCT → ^{3}LC → ^{3}MMLCT, red dotted arrow).

CVWP is the excited state wavepacket, its fast damping time of ~0.2 ps is largely due to the Pt–Pt contraction converting the Pt–Pt stretching frequency from 110 to 145 cm^{-1}. It should be noted that the TA kinetics analysis showed a decrease in the ^{1}MMLCT transition energy with a redshift of the SE feature on a similar time scale of 0.253 ps (Fig. 1d). Since the energy of the ^{3}LC state is expected to be unaffected by the Pt–Pt shortening, this suggests that the Pt–Pt contraction triggered by the ^{1}MMLCT excitation moves the ^{1}MMLCT population away from the FC state and the ^{3}LC state in the early time. Therefore, it can be concluded that the ultrafast ISC and the Pt–Pt stretching CVWP transfer to the ^{3}LC state have to occur within ~0.2 ps before completing the Pt–Pt contraction. Although the 17.5 fs pump pulse can instantly induce the shortening of the Pt–Pt bond, the close contact between the ligands (<3.5 Å) seems to slow down the bond contraction through the interligand structural reorganization, such as the changes of the distance and twisting angle between the ligand planes (Fig. 4b).

Our findings shed new light on the ISC dynamics in similar bimetallic Pt(II) dimers. Recently, Mewes *et al.* have reported coherent Pt–Pt stretching motions during the ISC in a pyrazolate-bridged Pt(II) dimer.[40] Interestingly, this complex has not shown the CVWP transfer to the ^{3}MMLCT state, although the excited state PES calculations predicted the presence of a high-lying triplet state adjacent to the FC state. Furthermore, in our previous report, the CVWP transfer was not observed in a Pt(II) dimer derivative with methyl substituents on the same bridging ligands of the present Pt(II) dimer.[27] For the Pt(II) dimer in the present study, a notable structural difference from the methyl substituted Pt(II) dimer is the phenyl substituents on the bridging ligands. Even though these Pt(II) dimers have similar Pt–Pt bond lengths in the ground and the excited MMLCT states, the molecular orbital calculations showed that interactions between the rings and Pt(II) atoms destabilized the dσ* orbital, thereby lowering the ^{1}MMLCT state relative to the ^{3}LC state and then decreasing the energy gap between the FC and the ^{3}LC states more than the ones of the methyl substituted Pt(II) dimer. Therefore, close proximity in the energies between the FC and ^{3}LC states is a requirement for the ultrafast ISC and the transfer of Pt–Pt stretching CVWPs.

It is impossible to identify the transient features of the ^{3}LC state from the TA kinetic analysis for the Pt(II) dimer due to the overlap of the broad SE and ESA of the MMLCT states. Hence, the exact time scale of the ISC pathway *via* the intermediate state is unknown in this work. Also, it is unclear that the Pt–Pt stretching vibrational coordinate is still the reaction coordinate for the subsequent internal conversion (IC) process from the ^{3}LC to the ^{3}MMLCT states. In the future, to prove the involvement of the intermediate state and the nuclear dynamics during the IC, time-resolved impulsive stimulated Raman spectroscopy will be done to track the CVWP motions from the FC to the intermediate and then the final ^{3}MMLCT states.

Conclusion

In this work, we investigated the ISC processes from the singlet to triplet MMLCT states using fs-BBTA spectroscopy. The ultrashort pump pulse allowed us to capture the CVWP motions associated with the photoinduced structural changes upon the ^{1}MMLCT excitation. The broadband probe spanning to 850 nm fully unraveled the SE spectral features corresponding to the ^{1}MMLCT state, which was unclear in our previous work.[27] With the aid of normal mode calculation, correlating the temporal behavior between the SE signals and the beating nodes confirmed that the short- and long-lived coherent Pt–Pt stretching vibrations with 145 cm^{-1} frequency originate from the ^{1}MMLCT and the ^{3}MMLCT states, respectively. This observation is strongly indicative of the ultrafast branching of the initially populated ^{1}MMLCT into the direct (^{1}MMLCT → ^{3}MMLCT) and indirect ISC pathways (^{1}MMLCT → ^{3}LC → ^{3}MMLCT), proposed in our previous work. Furthermore, the ~0.2 ps vibrational dephasing of the 110 cm^{-1} mode clearly resolved in the present study suggested that the branching dynamics occur before the Pt–Pt bond contraction in order to transfer a fraction of the Pt–Pt stretching CVWPs from the FC to the ^{3}LC states. This also implies that the Pt–Pt stretching motion is the key vibration for the ultrafast ISC from the ^{1}MMLCT to the ^{3}LC states. In conclusion, comprehensive interrogation of the vibrational wavepacket dynamics has provided new insights into the impact of the photoinduced Pt–Pt contraction on the ISC processes in the Pt(II) dimer. We believe that the methodology used to evaluate the vibrational wavepacket evolution in our study can contribute to a better understanding of ultrafast excited state and structural dynamics for other TMCs.

Conflicts of interest

There are no conflicts to declare.

Acknowledgements

This work has been supported by the National Science Foundation (CHE-1955806 to LXC and CHE-1955795 to FNC). The computational work and P. K. and L. X. C. were supported in part by the Ultrafast Initiative of the U. S. Department of Energy, Office of Science, Office of Basic Energy Sciences, through Argonne National Laboratory under Contract No. DE-AC02-06CH11357. The development of the computational method for simulating the excited state dynamics was supported by the National Science Foundation (CHE-1856210 to XL).

References

1 J. K. McCusker, *Science*, 2019, **363**, 484.

2 M. W. Mara, D. N. Bowman, O. Buyukcakir, M. L. Shelby, K. Haldrup, J. Huang, M. R. Harpham, A. B. Stickrath, X. Zhang, J. F. Stoddart, A. Coskun, E. Jakubikova and L. X. Chen, *J. Am. Chem. Soc.*, 2015, **137**, 9670–9684.

3 T. C. B. Harlang, Y. Liu, O. Gordivska, L. A. Fredin, C. S. Ponseca, P. Huang, P. Chábera, K. S. Kjaer, H. Mateos, J. Uhlig, R. Lomoth, R. Wallenberg, S. Styring, P. Persson, V. Sundström and K. Wärnmark, *Nat. Chem.*, 2015, **7**, 883–889.

4 J. Twilton, C. Le, P. Zhang, M. H. Shaw, R. W. Evans and D. W. C. MacMillan, *Nat. Rev. Chem.*, 2017, **1**, 0052.

5 D. M. Arias-Rotondo and J. K. McCusker, *Chem. Soc. Rev.*, 2016, **45**, 5803–5820.

6 O. S. Wenger, *J. Am. Chem. Soc.*, 2018, **140**, 13522–13533.

7 K. Kalyanasundaram and M. Grätzel, *Coord. Chem. Rev.*, 1998, **177**, 347–414.

8 M. Chaaban, C. Zhou, H. Lin, B. Chyi and B. Ma, *J. Mater. Chem. C*, 2019, 7, 5910–5924.

9 N. Huse, T. K. Kim, L. Jamula, J. K. McCusker, F. M. F. de Groot and R. W. Schoenlein, *J. Am. Chem. Soc.*, 2010, **132**, 6809–6816.

10 W. Zhang, R. Alonso-Mori, U. Bergmann, C. Bressler, M. Chollet, A. Galler, W. Gawelda, R. G. Hadt, R. W. Hartsock, T. Kroll, K. S. Kjær, K. Kubiček, H. T. Lemke, H. W. Liang, D. A. Meyer, M. M. Nielsen, C. Purser, J. S. Robinson, E. I. Solomon, Z. Sun, D. Sokaras, T. B. van Driel, G. Vankó, T.-C. Weng, D. Zhu and K. J. Gaffney, *Nature*, 2014, **509**, 345–348.

11 H. T. Lemke, K. S. Kjær, R. Hartsock, T. B. van Driel, M. Chollet, J. M. Glownia, S. Song, D. Zhu, E. Pace, S. F. Matar, M. M. Nielsen, M. Benfatto, K. J. Gaffney, E. Collet and M. Cammarata, *Nat. Commun.*, 2017, **8**, 15342.

12 T. Katayama, T. Northey, W. Gawelda, C. J. Milne, G. Vankó, F. A. Lima, R. Bohinc, Z. Németh, S. Nozawa, T. Sato, D. Khakhulin, J. Szlachetko, T. Togashi, S. Owada, S.-i. Adachi, C. Bressler, M. Yabashi and T. J. Penfold, *Nat. Commun.*, 2019, **10**, 3606.

13 S. Cho, M. W. Mara, X. Wang, J. V. Lockard, A. A. Rachford, F. N. Castellano and L. X. Chen, *J. Phys. Chem. A*, 2011, **115**, 3990–3996.

14 M. Iwamura, S. Takeuchi and T. Tahara, *Acc. Chem. Res.*, 2015, **48**, 782–791.

15 G. Auböck and M. Chergui, *Nat. Chem.*, 2015, **7**, 629–633.

16 J. D. Gaynor, J. Sandwisch and M. Khalil, *Nat. Commun.*, 2019, **10**, 5621.

17 Q.-C. Zhang, H. Xiao, X. Zhang, L.-J. Xu and Z.-N. Chen, *Coord. Chem. Rev.*, 2019, **378**, 121–133.

18 B. Ma, J. Li, P. I. Djurovich, M. Yousufuddin, R. Bau and M. E. Thompson, *J. Am. Chem. Soc.*, 2005, **127**, 28–29.

19 A. A. Rachford and F. N. Castellano, *Inorg. Chem.*, 2009, **48**, 10865–10867.

20 A. Chakraborty, J. C. Deaton, A. Haefele and F. N. Castellano, *Organometallics*, 2013, **32**, 3819–3829.

21 M. Han, Y. Tian, Z. Yuan, L. Zhu and B. Ma, *Angew. Chem., Int. Ed.*, 2014, **53**, 10908–10912.

22 C. Zhou, L. Yuan, Z. Yuan, N. K. Doyle, T. Dilbeck, D. Bahadur, S. Ramakrishnan, A. Dearden, C. Huang and B. Ma, *Inorg. Chem.*, 2016, **55**, 8564–8569.

23 S. E. Brown-Xu, M. S. J. Kelley, K. A. Fransted, A. Chakraborty, G. C. Schatz, F. N. Castellano and L. X. Chen, *J. Phys. Chem. A*, 2016, **120**, 543–550.

24 J. V. Lockard, A. A. Rachford, G. Smolentsev, A. B. Stickrath, X. Wang, X. Zhang, K. Atenkoffer, G. Jennings, A. Soldatov, A. L. Rheingold, F. N. Castellano and L. X. Chen, *J. Phys. Chem. A*, 2010, **114**, 12780–12787.

25 K. Haldrup, A. O. Dohn, M. L. Shelby, M. W. Mara, A. B. Stickrath, M. R. Harpham, J. Huang, X. Zhang, K. B. Møller, A. Chakraborty, F. N. Castellano, D. M. Tiede and L. X. Chen, *J. Phys. Chem. A*, 2016, **120**, 7475–7483.

26 P. Kim, M. S. Kelley, A. Chakraborty, N. L. Wong, R. P. Van Duyne, G. C. Schatz, F. N. Castellano and L. X. Chen, *J. Phys. Chem. C*, 2018, **122**, 14195–14204.

27 P. Kim, A. J. S. Valentine, S. Roy, A. W. Mills, A. Chakraborty, F. N. Castellano, X. Li and L. X. Chen, *J. Phys. Chem. Lett.*, 2021, **12**, 6794–6803.

28 A. Chakraborty, J. E. Yarnell, R. D. Sommer, S. Roy and F. N. Castellano, *Inorg. Chem.*, 2018, **57**, 1298–1310.

29 A. Grupp, A. Budweg, M. P. Fischer, J. Allerbeck, G. Soavi, A. Leitenstorfer and D. Brida, *J. Opt.*, 2018, **20**, 014005.

30 M. J. Frisch, G. W. Trucks, H. B. Schlegel, G. E. Scuseria, M. A. Robb, J. R. Cheeseman, G. Scalmani, V. Barone, G. A. Petersson, H. Nakatsuji, X. Li, M. Caricato, A. V. Marenich, J. Bloino, B. G. Janesko, R. Gomperts, B. Mennucci, H. P. Hratchian, J. V. Ortiz, A. F. Izmaylov, J. L. Sonnenberg, D. Williams-Young, F. Ding, F. Lipparini, F. Egidi, J. Goings, B. Peng, A. Petrone, T. Henderson, D. Ranasinghe, V. G. Zakrzewski, J. Gao, N. Rega, G. Zheng, W. Liang, M. Hada, M. Ehara, K. Toyota, R. Fukuda, J. Hasegawa, M. Ishida, T. Nakajima, Y. Honda, O. Kitao, H. Nakai, T. Vreven, K. Throssell, J. A. Montgomery Jr, J. E. Peralta, F. Ogliaro, M. J. Bearpark, J. J. Heyd, E. N. Brothers, K. N. Kudin, V. N. Staroverov, T. A. Keith, R. Kobayashi, J. Normand, K. Raghavachari, A. P. Rendell, J. C. Burant, S. S. Iyengar, J. Tomasi, M. Cossi, J. M. Millam, M. Klene, C. Adamo, R. Cammi, J. W. Ochterski, R. L. Martin, K. Morokuma, O. Farkas, J. B. Foresman and D. J. Fox, *Gaussian 16 Rev.B.01.*, Gaussian, Inc., Wallingford CT, 2016.

31 P. J. Hay and W. R. Wadt, *J. Chem. Phys.*, 1985, **82**, 299–310.

32 J.-D. Chai and M. Head-Gordon, *Phys. Chem. Chem. Phys.*, 2008, **10**, 6615–6620.

33 A. J. Atkins, F. Talotta, L. Freitag, M. Boggio-Pasqua and L. González, *J. Chem. Theory Comput.*, 2017, **13**, 4123–4145.

34 J. J. Snellenburg, S. P. Laptenok, R. Seger, K. M. Mullen and I. H. M. van Stokkum, *J. Stat. Softw.*, 2012, **49**, 1–22.

35 H. F. Shurvell and M. C. Southby, *Vib. Spectrosc.*, 1997, **15**, 137–146.

36 M. H. Vos, F. Rappaport, J.-C. Lambry, J. Breton and J.-L. Martin, *Nature*, 1993, **363**, 320–325.

37 S. D. McClure, D. B. Turner, P. C. Arpin, T. Mirkovic and G. D. Scholes, *J. Phys. Chem. B*, 2014, **118**, 1296–1308.

38 C. C. Jumper, P. C. Arpin, D. B. Turner, S. D. McClure, S. Rafiq, J. C. Dean, J. A. Cina, P. A. Kovac, T. Mirkovic and G. D. Scholes, *J. Phys. Chem. Lett.*, 2016, **7**, 4722–4731.

39 S. Rafiq and G. D. Scholes, *J. Phys. Chem. A*, 2016, **120**, 6792–6799.

40 L. Mewes, R. A. Ingle, S. Megow, H. Böhnke, E. Baranoff, F. Temps and M. Chergui, *Inorg. Chem.*, 2020, **59**, 14643–14653.

41 N. P. Weingartz, M. W. Mara, S. Roy, J. Hong, A. Chakraborty, S. E. Brown-Xu, B. T. Phelan, F. N. Castellano and L. X. Chen, *J. Phys. Chem. A*, 2021, **125**, 8891–8898.

Faraday Discussions

ROYAL SOCIETY OF CHEMISTRY

PAPER

On the use of vibronic coherence to identify reaction coordinates for ultrafast excited-state dynamics of transition metal-based chromophores

Bryan C. Paulus and James K. McCusker *

Received 17th May 2022, Accepted 17th May 2022
DOI: 10.1039/d2fd00106c

The question of whether one can use information from quantum coherence as a means of identifying vibrational degrees of freedom that are active along an excited-state reaction coordinate is discussed. Specifically, we are exploring the notion of whether quantum oscillations observed in single-wavelength kinetics data exhibiting coherence dephasing times that are intermediate between that expected for either pure electronic or pure vibrational dephasing are vibronic in nature and therefore may be coupled to electronic state-to-state evolution. In the case of a previously published Fe(II) polypyridyl complex, coherences observed subsequent to $^1A_1 \rightarrow {}^1$MLCT excitation were linked to large-amplitude motion of a portion of the ligand framework; dephasing times on the order of 200–300 fs suggested that these degrees of freedom could be associated with ultrafast (~100 fs) conversion from the initially formed MLCT excited state to lower-energy, metal-centered ligand-field excited state(s) of the compound. Incorporation of an electronically benign but sterically restrictive Cu(I) ion into the superstructure designed to interfere with this motion yielded a compound exhibiting a ~25-fold increase in the compound's MLCT lifetime, a result that was interpreted as confirmation of the initial hypothesis. However, new data acquired on a different chemical system – Cr(acac′)$_3$ (where acac′ represents various derivatives of acetylacetonate) – yielded results that call into question this same hypothesis. Coherences observed subsequent to $^4A_2 \rightarrow {}^4T_2$ ligand-field excitation on a series of molecules implicated similar vibrational degrees of freedom across the series, but exhibited dephasing times ranging from 340 fs to 2.5 ps without any clear correlation to the dynamics of excited-state evolution in the system. Taken together, the results obtained on both of these chemical platforms suggest that while identification of coherences can indeed point to degrees of freedom that should be considered as candidate modes for defining reaction trajectories, our understanding of the factors that determine the interplay across coherences, dephasing times, and electronic and geometric structure is insufficient at the present time to view

Department of Chemistry, Michigan State University, 578 South Shaw Lane, East Lansing, MI 48824, USA. E-mail: jkm@chemistry.msu.edu

this parameter as a robust metric for differentiating active *versus* spectator modes for ultrafast dynamics.

Introduction

Chemical processes – thermal and photochemical, intra- as well as intermolecular – proceed along complex potential energy surfaces that define pathways from reactants to products. Each step in this process will have an optimal trajectory that competes with alternate, non-productive pathways that serve as mechanisms for energy dissipation. Having *a priori* control of these trajectories – in effect dictating the reaction pathway along which the system evolves – has been a long-standing goal for chemists. Current theories that provide the conceptual basis for our understanding and chemical intuition about molecular processes generally assume that the kinetic energy of reactants is not stored as the system moves from peaks to valleys but rather is thermalized and therefore dissipated, leading to the exponential nature of barrier crossings for reaction dynamics. If we could alter this paradigm by programming the desired reaction pathway using information about the interplay between the electronic and vibrational degrees of freedom of a system, reactions could be choreographed at the molecular level with much greater precision and efficiency. To achieve this, one needs to exploit processes that involve strong coupling of electronic and nuclear degrees of freedom, which in turn requires the identification of trajectories that involve phase coherent vibrational motions and nonadiabatic mechanisms. By designing molecules with vibrational modes that couple the desired reactant and product states, one could, in theory, optimize the efficiency of a given chemical process by making dissipative reaction pathways less favorable. In effect, such an advance would enable chemists to control chemical dynamics through synthetic design driven by the informed application of quantum information about the system reflected in a phenomenon known as coherence.[1,2]

The concept of coherence is one that we are at once very familiar with and wholly unfamiliar with. Sitting at a stop light, we all notice when our turn signal and the one on the car in front of us is slightly out-of-phase. The specifics of this phase relationship ultimately leads to a beat pattern where, for a brief period of time, the two blinkers are fully synchronized: this is one form of coherence. In quantum mechanics, coherence refers to a correlation between different degrees of freedom in a quantum object such that their properties can only be represented as a superposition. A simple example of this is the idea of resonance.[3] Benzene, for instance, has two main resonance structures denoted by ψ_1 and ψ_2 (Fig. 1); neither of these two Kekulé structures correspond to the true wavefunction of benzene. Furthermore, the molecule is not flickering back and forth between them. The true wavefunction is a superposition, *i.e.*, a linear combination of each Kekulé wavefunction described by $a\psi_1 + b\psi_2$ and is a form of quantum mechanical coherence.

When considering coherence in the context of excited-state dynamics, ultrashort laser pulses with bandwidths that are broad enough to span multiple vibrational levels on an electronic excited state can excite several vibrational levels simultaneously. The resulting wavefunction is represented by a linear combination of each of the constituent wavefunctions as shown in eqn (1),

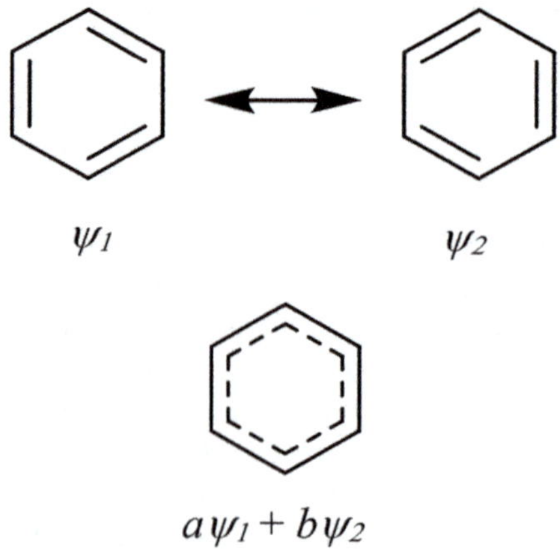

Fig. 1 Resonance structures of benzene. The actual wavefunction describing the electronic ground state of benzene is not represented by ψ_1 or ψ_2, but rather the superposition of these two wavefunctions reflected in the linear combination $a\psi_1 + b\psi_2$. This is an example of quantum mechanical coherence.

$$\psi(R,t) = \sum_n c_n \psi_n(R)\ \exp\left(-\frac{iE_n t}{h}\right) \tag{1}$$

where $\psi(R,t)$ is the net wavefunction, c_n is the population coefficient, $\psi_n(R)$ is the wavefunction of the nth vibrational level, and E_n is the energy of the nth vibrational level;[4] the wavefunction represented by eqn (1) is conceptually no different than the linear combination we use to represent the ground state wavefunction of benzene. A pictorial representation of this is shown in Fig. 2, in which the net wavefunction (a so-called wavepacket represented by $|\psi_{\text{tot}}|^2$) is, in this case, the

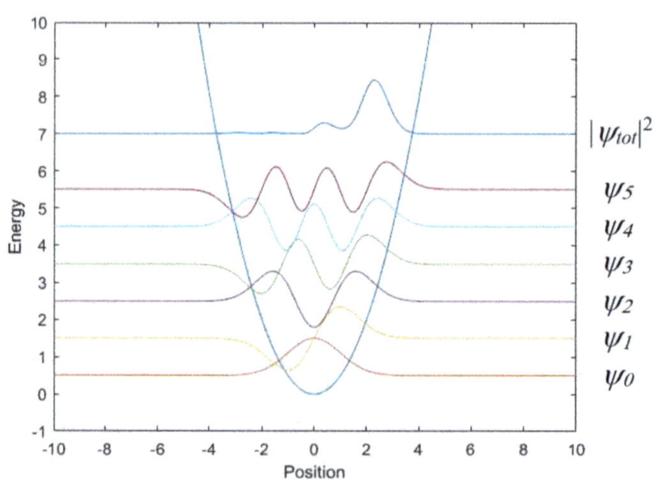

Fig. 2 Pictorial representation of a coherent superposition state – more commonly referred to as a wavepacket – resulting from a linear combination of oscillatory wavefunctions as described by eqn (1).

sum of each of the first five vibrational wavefunctions. An important difference in the case of the excited-state superposition is that, in general, the Franck–Condon state produced upon excitation does not correspond to the structurally equilibrated excited state (*i.e.*, $\Delta Q \neq 0$), so the excited-state potential will be exerting a force on the system. This situation coupled with the combined effects of constructive and destructive interferences has two important consequences: (1) a much more localized amplitude of the wavefunction, and (2) a time-dependent shift of the interference pattern, causing the wavepacket to migrate back and forth across the potential surface at the frequency of the activated vibration.

The time-dependent evolution of the wavepacket just described ultimately provides us with the chemical insight we need to take advantage of coherence in terms of synthetic design. Excited-state absorption spectra are no different than ground-state profiles insofar as they reflect a superposition of Franck–Condon overlap factors that arise between the two electronic states involved in the transition. In the case of transient absorption spectroscopy, the two states involved are the excited state that is formed upon photoexcitation and higher-lying excited states that the probe beam samples to yield the excited-state absorption spectrum. The critical difference is that the oscillatory nature of the wavepacket on the excited-state potential surface results in changes in the Franck–Condon factors as the wavepacket evolves in time. This manifests as fine structure superimposed on the kinetics associated with excited-state evolution (Fig. 3), fine structure that contains information on the coupling that exists between the electronic and nuclear degrees of freedom of the molecule and thus contains a blueprint of how a given excited-state process depends on and/or is dictated by the electronic and geometric properties of the system.

Despite the wealth of information these coherence signatures possess, there is a significant problem that arises: how does one differentiate coherences associated with degrees of freedom that are relevant to the excited-state process of

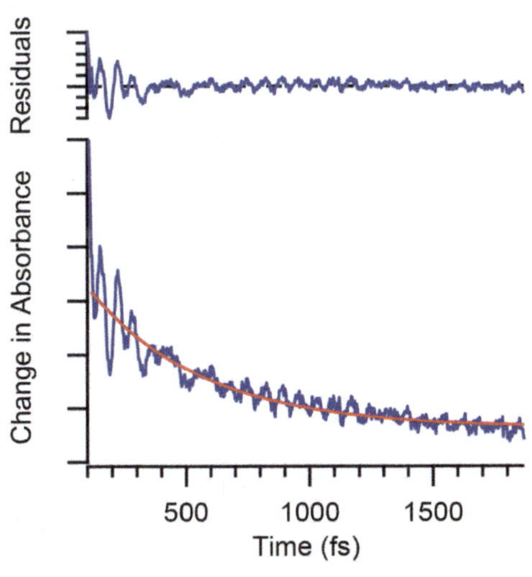

Fig. 3 Time-resolved absorption data for $Cr(acac)_3$ following *ca.* 40 fs excitation into the $^4A_2 \rightarrow {}^4T_2$ ligand-field absorption at ~550 nm. The oscillatory features are due to quantum oscillations associated with electronic state evolution in this compound.

interest (*e.g.*, intersystem crossing, electron transfer, energy transfer, *etc.*) – so-called "active modes" – from coherences that can arise along coordinates having nothing to do with that process ("spectator modes")? A non-linear molecule has $3N - 6$ vibrational degrees of freedom, so in order for vibronic coherence to be of use as a design principle, it's necessary to figure out which subset of those $3N - 6$ degrees of freedom are actually relevant from the perspective of photo-functionality. With this report, we present results on two different chemical systems that we have explored that highlight the nature of this difficulty with the goal of stimulating a dialog on how to transform coherence from an observational phenomenon to one that can (potentially) guide chemical innovation.

Experimental section

General

All reagents were purchased from Sigma-Aldrich or Strem Chemical and used as received unless otherwise indicated. The syntheses of tris(2,4-pentanediono)chromium(III) ($Cr(acac)_3$), tris(3-chloro-2,4-pentanediono)chromium(III) ($Cr(3\text{-}Cl\text{-}acac)_3$), tris(3-bromo-2,4-pentanediono)chromium(III) ($Cr(3\text{-}Br\text{-}acac)_3$), tris(3-iodo-2,4-pentanediono)chromium(III) ($Cr(3\text{-}I\text{-}acac)_3$), and tris(2,2,6,6-tetramethyl-3,5-heptanediono)chromium(III) ($Cr(TMHD)_3$) were prepared as described elsewhere.[5]

Computational methods

All calculations were performed with Gaussian 09 software[6] on the Michigan State University High Performance Computing Center servers. Previously optimized geometries for $Cr(acac)_3$ and $Cr(TMHD)_3$ were used as starting geometries for $Cr(acac)_3$ and $Cr(TMHD)_3$ ground state optimizations.[7] Optimizations were performed using the unrestricted B3LYP functional, 6-311G** basis set, tight convergence criteria; calculations for $Cr(3\text{-}I\text{-}acac)_3$ were unique in that the LANL2DZ basis set was used. Choice of solvent used for the polarizable continuum model had little effect on the results in the frequency region of interest, but $Cr(acac)_3$ calculations were performed in acetonitrile, $Cr(TMHD)_3$ in tetrahydrofuran, $Cr(3\text{-}Cl\text{-}acac)_3$ and $Cr(3\text{-}I\text{-}acac)_3$ in dichloromethane, and $Cr(3\text{-}Br\text{-}acac)_3$ in toluene. Mass weighted vector displacement diagrams of each pertinent vibrational mode were created in the Visual Molecular Dynamics program.[8]

Time resolved spectroscopy

The ultrafast time-resolved absorption spectrometer used for these studies has been described in detail previously.[9,10] Briefly, a Ti:sapphire oscillator (Coherent: Mantis) provides a ~400 mW, 80 MHz seed beam centered at ~800 nm for use in a regenerative amplifier (Coherent: Legend Elite) with a 527 nm Nd:YLF pump (Coherent: Evolution) producing a 1 KHz pulse train of approximately 35 fs 1 mJ pulses at 800 nm. This output is split using a 70/30 beam splitter with the higher energy portion used to pump a two-stage optical parametric amplifier (OPA) (Coherent: OperA Solo) to generate a visible pump beam which is used as the excitation source during the transient absorption experiment. This pump beam is again split with a 90 : 10 beam splitter with the lower energy portion being used as a probe beam for degenerate pump–probe experiments. The remaining 30% of the 800 nm output from the regenerative amplifier is either used to pump a second OPA (Coherent: Opera Solo) to generate a probe beam for single wavelength nondegenerate pump–probe experiments or used to make a white light continuum by attenuating the power and focusing ~1 μJ pulses into a 3 mm sapphire window. This white light continuum can again be used for single wavelength nondegenerate pump–probe experiments or a large swath of the continuum can be measured using full spectrum measurements. Group velocity dispersion for each OPA output is corrected using a folded prism compressor. Pump–probe delay is adjusted by routing the pump OPA output into a computer-controlled translation stage (Aerotech). Pump and probe polarizations are set to ~54.7° relative to each other to eliminate anisotropic contributions to the signal and focused into a 1 mm quartz cuvette containing the sample. The pump pulse energies were set in the range of 3–6 μJ and the signal was checked for linearity by halving the pump power using a variable neutral density filter wheel. Probe pulse energies were set to $1/10^{th}$ of the pump.

Single-wavelength measurements were obtained by passing the pump beam through a 446 Hz mechanical chopper. A monochromator (Jarrell Ash: MonoSpec 18; 1200 grooves per mm grating, blaze 500 nm) was placed after the sample and used to select a 3.8 nm FWHM slice of the probe beam spectrum to be measured on an amplified silicon photodiode (Thor laboratories: PDA55). The center wavelength of this slice of the probe spectrum is specified for each single wavelength TA trace. A reference beam is produced from a reflection of the probe beam off a microscope slide cover slip inserted into the beam path prior to the sample. This reference beam is measured on another amplified silicon photodiode and attenuated with neutral density filters and an iris until its intensity matches that of the signal beam. Both the signal and reference beams are coupled to a lock-in amplifier (Stanford Research: SR810) which is synchronized to the chopper modulating the pump beam. Connection of the lock-in amplifier to a data acquisition card allows the signal to be converted from analogue to digital. Data were acquired and analyzed using a locally written LabView program.

Full spectra were acquired by collimating the white light continuum probe and focusing it into the sample using spherical mirrors. A lens placed after the sample was then used to focus the transmitted continuum probe into a liquid light guide which was coupled to a spectrometer (Spex 270 M) with 1 mm entrance slits. A grating (300 grooves per mm; blaze 600 nm) within the spectrometer dispersed the probe beam onto a diode array detector (Hamamatsu: HC233-0900; C5964 NMOS, 1 × 512 pixel array) affording a roughly 300 nm spectral window. Data were acquired using a locally written LabView program by first collecting "dark counts", which is effectively a measurement of the pump beam scatter reaching the detector in the absence of the probe beam. Next, a background is collected by measuring the signal with both pump and probe incident on the sample at negative time delay. The dark counts spectrum is then subtracted from this background to provide a baseline (I_0) for calculating the change in absorbance as a function of pump–probe delay.

Results and discussion

Transition metal complexes represent an excellent conduit for examining the interplay between electronic and vibrational dynamics due to the nature of their electronic structures. Specifically, the spin-state variability associated with ions of the d-block, the co-existence of both charge-transfer and ligand-field excited states – each of these being defined by distinct differences in intramolecular charge distribution and degree of coupling to the structure of the molecule – as well as chemical tunability allow for the kind of compositional flexibility that is necessary in order to connect excited-state evolution to electronic structure, geometric structure, and molecular functionality.

As indicated in the Introduction, one of the difficulties associated with identifying the degrees of freedom relevant for defining excited-state reaction trajectories is the sheer number of vibrational modes that large molecules possess. A variable we have proposed that *may* serve as a means of differentiating active *versus* spectator modes is the dephasing time of the coherence.[11–13] Upon excitation with a sufficiently short laser pulse, the initial excited state is characterized by a coherent superposition of vibrational states that are all phase-locked; the motion of this wavepacket is what gives rise to the oscillations shown in Fig. 3 and described by eqn (1). This phase relationship will eventually degrade due to any number of reasons, *e.g.*, interactions with solvent, excited-state processes within the chromophore, *etc.*, leading to a damping of the oscillations and eventual loss of the characteristic oscillatory pattern. For a pure vibrational mode, *i.e.*, one that would be observed following direct excitation of a ground-state vibration, dephasing times on the order of 1–10 ps are routinely observed.[14] In contrast, electronic coherences dephase much more rapidly, *i.e.*, on the order of a few tens to a hundred femtoseconds.[15] We have postulated that modes exhibiting dephasing times between these two extremes may indicate a vibronic nature to that mode, thereby linking it to the coupled evolution of the electronic and vibrational degrees of freedom following photoexcitation. If true, one could potentially use dephasing times as a means of reducing the dimensionality of the problem of identifying active modes from $3N - 6$ to a much smaller subset of trajectories. This in turn would provide physical insight into the nuclear motions of the molecule that (potentially) define the reaction coordinate and provide a roadmap for imparting synthetic control over ultrafast excited-state dynamics through targeted chemical design.

As far as we are aware, there is no robust theory to support (or refute) the notion of linking dephasing times of observed coherences to that corresponding vibration's role in the excited-state process(es) associated with the population dynamics. Herein we will present results on two chemical systems that we have studied – one for which this correlation appears to hold and another where its validity is questionable – with the goal of stimulating discussion around this issue.

Dephasing times as an indicator of reaction coordinate: the photophysics of an Fe(II) polypyridyl complex and manipulation of metal-to-ligand charge transfer-state lifetimes

A recent report from our lab represents a case study in which synthetic redesign targeting vibrational degrees of freedom with coherence dephasing times in the range of 100s of femtoseconds appears to have been successful.[16] The chemical platform is based on polypyridyl complexes of Fe(II). Chromophores in this general class are of interest as potential earth-abundant alternatives to compounds based on rare metals such as ruthenium and iridium for applications reliant on photo-induced electron transfer chemistry (*e.g.*, solar energy conversion, photo-redox catalysis, *etc.*).[17] A critical problem with such systems is the sub-100 fs lifetime associated with the MLCT excited states of compounds like $[Fe(bpy)_3]^{2+}$ (where bpy is 2,2′-bipyridine).[18] This corresponds to a reduction in lifetime by 6–7 orders of magnitude relative to the corresponding state in a compound such as $[Ru(bpy)_3]^{2+}$ and effectively undercuts the ability to leverage the chemical potential stored in this type of charge-separated excited state. The origin of the problem is tied to the difference in the relative energies of the charge-transfer and ligand-field (*i.e.*, "d–d") excited-state manifolds between compounds of the first- *versus* second- and third-row transition series.[19] Specifically, the limited radial extension of the 3d orbitals (as compared to 4d and 5d) due to the primogenic effect[20] attenuates metal–ligand overlap, thereby reducing the magnitude of ligand field-induced splitting among those orbitals in a given compound. This positions the ligand-field excited states lower in energy than what one finds in corresponding second- and third-row metal complexes to where the lowest energy ligand-field excited states now sit below the MLCT states, in contrast to the situation one finds for their heavier congeners (Fig. 4). This energetic situation, combined with the large difference in equilibrium geometry that characterizes these excited ligand-field states, is the main driver for the ultrafast charge-transfer to ligand field-state conversion seen in these compounds.

The most successful approach thus far toward circumventing this problem has been through the use of carbene-based ligands,[21] which present significantly stronger ligand fields to a given metal center due to their strongly σ-donating character and has the effect of destabilizing the ligand-field excited states: if pushed sufficiently high in energy, one can realize an electronic structure that more closely resembles that of a second- or third-row transition metal complex. This strategy amounts to manipulation of the *y*-axis in the potential energy surface drawings in Fig. 4, which raised a question in our minds as to whether one could also tackle the problem through manipulation of the *x*-axis, *i.e.*, the

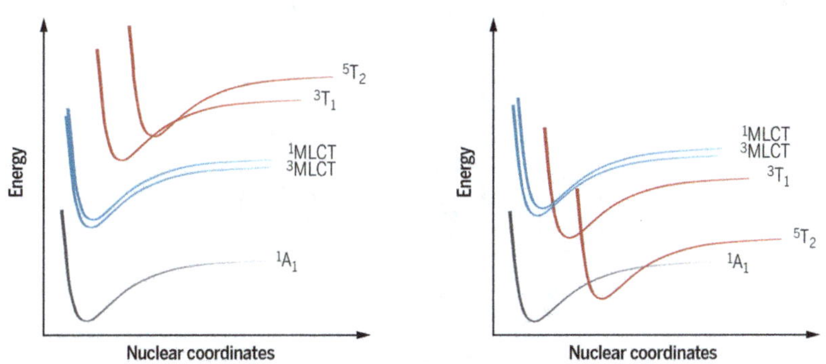

Fig. 4 Qualitative potential energy surface diagrams appropriate for an octahedral d^6 metal ion of the second or third transition series (left) and the first-row (right). The inversion in relative energies between the MLCT and ligand-field states contributes to ultrafast non-radiative decay in the case of the latter, which limits the utility of compounds like $[Fe(bpy)_3]^{2+}$ for photo-induced electron transfer chemistry. Adapted from ref. 19.

reaction coordinate. We posited that identification of vibronic coherences could aid in this regard by allowing for the visualization of the structural degrees of freedom that couple to the MLCT-to-LF state conversion, and then use that information as the basis of synthetically tailoring ligands to interfere with evolution along those degrees of freedom. We therefore conceived of the chemical platform shown in Scheme 1, which allows for modifications of the structural integrity of the ligand framework along the metal–ligand bond stretch coordinate as well as torsional motion about the primary coordination sphere; both of these degrees of freedom have been implicated in the dynamics of interconversion between the low-spin ($^1A_1$) and high-spin ($^5T_2$) forms of Fe(II) complexes.[22]

A full report of this work has been published elsewhere,[16] so for the purposes of this discussion we will merely summarize the key observations and results. Excitation into the $^1A_1 \rightarrow {}^1$MLCT absorption of the mononuclear Fe(II) complex with sub-50 fs pulses revealed an MLCT lifetime of 110 ± 30 fs, a value that is typical for Fe(II) polypyridyl complexes. The bandwidth of the excitation pulse was sufficient to create a wavepacket on the MLCT-state surface, producing quantum oscillations that were superimposed on the single-wavelength kinetic traces (Fig. 5). Analysis of the data indicated the presence of vibrational signatures corresponding to energies of 127 ± 5 cm^{-1}, 156 ± 2 cm^{-1}, 173 ± 3 cm^{-1}, and 215 ± 12 cm^{-1}. All of these modes exhibited dephasing times between the two extremes mentioned above, with the 156 ± 2 cm^{-1} mode being the most well-defined in terms of error bars at 300 ± 30 fs. This intermediate dephasing time led us to speculate that some (or all) of these modes could be vibronic in nature and therefore coupled to the MLCT-to-LF conversion process that we were seeking to suppress. DFT calculations enabled us to visualize these modes, all of which indicated large amplitude motion associated with the aliphatic portions of the molecule (Fig. 6). We postulated that incorporation of an electronically benign

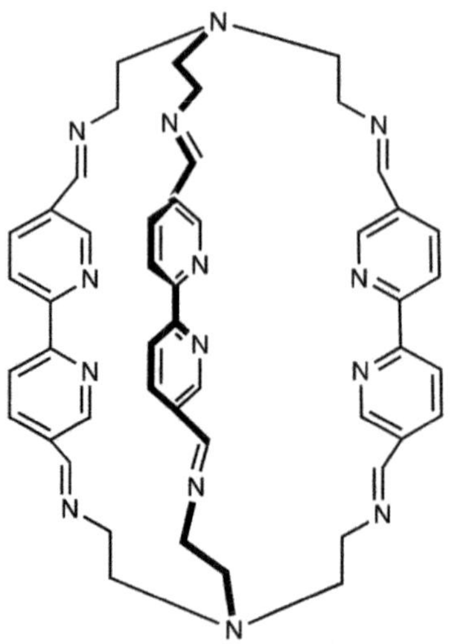

Scheme 1 Multidentate ligand (L) synthesized for the study of the use of vibronic coherence for probing the reaction coordinate for ultrafast MLCT-state relaxation in an Fe(II) polypyridyl complex (from ref. 16).

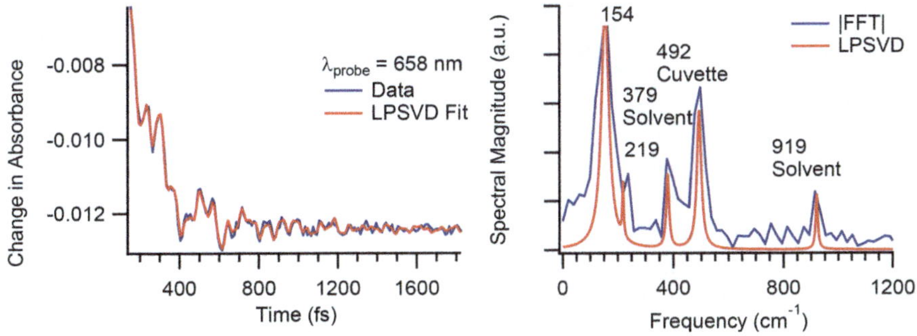

Fig. 5 Time-resolved absorption data acquired on $[Fe(L)]^{2+}$ in CH_3CN solution following *ca.* 40 fs excitation at 600 nm. The plot on the left shows single-wavelength kinetics measurements collected at 658 nm; the red line superimposed on the kinetics data corresponds to a simulation of the data derived from a linear predictive single-valued decomposition analysis based on eqn (1) (right). The mode at 154 cm^{-1} is pictured in Fig. 6. (Adapted from ref. 16.)

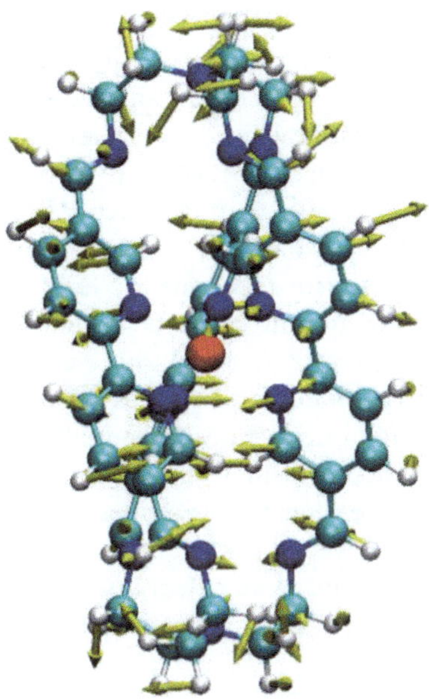

Fig. 6 DFT-based vector displacement diagram of the 154 cm^{-1} mode identified from the LPSVD analysis of the data plotted in Fig. 5. The large-amplitude motion of the ligand framework evident above and below the Fe(II) coordination site (red) was targeted for chemical modification *via* binding to Cu(I).

metal ion in the N_4 coordination site on both ends of the macrocycle would attenuate this motion and possibly decouple it from the reaction coordinate, leading to an increase in the MLCT-state lifetime. Photophysical data were therefore collected on the $[FeCu_2(L)]^{4+}$ under identical conditions as the iron-only system. The amplitude of the quantum oscillations observed for this compound were attenuated relative to $[Fe(L)]^{2+}$, and we noted a small but discernible increase

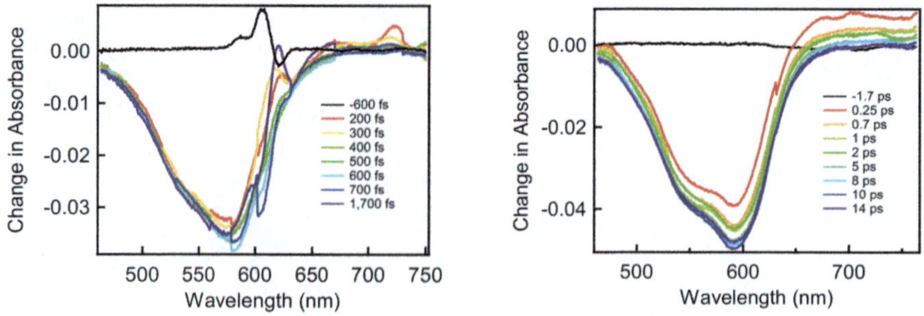

Fig. 7 Full-spectrum time-resolved absorption data acquired on $[Fe(L)]^{2+}$ (left) and $[FeCu_2(L)]^{4+}$ (right) in CH_3CN solution following *ca.* 40 fs excitation at 600 nm. Positive absorptions to the red of 625 nm are associated with the bpy^- species that is only present in the MLCT excited state of the compound. The appearance of this signal in $[FeCu_2(L)]^{4+}$ indicates that this excited state persists for a much longer period of time, indicating that the incorporation of the Cu(I) ions into the structure has caused a ~25-fold increase in the MLCT-state lifetime relative to $[Fe(L)]^{2+}$. Adapted from ref. 16.

in the dephasing times of the relevant modes. While these observations were suggestive that we had achieved the desired result, proof was obtained in the form of transient absorption spectra which clearly showed enhanced persistence of the optical signature associated with the bpy^- radical anion in the MLCT excited state (Fig. 7). The measured lifetime of the ^{3}MLCT state of the $[FeCu_2(L)]^{4+}$ complex was 2.6 ± 0.1 ps, corresponding to a *ca.* 25-fold increase in the lifetime of that state.

We viewed the results on this Fe(II)-based system as a successful, albeit empirical demonstration of the idea that one can leverage information contained in coherences as a means of guiding synthetic design. In this case, the goal was identifying vibrational degrees of freedom that served to define the reaction coordinate along which the compound evolves from the MLCT state to lower energy, metal-centered ligand-field states (essentially the *x*-axis in Fig. 4). This result notwithstanding, data on one molecule is simply not sufficient to argue that intermediate dephasing times constitutes a robust criterion upon which to rely for the identification of active modes along a given reaction coordinate. In order to probe this question in greater detail, we turn to a chemical platform that we had previously explored over a decade ago but can now examine in far greater detail.

The basic photophysics of $Cr(acac)_3$

One of the motivations underpinning the study of Fe(II)-based charge-transfer chromophores is their potential utility for facilitating photo-induced electron transfer chemistry. The strong visible absorption cross-section of these compounds is an important attribute in this regard, but unravelling details concerning excited-state dynamics is challenging due to the complexity of their electronic structures. It is for this reason that we began examining the ultrafast photophysics of $Cr(acac)_3$ (where acac is the mono-deprotonated form of acetylacetone). Chromium(III), possessing a d^3 valence configuration, has significantly fewer metal-centered electronic excited states than a low-spin d^6 ion like Fe(II). In particular, the isolation of the lowest-energy $^4A_2 \rightarrow {}^4T_2$ spin-allowed d–

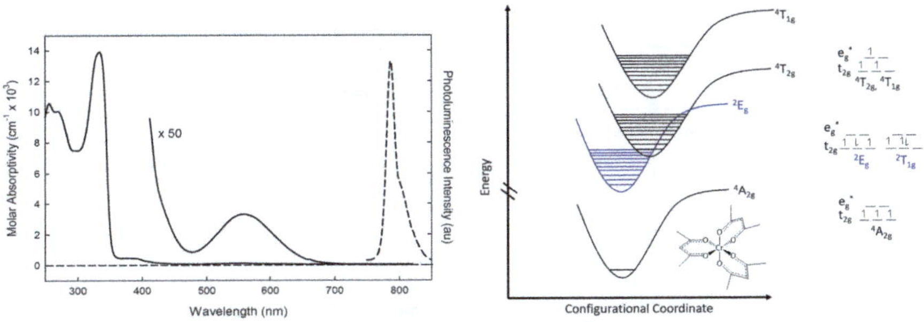

Fig. 8 (Left) Solution-phase absorption (solid line) and 90 K emission (dashed) spectra of Cr(acac)$_3$. The broad absorption centered at 560 nm corresponds to the lowest energy $^4A_2 \rightarrow {}^4T_2$ ligand-field absorption, whereas the emission at 790 nm arises from the ^2E excited state that is formed in <100 fs following excitation. Adapted from ref. 23. (Right) Qualitative potential energy surface diagrams appropriate for Cr(acac)$_3$, with the corresponding ligand-field splitting diagrams depicting the valence d-orbital based electronic structure. The PES for the $^2T_{1g}$ state has been omitted for clarity.

d absorption at 560 nm allows for selective excitation into a single electronic state; with only the lower energy ^2E state and ground-state energetically accessible, the number of possible relaxation pathways is extremely limited (Fig. 8). Finally, the small size of the ligand framework translates to a reduction in the size of the compound's vibrational partition function, which in turn will simplify analysis of vibronic coherences that may be created in an ultrafast time-resolved absorption measurement.

Our initial ultrafast time-resolved absorption measurements[23] led to a model whereby initial formation of the $^4T_2$ state was followed by intersystem crossing to the lower-energy ^2E state with a time constant that exceeded our temporal resolution at that time (<100 fs). Longer time-scale measurements indicated a ground-state recovery time constant of $\sim$800 ps: given this, a process with a 1.1 ± 0.2 ps time constant that appeared to correspond to a narrowing and slight shift of a transient absorption profile that was virtually indistinguishable to the known differential absorption spectrum of the ^2E state was assigned to vibrational relaxation in the ^2E excited state. Subsequent studies by Kunttu and co-workers focused on the use of ultrafast infrared absorption spectroscopy,[24,25] which is expected to be much more discriminating for vibrational relaxation dynamics than what can be inferred from electronic absorption spectra.[26] Their data were consistent with an initial, sub-100 fs $^4T_{1,2} \rightarrow {}^2E$ intersystem crossing process, but biphasic kinetics associated with bleach recovery of the C–O and C–C stretching modes of the ligand of 15 ps and 760 ps indicated a more complex process than we initially suggested.

A more detailed discussion and analysis of the population dynamics of this system can be found elsewhere,[27] but a basic picture of what is most broadly consistent with all of the experimental and computational[28] studies on this system of which we are aware is summarized in Fig. 9. This model retains our initial hypothesis of impulsive intersystem crossing to the ^2E state immediately following $^4A_2 \rightarrow {}^4T_2$ excitation. This is followed by (thermally activated) back-intersystem crossing to the $^4T_2$ state that occurs in competition with evolution

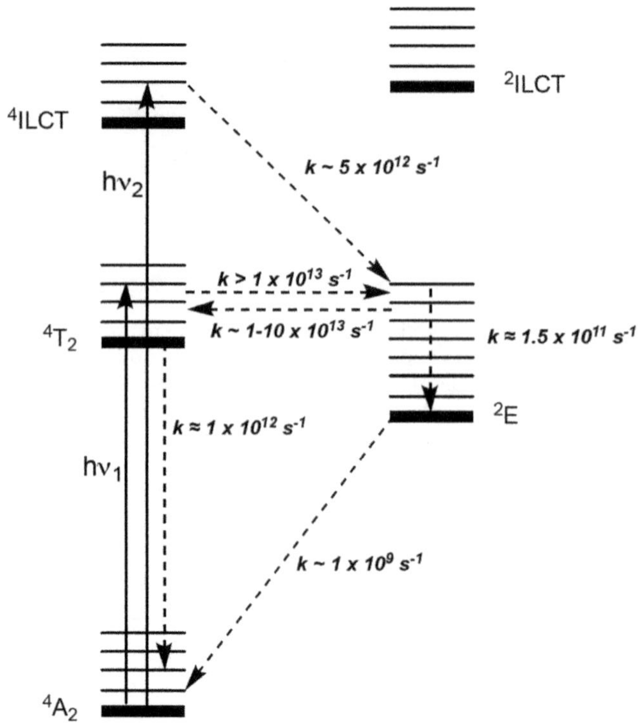

Fig. 9 Jablonski diagram for $Cr(acac)_3$, summarizing results from published femtosecond time-resolved absorption studies.

on the ^2E surface, resulting in a bifurcation of the excited-state population density. The back-intersystem crossing process accesses an internal conversion process from the $^4T_2$ state back to the ground state (*ca.* 1 ps), which then cools ($\tau \sim$ 10–15 ps). The fraction of population that does not undergo back-intersystem crossing cools on the ^2E surface with a time constant of ~7 ps, whereupon it also relaxes back to the ground state but on the much longer time scale of ~800 ps identified by both our group and that of Kunttu.

The role of vibronic coherence in this system begins with a study we published in 2010.[29] The initial motivation for this work was to take advantage of improved temporal resolution to try and quantify the time constant for the $^4T_2 \rightarrow {}^2E$ intersystem crossing process that eluded us in the earlier study. While the ISC process could not be definitively identified, what we did observe was the first example of excited-state coherence in a coordination complex. The oscillation was fit to a *ca.* 164 cm^{-1} mode and assigned to a torsional motion about the primary coordination sphere of the compound. The nature of this motion coupled with the fact that (1) the coherence appears instantaneously upon excitation and persists through what was known to be the time scale of $^4T_2 \rightarrow {}^2E$ intersystem crossing, and (2) a dephasing time of ~70 fs led to the speculation that this mode might define the reaction coordinate along which ISC proceeds. Inspection of the motion associated with the 164 cm^{-1} coherence reveals large amplitude displacement on the periphery of the acac ligand, so a derivative was synthesized that replaced the terminal CH_3 groups with *tert*-butyl groups (TMHD) in order to increase the steric bulk of this portion of the ligand framework. A time-dependent change in the spectral profile was interpreted as being associated with the $^4T_2 \rightarrow {}^2E$ intersystem crossing process, suggesting a slowing in these dynamics of over an order of magnitude in response to this steric modification of the ligand. It should be noted that this study was published prior to the work of Kunttu and co-workers, so the interpretation of these data was based on what now must be viewed as an incorrect model for the overall kinetics of this class of compounds. We must therefore revisit this analysis in light of Kunttu's work and the higher quality data we have been able to obtain on these chromophores since our previous report.

Because there are many mechanisms by which a coherence can be created in a sample during a pump–probe transient absorption experiment, it is important to assess the specific origin of each oscillatory component carefully. As mentioned in the Introduction, one can often exploit the disparity of time scales for dephasing between purely electronic and purely vibrational coherences to identify the type of coherence being monitored, *i.e.* electronic or vibrational. Electronic coherences typically dephase on sub-100 fs timescales,[30,31] whereas vibrational coherences can require several picoseconds to completely dephase.[32–35] In the system of interest, oscillations persist from several hundreds of femtoseconds to several picoseconds,[7,36] excluding purely electronic coherences as a likely component of the signal. This leaves three possible origins for the oscillatory components in the data: (1) ground state vibrational coherence in the solvent, (2) ground state vibrational coherence in the solute, or (3) vibrational coherence in the excited electronic state of the solute. Solvent contributions can often be identified by performing the pump–probe experiment on neat solvent or performing a frequency domain Raman experiment and are therefore easily accounted for.

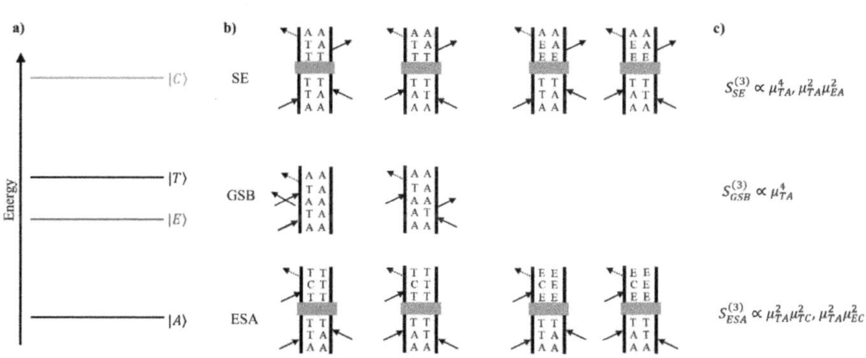

Fig. 10 (a) Qualitative energy level diagram depicting the ground $^4A_2$ state (|A⟩), ^2E state (|E⟩), $^4T_2$ state (|T⟩), and charge transfer manifold (|C⟩) that the ^2E and $^4T_2$ states can be promoted to through resonance with the probe beam. (b) Double sided Feynman diagrams depicting stimulated emission (SE), ground-state bleaching (GSB), and excited-state absorption (ESA). Following excitation to the $^4T_2$ state, the system may evolve. Depending on the time delay between pump and probe, a population may exist in the $^4T_2$ or ^2E states. This relaxation pathway is denoted by a gray box. (c) The relationship between signal amplitude (third-order response function) and transition dipole moments for relevant field interactions for SE, GSB, and ESA.

Discerning the nature of the solute coherence is in general more difficult. For this, it is helpful to consult a Feynman diagram: one appropriate for the $Cr(acac)_3$ system is shown in Fig. 10. The processes depicted are stimulated emission (SE), ground state bleaching (GSB) and excited state absorption (ESA).[37] The magnitude of the signal from each of these processes is a function of the probability of each necessary field interaction occurring, which is proportional to the transition dipole of the transition it induces. In the system of interest, both SE and GSB involve transitions solely between ligand field states, either between the $^4A_2$ and $^4T_2$ states or the $^4A_2$ and ^2E states. Due to their Laporte and/or spin forbidden nature, the transition dipoles associated with these features – essentially those coupling to the ground state – are small. In contrast, transitions arising from the ^2E and/or $^4T_2$ states will couple to charge-transfer states at higher energy, which are Laporte- and spin-allowed and will therefore possess substantially larger transition dipoles. Based on these considerations, the amplitudes of ESA signals will be much greater than for SE or GSB. Furthermore, we can acquire a rough estimate of the magnitude of this disparity by approximating the squared transition dipole for the $^4A_2 \leftrightarrow {}^4T_2$ and the ligand field $\leftrightarrow$ charge transfer manifold transitions as the extinction coefficients in the ground state absorption spectra of $Cr(acac)_3$ ($\sim$100 M^{-1} cm^{-1})[23] and $[Fe(bpy)_3]^{2+}$ ($\sim$10 000 M^{-1} cm^{-1}),[16] respectively. From the third order response functions listed in Fig. 10, these values imply that the ESA signal is approximately 100-fold larger than those for GSB or SE. This is further supported by the observation that the TA spectrum for the systems of interest are positive throughout the visible spectrum.[5,36,38] We can therefore conclude that any observed coherences in the time-resolved excited-state spectroscopy of $Cr(acac)_3$ or its derivatives must be associated with an electronic excited state.

Now that we have established the nature of the signals that will be observed during a transient absorption experiment, it is important to address the mechanisms by which coherences dephase. In the optical Bloch picture, dephasing occurs primarily by population relaxation (T_1) and pure dephasing (T_2^*) caused by random environmental fluctuations that cause a loss of phase information. The contributions of these two processes to the observed homogenous dephasing time (T_2) are additive as seen in eqn (2),

$$\frac{1}{T_2} = \frac{1}{2}\left(\frac{1}{T_{1,a}} + \frac{1}{T_{1,b}}\right) + \frac{1}{T_2^*} \tag{2}$$

where $T_{1,a}$ and $T_{1,b}$ are the population relaxation times of the states involved in the coherence. Vibrational coherences typically show dephasing times on the order of picoseconds[32–35] and in solution phase this is often dominated by pure dephasing processes.[39] It is often observed that excited state vibrational coherences dephase faster than their ground state analogues due to either higher anharmonicity of the potential surface or an increased density of states leading to faster population relaxation. An alternative view of coherences involves their description as localized wavepackets. In this picture, dephasing can in part be thought of as a spatial dispersion of the wavepacket. This can occur in a highly anharmonic portion of the potential surface or as a result of multiple population transfer and back-transfer events between potential surfaces.

The vibrational coherences of $Cr(acac)_3$ and their corresponding dephasing times following $^4A_2 \rightarrow {}^4T_2$ excitation were monitored using transient absorption

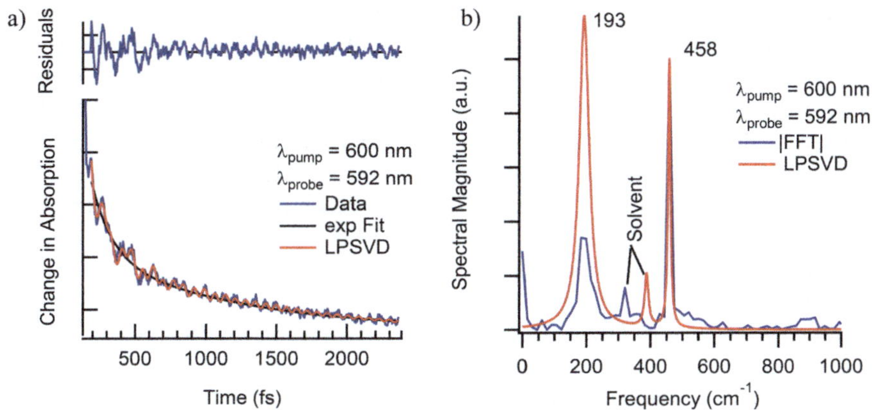

Fig. 11 (a) Single-wavelength kinetics of excited-state evolution in $Cr(acac)_3$ in CH_3CN solution at 592 nm following ~40 fs excitation into the $^4A_2 \rightarrow {}^4T_2$ absorption at 600 nm. The blue line corresponds to the experimental data, whereas the black and red lines correspond to a fit to a single-exponential decay model and an LPSVD analysis of the oscillatory features. (b) LPSVD (red) and fast Fourier transform (FFT, blue) analyses of the oscillatory components of the data shown in part (a). It should be noted that the data shown here are from a single measurement, whereas those discussed more generally in the text (and in Fig. 12) reflect an average of multiple data sets, hence the slight discrepancy in the vibrational frequencies.

spectroscopy. Studies were conducted in acetonitrile, 1,4-dioxane, and tetrahydrofuran while probing on the red and blue edges of the excited state absorption feature. For the purposes of the present discussion, we will focus only on vibrational modes at 193 cm^{-1} and 461 cm^{-1}.[40] These appear robustly in data acquired in all three solvents and are largely independent of pump wavelength across the $^4A_2 \rightarrow {}^4T_2$ absorption envelope and probe wavelengths spanning the excited-state absorption spectrum in the visible region and are therefore clearly associated with the electronic excited state(s) of the solute molecule (Fig. 11).

The lower frequency 193 cm^{-1} mode observed here is likely a more accurate value of the frequency observed previously.[29] The dephasing time observed in this study (350 $\pm$ 130 fs) is also significantly longer than reported originally (70 fs) but is still substantially shorter than the ~ps dephasing times typically thought of for vibrational coherences. The 461 cm^{-1} mode, on the other hand, shows a dephasing time of 1130 $\pm$ 500 fs that fits well with traditional values for pure vibrational coherences. Using the vibrational relaxation time constants of 7 ps and 12 ps measured by Kunttu and co-workers for the excited- and ground-electronic states, respectively,[25] the ground-state recovery time of ~1 ns is far too long for this electronic process to contribute to the observed dephasing time. Rather, the time scale for vibrational relaxation implies that the bulk of the dephasing occurs *via* T_2^* with a time constant of 1.3 ps, in agreement with the standard observation that most dephasing in liquids is due to pure dephasing.[33] Based on these and other considerations arising from resonance Raman linewidth analyses, we conclude that the 461 cm^{-1} mode is most likely a pure vibrational mode evolving on the ^2E excited state of the compound and therefore not associated with the reaction coordinate for the $^4T_2 \rightarrow {}^2E$ intersystem crossing process.

A corresponding analysis of the 193 cm^{-1} mode was inconclusive due in part to a significantly lower amplitude of the corresponding feature in the experimental Raman spectrum and distortion of the spectral shape from background signals that could not be adequately subtracted from the data. So, the presence of this feature in the coherence signal coupled with its intermediate dephasing time means that a link between dephasing time and the identification of an active mode along the reaction coordinate is still viable. Using density functional theory (DFT) calculations, we can start to acquire a better idea of the nature of each vibrational mode observed in the TA experiments and further define the type of motions active during the ISC process. It is important to reiterate that the vibrational modes in $Cr(acac)_3$ are predominantly sampled from the ^2E state which has the same electron configuration as the ground state and as such, is expected to have similar vibrational frequencies. The vibrational modes calculated for the ground electronic state should thus serve as an adequate substitute for vibrational modes in the excited state of $Cr(acac)_3$ and simplify the calculations. On the basis of frequency matching, the 193 cm^{-1} and 461 cm^{-1} modes can be assigned to the pair of metal–ligand bond stretching vibrations calculated at 185 cm^{-1} and 453 cm^{-1}, respectively (Fig. 12). Given the σ-antibonding character of the initially generated $^4T_2$ state and the significant displacement along these types of coordinates on the $^4T_2 \rightarrow {}^2E$ trajectory, it also makes intuitive sense that these modes involve significant distortion of the metal–ligand bond distance. The lower frequency 193 cm^{-1} mode is characterized by the entire ligand pushing toward the metal center while the two oxygen atoms splay outwards. The 461 cm^{-1} mode shows a slightly different motion, where each ligand compresses inward to make a smaller volumetric footprint through a decrease in the bite angle of the ligand. Neither of these vibrational modes involve the oxygen atoms moving in a linear trajectory with respect to the metal–oxygen bond vector, but rather appear more torsional in nature.

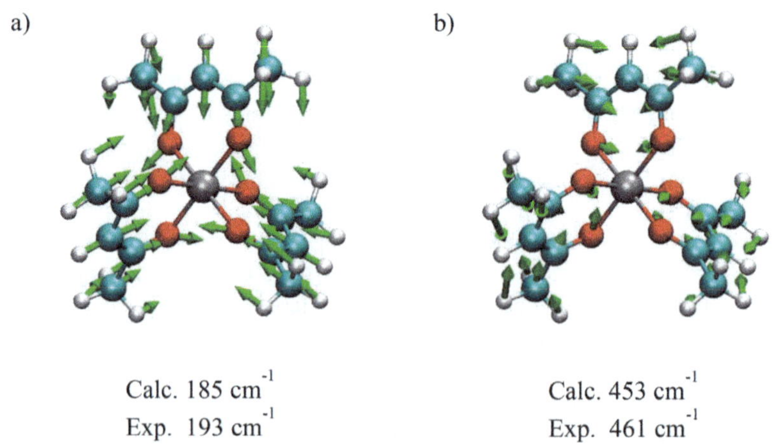

Fig. 12 Nuclear displacements associated with the coherent oscillations shown in Fig. 11 at (a) 193 cm^{-1} and (b) 461 cm^{-1}. The modes were visualized based on DFT calculations that yielded the energies indicated, whereas the experimental numbers reflect averages over multiple data sets.

It seems likely that a twisting motion would be effective in relaxing the Laporte selection rule and thus enhance the efficiency of an interconfigurational process such as the $^4T_2 \rightarrow {}^2E$ conversion. Indeed, evidence of the importance of torsional motion in transition metal complexes has been observed in several systems. Endicott and co-workers have shown trigonal twisting distortions to facilitate $^2E \rightarrow {}^4A_2$ ISC in octahedral Cr(III) complexes.[41,42] In the case of Fe(II) complexes, Purcell used an angular overlap-based analysis to propose that Bailar and Ray–Dutt twisting deformations may be involved in the racemization dynamics of $[Fe(phen)_3]^{2+}$ (where phen = 1,10-phenanthroline) by facilitating interconversion between its low-spin ($^1A_1$) and high-spin ($^5T_2$) configurations,[43] a model that was supported theoretically by Vanquickenborne and Pierloot[44] and experimentally by McCusker *et al.*, who showed that the kinetics of $^5T_2 \rightarrow {}^1A_1$ relaxation appeared to depend on synthetic perturbations along torsional degrees of freedom.[45] Hörner and co-workers later showed that both trigonal twisting and ligand breathing motions are likely coupled to the $^1A_1 \leftrightarrow {}^5T_2$ conversion in constrained Fe(II)-based spin-crossover systems.[46] So, it is certainly reasonable that the types of motion identified in the coherence signatures observed for $Cr(acac)_3$ could be coupled with the dynamics of $^4T_2 \rightarrow {}^2E$ intersystem crossing. The 340 fs dephasing time of the 193 cm^{-1} mode is certainly something we would have described as the type of intermediate dephasing time that proved successful (at least empirically) for identifying a viable reaction coordinate in the case of the Fe(II) polypyridyl complex. That stated, the 461 cm^{-1} mode, which has been assigned as a pure vibrational mode in part due to its longer dephasing time, is strikingly similar to the 193 cm^{-1} mode in terms of its nuclear displacement. It therefore remains to be determined what significance, if any, do we ascribe the different dephasing times associated with these modes as it pertains to the $^4T_2 \rightarrow {}^2E$ intersystem crossing reaction coordinate (and, more broadly, as a diagnostic tool for synthetic control of ultrafast dynamics)?

Chemical perturbations and their effect on vibronic coherences

In an effort to answer the question just posed, we consider transient absorption data we have acquired on several additional compounds: $Cr(3\text{-}R\text{-}acac)_3$ (where R = Cl, Br, and I) and $Cr(TMHD)_3$. These compounds represent different chemical perturbations to the basic $Cr(acac)_3$ framework. Incorporating substituents in the 3-position of the acac ligand (*i.e.*, the carbon atom between the two C–O groups) induces changes that are more electronic in nature rather than structural by affecting both the basicity of the oxygen donor atoms as well as the π interactions between the ligand and the t_{2g}-symmetry orbitals of the metal center (the so-called nephelauxetic effect). Modifications at the 1- and 5-positions (*e.g.*, *tert*-butyl groups replacing CH_3 groups in TMHD), on the other hand, primarily affect the sterics of the system.

The $^4A_2 \rightarrow {}^4T_2$ absorption band of $Cr(3\text{-}Cl\text{-}acac)_3$ (564 nm, 79 M^{-1} cm^{-1}) is slightly redshifted from the 560 nm maximum in $Cr(acac)_3$, consistent with the increased π-basicity of the chloro-substituted ligand raising the energy of the t_{2g} orbital set and thus marginally decreasing the ligand field strength of the compound by ~125 cm^{-1}. Based on analogous assignments on $Cr(acac)_3$, the bands centered at 272 and 358 nm can be assigned to intraligand $\pi \rightarrow \pi^*$ and $n \rightarrow \pi^*$ transitions, respectively.

Early time full spectrum TA data on Cr(3-Cl-acac)$_3$ following 600 nm excitation into the red edge of the $^4A_2 \rightarrow {}^4T_2$ ground state absorption band showed a clear decay in the blue and red edges of the transient absorption band similar to what was observed for the unsubstituted Cr(acac)$_3$ parent molecule. Global fitting of the data to two sequentially decaying kinetic components revealed a 1.07 ps kinetic component responsible for the spectral changes at early times, again similar to what was observed for Cr(acac)$_3$. Given the similar kinetics and observations between the two compounds, the ~1.1 ps kinetic component of Cr(3-Cl-acac)$_3$ can be assigned in an analogous manner to the parent compound, *i.e.*, a convolution of processes involving $^4T_2 \rightarrow {}^4A_2$ internal conversion subsequent to impulsive $^4T_2 \rightarrow {}^2E$ intersystem crossing and back-conversion as the system thermalizes. Single wavelength datasets show an average time constant of 1.24 ± 0.07 ps, in good agreement with the full spectrum fit and quite similar to the time constant observed in the unsubstituted Cr(acac)$_3$ compound.

Coherent oscillations were detected following excitation of Cr(3-Cl-acac)$_3$ similar to that described for Cr(acac)$_3$. The most robustly observed feature was a 149 cm^{-1} mode which was observed at all pump–probe combinations and confirmed to be associated with the chromophore. The nuclear displacements that characterize this degree of freedom are virtually identical to the 193 cm^{-1} mode observed for Cr(acac)$_3$, with the reduction in energy likely the result of the increased reduced mass of the vibration due to the incorporation of the chloro group in the ligand framework of Cr(3-Cl-acac)$_3$. Strikingly similar results were obtained for Cr(3-Br-acac)$_3$ and Cr(3-I-acac)$_3$, with coherent oscillations observed at energies of 123 cm^{-1} and 101 cm^{-1}, respectively. Vector displacement representations of these modes for all three of these new derivatives (as well as the parent Cr(acac)$_3$ complex) are illustrated in Fig. 13.

With these results in hand, we can gain further insights by comparing the results on each compound. First, it is clear by the qualitatively similar spectral evolution across the series that the excited state decay mechanism is likely the same for each member, *i.e.*, following excitation into the $^4T_2$ state, the molecule rapidly undergoes impulsive, ultrafast ISC to the doublet manifold followed by

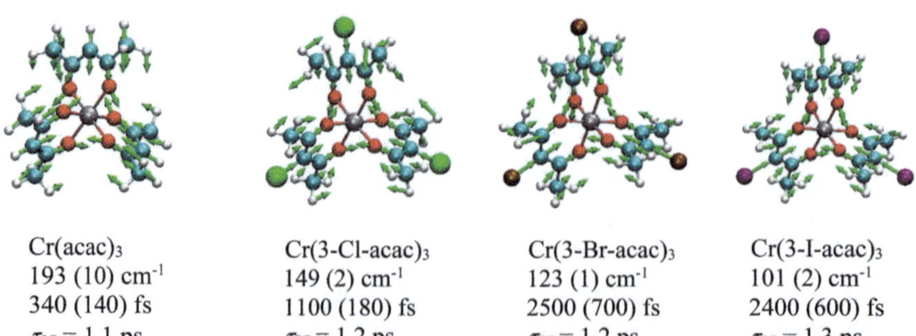

Fig. 13 Nuclear displacements associated with the coherent oscillations identified for the Cr(3-R-acac)$_3$ derivatives. Below each compound's name is listed the DFT calculated frequency associated with the coherence feature, the measured dephasing time, and the time constant corresponding to the observed kinetics for excited-state evolution based on the model shown in Fig. 9.

back-intersystem crossing to create a quasi-equilibrium between the $^4T_2$ and ^2E states. This provided a pathway whereby a portion of the $^4T_2$ population can relax back to the ground state with a time constant of ~1 ps while leaving some population to thermalize in the ^2E state. Perhaps unsurprisingly given the identical decay mechanisms across the Cr(3-R-acac)$_3$ series, the coherent vibrational mode that is active in each compound is also very similar. What is striking is the wide variation in the dephasing times for this mode, ranging from an "intermediate" value of *ca.* 350 fs in Cr(acac)$_3$ to ~2.5 ps for both Cr(3-Br-acac)$_3$ and Cr(3-I-acac)$_3$. From the Feynman diagram shown in Fig. 10, we know that the coherences sampled during the TA experiments must be from an electronic excited state ($^4T_2$ or ^2E). Given that the internal conversion time constant of *ca.* 1 ps is significantly smaller than the *ca.* 2.5 ps dephasing time for the heavier halogen-substituted derivatives (*i.e.*, R = Br and I), it is clearly the case that this vibrational mode is not sampled from the $^4T_2$ state in these systems. That stated, one cannot use this line of reasoning as definitively in the cases of either Cr(3-Cl-acac)$_3$ or Cr(acac)$_3$ since the dephasing times in these compounds are comparable to or shorter than the population dynamics associated with interconversion between the $^4T_2$ and ^2E states.

Data acquired on Cr(TMHD)$_3$ provides one final component to this discussion. In the previously published study,[29] a pronounced change in the evolution of the spectral profile was attributed to a decrease in the rate of the initial intersystem crossing from the $^4T_2$ state to the lower-lying ^2E state relative to what was inferred for Cr(acac)$_3$. This interpretation must be questioned in light of the work published by Kunttu subsequent to the 2010 study,[24,25] but the fact that there is an obvious difference in the time-dependent characteristics of the differential absorption profile between Cr(acac)$_3$ and Cr(TMHD)$_3$ indicates that the kinetics of the latter have changed more dramatically than what was observed due to compositional variations across the Cr(3-R-acac)$_3$ series. So, the introduction of steric bulk on the periphery of the ligand framework is clearly altering the excited-state dynamics of the system, but the question now is which process is being affected and does the dephasing time provide an indicator as to whether a given mode is coupled to this process.

Time-resolved absorption data were acquired on Cr(TMHD)$_3$ as a function of pump and probe wavelength in dioxane, benzene, and THF solutions; a representative example is shown in Fig. 14. Modes at 125 $\pm$ 10 cm^{-1}, 173 $\pm$ 12 cm^{-1}, and 249 $\pm$ 6 cm^{-1} with dephasing times of 680 $\pm$ 420 fs, 770 $\pm$ 450 fs, and 960 $\pm$ 700 fs were identified consistently across all sample conditions. Vector projections of the 173 cm^{-1} and 249 cm^{-1} modes bear resemblance to modes identified in the other compounds that have been discussed in this report, and while it appears that the more sterically encumbered Cr(TMHD)$_3$ does exhibit dephasing times that are intermediate between the unsubstituted Cr(acac)$_3$ (340 fs) and the halogenated 3-substituted analogs, the error bars derived from sampling data across all of the various sample conditions makes it difficult to reach any firm conclusions. It does seem to be the case that the Cr(3-R-acac)$_3$ series of compounds gave rise to a reduction in the vibrational frequency of the dominant mode and a concomitant increase in the dephasing time, whereas an opposite effect is evident in response to increased steric bulk on the periphery of the ligand framework. This trend prompts the question of whether there is an inverse relationship between the vibrational frequency (and by extension the vibrational period) and the susceptibility of

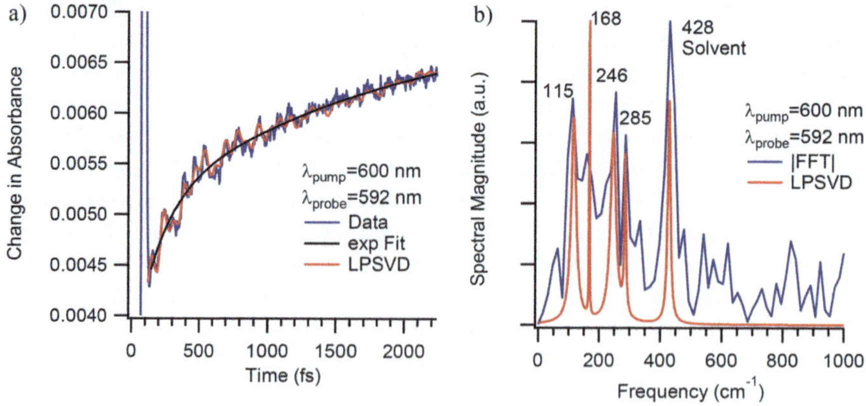

Fig. 14 Ultrafast time-resolved absorption data for $Cr(TMHD)_3$ in a 1,4-dioxane solution. (a) Single-wavelength kinetics data acquired at 592 nm following *ca.* 40 fs excitation into the $^4A_2 \rightarrow {}^4T_2$ ligand-field absorption band at 600 nm. The blue line corresponds to the experimental data, the black line a fit of the population dynamics to a single exponential kinetic model, and the red to the results of an LPSVD analysis. (b) LPSVD (red) and FFT (blue) analysis of the quantum oscillations observed in the experimental data shown in part (a). Features appearing at 115 cm^{-1}, 168 cm^{-1}, and 246 cm^{-1} correspond to the three vibrational signatures that were consistently observed across variations in solvent as well as pump/probe wavelength combinations.

a vibrational mode to be dephased by various excited-state processes (*e.g.*, vibrational relaxation, intersystem crossing, internal conversion) or if there is a more subtle relationship between the collection of vibrational frequencies of a molecule and these excited-state dynamics at work. At this stage, the data we have in hand (as well as the current state of theoretical treatments of this issue) does not allow us to draw any firm conclusions at this time.

Concluding comments

We have presented an overview of two chemical systems in which excited-state coherences were used in an effort to gain insight into the vibrational degrees of freedom that served to define the reaction coordinate of ultrafast excited-state dynamics. In particular, it was postulated that coherence dephasing times that were intermediate between that expected for pure electronic *versus* pure vibrational processes were vibronic and therefore could constitute a metric by which certain modes could be singled out as likely candidates for synthetic manipulation. If successful, such an approach would provide a roadmap for how one could use ultrafast spectroscopy to guide synthetic design. In the case of an Fe(II) polypyridyl complex, identification of large-amplitude motions in the experimentally observed excited-state coherences suggested a mechanism for altering the trajectory of MLCT-to-LF state evolution. The incorporation of an electronic benign but sterically impactful metal ion in this portion of the superstructure yielded a compound whose MLCT excited-state lifetime was indeed extended by more than an order of magnitude. We have also presented a counterexample in the form of the electronically and vibrationally much simpler $Cr(acac)_3$ platform. In this system, photoexcitation into the lowest energy $^4A_2 \rightarrow {}^4T_2$ ligand-field absorption leads to an ultrafast intersystem crossing process to the lower lying ^2E state, followed by an excited-state evolution process that bifurcates between back-intersystem crossing to the $^4T_2$ state (with subsequent internal conversion back to the ground state) and vibrational relaxation on the ^2E surface. Quantum oscillations superimposed on the observed kinetics suggested common vibrational degrees of freedom across a series of derivatives of the parent $Cr(acac)_3$ complex, but wide variations in dephasing times from 340 fs to over 2 ps did not reveal any clear correlation with evolution along the operative reaction coordinates. So, while the results on the Fe(II) system reveal an instance where this correlation appears to have worked, the $Cr(acac)_3$ results suggest that such a metric is far from robust as a predictive element for synthetic design.

We are therefore left with the same question with which we began, *i.e.*, can one use information from quantum coherences to inform on the degrees of freedom that couple to ultrafast excited-state dynamics? From the limited information available on this issue (much of which is contained in the results we have just presented), the only thing that seems clear at this stage is that the answer to this question is *not* clear. At present, dephasing times can suggest which modes *might* be of interest, but definitive proof that a given mode is, in fact, relevant to the reaction coordinate must still rely on an approach along the lines of what we demonstrated for the Fe(II) system, *i.e.*, synthetic modification and subsequent experimental measurements on that new compound to determine if the synthetic change has had the desired effect. We believe that resolution of this issue – and an answer to what we believe is a potentially transformative approach to controlling excited-state dynamics *a priori* – will require a synergistic effort across teams with synthetic, spectroscopic, and theoretical expertise. We would argue that transition metal complexes provide a uniquely suitable platform for such an effort due to the fact that their electronic structures are intrinsically vibronic in nature and possess a degree of compositional flexibility to allow for the type of structure–property correlations that will be required. Efforts along these lines are currently underway.

Conflicts of interest

The authors confirm that there are no conflicts of interest to declare.

Acknowledgements

The authors wish to acknowledge contributions from Dr Joel N. Schrauben and Dr Eileen D. Foszcz for their efforts in the initial development of the $Cr(acac)_3$-based project. The research on the Fe(II) polypyridyl complexes was supported through a grant from the Chemical Sciences, Geosciences, and Biosciences Division, Office of Basic Energy Science, Office of Science, U.S. Department of Energy under Grant No. DE-FG02-01ER15282. The $Cr(acac)_3$ was previously supported by the U.S. National Science Foundation under grant no. CHE-1300096, with continuing work now supported under CHE-2154233.

Notes and references

1 G. D. Scholes, G. R. Fleming, L. X. Chen, A. Aspuru-Guzik, A. Buchleitner, D. F. Coker, G. S. Engel, R. van Grondelle, A. Ishizaki, D. M. Jonas,

J. S. Lundeen, J. K. McCusker, S. Mukamel, J. P. Ogilvie, A. Olaya-Castro, R. A. Ratner, F. C. Spano, B. K. Whaley and X. Zhu, Using Coherence to Enhance Function in Chemical and Biological System, *Nature*, 2017, **543**, 647–656.
2 M. R. Wasielewski, M. D. E. Forbes, N. L. Frank, K. Kowalski, G. D. Scholes, J. Yuen-Zhou, M. A. Baldo, D. E. Freedman, R. H. Goldsmith, T. Goodson III, M. L. Kirk, J. K. McCusker, J. P. Ogilvie, D. A. Shultz, S. Stoll and B. K. Whaley, Exploiting chemistry and molecular systems for quantum information science, *Nat. Rev. Chem.*, 2020, **4**, 490–504.
3 An excellent discussion of how one can link quantum coherence to a range of well-known chemical phenomena can be found in the information boxes contained in ref. 1.
4 A. H. Zewail, Femtochemistry: Atomic-Scale Dynamics of the Chemical Bond, *J. Phys. Chem. A*, 2000, **104**, 5660–5694.
5 J. N. Schrauben, *Electronic Structure and Excited State Dynamics of Chromium(III) Complexes*, PhD thesis, Michigan State University, 2010.
6 M. J. Frisch, G. W. Trucks, H. B. Schlegel, G. E. Scuseria, M. A. Robb, J. R. Cheeseman, G. Scalmani, V. Barone, G. A. Petersson, H. Nakatsuji, *et al.*, *Gaussian 09*, Gaussian, Inc., Wallingford CT, 2009.
7 E. D. Foszcz, *Understanding the Interplay Between Geometry and Ultrafast Dynamics in Ligand Field Excited States of Inorganic Chromophores*, PhD thesis, Michigan State University, 2015.
8 W. Humphrey, A. Dalke and K. Schulten, VMD: Visual Molecular Dynamics, *J. Mol. Graphics Modell.*, 1996, **14**, 33–38.
9 A more complete discussion of the experimental set-up as well as pulse characterization, data analysis procedures, *etc.* can be found in ref. 5, 7, 10 and 27.
10 A. M. Brown, C. E. McCusker, M. C. Carey, A. M. Blanco-Rodríguez, M. Towrie, I. P. Clark, A. Vlček and J. K. McCusker, Vibrational Relaxation and Redistribution Dynamics in Ruthenium(II) Polypyridyl-Based Charge-Transfer Excited State: A Combined Ultrafast Electronic and Infrared Absorption Study, *J. Phys. Chem. A*, 2018, **122**, 7941–7953.
11 L. Wang, M. A. Allodi and G. S. Engel, Quantum coherences reveal excited-state dynamics in biophysical systems, *Nat. Rev. Chem.*, 2019, **3**(8), 477–490.
12 Y. Nagasawa, Ultrafast photon echo experiments in condensed phase: detection of solvation dynamics, coherent wavepacket motions and static inhomogeneity, *J. Photochem. Photobiol., C*, 2011, **12**(1), 31–45.
13 (*a*) F. D. Fuller, J. Pan, A. Gelzinis, V. Butkus, S. S. Senlik, D. E. Wilcox, C. F. Yocum, L. Valkunas, D. Abramavicius and J. P. Ogilvie, Vibronic coherence in oxygenic photosynthesis, *Nat. Chem.*, 2014, **6**(8), 706–711; (*b*) J. D. Gaynor, J. Sandwisch and M. Khalil, Vibronic coherence evolution in multidimensional ultrafast photochemical processes, *Nat. Commun.*, 2019, **10**(1), 5621; (*c*) P. Karak, K. Ruud and S. Chakrabarti, Demystifying the Origin of Vibrational Coherence Transfer Between the S_1 and T_1 States of the Pt-pop Complex, *J. Phys. Chem. Lett.*, 2021, **12**(40), 9768–9773.
14 J. C. Dean and G. D. Scholes, Coherence Spectroscopy in the Condensed Phase: Insights into Molecular Structure, Environment, and Interactions, *Acc. Chem. Res.*, 2017, **50**(11), 2746–2755.

15 It should be noted that, to our knowledge, there has never been any report describing the observation of an electronic coherence signal for a transition metal-containing complex.

16 B. C. Paulus, S. L. Adelman, L. L. Jamula and J. K. McCusker, Leveraging excited-state coherence for synthetic control of ultrafast dynamics, *Nature*, 2020, **582**(7811), 214–218.

17 (*a*) O. S. Wenger, Is Iron the New Ruthenium?, *Chem.–Eur. J.*, 2019, **25**(24), 6043–6052; (*b*) C. Wegeberg and O. S. Wenger, Luminescent First-Row Transition Metal Complexes, *JACS Au*, 2021, **1**, 1860–1876.

18 K. J. Gaffney, Capturing photochemical and photophysical transformations in iron complexes with ultrafast X-ray spectroscopy and scattering, *Chem. Sci.*, 2021, **12**, 8010–8025, and references therein.

19 J. K. McCusker, Electronic structure in the transition metal block and its implications for light harvesting, *Science*, 2019, **363**(6426), 484–488.

20 (*a*) P. Pyykko, Relativistic Effects in Structural Chemistry, *Chem. Rev.*, 1988, **88**, 563–594; (*b*) M. Kaupp, The Role of Radial Nodes of Atomic Orbitals for Chemical Bonding and the Periodic Table, *J. Comput. Chem.*, 2007, **28**, 320–325.

21 (*a*) Y. Liu, T. Harlang, S. E. Canton, P. Chábera, K. Suárez-Alcántara, A. Fleckhaus, D. A. Vithanage, E. Göransson, A. Corani, R. Lomoth, V. Sundström and K. Wärnmark, Towards Longer-Lived Metal-to-Ligand Charge Transfer States of Iron(II) Complexes: An N-Heterocyclic Carbene Approach, *Chem. Commun.*, 2013, **49**, 6412; (*b*) T. Reuter, A. Kruse, R. Schoch, S. Lochbrunner, M. Bauer and K. Heinze, Higher MLCT Lifetime of Carbene Iron(II) Complexes by Chelate Ring Expansion, *Chem. Commun.*, 2021, **57**, 7541–7544; (*c*) J. Steube, L. Burkhardt, A. Päpcke, J. Moll, P. Zimmer, R. Schoch, C. Wölper, K. Heinze, S. Lochbrunner and M. Bauer, Excited-State Kinetics of an Air-Stable Cyclometalated Iron(II) Complex, *Chem.–Eur. J.*, 2019, **25**, 11826–11830.

22 (*a*) D. C. Ashley and E. Jakubikova, Ray-Dutt and Bailar Twists in Fe(II)-Tris(2,2′-Bipyridine): Spin States, Sterics, and Fe–N Bond Strengths, *Inorg. Chem.*, 2018, **57**, 5585–5596; (*b*) J. Nance, D. N. Bowman, S. Mukherjee, C. T. Kelley and E. Jakubikova, Insights into the Spin-State Transitions in $[Fe(tpy)_2]^{2+}$: Importance of the Terpyridine Rocking Motion, *Inorg. Chem.*, 2015, **54**, 11259–11268; (*c*) H.-Y. Kwon, D. C. Ashley and E. Jakubikova, Halogenation Affects Driving Forces, Reorganization Energies and "Rocking" Motions in Strained $[Fe(Tpy)_2]^{2+}$ Complexes, *Dalton Trans.*, 2021, **50**, 14566–14575. See also ref. 43 and 45.

23 E. A. Juban and J. K. McCusker, Ultrafast Dynamics of 2E State Formation in $Cr(acac)_3$, *J. Am. Chem. Soc.*, 2005, **127**, 6857–6865.

24 E. M. S. Maçôas, R. Kananavicius, P. Myllyperkiö, M. Pettersson and H. Kunttu, Relaxation Dynamics of $Cr(acac)_3$ Probed by Ultrafast Infrared Spectroscopy, *J. Am. Chem. Soc.*, 2007, **129**, 8934–8935.

25 E. M. S. Maçôas, S. Mustalahti, P. Myllyperkiö, H. Kunttu and M. Pettersson, Role of Vibrational Dynamics in Electronic Relaxation of $Cr(acac)_3$, *J. Phys. Chem. A*, 2015, **119**, 2727–2734.

26 While time-resolved vibrational studies represent the most direct way for probing vibrational relaxation dynamics, transient absorption spectra do provide some insights along these lines. See ref. 10 for an example.

27 B. C. Paulus, *Insights into the Photophysical Reaction Coordinate of 1st Row Transition Metal Chromophores from Vibrational Coherences*, PhD thesis, Michigan State University, 2020.

28 H. Ando, S. Iuchi and H. Sato, Theoretical Study on Ultrafast Intersystem Crossing of Chromium(III) Acetylacetonate, *Chem. Phys. Lett.*, 2012, **535**, 177–181.

29 J. N. Schrauben, K. L. Dillman, W. F. Beck and J. K. McCusker, Vibrational Coherence in the Excited State Dynamics of $Cr(acac)_3$: Probing the Reaction Coordinate for Ultrafast Intersystem Crossing, *Chem. Sci.*, 2010, **1**, 405–410.

30 S. De Silvestri, G. Cerullo and G. Lanzani, *Coherent Vibrational Dynamics*, CRC Press, Boca Raton, FL, 2008.

31 P. J. Reid, C. Silva and P. F. Barbara, Electronic Coherence, Vibrational Coherence, and Solvent Degrees of Freedom in the Femtosecond Spectroscopy of Mixed-Valence Metal Dimers in H_2O and D_2O, *J. Phys. Chem.*, 1995, **99**, 2609–2616.

32 P. Hamm and M. Zanni, *Concepts and Methods of 2D Infrared Spectroscopy*, Cambridge University Press, Cambridge, 2011.

33 R. Monni, G. Auböck, D. Kinschel, K. M. Aziz-Lange, H. B. Gray, A. Vlček and M. Chergui, Conservation of Vibrational Coherence in Ultrafast Electronic Relaxation: The Case of Diplatinum Complexes in Solution, *Chem. Phys. Lett.*, 2017, **683**, 112–120.

34 Q. Wang, R. W. Schoenlein, L. A. Peteanu, R. A. Mathies and C. V. Shank, Vibrationally Coherent Photochemistry in the Femtosecond Primary Event of Vision, *Science*, 1994, **266**, 422–424.

35 (*a*) S. Rafiq and G. D. Scholes, Slow Intramolecular Vibrational Relaxation Leads to Long-Lived Excited-State Wavepackets, *J. Phys. Chem. A*, 2016, **120**, 6792–6799; (*b*) S. Rafiq and G. D. Scholes, From Fundamental Theories to Quantum Coherences in Electron Transfer, *J. Am. Chem. Soc.*, 2019, **141**, 708–722.

36 Additional details can be found in the ESI associated with ref. 29.

37 It should be noted that the mechanism by which ground state vibrational coherence in the solute is generated (impulsive stimulated Raman scattering) can be grouped with GSB in terms of the sign of the signal.

38 Additional details can be found in ref. 23.

39 E. Gershgoren, Z. Wang, S. Ruhman, J. Vala and R. Kosloff, Investigating Pure Vibrational Dephasing of I_3^- in Solution: Temperature Dependence of T_2^* for the Fundamental and First Harmonic of ν_1, *J. Chem. Phys.*, 2003, **118**, 3660–3667.

40 There are a number of additional signals that are accounted for through resonance Raman specrtra, bandwidth analyses, *etc.* Details can be found in ref. 27.

41 M. W. Perkovic and J. F. Endicott, Stereochemical Tuning of Chromium(III) Photophysics With N,N′,N″-tris(alkylamine)-1,4,7-triazacyclononane complexes, *J. Phys. Chem.*, 1990, **94**, 1217–1219.

42 M. W. Perkovic, M. J. Heeg and J. F. Endicott, Stereochemical Perturbations of the Relaxation Behavior of ^2E Chromium(III). Ground-State X-Ray Crystal Structure, Photophysics, and Molecular Mechanics Simulations of the Quasi-Cage Complex [4,4′,4″-Ethylidynetris(3-Azabutan-1-Amine)]Chromium Tribromide, *Inorg. Chem.*, 1991, **30**, 3140–3147.

43 K. F. Purcell, Pseudorotational Intersystem Crossing in d^6 Complexes, *J. Am. Chem. Soc.*, 1979, **101**, 5147–5152.

44 L. G. Vanquickenborne and K. Pierloot, Role of Spin Change in the Stereomobile Reactions of Strong-Field d^6 Transition-Metal Complexes, *Inorg. Chem.*, 1981, **20**, 3673–3677.

45 J. K. McCusker, A. L. Rheingold and D. N. Hendrickson, Variable-Temperature Studies of Laser-Initiated $^5T_2 \rightarrow {}^1A_1$ Intersystem Crossing in Spin-Crossover Complexes: Empirical Correlations between Activation Parameters and Ligand Structure in a Series of Polypyridyl Ferrous Complexes, *Inorg. Chem.*, 1996, **35**, 2100–2112.

46 P. Stock, E. Deck, S. Hohnstein, J. Korzekwa, K. Meyer, F. W. Heinemann, F. Breher and G. Hörner, Molecular Spin Crossover in Slow Motion: Light-Induced Spin-State Transitions in Trigonal Prismatic Iron(II) Complexes, *Inorg. Chem.*, 2016, **55**, 5254–5265.

Faraday Discussions

PAPER

Ultrafast manipulation of the NiO antiferromagnetic order *via* sub-gap optical excitation

Xiaocui Wang, [a] Robin Y. Engel,[bc] Igor Vaskivskyi,[d] Diego Turenne,[a] Vishal Shokeen,[a] Alexander Yaroslavtsev,[a] Oscar Grånäs, [a] Ronny Knut,[a] Jan O. Schunck, [bc] Siarhei Dziarzhytski, [b] Günter Brenner,[b] Ru-Pan Wang, [c] Marion Kuhlmann,[b] Frederik Kuschewski,[b] Wibke Bronsch, [e] Christian Schüßler-Langeheine,[f] Andriy Styervoyedov,[g] Stuart S. P. Parkin,[g] Fulvio Parmigiani,[e] Olle Eriksson,[a] Martin Beye [bc] and Hermann A. Dürr *[a]

Received 10th January 2022, Accepted 15th February 2022
DOI: 10.1039/d2fd00005a

Wide-band-gap insulators such as NiO offer the exciting prospect of coherently manipulating electronic correlations with strong optical fields. Contrary to metals where rapid dephasing of optical excitation *via* electronic processes occurs, the sub-gap excitation in charge-transfer insulators has been shown to couple to low-energy bosonic excitations. However, it is currently unknown if the bosonic dressing field is composed of phonons or magnons. Here we use the prototypical charge-transfer insulator NiO to demonstrate that 1.5 eV sub-gap optical excitation leads to a renormalised NiO band-gap in combination with a significant reduction of the antiferromagnetic order. We employ element-specific X-ray reflectivity at the FLASH free-electron laser to demonstrate the reduction of the upper band-edge at the O 1s–2p core–valence resonance (K-edge) whereas the antiferromagnetic order is probed *via* X-ray magnetic linear dichroism (XMLD) at the Ni 2p–3d resonance (L_2-edge). Comparing the transient XMLD spectral line shape to ground-state measurements allows us to extract a spin temperature rise of 65 ± 5 K for time delays longer than 400 fs while at earlier times a non-equilibrium spin state is formed. We identify transient mid-gap states being formed during the first 200 fs accompanied by a band-gap

[a]*Department of Physics and Astronomy, Uppsala University, 75120 Uppsala, Sweden. E-mail: hermann.durr@physics.uu.se*
[b]*Deutsches Elektronen-Synchrotron DESY, Notkestr. 85, 22607 Hamburg, Germany*
[c]*Department of Physics, Universität Hamburg, Luruper Chaussee 149, 22761 Hamburg, Germany*
[d]*Department for Complex Matter, Jozef Stefan Institute, Jamova cesta 39, 1000 Ljubljana, Slovenia*
[e]*Elettra-Sincrotrone Trieste S.C.p.A., 34149 Basovizza, Italy*
[f]*Helmholtz-Zentrum Berlin für Materialien und Energie, Albert-Einstein-Str. 15, 12489 Berlin, Germany*
[g]*Max-Planck Institut für Mikrostrukturphysik, Weinberg 2, 06120 Halle, Germany*

reduction lasting at least up to the maximum measured time delay of 2.4 ps. Electronic structure calculations indicate that magnon excitations significantly contribute to the reduction of the NiO band gap.

1 Introduction

Antiferromagnets (AFMs) have gained increasing interest and attention in both fundamental research and technological development. Since AFMs have zero net magnetization in their ground state, they are robust against external magnetic fields and have the promising potential to facilitate spintronic devices with ultrahigh speeds in the THz range.[1,2] Despite intense experimental studies and theoretical modelling efforts that have been dedicated to the coupling mechanisms between charge, spin and lattice degrees of freedom, the important interaction pathways remain elusive.

Owing to the development of ultrafast optical laser and free-electron laser (FEL) radiation sources that improve the temporal resolution even towards the few-femtosecond regime and the development of computing algorithms, scholars in this field have been able to unambiguously detect and model the complex dynamics in strongly correlated materials. With scattering techniques, using electrons,[3] X-ray photons[4] or neutrons,[5] one can investigate the ordering of the lattice or the AFM order. The diffraction signal of multiple-integer Bragg peaks is assigned to the structural re-arrangement, while the multiples of half-integer peaks are assigned to the magnetic ordering.[6] With spectroscopic techniques, one can study the electronic configurations with element specificity[7–9] and spin excitations in magnetic systems.[10]

Nickel oxide (NiO), as a prototype AFM, is a good candidate for studies in strongly correlated materials due to its well separated intra-gap states,[11] large spin density[5,12,13] and a high Néel temperature ($T_N \sim 523$ K).[7,14,15] Above T_N, NiO has a rock-salt structure (point group $Fm\bar{3}m$) with a lattice constant of 4.176 Å. Below T_N, the Ni^{2+} spins are ferromagnetically aligned along $\langle 11{-}2\rangle$ directions within the (111) plane, and adjacent (111) planes are coupled antiferromagnetically, which composes two sublattice AFM systems. Due to exchange striction, the AFM ordering of the spins induces a small contraction along the $\langle 111\rangle$ direction, resulting in a rhombohedral distortion of the crystalline structure. This distortion induces a reduction of the crystallographic symmetry from point group $Fm\bar{3}m$ of the cubic structure to point group $C2/m$ of the rhombohedral structure. Since there are four energetically degenerate $\langle 111\rangle$ directions, the distortion can occur along any of the four equivalent directions. This results in four types of twin-domains (T-domain). For each T-domain, there are three possible spin orientations (S-domain). In total, there can be 12 orientational domains in NiO crystals.

In various static measurements, the typical X-ray magnetic linear dichroic (XMLD) line shape, which is directly linked to the long-range ordering in AFMs, has been observed in NiO by varying the temperature,[7] the experimental geometry[7,16] or measuring XMLD from different AFM domains in microscopy.[17,18] There have also been a large number of time-resolved measurements carried out in NiO in order to disentangle the coupling between different degrees of freedom. With time-resolved optical measurements, oscillations with THz frequencies have been observed and assigned to magnon excitation.[14,19,20] In a recent experiment, researchers have investigated the lattice dynamics of NiO using femtosecond electron diffraction,[3] and the origin of the rhombohedral lattice distortion was assigned to the weakening of the AFM order. However, there is so far no direct evidence of a laser-induced change in the AFM order driven by femtosecond laser pulses in time-resolved magnetic diffraction measurements.[6]

In this article, we report the investigation of the ultrafast manipulation of the AFM order in NiO films *via* sub-gap optical excitation. Using time-resolved resonant X-ray reflectivity, we observe XMLD line shape changes at the nickel L_2-edge, *i.e.*, for $2p_{1/2}$–3d transitions, and band gap renormalisation at the oxygen K-edge (1s–2p transitions). By comparing with static XMLD measurements at different temperatures,[7] we determine the laser-induced temperature change of the spin system to be 65 ± 5 K. In order to explain the band gap renormalisation observed at the oxygen K-edge, we use first principles calculations to investigate the influence of magnetic excitations on the band structure with the density-functional theory + Hubbard U formalism.

2 Methods

2.1 Sample preparation and characterization

NiO(001) crystalline films were used as our samples. The NiO films were epitaxially grown with surface normal along the [001] direction on a single crystal MgO(001) substrate. The MgO substrate was polished on both sides. A 2 nm thick MgO underlayer was first deposited by radio frequency magnetron sputtering in 3 mTorr argon at a temperature below 100 °C. The NiO layer was then deposited at 700 °C in a gas mixture of Ar(90%)/O_2(10%) at a pressure of 3 mTorr and was annealed *in situ* for 15 min with the same temperature and gas mixture condition. The thickness of the NiO film was 40 nm.

The ground-state electronic and magnetic structure of NiO crystalline films was characterised by performing X-ray absorption spectroscopy (XAS) measurements in drain current mode at the PM3 beamline at the BESSY II electron storage ring operated by Helmholtz-Zentrum Berlin.[21] Static XAS spectra were acquired for various incidence angles at room temperature.

The previous optical measurements of NiO,[22,23] show a variation in the optical absorption coefficient. Therefore, optical spectroscopy measurements were carried out prior to the time-resolved X-ray reflectivity experiment in order to evaluate the optical absorption properties of the NiO film during the optical pump–X-ray probe experiments, as shown in Fig. 1. The symbols in Fig. 1 represent the measured optical absorption between 450 nm and 1450 nm, and the dashed lines represent the calculated absorption for a 40 nm NiO film using the tabulated absorption coefficient.[22,23] The sharp drop at 860 nm in the experimental data is due to the change of photodetector in the optical measurements. From experimental data one can see two absorption bands that are centred at 720 nm and 1150 nm. The peak position of these two absorption bands is close to the values reported in previous work.[22–24] The origin of these two absorption bands has been assigned to the crystal-field d–d transitions.[25,26] We note that the dashed red line is truncated at 500 nm because the short wavelength limit reported in ref. 23 is 500 nm. Overall, the absorption bands shown in the experimental data indicate the good quality of our sample.

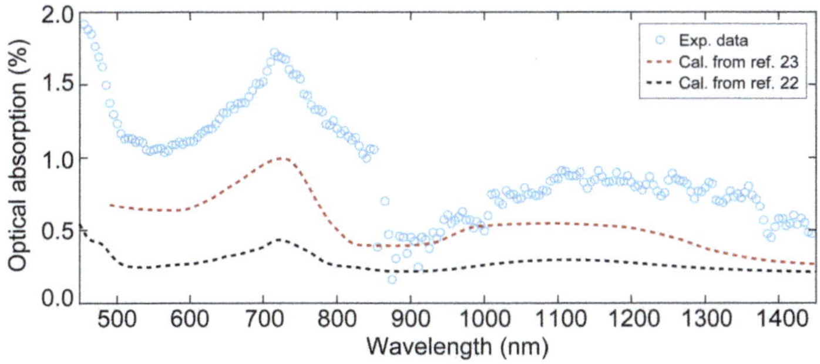

Fig. 1 The optical absorption of the NiO film with a thickness of 40 nm at wavelengths between 450 nm and 1450 nm. The circles represent the experimental data obtained from visible-infrared spectroscopy measurements. The sharp drop at 860 nm is due to the change of photodetector in the measurement. The dashed black and red lines denote the calculated optical absorption using the reported absorption coefficient in ref. 22 and 23, respectively.

2.2 Time-resolved X-ray reflectivity measurements at grazing angles

The time-resolved X-ray reflectivity experiments were carried out using the MUSIX end-station at beamline FL24 at the free-electron laser, FLASH, in Hamburg.[27,28] Experiments were carried out in reflection geometry as illustrated in Fig. 2(a). The NiO film was excited by a femtosecond laser pulse with a central wavelength around 800 nm and a pulse duration of 50 fs. The temporal evolution of the X-ray reflectivity following laser excitation was probed at both the oxygen K-edge and the nickel $L_{2,3}$-edges. The angle between the incident pump and the probe beams was 0.75°, *i.e.* nearly collinear.

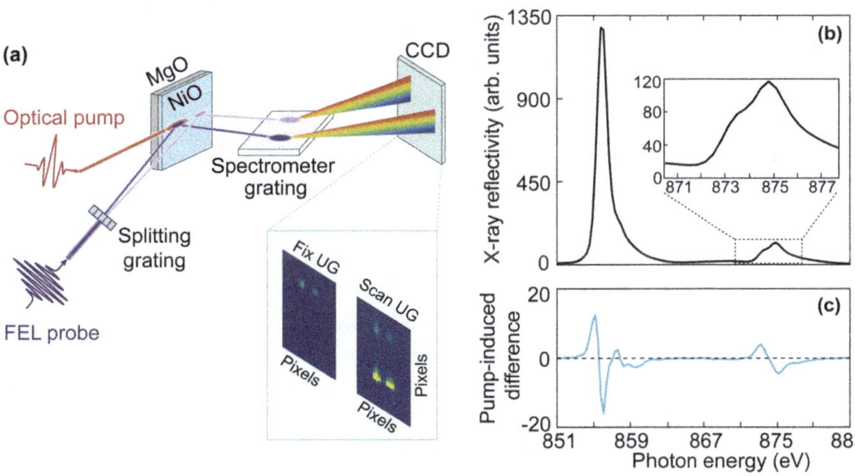

Fig. 2 (a) Schematic of experimental set-up for time-resolved X-ray reflectivity measurements. The inset shows two representative CCD images for measurements with fixed undulator gap (UG) and scanning UG. (b) Static X-ray reflectivity spectrum of NiO at the nickel $L_{2,3}$-edges. The inset shows the enlarged spectrum at the nickel L_2-edge. (c) Pump-induced difference in the X-ray reflectivity spectrum for time delays longer than 400 fs.

FLASH was operated in a 10 Hz burst mode with each burst consisting of 40 X-ray pulses. The FEL was tuned for the third harmonic radiation to reach the oxygen K-edge and the nickel $L_{2,3}$-edges. Reflectivity spectra were recorded utilizing a split-beam normalisation scheme.[28] As illustrated in Fig. 2(a), the FEL probe beam propagated through a beam splitting grating and was horizontally split such that the zeroth and first diffraction orders hit the sample. The zeroth order beam overlapped both spatially and temporally with the optical pump beam to record the transient X-ray reflectivity. The zeroth and first diffraction orders were separated by 3 mm on the sample, ensuring that the first diffraction order of the FEL beam had no overlap with the optical laser beam and could be used to normalise fluctuations of the FEL spectral intensity stemming from the self-amplified spontaneous emission (SASE) process. X-ray energy resolution was obtained by directing both the zeroth and first diffraction orders onto a down-stream spectrometer grating and consecutively, a CCD camera. The dispersion of the spectrometer was calibrated and found to be 0.33 eV per pixel. To measure reflectivity spectra around the absorption edges, the photon energy was scanned in steps of 0.5 eV at the fundamental wavelength by varying the undulator gap, which corresponds to 1.5 eV at the third harmonic in the time-resolved X-ray reflectivity measurement. As each FEL spectrum covers a bandwidth of about 1% of the total photon energy, this ensures a seamless spectrum of the scanned range. In reflection geometry, by varying the angle of incidence, one can access different parts of the Brillouin zone. The size of the X-ray beam was 50 μm × 75 μm. The X-ray angle of incidence was set to 7.23° at the nickel $L_{2,3}$-edges and 9.00° at the oxygen K-edge in order to access a similar q-range in reciprocal space and match the X-ray penetration depth with the sample thickness. The X-ray penetration depth is 50 nm at the oxygen K-edge and 60 nm at the nickel $L_{2,3}$-edges.[29] The transferred wave-vector q was 0.85 nm^{-1} at oxygen K-edge and 1.09 nm^{-1} at the nickel $L_{2,3}$-edges, which corresponds to 0.06 and 0.07 reciprocal lattice units, respectively. Therefore, the q-range lies close to the Γ point of the Brillouin zone. Fig. 2(b) shows a static X-ray reflectivity spectrum of NiO at the nickel $L_{2,3}$-edges, and the enlarged nickel L_2-edge is displayed in the inset.

The size of the optical pump beam was characterised both by imaging a virtual focus on an equivalent plane and by a knife-edge measurement. The size of the pump beam was (270 ± 10) μm × (257 ± 2) μm. The incident pump fluence was 60 mJ cm^{-2} and 75 mJ cm^{-2} due to the different angles of incidence at the nickel $L_{2,3}$-edges and the oxygen K-edge, respectively. Fig. 2(c) shows the pump-induced change of the X-ray reflectivity spectrum at the nickel L_3- and L_2-edges averaged over time delays between 400 fs and 2.4 ps. We will describe in the following Section 2.3 the data analysis procedure and involved uncertainties. The temporal resolution and pump–probe temporal overlap were characterised at regular intervals during the measurements by probing the transient optical reflectivity change of a GaAs single crystal induced by the FEL pulse.[30] The temporal resolution was found to be 300 ± 100 fs full-width half-maximum (FWHM), which mainly arises from the jitter between the pump and probe pulses. The temporal resolution evaluated from the GaAs single crystal is overestimated because the measurements at the oxygen K-edge reveal dynamics faster than 300 fs. Therefore, the GaAs single crystal was mainly used to monitor the temporal overlap, *i.e.*, time zero. During the experiments, the delay between the optical laser and the FEL was varied by a delay stage in the optical path. The delay was scanned across a 3 ps time window in steps of 100 fs.

2.3 Data analysis for the normalisation of the SASE FEL fluctuations and slow drifts

In order to quantitatively extract the laser-induced reflectivity changes from spectra acquired at SASE FELs in an unambiguous manner, one needs to account for the shot-to-shot fluctuations in spectral intensity caused by the SASE process. Since the measured reflectivity spectrum is proportional to the product of the FEL spectrum, the spectral reflectivity of the sample, and the efficiency of the diffraction grating, the direct measure of the laser-induced reflectivity change is the ratio of the measured spectra without and with laser excitation. To compensate for the slow drifts between the spectra acquired with and without laser excitation, spectral changes observed from the signal sample spot, which is illuminated by both the optical laser beam and the FEL beam, are normalised by the changes observed from the reference sample spot illuminated only by the FEL beam. The normalised ratio eventually gives the real pump–probe effect.

Fig. 3(a) and (b) show the X-ray reflectivity spectra at the nickel $L_{2,3}$-edges, respectively. The dashed black and the solid red lines represent the spectra acquired without and with laser excitation, respectively. From the comparison in Fig. 3(a) and (b), one can directly see that the change at the nickel L_3-edge is significantly smaller than at the nickel L_2-edge. To quantify the relative change induced by the optical laser, we calculated the ratio of the spectra with and without laser excitation. Fig. 3(c) and (d) show the ratio plots as a function of delay and photon energy at the nickel $L_{2,3}$-edges, respectively. In Fig. 3(c) and (d), it can be seen that the laser-induced change at the nickel L_2-edge is less noisy than that at the nickel L_3-edge. Therefore, in the results and discussion about the nickel L-edge measurements, we will focus mainly on the nickel L_2-edge.

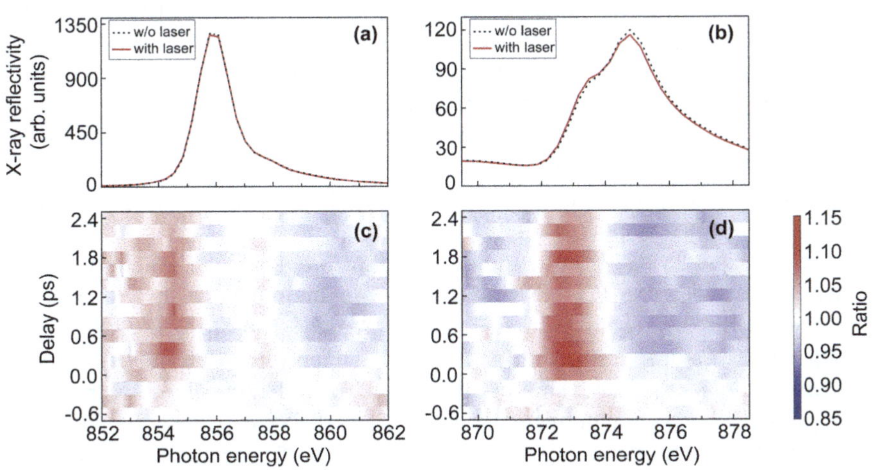

Fig. 3 (a and b) X-ray reflectivity spectra of NiO at the nickel $L_{2,3}$-edges. The dashed black and solid red lines represent the spectra acquired at time delays longer than 400 fs without and with laser excitation, respectively. (c and d) The ratio map as a function of delay and photon energy at the nickel $L_{2,3}$-edges.

3 Results and discussion

In this section we present time-resolved measurements at the O K-edge and Ni $L_{2,3}$-edges that probe O 1s–2p and Ni 2p–3d core–valence excitations, respectively. These orbitally resolved measurements enable us to extract information about the temporal evolution of the band-gap renormalisation and the modification of the AFM order, respectively. We will finally present a model that relates the changes of the upper band-edge, composed of O 2p and Ni 4s orbitals, to canted AFM moments that typically occur during magnon excitation. We start the section by describing the measured X-ray reflectivity spectra and relating them to XAS measurements *via* Kramers–Kronig analysis.

3.1 Comparison of X-ray reflectivity and XAS spectral line shape

Fig. 2(b) shows a Ni L-edge reflectivity spectrum obtained without laser excitation. The spectral line shape compares well with literature measurements on similar NiO/MgO(001) samples.[7,28,31] The dominant feature is a sharp peak at the Ni L_3-edge and a double-peak structure at the Ni L_2-edge. We note that the width of the sharp L_3-edge peak is limited by the energy resolution of the down-stream spectrometer grating. As mentioned above, the energy resolution of the spectrometer grating was 0.33 eV per pixel and the corresponding energy step size can be seen from the flat top of the L_3-peak. In the following we focus mainly on the L_2-edge since (i) the spectral features are broader and, therefore, contain a sufficient amount of energy steps, and (ii) XMLD measurements have mainly been done using the double-peak structure that gives rise to a characteristic derivative-like XMLD line shape.[7] Since XMLD measurements are typically performed in absorption, we performed ground-state reference XAS measurements monitoring the sample drain current for various grazing incidence angles between 7–26°, as shown in Fig. 4(a). In all the XAS spectra, one can see the crystal field induced peak splitting at the nickel L_2-edge whose energy positions compare well with the shoulder at 873.3 eV and the main peak at 874.5 eV observed in the reflectivity spectra. However, the respective peak intensities differ for XAS and reflectivity. While the XAS spectra correspond to the imaginary part of the X-ray dielectric constant, both imaginary and real parts of the X-ray dielectric constant contribute to the reflectivity. In principle, the real part of the dielectric constant can be computed from the imaginary part using Kramers–Kronig transformation, and consequently, the X-ray reflectivity spectra can be calculated using Fresnel's equation for a specific polarization.[31]

The XAS spectra in Fig. 4(a) show the typical line shape variations that are expected for L_2-edge XMLD of NiO/MgO(001).[7] As the incidence angle increases, the intensity of the second peak decreases gradually. The dip between the two peaks slightly shifts towards higher photon energy as the incident angle increases. There is also a red shift of the slope on the right side of the second peak along with increasing incident angles. To calculate the X-ray reflectivity spectrum, we chose the XAS spectrum at a grazing angle of 7° in Fig. 4(a), as it is the closest angle to the one in our time-resolved reflectivity measurement at FLASH. We follow ref. 31 to calculate the reflectivity spectrum. Briefly, (i) use the XAS spectrum as the input to calculate the extinction coefficient k, which is the imaginary part of the complex refractive index; (ii) compute the real part n of the complex

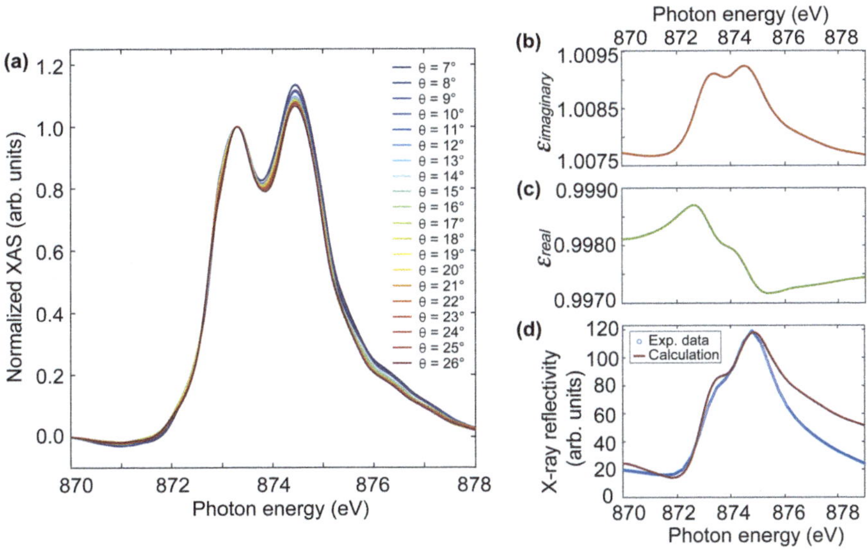

Fig. 4 (a) Static XAS spectra measured at the nickel L_2-edge at different grazing angles. Each XAS spectrum was normalised to the first peak at 873.3 eV. (b) The imaginary part of the dielectric constant, which was calculated using the XAS spectrum at a grazing angle of 7°. (c) The real part of the dielectric constant, which was calculated from the imaginary part in (b) using Kramers–Kronig transformation. (d) Comparison of the experimental X-ray reflectivity spectrum measured at FLASH (blue symbols) and the calculated reflectivity (red line).

refractive index from the imaginary part k using Kramers–Kronig transformation; (iii) calculate the reflectivity R using the Fresnel equation for p-polarized light, which is the polarisation in our reflectivity experiment. The Kramers–Kronig transformation is performed using Maclaurin's formula.[32,33] This results in similar line shapes of the X-ray dielectric constant compared to the reported values in ref. 31. However, their magnitude was rescaled to the literature values to take the different detection efficiency into account. Fig. 4(b) and (c) show the imaginary and real parts of the X-ray dielectric constant, respectively. From the comparison in Fig. 4(d), one can see reasonable agreement between the measured (blue symbols) and the calculated (red line) X-ray reflectivity spectra. The lower intensity in the measured reflectivity around 878 eV can occur due to lower flux from the FEL since this energy is close to the end of the scan range.

3.2 Temporal evolution of AFM order visualised at the nickel L_2-edge

The double-peak multiplet structure of the nickel L_2-edge has been used extensively to probe the NiO AFM order in thermal equilibrium. The interplay between AFM order, crystalline field effects and spin–orbit coupling gives rise to a characteristic XMLD line shape depending on the alignment of the X-ray polarisation and the AFM moment orientation.[7,16,17] In static measurements, the sign of the XMLD line shape can be reversed due to the angle between the X-ray polarisation and the crystal orientation[16] or the subtraction of the X-ray signal between different AFM domains.[17] Here we describe the first NiO XMLD measurement in the time-domain caused by pump-induced changes to the AFM order.

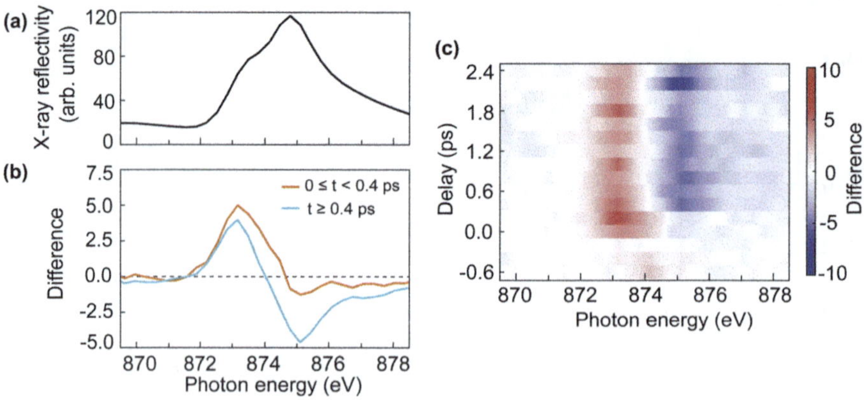

Fig. 5 (a) The static X-ray reflectivity spectrum at the nickel L_2-edge. (b) The difference spectra of the X-ray reflectivity with laser and without laser at transient time ($0 \leq t < 400$ fs, orange line) and at longer delays ($t \geq 400$ fs, cyan line). (c) The map of the difference spectra as a function of delay and photon energy.

Fig. 5(a) shows the static Ni L_2-edge X-ray reflectivity of NiO. The difference in reflectivity between laser on and laser off is shown in Fig. 5(b) for two time-delay regions, averaged for the first 400 fs (orange line) and between 400 fs and 2.4 ps (cyan line). The difference reflectivity spectra as a function of time delay and photon energy are shown as a colour map in Fig. 5(c). From Fig. 5(b), one can see that the spectral difference line shape evolves with time delay. After 400 fs the displayed difference spectra in Fig. 5(b) are essentially indistinguishable from what is expected for XMLD in thermal equilibrium.[7,16,17] Moreover, the observed positive–negative spectral difference line shape with increasing photon energy demonstrates that a reduction of AFM order takes place.[7] Interestingly, the spectral difference line shape is very different for the first 400 fs. It is characterised by a positive peak at lower photon energy and a negligible negative tail for the high photon-energy multiplet peak. This is an indication of a non-thermal state of the laser-excited spin system, which means the spin system has not reached a local thermal equilibrium. Around 400 fs the difference spectrum changes its line shape which remains unaltered afterwards, as can be clearly seen in the difference map of Fig. 5(c). This implies that the system reaches a local metastable equilibrium after 400 fs. It is conceivable that the XMLD line shape during the first 400 fs is also influenced by electronic excitations that could alter the effective local crystalline field experienced by the nickel ions. Crystalline field changes in monolayer thin NiO films have been reported to lead to changes in the L_2 multiplet peaks different from XMLD.[34]

Since the spin system reaches a local equilibrium after 400 fs, it is of interest to evaluate the temperature change of the spin system. It has been observed that the temperature dependence of the ratio between the two multiplets peaks at the nickel L_2-edge follows a Brillouin–Langevin function.[7] We observe a 7% change in the cyan XMLD spectrum of Fig. 5(b). This implies that we can use a linear approximation to evaluate the spin-temperature-change of the system. We convert the L_2 multiplet peak ratio measured in reflectivity to absorption as outlined in Section 3.1 and use the measured temperature dependence in ref. 7 to obtain an effective spin temperature increase of 65 ± 5 K, relative to the room temperature starting condition before optical excitation. The uncertainty is estimated from the photon statistics.

This observation of a significant reduction of the AFM order in NiO thin films is surprising especially since a negligible change was reported in a bulk sensitive magnetic scattering experiment performed under very similar excitation conditions.[6] In the following we discuss if optical excitation can indeed deposit enough energy to drive a comparable temperature increase. We evaluated the absorption coefficient of our NiO film at 800 nm to be around 1860 cm^{-1}. The refractive index of NiO at the excitation wavelength is 2.35 and the reflection of the sample is about 20% for p-polarized light.[35] Using the Beer–Lambert law and the magnetic specific heat of NiO,[36] we estimated the maximum temperature rise of the spin system to be ~150 K. This represents an upper limit since the magnetic specific heat increases rapidly as the temperature approaches T_N.[36] We alternatively estimated the lattice temperature rise base on a previous ultrafast electron diffraction measurement.[3] Assuming that at the slightly higher pump photon energy used in ref. 3 the optical absorption is of at least similar strength, we can extrapolate the measured lattice temperature in ref. 3 to ~85 K given our pump fluence. We note that in ref. 3 no optical absorption data for the measured sample was available and the available literature data provided an insufficient heat deposition leading the authors in ref. 3 to conclude that two-photon above-gap excitation was the actual driving force. However, our lower photon energy would require three-photon events for above-gap absorption. We can conclude that the enhanced optical absorption in thin NiO films is sufficient to explain our observed level of AFM order reduction.

3.3 Band gap renormalisation visualised at the oxygen K-edge

Fig. 6 displays the results of transient reflectivity spectra measured at the oxygen K-edge. We focus on the lowest photon energy feature as this is the one that defined the upper gap-edge. However, we corroborated in static measurements that our samples display the typical three-peak structure between 525 eV and 545 eV expected for NiO.[37–39] Fig. 6(a) shows the static X-ray reflectivity as a reference for the map of the difference spectra between laser on and laser off in Fig. 6(b). The pre-edge dip at 531.5 eV in Fig. 6(a) is related to the reflection geometry, which can also be seen from the calculated spectrum in Fig. 4(d). The evolution of the difference spectra shown in Fig. 6(b) is plotted using a temporal binning of 200 fs to improve the signal-to-noise ratio (SNR). The positive signal region centred around 532 eV in Fig. 6(b) is an indication of a red-shift (≈60 meV) of the oxygen K-edge. The amount of the red-shift will be discussed in detail in the next section in combination with first-principles modelling.

The most prominent feature visible in the difference map of Fig. 6(b) is clearly the red-shifted absorption edge, indicated by the red intensity increase near 532 eV photon energy and a broader blue decrease above the edge (533–534 eV). A gap reduction following sub-gap optical excitation has been observed in the charge-transfer insulator, La_2CuO_4,[40] and it is tempting to relate our observation to the same effect. However, strictly speaking we observe in the present experiment only the red-shifted upper gap-edge while a lower gap-edge shift would have to be assessed *e.g.* in complementary time-resolved photoemission experiments. The

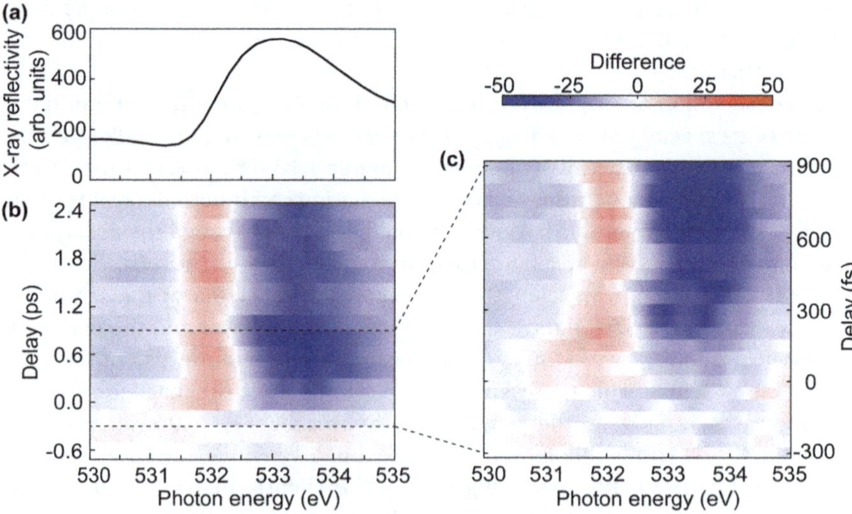

Fig. 6 (a) The static X-ray reflectivity spectrum at the oxygen K-edge. (b) The map of the difference spectra of the X-ray reflectivity with laser and without laser as a function of delay and photon energy. (c) The map of laser-induced difference in X-ray reflectivity as a function of delay and photon energy at a fixed undulator gap. The delay step size is 50 fs.

gap renormalisation in ref. 40 was assigned to the coupling of optical excitation to a bosonic field, although it remained unclear if this is composed of phonons or magnons. A possible magnetic origin of the red-shifted NiO upper gap-edge becomes apparent when we analyse the observed oscillations visible in Fig. 6(b).

The oscillation of the positive difference signal centred around 532 eV is compatible with a period of ~1 ps, which corresponds to a frequency of ~1 THz. This frequency is close to the eigenfrequency of the out-of-plane magnon mode, which has been extensively studied in both optical and THz spectroscopy measurements.[14,19,41–49] There are two magnon eigenmodes: in-plane and out-of-plane modes. For the in-plane mode, the AFM vector between two adjacent ferromagnetic planes is modulated along the $[1\bar{1}0]$ direction, which lies in the (111) plane. For the out-of-plane mode, the modulation of the AFM vector is along the [111] direction, which is perpendicular to the ferromagnetic plane.[20] Due to the large magnetic anisotropy along the easy axis (*i.e.* [111] direction), the out-of-plane mode has a much higher eigenfrequency than the in-plane mode. Both of these two magnon modes are excitable by ultrashort optical laser pulses and have been detected at around 0.14 THz (ref. 20 and 46) and 1.07 THz (ref. 20, 42, 45–47 and 49) for the in-plane mode and the out-of-plane mode, respectively. The mode also displays a temperature dependence with a frequency reduction as the Néel temperature is approached.[14,42,49] We note that also the observed phase of the oscillation in Fig. 6(b) matches that reported for the out-of-plane magnon mode.[20]

Fig. 6(c) shows the oxygen K-edge difference reflectivity signal up to 900 fs with 50 fs step size. The measurements were performed at a fixed undulator gap corresponding to a nominal photon energy of 533 eV. The displayed photon energy region therefore represents the full bandwidth of the undulator emission. The line shape of the X-ray reflectivity (not shown) is similar to the one displayed in Fig. 6(a), however, the best SNR is obtained for the undulator maximum (533 eV).

Fig. 6(c) shows, with a much better time resolution than Fig. 6(b), that the red-shift of the upper gap-edge is preceded by the appearance of transient states much further into the band-gap. The temporal duration of such transient mid-gap states seems to essentially be determined our temporal resolution, *i.e.* the optical and FEL pulse durations and the jitter between them. This reflects to some degree earlier measurements at pump photon energies off-resonant to any d–d transitions in the NiO band-gap.[9] It is conceivable that the appearance of transient mid-gap states are a consequence of the strong optical driving field. However, with off-resonance excitation no significant lasting band-gap renormalisation was observed.[9] The red-shift of the upper gap-edge is therefore likely a consequence of the energetic proximity of our excitation frequency to d–d transitions that may facilitate the coupling to spin excitations.[50]

3.4 A simple model for the band gap renormalisation in NiO

With the evidence of the laser-induced XMLD line shape changes at the nickel L_2-edge (Section 3.2) and the observation of the 1 THz magnon oscillation in the NiO upper gap-edge red-shift (Section 3.3), it is of interest to investigate how changes in the spin orientation influence the band gap in NiO. Therefore, we performed first-principles calculations of the NiO band structure with different canting angles of the Ni^{2+} spins, to mimic various spin temperatures. The super-exchange and correlation in the calculations were taken into account using local density approximation together with an on-site Hubbard U term in the Hamiltonian. Although dynamical correlations have been pointed out to be important for NiO, we adopt for simplicity here the LDA+U formalism, that is based on single determinant states. A detailed description of the theoretical modelling can be found in ref. 9. The canting angles at different temperatures were derived from the reduction of the magnetic moment of the Ni^{2+} spins.[15]

Fig. 7(a) shows the band structure calculated with Ni^{2+} spins oriented along the $[11\bar{2}]$ directions. The conduction band is composed of essentially dispersionless bands of Ni 3d character and dispersive bands consisting of hybridised O 2p and Ni 4s states. The upper gap-edge is composed of the latter and is marked by the light blue shaded rectangular in Fig. 7(a). The green and orange dots in Fig. 7(a) represent the experimental q values probed in our time-resolved X-ray reflectivity measurements. Fig. 7(b) shows the influence of canted magnetic moments on adjacent Ni atoms along the [111] direction on the dispersive band near the upper gap-edge. The chosen spin canting pattern is that of the 1 THz magnon mode[20] (see Fig. 7(c)). Initially degenerate bands are split in energy once the spins are canted. This splitting increases as the canting angle becomes larger. The band shifting downwards in energy at the Γ point [000] of the Brillouin zone reduces the NiO band gap as probed by the oxygen K-edge. Although the band structure in Fig. 7 was calculated for frozen magnetic moments, it is straightforward to see that for oscillating spins the upper gap-edge position would also be modulated by the magnon frequency.

In the following we quantify the redshift of the upper gap-edge and separate the coherent oscillatory part from an incoherent redshift. Fig. 8(a) compares the measured ground-state X-ray reflectivity spectrum (solid line) to the redshifted experimental spectra for 60 (blue dashed line) and 80 meV (red dashed line) shifts. In addition to the redshift, we also observe experimentally a 5% reduction

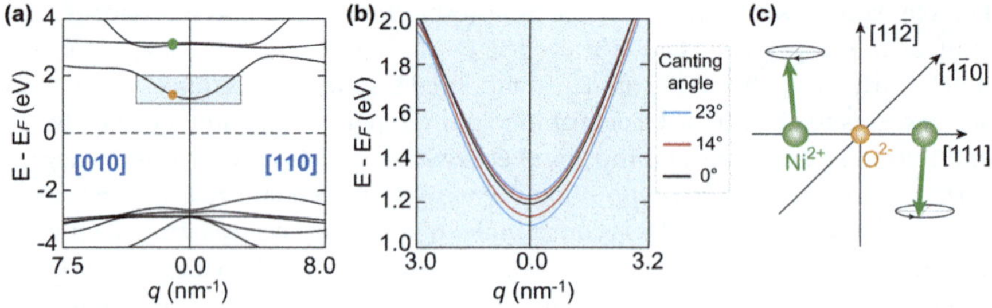

Fig. 7 (a) First-principles calculations of the NiO band structure along two high symmetry lines:[9] [010] and [110]. The green and orange dots along the [010] direction denote the experimental wavevectors that were probed in the time-resolved X-ray reflectivity measurement. (b) Zoom-in view of the band structure marked within the light blue shaded rectangle in (a). The band structure was calculated for different canting angles of the Ni^{2+} spins. (c) Illustration of the canting angle of the Ni^{2+} spins in the NiO unit cell that was employed in the calculations. The canting angle is defined as the angle between the Ni^{2+} spins and the plane that is spanned by [11$\bar{2}$] and [111] directions. The spin canting pattern is the same as the out-of-plane mode in ref. 20. The Ni^{2+} and the O^{2-} ions are shown with green and orange spheres, respectively.

to the spectral peak intensity likely related to the blueshift of some initially degenerate bands as seen in Fig. 7(b). This results in good agreement with the experiment for a redshift of 80 $\pm$ 10 meV at 0.6 ps time delay (see Fig. 8(b)) and 60 $\pm$ 10 meV at 1.0 ps time delay (see Fig. 8(c)). We note that for increasing photon energy the agreement becomes less good, possibly due to the influence of the additional O K-edge spectral peaks neglected here. From Fig. 6(b) one can see that the positive signal around 532 eV is composed of an overall intensity increase and

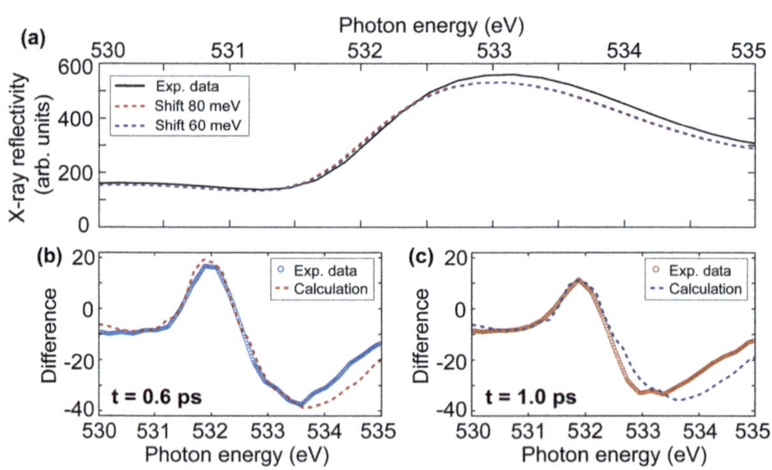

Fig. 8 (a) The black line represents the static X-ray reflectivity spectrum acquired at FLASH. The dashed red and blue lines represent the shifted spectra by 80 meV and 60 meV, respectively. (b) Comparison of the experimental difference spectrum at 0.6 ps after laser excitation and the calculated spectrum that is obtained by subtracting the black line from the dashed red line in (a). (c) Comparison between the experimental data at 1.0 ps after laser excitation and calculated difference spectrum; that is the subtraction of the black line and the dashed blue line in (a).

an oscillatory intensity variation. The overall intensity increase corresponds to an incoherent redshift of 70 meV. The amplitude of the coherent oscillatory redshift related to the 1 THz magnon mode is approximately 10 meV. Using the calculations in Fig. 7(b) we can estimate the corresponding magnon precession angle to $\sim 2.5°$. We note that the reduction of AFM due to such large spin precession amplitude would amount to about 28% of the spin temperature increase we observed in Section 3.2.

4 Conclusions

In conclusion, we performed time-resolved resonant X-ray reflectivity measurements in NiO following femtosecond 1.5 eV sub-gap excitation. The temporal evolution of the X-ray reflectivity spectra was monitored at the Ni L_2-edge and the O K-edge to probe the dynamics of the AFM order and the band gap renormalisation, respectively.

At the Ni L_2-edge, we observed laser-induced XMLD spectral line shape for time delays longer than ~ 400 fs after laser excitation. By comparing the transient XMLD spectra with temperature-dependent ground-state measurements, we extracted a spin temperature rise of 65 ± 5 K. The transient XMLD line shape demonstrates a significant reduction of the AFM order in NiO. The change of the spin temperature is consistent with what is expected from optical absorption measurements. At the O K-edge, we observed the red-shift of the upper band gap edge across the measured time delay range up to 2.4 ps. This upper gap-edge renormalisation is accompanied by a coherent oscillation with ~ 1 THz frequency. During the optical driving field we find evidence for the formation of transient mid-gap states that were also observed previously for optical frequencies off-resonant to d–d excitations.[9]

We finally discuss a ground-state model of coupling Ni spin precession to the electronic band-structure. Using first-principles calculations, we find that the O 2p dispersive states forming the upper gap-edge shift in energy as the Ni magnetic moments precess as given by the 1 THz magnon mode. This even allows us to model the observed O K-edge measurements.

The high-quality X-ray spectroscopy measurements presented here have become possible due to the implementation of a multiple X-ray beams experimental setup that enables us to significantly reduce the intensity and photon-energy fluctuations inherent to SASE FELs. This opens up new possibilities to investigate the ultrafast dynamics of electronic structure and the magnetic ordering phenomena across a wide range of materials.

Author contributions

M. B. and H. A. D. designed the experiments. X. W., R. Y. E., I. V., D. T., J. O. S., S. D., G. B., R.-P. W., M. K., F. K., W. B., M. B. and H. A. D. performed the time-resolved X-ray reflectivity experiment and on-site data analysis. R. K. and I. V. performed the optical spectroscopy experiment. C. S.-L. performed the static XAS experiment. X. W. conducted the post data analysis. O. G. conducted the theoretical calculations. X. W., R. Y. E., I. V., D. T., V. S., A. Y., O. G., F. P., O. E., M. B., and H. A. D. discussed the results. X. W. wrote the first draft of the manuscript. X. W. and H. A. D. revised the manuscript with inputs from all other authors.

Conflicts of interest

There are no conflicts to declare.

Acknowledgements

The authors acknowledge FLASH in Hamburg, Germany, for the provision of X-ray free-electron laser beamtime at beamline FL24 and thank the machine group and facility staff for their assistance during the beamtime. The beamtime was allocated for proposal F-20191553 EC. The research work at FLASH has been supported by the project CALIPSOplus under the Grant Agreement 730872 from the EU Framework Programme for Research and Innovation HORIZON 2020. X. W., D. T., V. S., and H. A. D. acknowledge support from the Swedish Research Council (VR), Grants 2017-06711 and 2018-04918. A. Y. acknowledges support from the Carl Trygger Foundation. O. G. acknowledges support from the Strategic Research Council (SSF) grant ICA16-0037 and the Swedish Research Council (VR) grant 2019-03901. The computations were enabled by resources provided by the Swedish National Infrastructure for Computing (SNIC), partially funded by the Swedish Research Council through grant agreement no. 2018-05973. R. Y. E., J. O. S., and M. B. were supported by the Helmholtz Association through grant VH-NG-1105. I. V. acknowledges support from the Slovene Ministry of Science (Raziskovalci-2.1-IJS-952005). R.-P. W. acknowledges support from the German Ministry of Education and Research, grant number 05K19GU2. We thank Prof. Arne Roos for his help with the optical spectroscopy measurements.

References

1 T. Jungwirth, X. Marti, P. Wadley and J. Wunderlich, *Nat. Nanotechnol.*, 2016, **11**, 231–241.

2 V. Baltz, A. Manchon, M. Tsoi, T. Moriyama, T. Ono and Y. Tserkovnyak, *Rev. Mod. Phys.*, 2018, **90**, 015005.

3 Y. W. Windsor, D. Zahn, R. Kamrla, J. Feldl, H. Seiler, C.-T. Chiang, M. Ramsteiner, W. Widdra, R. Ernstorfer and L. Rettig, *Phys. Rev. Lett.*, 2021, **126**, 147202.

4 M. A. Peck and M. A. Langell, *Chem. Mater.*, 2012, **24**, 4483–4490.

5 A. M. Balagurov, I. A. Bobrikov, S. V. Sumnikov, V. Y. Yushankhai, J. Grabis, A. Kuzmin, N. Mironova-Ulmane and I. Sildos, *Phys. Status Solidi B*, 2016, **253**, 1529–1536.

6 L. Huber, A. Ferrer, T. Kubacka, T. Huber, C. Dornes, T. Sato, K. Ogawa, K. Tono, T. Katayama, Y. Inubushi, M. Yabashi, Y. Tanaka, P. Beaud, M. Fiebig, V. Scagnoli, U. Staub and S. L. Johnson, *Phys. Rev. B*, 2015, **92**, 094304.

7 D. Alders, L. H. Tjeng, F. C. Voogt, T. Hibma, G. A. Sawatzky, C. T. Chen, J. Vogel, M. Sacchi and S. Iacobucci, *Phys. Rev. B*, 1998, **57**, 11623.

8 L.-C. Duda, T. Schmitt, M. Magnuson, J. Forsberg, A. Olsson, J. Nordgren, K. Okada and A. Kotani, *Phys. Rev. Lett.*, 2006, **96**, 067402.

9 O. Grånäs, I. Vaskivskyi, X. Wang, P. Thunström, S. Ghimire, R. Knut, J. Söderström, L. Kjellsson, D. Turenne, R. Y. Engel, M. Beye, J. Lu, A. H. Reid, W. Schlotter, G. Coslovich, M. Hoffmann, G. Kolesov,

C. Schüßler-Langeheine, A. Styervoyedov, N. Tancogne-Dejean, M. A. Sentef, D. A. Reis, A. Rubio, S. S. P. Parkin, O. Karis, J. Nordgren, J.-E. Rubensson, O. Eriksson and H. A. Dürr, 2020, arXiv:2008.11115.
10 D. Betto, Y. Y. Peng, S. B. Porter, G. Berti, A. Calloni, G. Ghiringhelli and N. B. Brookes, *Phys. Rev. B*, 2017, **96**, 020409(R).
11 G. Lefkidis and W. Hübner, *Phys. Rev. Lett.*, 2005, **95**, 077401.
12 A. K. Cheetham and D. A. O. Hope, *Phys. Rev. B*, 1983, **27**, 6964.
13 Y. Gao, Q. Sun, J. M. Yu, M. Motta, J. McClain, A. F. White, A. J. Minnich and G. K.-L. Chan, *Phys. Rev. B*, 2020, **101**, 165138.
14 J. Nishitani, K. Kozuki, T. Nagashima and M. Hangyo, *Appl. Phys. Lett.*, 2010, **96**, 221906.
15 N. Rinaldi-Montes, P. Gorria, D. Martínez-Blanco, A. B. Fuertes, I. Puente-Orench, L. Olivi and J. A. Blanco, *AIP Adv.*, 2016, **6**, 056104.
16 E. Arenholz, G. van der Laan, R. V. Chopdekar and Y. Suzuki, *Phys. Rev. Lett.*, 2007, **98**, 197201.
17 H. Ohldag, G. van der Laan and E. Arenholz, *Phys. Rev. B*, 2009, **79**, 052403.
18 J. Xu, C. Zhou, M. Jia, D. Shi, C. Liu, H. Chen, G. Chen, G. Zhang, Y. Liang, J. Li, W. Zhang and Y. Wu, *Phys. Rev. B*, 2019, **100**, 134413.
19 T. Satoh, S.-J. Cho, R. Iida, T. Shimura, K. Kuroda, H. Ueda, Y. Ueda, B. A. Ivanov, F. Nori and M. Fiebig, *Phys. Rev. Lett.*, 2010, **105**, 077402.
20 C. Tzschaschel, K. Otani, R. Iida, T. Shimura, H. Ueda, S. Günther, M. Fiebig and T. Satoh, *Phys. Rev. B*, 2017, **95**, 174407.
21 T. Kachel, F. Eggenstein and R. Follath, *J. Synchrotron Radiat.*, 2015, **22**, 1301–1305.
22 R. Newman and R. M. Chrenko, *Phys. Rev.*, 1959, **114**, 1507.
23 I. G. Austin, B. D. Clay and C. E. Turner, *J. Phys. C: Solid State Phys.*, 1968, **1**, 1418.
24 G. R. Rossman, R. D. Shannon and R. K. Waring, *J. Solid State Chem.*, 1981, **39**, 277–287.
25 T. Tsuboi and W. Kleemann, *J. Phys.: Condens. Matter*, 1994, **6**, 8625.
26 V. I. Sokolov, V. A. Pustovarov, V. N. Churmanov, V. Y. Ivanov, N. B. Gruzdev, P. S. Sokolov, A. N. Baranov and A. S. Moskvin, *Phys. Rev. B*, 2012, **86**, 115128.
27 M. Beye, R. Y. Engel, J. O. Schunck, S. Dziarzhytski, G. Brenner and P. S. Miedema, *J. Phys.: Condens. Matter*, 2018, **31**, 014003.
28 R. Y. Engel, P. S. Miedema, D. Turenne, I. Vaskivskyi, G. Brenner, S. Dziarzhytski, M. Kuhlmann, J. O. Schunck, F. Döring, A. Styervoyedov, S. S. P. Parkin, C. David, C. Schüßler-Langeheine, H. A. Dürr and M. Beye, *Appl. Sci.*, 2020, **10**, 6947.
29 B. L. Henke, E. M. Gullikson and J. C. Davis, *At. Data Nucl. Data Tables*, 1993, **54**, 181–342.
30 C. Gahl, A. Azima, M. Beye, M. Deppe, K. Döbrich, U. Hasslinger, F. Hennies, A. Melnikov, M. Nagasono, A. Pietzsch, M. Wolf, W. Wurth and A. Föhlisch, *Nat. Photonics*, 2008, **2**, 165–169.
31 D. Alders, T. Hibma, G. A. Sawatzky, K. C. Cheung, G. E. van Dorssen, M. D. Roper, H. A. Padmore, G. van der Laan, J. Vogel and M. Sacchi, *J. Appl. Phys.*, 1997, **82**, 3120–3124.
32 K. Ohta and H. Ishida, *Appl. Spectrosc.*, 1988, **42**, 952–957.
33 T. G. Mayerhöfer and J. Popp, *Spectrochim. Acta, Part A*, 2019, **213**, 391–396.

34 M. W. Haverkort, S. I. Csiszar, Z. Hu, S. Altieri, A. Tanaka, H. H. Hsieh, H.-J. Lin, C. T. Chen, T. Hibma and L. H. Tjeng, *Phys. Rev. B*, 2004, **69**, 020408(R).

35 M. M. El-Nahass, M. Emam-Ismail and M. El-Hagary, *J. Alloys Compd.*, 2015, **646**, 937–945.

36 R. J. Radwanski and Z. Ropka, *Acta Phys. Pol., A*, 2008, **114**, 213–218.

37 H. Y. Peng, Y. F. Li, W. N. Lin, Y. Z. Wang, X. Y. Gao and T. Wu, *Sci. Rep.*, 2012, **2**, 442.

38 R. J. O. Mossanek, G. Domínguez-Cañizares, A. Gutiérrez, M. Abbate, D. Díaz-Fernández and L. Soriano, *J. Phys.: Condens. Matter*, 2013, **25**, 495506.

39 J. Y. Zhang, W. W. Li, R. L. Z. Hoye, J. L. MacManus-Driscoll, M. Budde, O. Bierwagen, L. Wang, Y. Du, M. J. Wahila, L. F. J. Piper, T.-L. Lee, H. J. Edwards, V. R. Dhanak and K. H. L. Zhang, *J. Mater. Chem. C*, 2018, **6**, 2275–2282.

40 F. Novelli, G. De Filippis, V. Cataudella, M. Esposito, I. Vergara, F. Cilento, E. Sindici, A. Amaricci, C. Giannetti, D. Prabhakaran, S. Wall, A. Perucchi, S. Dal Conte, G. Cerullo, M. Capone, A. Mishchenko, M. Grüninger, N. Nagaosa, F. Parmigiani and D. Fausti, *Nat. Commun.*, 2014, **5**, 5112.

41 S. Kovalev, Z. Wang, J.-C. Deinert, N. Awari, M. Chen, B. Green, S. Germanskiy, T. V. A. G. de Oliveira, J. S. Lee, A. Deac, D. Turchinovich, N. Stojanovic, S. Eisebitt, I. Radu, S. Bonetti, T. Kampfrath and M. Gensch, *J. Phys. D: Appl. Phys.*, 2018, **51**, 114007.

42 T. Moriyama, K. Hayashi, K. Yamada, M. Shima, Y. Ohya and T. Ono, *Phys. Rev. Mater.*, 2020, **4**, 074402.

43 T. Kampfrath, A. Sell, G. Klatt, A. Pashkin, S. Mährlein, T. Dekorsy, M. Wolf, M. Fiebig, A. Leitenstorfer and R. Huber, *Nat. Photonics*, 2011, **5**, 31–34.

44 J. Nishitani, T. Nagashima and M. Hangyo, *Phys. Rev. B*, 2012, **85**, 174439.

45 S. Baierl, J. H. Mentink, M. Hohenleutner, L. Braun, T.-M. Do, C. Lange, A. Sell, M. Fiebig, G. Woltersdorf, T. Kampfrath and R. Huber, *Phys. Rev. Lett.*, 2016, **117**, 197201.

46 T. Kohmoto, T. Moriyasu, S. Wakabayashi, H. Jinn, M. Takahara and K. Kakita, *J. Infrared, Millimeter, Terahertz Waves*, 2018, **39**, 77–92.

47 Z. Wang, S. Kovalev, N. Awari, M. Chen, S. Germanskiy, B. Green, J.-C. Deinert, T. Kampfrath, J. Milano and M. Gensch, *Appl. Phys. Lett.*, 2018, **112**, 252404.

48 O. V. Chefonov, A. V. Ovchinnikov, C. P. Hauri and M. B. Agranat, *Opt. Express*, 2019, **27**, 27273–27281.

49 T. Moriyama, K. Hayashi, K. Yamada, M. Shima, Y. Ohya, Y. Tserkovnyak and T. Ono, *Phys. Rev. B*, 2020, **101**, 060402(R).

50 R. Gómez-Abal, O. Ney, K. Satitkovitchai and W. Hübner, *Phys. Rev. Lett.*, 2004, **92**, 227402.

Faraday Discussions

ROYAL SOCIETY OF CHEMISTRY

PAPER

Attosecond electronic delay response in dielectric materials

Husain Alqattan, † Dandan Hui, † Mohamed Sennary and Mohammed Th. Hassan *

Received 6th January 2022, Accepted 7th February 2022

DOI: 10.1039/d2fd00002d

The advancement in the attosecond field and the generation of XUV attosecond pulses has enabled the study of electron dynamics in the solid-state by high harmonic generation spectroscopy. Here, we introduce new all-optical attosecond metrology to study the light-field induced electron dynamics in dielectric systems. This new methodology is based on the phase transition of a dielectric material due to its interaction with a strong light field. Hence, the charge carriers undergo an inter- and intraband transition, causing a modification in the electronic structure, dielectric constant, and optical properties of the dielectric system. Consequently, the dielectric material experiences an adiabatic semi-metal phase transition due to the strong polarizability. Therefore, the reflectivity of the dielectric system changes, following the shape of the pump field. Accordingly, the time-resolved reflectivity change measurement provides direct access to the phase transition and the related electronic dynamics of the system in real-time. In the reported experiment, a strong light field (pump pulse) induces the phase transition and modifies the fused silica sample's reflectivity, which is probed by another weak light field (probe pulse). The reflected probe beam spectrum is acquired as a function of the time delay between the pump and probe pulses. This measurement shows that the real-time phase transition dynamic (and reflectivity change) follows the pump field shape. Moreover, the reflectivity measurements have been recorded at different pump field strengths performed under the same conditions. The reflectivity trace shows a retardation phase delay at a higher driver field strength. The delay response—retrieved from the recorded reflectivity traces—is in the order of a few hundred attoseconds. In addition, the results show that the delay response monotonically increases as the trigger field escalates. Furthermore, the reflectivity measurements have been acquired for another dielectric system (CaF_2). The electronic delay response shows the same linear behavior increase as that of the SiO_2 system. This work establishes a universal new attosecond metrology for measuring the electron dynamics and delay response in different materials.

Department of Physics, University of Arizona, Tucson, AZ 85721, USA. E-mail: mohammedhassan@email.arizona.edu

† These authors contributed equally to this work.

Introduction

Previously, it was thought that chemical processes were immeasurably fast. However, using ultrashort laser pulses in femtosecond time-resolved spectroscopy measurements provided access to the structural dynamics in chemical reactions.[1] This revolutionary advance provided information about the transition states in real-time and gave rise to the ultrafast science and femtochemistry fields.[2] Later, different ultrafast laser spectroscopic techniques were developed to trace the atomic motion of matter. However, capturing the electronic motion on a femtosecond or less timescale remained beyond reach due to the temporal resolution limit.

Studying electron dynamics in their native time scale is key to understanding and controlling the quantum mechanical underpinnings of the physical and chemical dynamics of matter.[3–5] In the last two decades, the generation of attosecond extreme ultraviolet (XUV) pulse spectroscopy[6–12]—based on a high harmonic generation process—has permitted the real-time observation of electron motion dynamics in atoms, molecules, and the solid-state.[4,5,10–15] However, the development of attosecond XUV spectroscopy allowed us to explore the electronic response of atoms and solids in the ionized state only. Recently, the development of synthesized attosecond laser pulses[16] enabled the determination of the nonlinear response of bound electron dynamics in the atomic system, which was determined to be in the order of a hundred attoseconds.[16–18] However, the electronic delay response of bound electrons in solid states has not been reported yet.

Here, we study the excited carrier dynamics and the related phase transition in dielectric systems to determine the relative electronic delay response in the presence of a strong light field. Previously, the underlying physics of strong-field interactions with dielectric solid-state systems and the associated electron and optical reflectivity dynamics were probed by XUV spectroscopy using a high-energy photon. It was demonstrated that the tracing of the optical reflectivity modulation would provide direct insights into the electronic response and the phase transition dynamics of the dielectric material.[19] The strong field-induced electron dynamics in dielectrics have been studied intensively theoretically and experimentally.[6–9,19–25] Based on these studies, the charge carriers are first excited from the valence band to the conduction band *via* multiphoton excitation through the virtual state.[21] Then, the excited electrons in the conduction band move in the reciprocal space by acquiring a time-dependent momentum from the driving field.[19–23,25,26] Thus, the electrons are accelerated and decelerated following the shape of the pump field's vector potential, altering the electronic structure and the optical properties of the system.[19–28]

In this work, we study the electronic delay response and its relation to the intensity of the external triggering field in different dielectric systems. Our study exploited the attosecond all-optical reflectivity modulation (atto-ARM) methodology[19] to measure the reflectivity modulation of the SiO_2 (band gap = 9 eV) dielectric system triggered by the strong field interaction of few-cycle visible laser pulses. From these measurements, we obtain the total reflectivity modulation (TRM) trace of the SiO_2 as a function of the time delay between the pump and probe pulses, which carries the signature of the transient phase-transition dynamics of the dielectric system following the vector potential of the pump field. Moreover, we measured the TRM traces at different field intensities and determined the relative phase delay retardation between these traces due to the electronic delay response of the SiO_2. Furthermore, we determined the relative electronic delay response of calcium fluoride (CaF_2), which has a higher bandgap (12 eV), using the same methodology. Finally, the presented measurements allowed us to compare the electronic dynamic behaviors in the two different dielectric materials.

Method

Experimental setup

In this work, we conducted two experiments using the atto-ARM[19] setup arrangement, illustrated in Fig. 1a, to study the transient phase transition dynamics, the reflectivity modulation in real-time, and the electronic response of two dielectric systems; SiO_2 and CaF_2. The experiment setup can be described as follows: high-power (1 mJ) multi-cycle pulses are generated from an optical parametric chirped-pulse amplification (OPCPA) laser system. These multi-cycle pulses are nonlinearly propagated inside a hollow-core fiber (HCF) filled with pressurized neon gas (2 bar). A powerful broadband spectrum that spans the visible and near-infrared spectral regions is generated. A chirped mirror compressor is used to compress the spectrum region from 500–700 nm (see Fig. 1b) to its Fourier-limit duration (6.5 fs). The carrier-envelope-phase (CEP) of these pulses is passively locked with an estimated phase jittering of ~100 mrad. Then, the output linearly polarized (p-pol) beam from the compressor is divided by a two-hole mask into two beams. The first beam that emerges from the first hole (diameter = 3 mm) has a high intensity (pump beam). In contrast, the second beam (probe beam) has a lower intensity, since it is emerging from a hole with a smaller diameter (1 mm). The second beam intensity is ~3.3% of the first beam intensity. The two beams are co-linearly propagated to two identical D-shape focusing mirrors (f = 100 mm) and then focused by a small incident angle on a 100 μm SiO_2 sample. One of the two D-shape mirrors is carried on a high-precise piezo positioning stage. This stage controls the arriving time (on the sample) and the delay between the two beams with attosecond resolution. The input beam intensity is controlled by a round continuously variable metallic neutral density filter introduced in the beam path (see Fig. 1a), by changing the rotatory angle of the filter so the propagation length inside the filter is constant.

The pump beam triggers the transient phase transition and the reflectivity of the SiO_2 system. The reflectivity modulation can be traced by recording the spectrum of the reflected probe beam after filtering out the pump beam. Hence, the reflected probe beam is focused into an optical spectrometer. The spectrum is recorded as a function of the time delay between the pump and probe pulses, which is controlled by the delay stage with a step size of 100 as. The intensity of the reflected spectrum at each instance of time is integrated to obtain the TRM trace. The TRM represents the transient phase transition of the material in real-time. Then, to study the electronic delay response in SiO_2, the TRMs are recorded at different pump beam intensity levels, ranging from $\sim 0.376 \times 10^{13}$ to 1.315×10^{13} W cm^{-2}. At higher intensities, damage on the sample is observed, and the intensity modulation signal disappears. Note that the intensity level is controlled

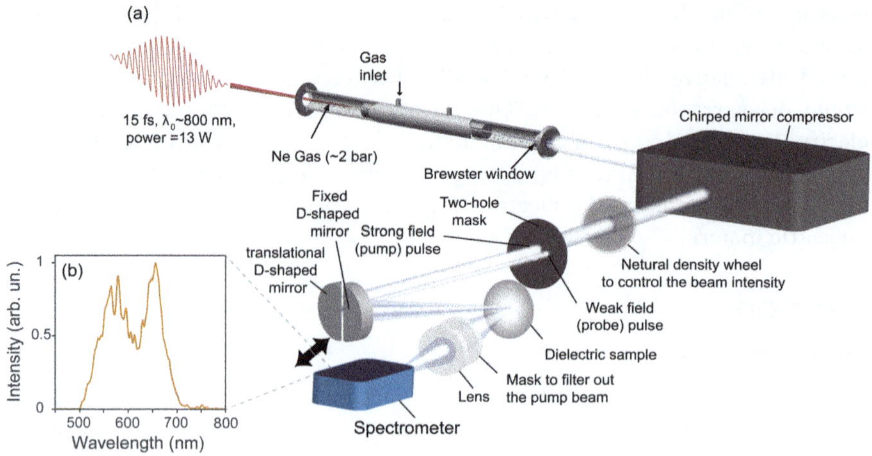

Fig. 1 (a) Attosecond all-optical reflectivity modulation (Atto-ARM) experimental setup to measure the reflectivity change in dielectric systems. The multi-cycle pulse spectrum is broadened by nonlinear propagation in a hollow-core fiber (HCF). The broad spectrum is compressed inside a chirped mirror compressor to 6.5 fs pulses. The output beam from the compressor is divided into the intense beam (pump) and the low power beam (probe) by a two-hole mask. A neutral density wheel filter is implemented in the beam path to control the input beam intensity. Two D-shape mirrors focus the two pump and probe beams onto the dielectric sample. One of the D-shape mirrors is attached to a delay stage which moves with a nanometer resolution. The reflected pump beam is filtered out, and only the probe beam spectrum is focused and measured by an optical spectrometer. (b) The normalized measured probe spectrum, which is identical to the pump beam spectrum (spanning from 500 to 700 nm).

by rotating the metallic filter in the input beam path, so the ratio between the pump and probe beams remains the same at all measurements. Also, no change of the spectral phase and the CEP of the input beam pulse occurred during the measurements at different input beam intensities. In the second experiment, we replaced the SiO_2 sample with a 200 μm CaF_2 sample and acquired the TRM of CaF_2 at different pump beam intensities, starting from $\sim 0.376 \times 10^{13}$ to 1×10^{13} W cm^{-2}; the sample is damaged at higher intensities. These measurements enabled the study of the relative electronic delay response of CaF_2 and to compare these results with the electronic delay response of the SiO_2 system.

Results and discussion

The transient photoinduced phase transition of fused silica in a strong field led to a change in the material's properties and its reflectivity, following the waveform of the pump field in real-time. Accordingly, the measurement of the TRM trace of fused silica provides access to the transient phase transition dynamics of SiO_2 in the strong field, which occur on the sub-femtosecond time scale. Moreover, the TRM represents the vector potential of the driver field. Previously, we exploited the transient phase transition dynamics in SiO_2 to demonstrate the all-optical field sampling methodology.[19] However, an important open question remains: are the induced transient phase transition dynamics of the dielectric system in a strong field material-dependent? To answer this question, we measured the

TRM of SiO_2 and CaF_2 and present these in Fig. 2a and b, respectively. The retrieved electric field, by calculating the derivative of the TRM of SiO_2 (Fig. 2a), is shown as a solid red line in Fig. 2c. The dashed blue line in Fig. 2c represents the electric field calculated from the derivative of the measured CaF_2 TRM (Fig. 2b). Remarkably, these results show that the transient phase transition dynamics are field-driven and independent of the dielectric system.

Correspondingly, the temporal profiles of the pulse, retrieved from the TRM in Fig. 2a and b, are shown by the dashed red and dashed blue lines, respectively, in Fig. 2d. In contrast, the measured temporal profile of the pump pulse by the

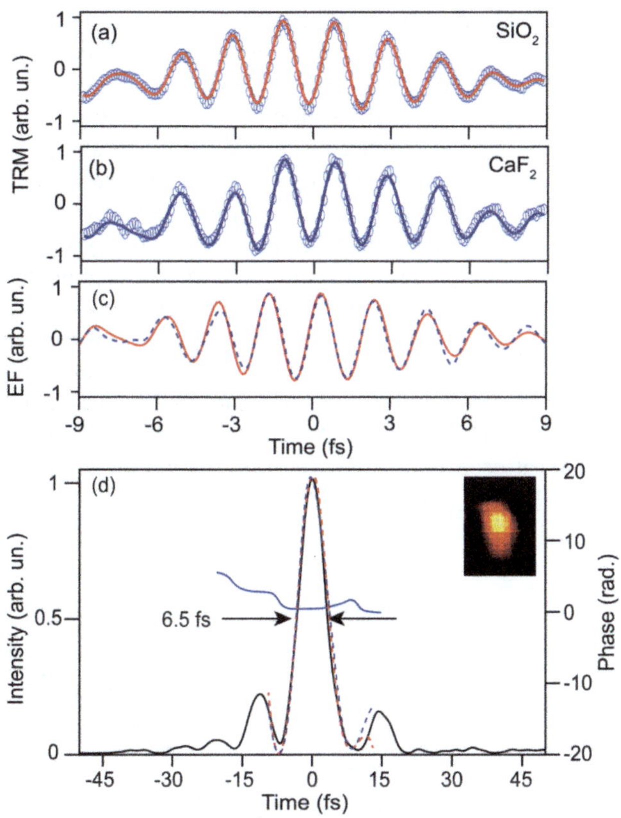

Fig. 2 (a) The measured total reflectivity modulation (TRM) trace of SiO_2. This trace is obtained by measuring the spectrum of the reflected probe beam from the front surface of SiO_2 as a function of the time delay between the pump and probe beams. Each point in the trace represents the integration of the probe spectrum at the corresponding time delay. (b) The measured TRM trace of CaF_2. (c) The electric fields calculated from the TRM trace in (a and b) are shown as solid red and dashed blue lines, respectively. The two calculated electric fields show significant agreement, indicating that the transient phase transition depends solely on the driver field. (d) The solid black line shows the temporal profile of the driver field pulse measured by TG-FROG. The full width at half maximum (FWHM) pulse duration is 6.5 fs. The TG-FROG trace is shown in the inset. The measured spectral phase is shown by the solid blue line. The temporal profiles calculated from the retrieved electric fields shown in (c) are depicted by the dashed red line (from the TRM of SiO_2) and the dashed blue line (from the TRM of CaF_2).

conventional transient grating frequency-resolved optical gating (TG-FROG) is depicted by the solid black line in Fig. 2d. The temporal profiles from the measured TRMs of SiO_2 and CaF_2 are in excellent agreement with the profile measured by TG-FROG, confirming the all-optical field sampling methodology, based on the transient phase transition in SiO_2, and shows that this methodology is universal.[19]

Nevertheless, the TRM measurements not only allowed the study of the phase transition dynamics but also gave access to the related electron dynamics in the dielectric system. In the strong light field, the carriers are excited to the conduction band. After excitation, the field-driven intraband transition of the electrons in the conduction band occurs and causes the change in the optical properties and the reflectivity of the dielectric system. Therefore, the measurements of TRM traces at different driver field intensities provide insights into the electronic delay response of the dielectric system to the external light field.

Accordingly, we measured the TRM traces of SiO_2 under the exact same conditions at different field intensities ranging from $\sim$0.376 to 1.315 $\times$ 10^{13} W cm^{-2}. The measured traces were normalized and plotted, as shown in Fig. 3a. As the intensity increases, the measured 11 TRM traces show a phase shift in the order of a few hundred attoseconds. The relative phase delay ($\Delta\tau$) is determined as the delay between the highest reflectivity peaks (marked by the red arrow in Fig. 3a) at a certain intensity and the maximum of the corresponding peak at the lowest intensity ($I \sim 0.376 \times 10^{13}$ W cm^{-2}). The retrieved phase delay is plotted as a function of the driver field intensity in Fig. 3b. These measurements indicate that the electronic delay response increases linearly as the pump field intensity increases.[27] Moreover, from the slope of the linear relation in Fig. 3b, we can estimate the delay response at the $\sim$0.376 $\times$ 10^{13} W cm^{-2} intensity to be roughly in the order of <800 attoseconds, under the assumption that the linear relation is maintained at lower field intensities. The increase of the delay response at

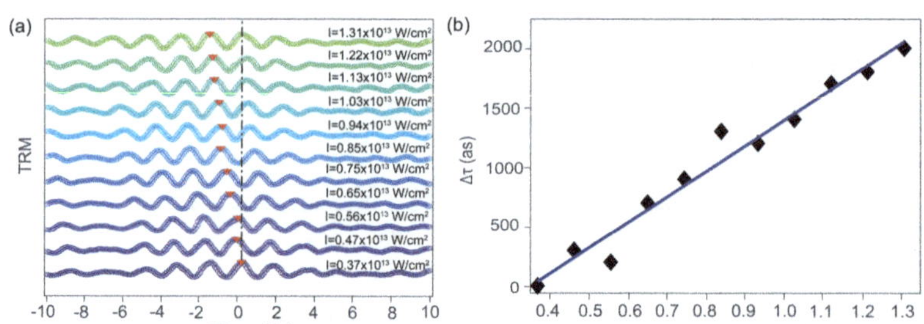

Fig. 3 (a) The measured TRM traces of SiO_2 at 11 different driver field intensities, ranging from 0.376 $\times$ 10^{13} to 1.31 $\times$ 10^{13} W cm^{-2}. The traces show a phase delay which increases at a higher pump field intensity. The maxima of the highest (same) peaks of the measured TRM traces at different intensities are indicated with red arrows. The relative phase delay ($\Delta\tau$) between the TRM traces measured at each intensity is calculated by determining the delay between the maximum of the highest peak at this certain intensity and the maximum of the same peak at $I = 0.376 \times 10^{13}$ W cm^{-2}. The black dashed line is a guide for the eye to show the relative delay increase. (b) The measured relative delay ($\Delta\tau$) as a function of the pump field intensity is shown in black points. The relative delay increases linearly as the pump intensity increases. The solid blue line shows the linear fitting of the measured relative delay.

a higher pump field may be attributed to the increase in the excitation carrier density, the AC Stark effect,[29] and the system's polarizability.[16,17,27]

Furthermore, we measured 8 TRM traces of CaF_2 at different pump field intensities (the results are shown in Fig. 4a). Also, the CaF_2 shows a phase retardation increase as the field intensity increases with a linear behavior, as shown in Fig. 4b. From the slope, we can estimate the delay response at $I \sim 0.376 \times 10^{13}$ W cm^{-2} intensity to be in the order of ~650 attoseconds (assuming that the delay response increase remains linear at lower intensity values), which is lower than the electronic delay response (~800 attoseconds) at the same intensity of SiO_2. This may be attributed to the difference in the bandgap value (~12 eV for CaF_2 and ~9 eV for SiO_2), the electronic structure, and the difference between the excited carrier densities of both systems at the same field intensity ($I \sim 0.376 \times$

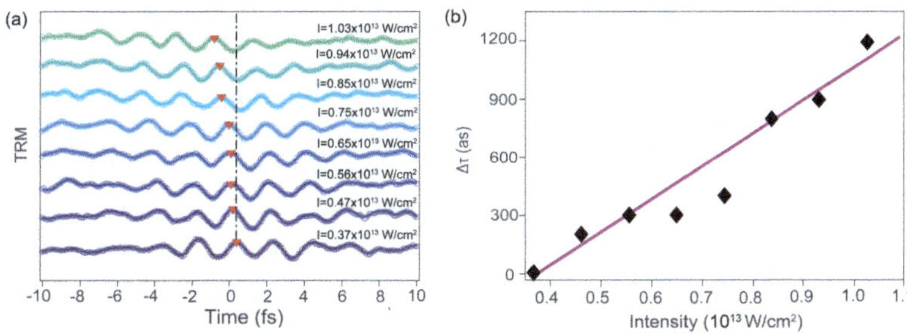

Fig. 4 (a) The measured TRM traces of CaF_2 at 8 different driver field intensities ranging from 0.376×10^{13} to 1.03×10^{13} W cm^{-2}. The traces show a phase delay which increases at higher pump field intensity. (b) The determined relative delay ($\Delta\tau$), between the maximum peak measured at the certain intensity with respect to the same peak at $I = 0.376 \times 10^{13}$ W cm^{-2}, as a function of the pump field intensity is presented by black points. The relative delay increases linearly as the pump intensity increases similar to SiO_2. The solid magenta line shows the linear fitting of the measured relative delay.

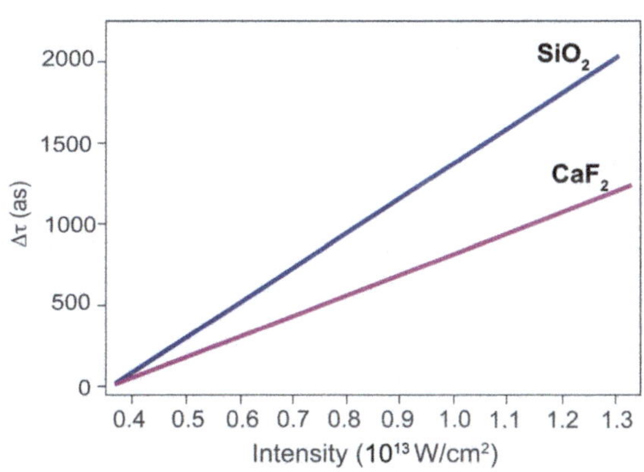

Fig. 5 The comparison between the linear behavior of the relative electronic delay response as a function of the pump field intensity measured for SiO_2 (in solid blue line) and CaF_2 (in solid magenta line). The two linear lines show different slops, indicating that the rise in the electronic delay response in SiO_2 at higher intensity is more than the increase in the electronic delay response in CaF_2.

10^{13} W cm^{-2}), which are theoretically calculated to be 6.58×10^{-10} nm^{-3} for CaF_2 and 7.3×10^{-7} nm^{-3} for SiO_2.

To compare the behavior of the electronic delay repose of SiO_2 and CaF_2, we plot the fitted lines in Fig. 3b and 4b together in the same plot in Fig. 5. These results show that the electronic delay response of SiO_2 increases more at the same field intensity than that of CaF_2. Accordingly, we can conclude that the electronic delay response behavior depends on the external field intensity, not on the type of the dielectric system (both systems show linear trends). However, the electronic delay response value and the increasing increment as the driver field increases depend on the dielectric's bandgap and electronic structure.

Conclusions

We studied the transient phase transition dynamics of the SiO_2 and CaF_2 dielectric systems by tracing the optical reflectivity of the two systems in a strong field. The reported results confirm that the phase transition dynamics depend solely on the shape of the driver field waveform and not the type of the material. Moreover, we determined the relative electronic delay responses of the two dielectric systems under different pump field intensities. The results show that the electronic delay response in the two dielectric systems is linearly increased by increasing the driver field intensity. However, the value of the electronic delay response and its increase at a higher intensity pump field depends on the dielectric system and its electronic structure.

Author contributions

H. A., D. H., and M. S. conducted the experiments and analyzed the data. M. H. conceived of, supervised, and directed the study. All authors discussed the results and their interpretation and wrote the manuscript.

Conflicts of interest

There are no conflicts of interest to declare.

Acknowledgements

This project was funded by the Air Force Office of Scientific Research under award number FA9550-19-1-0025 and partially funded by the Gordon and Betty Moore Foundation Grant GBMF7938. We are grateful for the W. M. Keck Foundation support. We thank Enam Chowdhury and Simin Zhang for the constructive discussion.

References

1 A. H. Zewail, *J. Phys. Chem. A*, 2000, **104**, 5660–5694.
2 A. H. Zewail, *Angew. Chem., Int. Ed.*, 2000, **39**, 2586–2631.
3 M. Nisoli and G. Sansone, *Prog. Quantum Electron.*, 2009, **33**, 17–59.

4 F. Calegari, D. Ayuso, A. Trabattoni, L. Belshaw, S. De Camillis, S. Anumula, F. Frassetto, L. Poletto, A. Palacios, P. Decleva, J. B. Greenwood, F. Martín and M. Nisoli, *Science*, 2014, **346**, 336–339.

5 F. Calegari, G. Sansone, S. Stagira, C. Vozzi and M. Nisoli, *J. Phys. B: At., Mol. Opt. Phys.*, 2016, **49**, 062001.

6 A. Schiffrin, T. Paasch-Colberg, N. Karpowicz, V. Apalkov, D. Gerster, S. Mühlbrandt, M. Korbman, J. Reichert, M. Schultze and S. Holzner, *Nature*, 2013, **493**, 70–74.

7 M. Schultze, E. M. Bothschafter, A. Sommer, S. Holzner, W. Schweinberger, M. Fiess, M. Hofstetter, R. Kienberger, V. Apalkov and V. S. Yakovlev, *Nature*, 2013, **493**, 75–78.

8 T. Paasch-Colberg, S. Y. Kruchinin, Ö. Sağlam, S. Kapser, S. Cabrini, S. Muehlbrandt, J. Reichert, J. V. Barth, R. Ernstorfer and R. Kienberger, *Optica*, 2016, **3**, 1358–1361.

9 T. T. Luu, M. Garg, S. Y. Kruchinin, A. Moulet, M. T. Hassan and E. Goulielmakis, *Nature*, 2015, **521**, 498–502.

10 F. Krausz and M. Ivanov, *Rev. Mod. Phys.*, 2009, **81**, 163.

11 F. Krausz, *Electrons in Motion: Attosecond Physics Explores Fastest Dynamics*, World Scientific Publishing Company Pte. Limited, 2019.

12 P. Corkum and F. Krausz, *Nat. Phys.*, 2007, **3**, 381–387.

13 G. Sansone, F. Kelkensberg, J. Pérez-Torres, F. Morales, M. F. Kling, W. Siu, O. Ghafur, P. Johnsson, M. Swoboda and E. Benedetti, *Nature*, 2010, **465**, 763–766.

14 V. S. Yakovlev, S. Y. Kruchinin, T. Paasch-Colberg, M. I. Stockman and F. Krausz, in *Ultrafast Dynamics Driven by Intense Light Pulses: From Atoms to Solids, from Lasers to Intense X-rays*, ed. M. Kitzler and S. Gräfe, Springer International Publishing, Cham, 2016, pp. 295–315, DOI: 10.1007/978-3-319-20173-3_12.

15 V. S. Yakovlev, M. I. Stockman, F. Krausz and P. Baum, *Sci. Rep.*, 2015, **5**, 14581.

16 M. T. Hassan, T. T. Luu, A. Moulet, O. Raskazovskaya, P. Zhokhov, M. Garg, N. Karpowicz, A. M. Zheltikov, V. Pervak, F. Krausz and E. Goulielmakis, *Nature*, 2016, **530**, 66–70.

17 A. Wirth, M. T. Hassan, I. Grguraš, J. Gagnon, A. Moulet, T. T. Luu, S. Pabst, R. Santra, Z. A. Alahmed, A. M. Azzeer, V. S. Yakovlev, V. Pervak, F. Krausz and E. Goulielmakis, *Science*, 2011, **334**, 195–200.

18 M. T. Hassan, A. Wirth, I. Grguras, A. Moulet, T. T. Luu, J. Gagnon, V. Pervak and E. Goulielmakis, *Rev. Sci. Instrum.*, 2012, **83**, 111301.

19 D. Hui, H. Alqattan, S. Yamada, V. Pervak, K. Yabana and M. T. Hassan, *Nat. Photonics*, 2022, **16**, 33–37.

20 G. Wachter, C. Lemell, J. Burgdörfer, S. A. Sato, X.-M. Tong and K. Yabana, *Phys. Rev. Lett.*, 2014, **113**, 087401.

21 J. B. Khurgin, *J. Opt. Soc. Am. B*, 2016, **33**, C1–C9.

22 S. Yamada, M. Noda, K. Nobusada and K. Yabana, *Phys. Rev. B: Condens. Matter Mater. Phys.*, 2018, **98**, 245147.

23 V. Apalkov and M. I. Stockman, *Phys. Rev. B: Condens. Matter Mater. Phys.*, 2012, **86**, 165118.

24 K. Yabana, T. Sugiyama, Y. Shinohara, T. Otobe and G. Bertsch, *Phys. Rev. B: Condens. Matter Mater. Phys.*, 2012, **85**, 045134.

25 T. Paasch-Colberg, S. Y. Kruchinin, Ö. Sağlam, S. Kapser, S. Cabrini, S. Muehlbrandt, J. Reichert, J. V. Barth, R. Ernstorfer, R. Kienberger, V. S. Yakovlev, N. Karpowicz and A. Schiffrin, *Optica*, 2016, **3**, 1358–1361.
26 S. Sato, K. Yabana, Y. Shinohara, T. Otobe and G. F. Bertsch, *Phys. Rev. B: Condens. Matter Mater. Phys.*, 2014, **89**, 064304.
27 S. Sederberg, D. Zimin, S. Keiber, F. Siegrist, M. S. Wismer, V. S. Yakovlev, I. Floss, C. Lemell, J. Burgdörfer and M. Schultze, *Nat. Commun.*, 2020, **11**, 430.
28 A. Korobenko, K. Johnston, M. Kubullek, L. Arissian, Z. Dube, T. Wang, M. Kübel, A. Y. Naumov, D. M. Villeneuve, M. F. Kling, P. B. Corkum, A. Staudte and B. Bergues, *Optica*, 2020, **7**, 1372–1376.
29 S. H. Autler and C. H. Townes, *Phys. Rev.*, 1955, **100**, 703–722.

Faraday Discussions

DISCUSSIONS

Optical excitation processes: general discussion

Martin Appleby, Kyle Barlow, Martina Basini, Lara Benfatto, Yorrick Boeije, Andrew Burnett, Lin X. Chen, Eric Collet, Hermann A. Dürr, Graham Fleming, Aswathy V. Girija, Mohammed Th. Hassan, Hamoon Hedayat, Shinichiro Iwai, J. Olof Johansson, Steven L. Johnson, Kota Katsumi, Pyosang Kim, James K. McCusker, Ryan Phelps, Jan Michael Rost, Erica Sutcliffe, Julian Wagner, Julia Weinstein and Serhane Zerdane

DOI: 10.1039/d2fd90037h

James McCusker opened discussion of the paper by Serhane Zerdane: Have you been able to carry out simultaneous time-resolved X-ray scattering and emission measurements in order to more firmly establish the mechanistic sequence and spin–structure correlations being proposed?

Serhane Zerdane responded: Indeed, complementary experiments would provide valuable input, especially regarding correlated changes of spin-state and structure. When we performed the XANES experiment at LCLS, it was not possible to perform emission spectroscopy and scattering experiments simultaneously. We expect to perform these measurements during the next beamtime.

Yorrick Boeije asked: Considering the 80 fs temporal resolution of the optical pump–probe measurements, how do you distinguish between local lattice torsion activation induced by (1) charge-transfer (CT) or (2) a process *prior* to charge-transfer, either optical excitation or the spin transition?

Serhane Zerdane replied: For monitoring changes of electronic state and structure we used two complementary techniques: optical spectroscopy and X-ray spectroscopy (XANES). The time resolution was limited to 80 fs in time-resolved optical spectroscopy and this technique is indirectly sensitive to CT and spin transition through the associated electronic reorganizations. The time resolution of the XANES experiment was about 25 fs (ref. 1) and this technique is sensitive to both inter-metallic charge transfer (opposite shifts for Co and Fe K edges) and spin transition (shifts the first extended X-ray absorption fine structure above the Co edge). In this way we could show that the spin transition occurs on the Co first, and is followed by an intermetallic charge-transfer.[1]

1 M. Cammarata, S. Zerdane, L. Balducci, G. Azzolina, S. Mazerat, C. Exertier, M. Trabuco, M. Levantino, R. Alonso-Mori, J. M. Glownia, S. Song, L. Catala, T. Mallah, S. F. Matar and E. Collet, *Nat. Chem.*, 2021, **13**, 10–14.

Aswathy V. Girija enquired: Have you performed any electron spin resonance spectroscopic measurements? Would that be a complementary technique to understand the spin transitions in these systems better? Do you think such measurements would be relevant here?

Serhane Zerdane answered: This technique can help characterizing spin state indeed. But the challenge is to extend this technique to ultrashort time resolution, below 100 fs.

Lin Chen added: The time resolution of electron paramagnetic resonance (EPR) was much lower than the lifetime of the spin species created in this particular example so that there would be a mismatch in the time resolution required to resolve the transient species *via* EPR since the time resolved EPR had a time resolution of around 100 ns, much longer than the lifetime of the transient species.

Steven Johnson asked: Does the "torsion mode" that is coherently driven as an intermediate in this process connect in any way to angular momentum? Is this a possible way in which angular momentum is transferred between the electron states to the lattice?

Serhane Zerdane replied: In spin-crossover materials it was shown that spin–orbit coupling is similar to electron–phonon coupling. Our current understanding of the photoinduced charge-transfer process, coupled to spin transition, in Prussian blue analogues (PBAs) is that lattice torsion modes are activated due to the local bond elongation, in a unit cell which has no time to expend on the 50 fs timescale. This local torsion mode has some angular momentum. However, this is a low symmetry mode, which is therefore incoherently activated. Therefore, it seems to us that there is no global momentum transferred to the lattice due to local disorder in the structural trapping process.

James McCusker enquired: The pressure dependence of what you're looking at could be really interesting. Did you carry out such measurements (or are they planned)?

Serhane Zerdane responded: This is a really interesting point, as the lifetime of the photoinduced charge-transfer is related to global expansion of the lattice. Pressure can therefore destabilize the photoinduced state of higher volume. At the Institut de Physique de Rennes, there is a pressure cell which allows performing ultrafast pump–probe studies. We will therefore perform such measurements you are suggesting to track how the lifetime of the photoinduced charge transfer states changes with pressure.

Lin Chen asked: In your transient Co K-edge XANES spectra in Fig. 3 of the paper (https://doi.org/10.1039/d2fd00015f), you observe a significant change at the pre-edge region around 7719 eV, what are the kinetics at this energy? It seems to be sensitive at this energy to give spin or oxidation state change information, but the kinetics are not shown.

Serhane Zerdane answered: In our previous femtosecond X-ray spectroscopy study[1] we have shown the kinetics at 7717 eV, which is close to 7719 eV. Indeed, these kinetics show a sensitivity to charge-transfer dynamics and torsion modes of the lattice. In our current paper, we are comparing time traces for 11 nm and 70 nm particles and, unfortunately, we did not have time to collect such data for both types of nanoparticles.

1 M. Cammarata, S. Zerdane, L. Balducci, G. Azzolina, S. Mazerat, C. Exertier, M. Trabuco, M. Levantino, R. Alonso-Mori, J. M. Glownia, S. Song, L. Catala, T. Mallah, S. F. Matar and E. Collet, *Nat. Chem.*, 2021, **13**, 10–14.

Lin Chen questioned: In your manuscript (https://doi.org/10.1039/d2fd00015f) you said the signal was not sufficient to see the coherent oscillation signal in XANES; however on slide (11) of your presentation there seems to be a nice coherent phonon mode at the Co K-edge, but the Fe K-edge data did not show corresponding oscillations. Given that the two metal centers are strongly coupled electronically, how would you explain this?

Serhane Zerdane replied: The oscillating component is weak in the signal, due to the torsion nature of the mode involved. This is due to the fact that XANES is mainly sensitive to inter-atomic distances. In addition, the signal/noise ratio was weaker in our data measured at the Fe edge than at the Co edge. In some time traces of ref. 1, we could distinguish an oscillating component around the Fe edge. Therefore, this may be due to signal/noise limitation, but this may also be due to the fact that the torsion occurs mainly around the Co atom. We hope that future measurements will provide better signal/noise to track such lattice torsion.

1 M. Cammarata, S. Zerdane, L. Balducci, G. Azzolina, S. Mazerat, C. Exertier, M. Trabuco, M. Levantino, R. Alonso-Mori, J. M. Glownia, S. Song, L. Catala, T. Mallah, S. F. Matar and E. Collet, *Nat. Chem.*, 2021, **13**, 10–14.

Lin Chen queried: What fraction of excited states were you able to create during the experiments?

Serhane Zerdane responded: This is hard to say, but we estimate that a fraction of 10–30% of sites are photo-excited, based on XANES changes before and after (3 ps) laser excitation. Future measurements will allow more accurate tracking of the fraction of photoexcited sites.

Lin Chen asked: Have you looked into intervalence charge transfer absorption in these systems, which may give you some idea about electron hopping between the metal centers?

Serhane Zerdane answered: Indeed, we used intervalence charge-transfer absorption bands to track ultrafast dynamics, but we were lacking time resolution. New experiments are currently performed with 20 fs time resolution.

J. Olof Johansson requested: How do your results fit in the broader context of photomagnetism?

Serhane Zerdane responded: This would be our dream. What we can say is that the pump wavelength is an important parameter, as different types of electronic excitations can be launched by laser pumping. DFT and TD-DFT can also help understanding the nature of the initial excited state, which may more or less couple to the targeted photoinduced state. This point was recently discuss by G. Azzolina *et al.*[1] in a MnFe PBA.

1 G. Azzolina, H. Tokoro, K. Imoto, M. Yoshikiyo, S. Ohkoshi and E. Collet, *Angew. Chem., Int. Ed.*, 2021, **60**, 23267–23273.

J. Olof Johansson enquired: You mention in your paper (https://doi.org/10.1039/d2fd00015f) that the degree of crystallinity might be important in the decay dynamics. Do you know the difference in crystallinity between the small and large particles you have studied?

Serhane Zerdane answered: The nanoparticles we studied are core–shell particles.[1] The sample made of ≃11 nm particles has a 2 nm passive shell (made of Co(II)) and a 7 nm active core. The ratio of active core and passive shell is different in the 70 nm particles, but this was not accurately characterized yet.

1 L. Trinh, S. Zerdane, S. Mazérat, N. Dia, D. Dragoe, C. Herrero, E. Rivière, L. Catala, M. Cammarata, E. Collet and T. Mallah, *Inorg. Chem.*, 2020, **59**(18), 13153–13161.

Eric Collet commented: We observed the coherent activation of lattice torsion modes in the RbMnFe system, which is analogous to the CoFe system presented by Dr Zerdane. DFT calculation of lattice phonon modes clearly identified the modes at 1.5 THz as torsion modes of the cyanide bridge between the constituting metals. The structural trapping of the photoinduced charge-transfer (CT) state occurs through the elongation of metal–nitrogen bonds. A local lattice torsion is required to accommodate the local volume expansion, as on the bond elongation timescale (sub ps) the lattice doesn't have time to expand. Lattice expansion occurs in a second stage as a result of the structural trapping of the photoinduced CT state.

Kyle Barlow communicated: A question on the dephasing of the breathing of the entire nanocrystal. How big an effect does the distribution of nanoparticle size have on the dephasing you measured in the experiment? If all particles are a different size, there will also be a distribution of frequencies that could lead to dephasing as they all begin with the same phase but will slowly go out of phase with each other. Could this be a possible explanation for the different dephasing times?

Serhane Zerdane communicated in reply: Yes, dephasing strongly depends on the distribution of the size of the particles. For 11 nm particles, the size distribution is quite narrow (11.0 ± 1.5 nm) as shown in ref. 1. For 70 nm this distribution is broader and this can enhance dephasing in the signal. Another explanation can be a stronger damping in larger particles, due to higher elastic energy. Studying single particles with time-resolved electron microscopy could be interesting to track global dynamics of a single particle, to avoid dephasing due to size distribution, and to study deeply how size affects dynamics.

1 L. Trinh, S. Zerdane, S. Mazérat, N. Dia, D. Dragoe, C. Herrero, E. Rivière, L. Catala, M. Cammarata, E. Collet and T. Mallah, *Inorg. Chem.*, 2020, **59**(18), 13153–13161.

J. Olof Johansson addressed Steven Johnson: Do you see any similarity of the torsional vibrational modes identified by S. Zerdane *et al.* discussed in this session with your results on Fe demagnetisation?[1]
1 C. Dornes, *et al.*, *Nature*, 2019, **565**, 209–212.

Steven Johnson replied: In that work we observed the formation of transverse acoustic strain waves from the interfaces of rapidly demagnetized Ni, which can be seen as a consequence of angular momentum conservation. In that system the dynamics were strongly related to the boundaries of the magnetic sample. These torsional motions are different in that they appear to be intra-unit-cell motions that correspond to some linear combination of optical vibrational modes. It is not clear to me that they can carry angular momentum without knowing more about their eigenvector.

Eric Collet remarked: It is important to underline that in these systems there is no thermal expansion, and sometime a weak negative thermal expansion. The volume expansion observed after photoexcitation is therefore only due to the less bonding nature of the photoinduced charge-transfer state. We have estimated that when a few percent of active sites are photoexcited, their structural trapping with local bond elongation corresponds to an internal pressure of the order of 2 kbars within the crystalline lattice.

Shinichiro Iwai opened discussion of the paper by Hamoon Hedayat and Julian Wagner: Is polarization of the pump light linear or circular?

Julian Wagner replied: Pump polarization is linear and cross polarized with respect to the linearly polarized probe.

Shinichiro Iwai asked: What is the build up time of the signal – is it attributed to increasing temperature or another mechanism?

Julian Wagner answered: The build up time shows a strict fluence dependence. In this regard, we assign it to the amount of created holons and doublons after pump excitation. Further, the temperature dependent measurements of the magnetic linear dichroism signal at high excitation fluences (see Fig. 3(a) and (b) in the paper; https://doi.org/10.1039/d2fd00006g) support the interpretation that the increase in temperature due to the pump laser beam is negligible.

Shinichiro Iwai questioned: What is the timescale of build up? Does it depend on magnetic field?

Julian Wagner responded: As discussed in the manuscript there are different regimes for the build up time. At low fluences the build up time measures approximately 30 ps and above the threshold fluence reduces exponentially reaching the sub-ps timescale for high excitation fluences. As shown in Fig. 4(d) of the paper https://doi.org/10.1039/d2fd00006g) the build up time shows similar fluence dependence for all applied field strengths.

Lara Benfatto asked: You showed that at larger fluence, after the fast reduction, the signal is stuck for up to 4000 ps. What happens afterwards? Similar effects have been seen in other systems in a quantum state, as *e.g.* superconductors: at large fluence, the system seems to reach a metastable state where the system remains for ns. Did anyone look afterwards? What are the timescales which limit the detection?

Julian Wagner answered: To the best of our knowledge, no other study shows the dynamics for time delays larger than 4000 ps for α-$RuCl_3$. However, in the intermediate fluence regime, we observed a relatively short plateau lasting for 2–3 ns. As we discussed, we found a universal recovery behavior in this regime. Increasing the fluence, the plateau persists more prolonged, but the recovery dynamics probably follow a similar trend. However, we can only conclude that the plateau is a transient state and recovers within the timescale of 1/(30 kHz) due to the absence of any background signal at negative time delays.

Lara Benfatto asked: Related to my previous question: if I understood you always have an upper bound set by the repetition rate of the experiment, and one cannot say what happens afterwards. Is this correct?

Julian Wagner replied: Due to experimental limitations, the maximum accessible delay between pump and probe pulses is given by the inverse repetition rate of the laser 1/(30 kHz), which is approx. 0.03 ms. We see that at negative delays close to 1/(30 kHz) there is no background signal, *i.e.*, the signal has recovered ultimately between consecutive pulses. In addition, the long-lived plateau can only be tracked for the first 4000 ps in detail after time zero due to the finiteness of the motorized delay stage. We can only conclude that the long-lived state recovers within 1/(30 kHz).

Lin Chen queried: What is the maximum delay/repetition rate in your experiment?

Julian Wagner responded: The repetition rate of the laser was set to 30 kHz. The maximum time between consecutive pulses is hence given by 1/(30 kHz), which is approx. 0.03 ms. The motorized delay stage allows to change the time delay between pump and probe up to 4000 ps.

Erica Sutcliffe asked: Is there any dependence of the number of layers on the magnetic ordering?

Hamoon Hedayat and **Julian Wagner** replied: From theoretical calculations, the ground state for bulk and monolayer crystals of α-$RuCl_3$ is determined to be the zigzag antiferromagnetic order. However, a ferromagnetic ground state is predicted upon photo-doping in the monolayer limit. In our study, we investigated a bulk crystal, for which the zigzag antiferromagnetic order is known.[1]

1 Y. Tian, W. Gao, E. A. Henriksen, J. R. Chelikowsky and L. Yang, *Nano Lett.*, 2019, **19**, 7673–7680.

J. Olof Johansson enquired: What is the difference between a quantum and normal paramagnet?

Hamoon Hedayat and **Julian Wagner** responded: In a normal paramagnet, there are unpaired electrons in the material, which cause a finite magnetic moment but do not interact with one another. Further, they are randomly oriented in the absence of an external field due to thermal agitation, which randomizes the spin orientation resulting in zero net magnetic moment. However, in a quantum paramagnet, the underlying quantum fluctuations and anisotropic interactions cause frustration regarding any magnetic order and hence cause a quantum paramagnetic state.

J. Olof Johansson questioned: Can you extract binding energies of the doublon–holon pairs from your measurements?

Hamoon Hedayat and **Julian Wagner** answered: Since, in our experiment, we use a probe with single photon energy, we cannot extract the binding energies of excitons. We used ref. 1 and 2 to interpret our data.

1 L. J. Sandilands, C. H. Sohn, H. J. Park, S. Y. Kim, K. W. Kim, J. A. Sears, Y.-J. Kim and T. W. Noh, *Phys. Rev. B*, 2016, **94**, 195156.
2 D. Nevola, A. Bataller, A. Kumar, S. Sridhar, J. Frick, S. O'Donnell, H. Ade, P. A. Maggard, A. F. Kemper, K. Gundogdu and D. B. Dougherty, *Phys. Rev. B*, 2021, **103**, 245105.

J. Olof Johansson enquired: What evidence do you have to support the observed demagnetisation? Could it be that it is more an optical effect than magnetic?

Hamoon Hedayat and **Julian Wagner** replied: The measured magnetic linear dichroism (MLD) signal $\Delta\theta_{MLD}$ is related to the magnetic order and probes the zigzag antiferromagnetic phase. In detail, we demonstrated this relation in our previous work on α-$RuCl_3$ at equilibrium conditions.[1] The top panel of Fig. 1(d) in the paper (https://doi.org/10.1039/d2fd00006g) shows the phase diagram as a function of temperature T and in-plane applied magnetic field B obtained under equilibrium conditions. The variation of different temperatures $\Delta\theta$ and applied magnetic fields reconstructs the already observed B–T phase diagram of α-$RuCl_3$.[2] To check if the pump-induced transient $\Delta\theta(t)$ is also directly related to the zigzag order, we measured $\Delta\theta_{MLD}(t)$ after photoexcitation and plotted the integrated signal over the entire dynamics as a function of temperatures and fields. Fig. 1(d) compares the B–T phase diagrams of the transient $\Delta\theta_{MLD}(t)$ and static $\Delta\theta_{MLD}$ showing analogous behavior. This evidence demonstrates the sensitivity of $\Delta\theta(t)$ to the zigzag antiferromagnetic order. Further, the $\Delta\theta(t)$ saturation above a threshold fluence indicates a maximum pump-induced quench of the antiferromagnetic order parameter, which is related to the maximum rotation measured by static MLD. Hence, in our MLD experiment, the direct relationship between the signal and the optical effects is insignificant and we observed the demagnetization dynamics.

1 J. Wagner, A. Sahasrabudhe, R. Versteeg, L. Wysocki, Z. Wang, V. Tsurkan, A. Loidl, D. Khomskii, H. Hedayat and P. van Loosdrecht, *npj Quantum Mater.*, 2022, 7, 28.
2 Z. Ma, K. Ran, J. Wang, S. Bao, Z. Cai, S. Li and J. Wen, *Chin. Phys. B*, 2018, **27**, 106101.

Shinichiro Iwai said: I understood that a thermal effect can be negligible in this case. I guess that the timescale (*ca.* 30 ps) of the build up time reflects magnetic interactions. If so, I think that the dependence on magnetic field should be detected. Strength of magnetic field is not enough to modulate the build up time or does another reason exist?

Julian Wagner answered: The data shows that the fluence dependence of the extracted build up time is similar for all applied field strengths (see Fig. 4(d) in the paper; https://doi.org/10.1039/d2fd00006g). We interpret this observation as indicative that the demagnetization's driving mechanism is related to the creation of holons and doublons and Mott–Hubbard excitons depending on the excitation fluence as discussed in the manuscript. For sufficiently small excitation densities holon–doublon pair recombination *via* magnon emission appears to dominate on 10s of ps timescales, while the hopping of unbound holons and doublons in the intermediate fluence regime effectively perturbs the long-range zigzag order. Further, the antiferromagnetic order is only quenched for magnetic fields larger than approx. 7.5 T,[1] which is above our experimentally accessible field strengths.

1 Z. Ma, K. Ran, J. Wang, S. Bao, Z. Cai, S. Li and J. Wen, *Chin. Phys. B*, 2018, **27**, 106101.

Lin Chen enquired: Do you know what is the exciton size in this kind of material? How does this relate to magnetic domain in such materials, are you able to get such information from experiments? If so, how?

Hamoon Hedayat responded: To the best of our knowledge, there is no report on the exciton size of α-$RuCl_3$ and our experimental technique is not suitable to extract it. However, based on the binding energy of the α-$RuCl_3$ exciton, we estimate a small size in the order of a few nm. The magnetic domains in the equilibrium condition are not related to the excitons. As we discussed in the manuscript (https://doi.org/10.1039/d2fd00006g), in the out-of-equilibrium condition, the dynamics of magnetic order are related to the "breaking" of excitons formed by photoexcitation. Based on our model, holons and doublons play the primary role in the dynamics of magnetic order, and the excitons are involved indirectly. Therefore, we think the size of the excitons does not significantly affect the magnetization dynamics either.

Julia Weinstein opened discussion of the paper by Pyosang Kim: There is a large difference between the rate of intersystem crossing (ISC) in the complex discussed, and Pt-POP. How different are the structural changes observed in the excited state of your system, and those in Pt-POP? Could you also please comment on the difference in the nature of the excited states undergoing ISC in these two molecules, and whether this difference has any effect on the mechanism of the ISC.

Pyosang Kim answered: Although the lowest electronic transition of Pt-POP (dσ* → pσ/metal-centered (MC)) is different from the one of the Pt dimer (dσ* → π*/MMLCT), the photoexcitation of the lowest band for the two complexes commonly depletes one electron from the antibonding orbital and thus induces Pt–Pt bond contraction. The Chergui group has studied the excited-state and vibrational coherent dynamics of the Pt-POP complex in various solvents using femtosecond optical transient absorption (TA) and fluorescence upconversion spectroscopies. Their JACS paper[1] has reported the ISC from the singlet to the triplet MC state with time scales of 10–30 ps and the Pt–Pt stretching coherent vibrational wavepacket (CVWP) motions with ~2 ps damping time. Since the coherence dephasing time is much shorter than the ISC rate, they assigned the CVWP motion to the Pt–Pt stretching vibration in the singlet MC PES. In their follow-up paper,[2] Pt-POP dissolved in MeCN has exhibited a much shorter ISC rate of 700 fs. Furthermore, the Pt–Pt stretching CVWP populated in the singlet MC state transferred to the triplet MC state with a similar time scale to ISC. Based on MD simulations, they rationalized the vibrational wavepacket transfer as a result of the ISC process *via* an intermediate state which is stabilized by MeCN. The ISC processes in the Pt dimer and Pt-POP (in MeCN) are commonly accelerated by the high-lying triplet intermediate state. However, the initially populated ^{1}MMLCT state for the Pt dimer branches into the direct and indirect ISC pathways in contrast to Pt-POP. This indicate that the ISC rate between the ^{1}MMLCT and the intermediate state (^{3}LC) in the Pt dimer may change during the Pt–Pt bond contraction, but not for Pt-POP.

1 R. M. van der Veen, A. Cannizzo, F. van Mourik, A. Vlček, Jr. and M. Chergui, *J. Am. Chem. Soc.*, 2011, **133**(2), 305–315.
2 R. Monni, G. Capano, G. Auböck, H. B. Gray, A. Vlček, I. Tavernelli and M. Chergui, *Proc. Natl. Acad. Sci. U. S. A.*, 2018, **115**, E6396.

Julia Weinstein asked: The optical data have a 0.28 ps component, which is assigned to the coherent VWP in the paper. Does the model consider only sequential processes, or is it possible to envisage branching and two (or more) channels for the ISC? And, would it be possible to distinguish between two models using fluorescence upconversion spectroscopy?

Pyosang Kim replied: I only considered the sequential model in the TA fitting. Thus, the early evolution-associated difference spectra (EADSs) are a mixture of the TA spectra from two or more excited-states. To distinguish the two ISC pathways (direct: ^{1}MMLCT → ^{3}MMLCT; indirect: ^{1}MMLCT → ^{3}LC → ^{3}MMLCT), I compared the TA spectral changes with the temporal evolution of CVWP motions. Particularly, I could identify the short-lived CVWP (~0.9 ps damping time) originates from the ^{1}MMLCT state because the nodal point (phase change) corresponding to the vibration evolves along the spectral redshift of the stimulated emission (~650 → ~700 nm). Interestingly, we also observed the long-lived CVWP (~2 ps damping time) and its temporal evolution along the spectral redshift of the excited-state absorption signal (~600 → ~650 nm). If these two CVWP motions come from the same excited state, the damping time for these two vibrations would be similar. Therefore, we could assign the long-lived coherence to the CVWP motion from the ^{3}MMLCT state through the indirect ISC path. If the fluorescence upconversion technique can detect the phosphorescence from the ^{3}MMLCT state, it should be possible to identify the branching dynamics in the ISC processes. For example, if the rise dynamics of the phosphorescence show double exponential components, we may assign the short- and long-time constants to the ISC rates for the indirect and direct ISC pathways, respectively. Currently, we are working on the experiment.

Lin Chen added: We planned to do fluorescence upconversion, working to improve time resolution to ~100 fs. Regarding the difference between Pt-POP and the Pt dimer, the flexibility of having different cyclometallic ligands is a distinct feature differing from PtPOP. Because the lowest energy transition is MMLCT in nature, the LUMO is dominated by the π^* orbital of the ligand. Thus, changing the ligand energy level will tune the conical intersection along the key reaction coordinate close to that of the Franck–Condon state, allowing ultrafast ISC from the molecular singlet MMLCT and to undergo ISC within tens of fs.

James McCusker enquired: I really liked the use of stimulated emission to identify the intersystem crossing event (very important measurement). That stated, since there is sufficient radiative coupling between the ground and excited states to produce stimulated emission, how do you differentiate excited- *versus* ground-state coherences?

Pyosang Kim responded: In this work, I compared the spectral evolution of the stimulated emission with the temporal evolution of the CVWP. Interestingly, the nodal point (phase change) corresponding to the 145 cm^{-1} CVWP redshifts from ~650 to ~700 nm, which is similar to the spectral redshift in the stimulated emission. Generally, the phase change occurs at the equilibrium of PES and thus, the node appears around the maximum of the TA spectra such as the stimulated emission in this case. Furthermore, the dephasing time for this CVWP (~0.9 ps) is similar to the direct ISC time scale of 1.25 ps. Thus, I could assign the 145 cm^{-1} CVWP to the excited-state coherence. Also, I observed the 110 cm^{-1} CVWP motion mostly around the ground-state absorption region. The structural optimization exhibited by the Pt–Pt contraction increases the Pt–Pt stretching frequency from ~115 to ~150 cm^{-1}, suggesting the 110 cm^{-1} originates from the ground state. However, since the dephasing time for this 110 cm^{-1} mode (<0.2 ps) is much shorter than the excited-state coherence (~0.9 and ~2 ps), there is another possibility that the CVWP could be the initial wavepacket motion populated in the FC state. As I mentioned in the manuscript (https://doi.org/10.1039/d2fd00009a), I plan to perform time-resolved impulsive stimulated Raman spectroscopy to identify the origin of the 110 cm^{-1} mode, because this technique can probe selectively the excited-state coherence.

James McCusker asked: Were there significant differences in dephasing times for the various coherences you identified?

Pyosang Kim answered: For the 145 cm^{-1} CVWP motions, I found two damping constants of ~0.9 and ~2 ps. The node corresponding to the wavepacket motions with ∼0.9 ps dephasing time appeared around ∼650 nm and then redshifted to ∼700 nm. This temporal behavior followed the spectral redshift of stimulated emission in the TA data and hence I assigned the short-lived CVWP to the Pt–Pt stretching vibration in the singlet MMLCT state. In contrast, the CVWP motion corresponding to the other node appearing around 620 nm and then redshifting to ∼650 nm showed ∼2 ps damping time. Although the ESA feature is structureless, the ESA signal in 600–650 nm exhibited the spectral redshift, which is similar to the temporal behavior of the other node. Therefore, I could assign the long-lived CVWP motion to the Pt–Pt stretching vibrational coherence generated in the triplet MMLCT state through ultrafast ISC *via* the high-lying intermediate state. Meanwhile, I also observed the 110 cm^{-1} CVWP motion mostly around the bleaching signal. The normal mode analysis suggested that the Pt–Pt stretching vibration in the ground-state has lower frequency than that in the excited-state, because of the longer Pt–Pt distance. However, the dephasing time for this CVWP motion displayed a much shorter time scale, <0.2 ps, than the excited-state CVWP motions. Therefore, I suggested in this work that the 110 cm^{-1} CVWP could be the CVWP motion around the Franck–Condon (FC) region before the Pt–Pt contraction. As I mentioned in the manuscript (https://doi.org/10.1039/d2fd00009a), I plan to perform time-resolved impulsive stimulated Raman spectroscopy to confirm this suggestion, because this technique can probe selectively the excited-state coherence.

Yorrick Boeije questioned: What is the branching ratio for the indirect : direct ISC pathways and what is it controlled by?

Pyosang Kim responded: I have not calculated the branching ratio in this compound. However, in our previous paper,[1] we found the branching dynamics are strongly affected by the existence of bulky substituents on the bridging ligands. We observed the long-lived coherence from the ^{3}MMLCT state only in the Pt dimer with the phenyl substituents on the bridging ligands (studied in the current manuscript; https://doi.org/10.1039/d2fd00009a) while the Pt dimer with the methyl substituents does not exhibit the long-lived coherence but a short-lived coherence from the ^{1}MMLCT state. In the excited-state PES calculations, the bulkier substituents (phenyl rings) stabilized the ^{1}MMLCT state relative to the triplet intermediate state (^{3}LC) and then the curve crossing between the ^{1}MMLCT and ^{3}LC gets closer to the Franck-Condon (FC) state in comparison to the other Pt dimer which has smaller substituents (methyl). Therefore, we could suggest that the indirect ISC pathway can be open when the FC state is energetically close to the intermediate state. Furthermore, this suggests the Pt–Pt contraction may change the ISC rate from the ^{1}MMLCT to the intermediate state and thus the branching ratio, because the energy of the ^{1}MMLCT state is rapidly stabilized out of the FC region due to the photoinduced Pt–Pt contraction.

1 P. Kim, A. J. S. Valentine, S. Roy, A. W. Mills, A. Chakraborty, F. N. Castellano, X. Li and L. X. Chen, *J. Phys. Chem. Lett.*, 2021, **12**(29), 6794–6803.

Aswathy V. Girija enquired: As a follow-up question, if we consider two structural parameters, one being Pt–Pt bond distance and the other being different ligand substitutions, which of these affects intersystem crossing the most? What are the major structural parameters that contribute to processes such as intersystem crossing in such cyclometallic complexes?

Pyosang Kim replied: In our previous work,[1] we observed the long-lived CVWP motion (~2 ps) from the ^{3}MMLCT state only in the Pt dimer with the phenyl substituents on the bridging ligands (studied in the current manuscript; https://doi.org/10.1039/d2fd00009a) while the other Pt dimer with methyl substituents does not exhibit the long-lived coherence but a short-lived coherence (~0.7 ps) from the ^{1}MMLCT state. However, when we modified the cyclometalating ligands from phenylpyridine to benzoquinoline, the CVWP motion of the Pt dimer also displayed the long-lived coherence (~2 ps) from the ^{3}MMLCT state. The excited-state PES calculations have shown that the bulkier substituents (phenyl rings) stabilized the ^{1}MMLCT state relative to the triplet intermediate state (^{3}LC) and then the curve crossing between the ^{1}MMLCT and ^{3}LC gets closer to the Franck–Condon (FC) state in comparison to the other Pt dimer which has smaller substituents (methyl). Therefore, we rationalize that the indirect ISC pathway can be open when the FC state is energetically close to the intermediate state. Furthermore, this suggests the Pt–Pt contraction may change the ISC rate from the ^{1}MMLCT to the intermediate state and thus the branching ratio, because the energy of the ^{1}MMLCT state is rapidly stabilized out of the FC region due to the photoinduced Pt–Pt contraction. In summary, our studies for hydroxypyridyl-bridged Pt dimers show that the substituents on the bridging ligands have a more significant effect on the ISC than the cyclometalating ligands.

1 P. Kim, A. J. S. Valentine, S. Roy, A. W. Mills, A. Chakraborty, F. N. Castellano, X. Li and L. X. Chen, *J. Phys. Chem. Lett.*, 2021, **12**(29), 6794–6803.

J. Olof Johansson asked: Given that you have a 1 mm path length, how long is the cross phase modulation (XPM) and can you rule out that the short decay component is not influenced by the XPM?

Pyosang Kim responded: The XPM was finished before 50 fs in the broadband transient absorption (BBTA) experiment because the pump (NOPA) and probe (WLC) pulses were well compressed using chirped mirror pairs. Moreover, the chirp of WLC is in ~30 fs along the wavelength axis. Thus, I was able to assign the short component, ~200 fs. Furthermore, I truncated the TA data up to 50 fs to avoid the contribution of the XPM to the TA kinetics and CVWP dynamics.

Martin Appleby communicated: The TA data in Fig. 1c of the paper (https://doi.org/10.1039/d2fd00009a) is very interesting. In the wavelength-resolved residual map spectra shown in the talk and in Fig. 2c. It looks as if the data was chirped. Is this from the same measurements as Fig. 1c, as this data does not appear to be chirped?

Pyosang Kim communicated in reply: The residual map was obtained from the same TA data and was drawn and analyzed after 100 fs to avoid the contribution of cross-phase modulation. As you mentioned, there is a chirp along the wavelength axis (550–850 nm), but the chip is within ~30 fs. The chirp occurring in the WLC probe was corrected by a chirped mirror pair and a post-processing to correct the chirp was not done in this work.

Martin Appleby communicated: Could the chirp be imprinting on the signal and be affecting the perceived dynamics of the two wave packets?

Pyosang Kim communicated in reply: Yes, but the chirp in the probe pulse does not impact on the damping dynamics and frequency of the vibrational wavepacket motions. Furthermore, the ~30 fs chirp in the WLC probe pulse is much smaller than the period of the Pt–Pt stretching vibration (~220 fs), so the chirp is no matter in the kinetic analysis of the nodal point in Fig. 2b of the paper (https://doi.org/10.1039/d2fd00009a). However, if the chirp in the pump pulse is longer than the half period of the vibrational wavepackets, it can affect the intensity and, maybe, the damping dynamics of the wavepacket motions.

Martin Appleby communicated: Does the chirp have a greater effect on the wavelength-resolved residual map compared to the fs-BBTA data map?

Pyosang Kim communicated in reply: The chirp in the probe pulse can affect both data maps. However, as I mentioned before, the probe chirp does not change the wavepacket dynamics, but the chirp should be corrected to analyze the TA spectral changes, because the time-zero would be different along the probe wavelength axis. Of course, you may correct the chirp mathematically. In my work, I compress the probe pulse using a chirped mirror pair for the convenience in TA and wavepacket data analysis.

Martin Appleby communicated: In the paper, you mention that the ISC occurs on a timescale faster than the Pt–Pt shortening/contraction. Have you looked at modifying the system to prevent the shortening and do you have any thoughts on what the effects of restricting the contraction or lengthening the timescale it takes place over may have on the ultrafast dynamics.

Pyosang Kim communicated in reply: In fact, we are co-working with a synthetic group to modify chemically the excited-state properties of the Pt dimers including the Pt–Pt contraction. This future work may answer how the Pt–Pt bond contraction affect the ISC dynamics in the vicinity of the Franck–Condon region.

Ryan Phelps opened discussion of the paper by James McCusker: The reported dephasing times were taken from experiments conducted in different solvents. As a result, it could be expected that normal modes of $Cr(acac)_3$ and its derivatives become shifted, altering intramolecular vibrational couplings. Similarly, the coupling to the solvent bath will also change. Could changes in the solvent properties account for any differences in the dephasing times reported?

James McCusker answered: The vibrational spectra of compounds of this type do not change significantly as a function of solvent; the fact that (a) the $Cr(acac)_3$ complexes are neutral, and (b) the excited-state dynamics are associated with ligand-field states that are fully localized on the metal means that the degree of solvent–solute coupling is significantly attenuated as compared to charge-transfer complexes, for example. That stated, we did look at a few compounds as a function of solvent (in part to differentiate solvent *versus* solute signals) and found that variations in the dephasing times for the modes we focused were within the error bars we cite.

Martina Basini queried: Can these coherent vibronic states be thought of as phonons?

James McCusker replied: In a manner of speaking, yes: the vibrational degrees of freedom of the compounds presented can be thought of as molecular analogs of phonons that one might identify for a solid-state material. I would add that this analogy is valid regardless of whether a given mode(s) forms the basis of a coherence signature.

Graham Fleming said: 2D Electronic vibrational (2DEV) spectroscopy measures cross correlations rather than autocorrelations which are obtained for degenerate four-wave mixing experiments such as 2D electronic spectroscopy. This makes it very sensitive to vibronic mixing and would be a good way of disentangling the interactions in your molecules.

James McCusker responded: Agreed – will be in touch with your group to get you a few initial samples.

Lin Chen asked: Are there any electronic energy changes when you attached halogen onto the $Cr(acac)_3$ ligands?

James McCusker replied: There are detectable changes to the absorption and emission maxima of the compounds, but they're pretty minor (maybe 5–10 nm). The main reason for this is that substitution at the 3 position of acac primarily tweaks the π interactions between the acac ligand and the metal, and the amount of π bonding in $Cr(acac)_3$ is pretty small.

J. Olof Johansson questioned: Does the chemical perturbation of the $Cr(acac)_3$ derivatives change the molar extinction coefficient of the compounds?

James McCusker answered: It likely does for the intraligand absorptions in the ultraviolet (I'd have to go back and check), but there is minimal effect (<10%) on the molar absorptivity of the d–d absorption in the visible where all of our measurements were focused.

J. Olof Johansson asked: Will the measurement of the wavepacket dephasing time not be affected by the noise in the experiment?

James McCusker responded: We do see a fair amount of variability in dephasing times that we obtain from the LPSVD analysis, and part of this (perhaps most) is no doubt due to the noise in the experiment. In order to mitigate this, we typically collect multiple data sets at a given pump–probe combination, analyze them individually, and then average both the frequencies as well as the dephasing times in order to provide what we believe is a reasonable representation of the error bars associated with those measurements under those conditions. We still see some variability, so we try to be conservative in terms of identifying what we think does and does not represent a significant enough difference to warrant attention.

Eric Collet enquired: There are two origins for decoherence of molecular vibration during ultrafast photoinduced change of electronic state: phonon–phonon coupling, damping the oscillation and decoherence through an intermediate state. In the two types of Fe systems shown in your presentation, where on the ligand is modified, are there different vibrational coherence or decoherence dynamics? Is lifetime changing a lot? Should this reduce amplitude of oscillation?

James McCusker replied: What we noticed when we introduced the Cu(I) ions into the N_4 coordination sites above and below the Fe-bpy core was (a) a dampening of the amplitude of the oscillatory features, and (b) an increase in the dephasing time of the coherence (although this increase was just outside of our error bars, so our confidence level on this point is not high). We interpreted these observations as an indication that the mode was still coupled to electronic evolution of the system (*i.e.*, the MLCT-to-LF conversion), but that coupling was being attenuated. In that study our focus was on the effect interference with that degree of freedom had on the MLCT-state lifetime, so we did not pursue a more sophisticated analysis of the decoherence mechanism in that case. We are looking into this issue now but using ligand constructs that are not quite as complex as the one presented so we can address the sorts of points you're raising in greater detail.

Jan Rost asked: In systems we deal with, we wouldn't necessarily expect reaction coordinate and dephasing to be linked. The reaction coordinate is often decided initially and very fast by absorption – then the system finds itself on the upper energy surface in a very different environment which determines dephasing. Is there a fundamental difference for the systems where you have found a correlation between reaction coordinate and dephasing time?

James McCusker responded: Nothing specific, but I would argue that the degree of vibronic coupling one finds in transition metal-based systems (particularly those involving metals of the first transition series like Fe and Cr) is generally far greater than what one encounters for organic systems. For example, the primary coordination sphere between low-spin Fe(II) (which is the ground state of one of the compounds discussed) and high-spin Fe(II) (which is the form produced upon relaxation out of the MLCT state) involves a *ca.* 0.2 Å increase in metal–ligand bond lengths, large angular deformations, and a molecular volume increase in the range of 20–25 $cm^3 mol^{-1}$. Similar (though not as large) changes are associated with the excited-state dynamics sampled in the Cr(III) compounds. The structural changes are therefore intimately linked with electronic state changes, and my understanding is that the latter are often identified as dephasing mechanisms. This was the origin of our hypothesis, but the data that we have acquired thus far indicates that such a correlation is far from robust (if it even exists at all).

Yorrick Boeije queried: Are the halide substitutions in $Cr(acac)_3$ electronically inert? So, do they only affect the vibrations and not the energetics of the excited states?

James McCusker replied: Substituting halides at the 3 position of the acac ligand has surprisingly little effect on the energetics of the electronic states of the metal center. The absorption and emission features (arising from the $^4T_2$ and ^2E ligand-field states, respectively) do shift by a few nanometers in terms of their maxima, but the shift represents a very small perturbation to the energetics of the system in terms of electronics. The reason for this is that substitutions at this position predominantly affect π interactions, which in the first transition series are relatively minor contributions to the effective ligand-field strength imposed upon the metal ion's electronic structure.

Yorrick Boeije questioned: As the rate of intramolecular vibrational relaxation scales with the frequency of the excited mode due to a larger Franck-Condon overlap, can we rationalize the trend of longer dephasing times with heavier halide substitutions (and therefore lower frequency) in the $Cr(acac')_3$ series by a less efficient vibrational relaxation?

James McCusker answered: I would think that the decrease in frequency (and concomitant increase in the vibrational density of states) would increase the efficiency of vibrational relaxation, not decrease it. In this context, I do think that could be linked to the longer dephasing time as the system gets established on the electronic excited-state potential more rapidly.

Pyosang Kim asked: For iron complexes, what was the frequency and dephasing time in different probe wavelengths, such as ground state bleaching signal (~550 nm) or ESA signal (700–750 nm), or any phase changes for those frequencies? I am just wondering how to distinguish the excited-state coherence from the ground-state coherence in this particular system showing a very strong bleaching signal compared to excited-state absorption and nothing in stimulated-emission.

James McCusker replied: For the Fe(II) complexes we didn't observe any coherences in the region of the ground-state bleach – only in regions of excited-state absorption – so that made it relatively easy in that case to ascribe the signal to an excited-state coherence. In the case of the Cr(III) complexes it was even easier due to the low molar absorptivity of the band into which we were exciting (*ca.* 50 M^{-1} cm^{-1}): the radiative coupling is far too low to support stimulated emission back to the ground state.

J. Olof Johansson queried: Do you think that torsional twisting modes are also important in the fast ISC rate observed in $Cr(acac)_3$, just like was discussed by K. F. Purcell[1] and yourself[2] in the case for Fe(II) ISC?

1 K. F. Purcell, *J. Am. Chem. Soc.*, 1979, **101**, 5147–5152.
2 J. K. McCusker, A. L. Rheingold and D. N. Hendrickson, *Inorg. Chem.*, 1996, **35**, 2100–2112.

James McCusker responded: I do think that twisting modes are significant, but our evidence to support this is still pretty circumstantial. If you look at the vector displacements associated with the modes we've identified in the $Cr(acac)_3$ system, the motion is clearly more complex than a simple variation in Cr–O bond distance. And certainly the work by Perkovic and Endicott[1,2] is very suggestive of torsional modes playing an important role in ISC for ground-state recovery. Ultimately, I think the best way to address this question is probably an ultrafast circular dichroism (CD) measurement along the lines that Chergui and co-workers just published in *Nat. Chem.* for $[Fe(bpy)_3]^{2+}$ (ref. 3).

1 M. W. Perkovic and J. F. Endicott, *J. Phys. Chem.*, 1990, **94**, 1217–1219.
2 M. W. Perkovic, M. J. Heeg and J. F. Endicott, *Inorg. Chem.*, 1991, **30**, 3140–3147.
3 M. Oppermann, F. Zinna, J. Lacour and M. Chergui, *Nat. Chem.*, 2022, **14**, 739–745.

Andrew Burnett enquired: How do you decide which calculated normal mode corresponds to the experimentally observed motion, is it simply the one with the closest calculated frequency – these low frequency modes can be tricky to calculate accurately and can be very sensitive to calculation choice (basis set/functional/dispersion/solvent/periodicity).

James McCusker answered: That's a great question and one that we constantly struggle with. Initially we focus on what we think is the best match in terms of frequency, but as you point out this can be tricky for low-frequency modes. The situation gets a bit easier when you get above *ca.* 100 cm^{-1} because the density of vibrational states tends to drop off a bit. What we try to do is consider the nature of the motion – that is to say the actual atomic displacements – across a series of related complexes and target the physical motion as opposed to the specific frequency (for example, the halogen-substituted acac complexes). This at least provides us a degree of internal consistency. We also routinely do these calculations using several different functionals and try to focus on modes that appear to be (relatively) independent of which functional we choose. We can also consider the electronic state in which the system resides and try to use some degree of chemical intuition in terms of how we expect a given mode to differ in that state relative to the ground state. But it is an imperfect process to say the least.

Eric Collet asked: In the chromium compound, where the ligand is substituted by different halogens, is the frequency of the breathing mode changing with the square of the reduced mass of the mode or of the mass of the ligand? In other words, is the strength of the metal–ligand bond unaffected by the substitution?

James McCusker responded: We didn't check for that specifically but my guess is, given the complexity of the motion indicated by the vector displacement representations, a simple correlation such as the one you suggest might be difficult to discern. If the relative invariance of the energy and oscillator strength of the d–d band is any indication, any perturbations to the Cr–O bond strength are probably pretty minor. This would be consistent with the fact that substitution in the 3 position of acac primarily couples to the π-symmetry orbitals of the metal (hence the nephelauxetic effect on the ^2E emission), and the degree of π contributions to the Cr–O bond is not that significant.

Yorrick Boeije communicated: On the differentiation between active and spectator modes, isn't excited-state vibrational coherence not always to some extent active (and involved in the reaction coordinate) as the amplitude of the excited-state vibrational oscillation depends on how much the excited-state is displaced with respect to the ground-state through the FC factor, which is already driving the wavepacket along the direction of the reaction coordinate. Or does this reasoning fail when we consider anharmonic coupling which may activate initially FC-inactive modes, resulting in, potentially, spectator vibrational coherence?

James McCusker communicated in reply: The amount of information available on coherence in transition metal-based compounds is relatively limited as compared to organic or biological systems, so I would be reluctant to make any generalizations along these lines. That stated, significant differences between the electronic structure of, say, a typical coordination complex *versus* an organic chromophore involve (a) a wider variety of electronic excited states in terms of their character (*e.g.*, charge-transfer, ligand-field, and intraligand) and (b) different degrees of vibronic coupling associated with these different states. I do think you are correct that the introduction of anharmonicity (which will be particularly important for the sorts of low-frequency modes that represent a significant fraction of the vibrational partition function for a typical transition metal complex) can result in the transference of coherence to modes that were not directly excited in the Franck–Condon state. Presumably this will have some correlation with the time scale of intramolecular vibrational redistribution, which for a metal complex is comparable to (or shorter than) the dephasing times we're talking about (*e.g.*, 100–200 fs).

Kyle Barlow communicated: I have some questions about the mechanisms of dephasing and the contributions of pure dephasing and populations relaxation. In the paper (https://doi.org/10.1039/d2fd00106c), it says "Vibrational coherences typically show dephasing times on the order of picoseconds and in solution phase this is often dominated by pure dephasing processes." The reader is referred to ref. 39 in the paper (copied below as ref. 1) which is an experiment on I_3^-, another reference used to support this is ref. 33 in the paper (copied below as ref. 2) which states in the article "…this observation indicates that the structure does not affect dephasing times. The solute–solvent interactions are also not a source of dephasing … We conclude that the origin of dephasing must be from anharmonic coupling of the Pt–Pt oscillator to internal low-frequency modes of the molecule." Ref. 39 is on a very small molecule with only a handful of internal vibrational degrees of freedom and ref. 33 suggests it is due to intermolecular vibrational coupling to low-frequency modes. Considering transition metal complexes are much larger molecules often with 100+ vibrational degrees of freedom is the idea that pure dephasing dominates the dephasing mechanism still true? If this isn't the case, how do we approach separately measuring pure dephasing and population relaxation and how they influence electronic relaxation?

1 E. Gershgoren, Z. Wang, S. Ruhman, J. Vala and R. Kosloff, Investigating Pure Vibrational Dephasing of I_3^- in Solution: Temperature Dependence of T_2* for the Fundamental and First Harmonic of ν_1, *J. Chem. Phys.*, 2003, **118**, 3660–3667 (ref. 39 in the paper).

2 R. Monni, G. Auböck, D. Kinschel, K. M. Aziz-Lange, H. B. Gray, A. Vlček and M. Chergui, Conservation of Vibrational Coherence in Ultrafast Electronic Relaxation: The Case of Diplatinum Complexes in Solution, *Chem. Phys. Lett.*, 2017, **683**, 112–120 (ref. 33 in the paper).

James McCusker communicated in reply: The fact that the transition metal complexes we are studying have many more vibrational degrees of freedom than the simpler triatomics where these phenomena were first studied is an excellent point. I think it is likely that more degrees of freedom, combined with anharmonic coupling among various vibrational modes, can both contribute to dephasing but also act as a conduit for coherence transfer. There may be some information along these lines from earlier studies of metal carbonyls, but these compounds have much simpler vibrational structures than the types of compounds we are considering (and I'm not aware of any work where this question has been directly addressed in the context of coordination complexes). In my opinion, there is simply not enough known at this point in terms of how coherence manifests in this class of compounds to make any definitive statements along the lines of the question you're asking (*i.e.*, more research is needed).

Steven Johnson opened discussion of the paper by Hermann Dürr: You report that you observe the 1 THz magnon mode as modulating the NiO band gap, which you see as a time-dependent shift of the O K-edge. By the symmetry of the mode and your calculations as presented in Fig. 7 of the paper (https://doi.org/10.1039/d2fd00005a), is it correct to say that this is actually a splitting of the conduction band into two bands with reduced symmetry rather than a pure gap shift? In optics this would show up as a change in the birefringence (Cotton–Mouton effect). Is this analogous to that, in that the K-edge absorption should show linear birefringence, with another polarisation perhaps showing oscillations with an opposite phase?

Hermann Dürr responded: Thank you for the questions. Regarding the first question, it is indeed a band splitting that occurs. When the two Ni^{2+} spins are canted, there will be a net ferromagnetic vector (sum of the Ni^{2+} spin vectors) pointing along the $[1\ \bar{1}\ 0]$ direction (see ref. 1 below and Fig. 7 in our paper; https://doi.org/10.1039/d2fd00005a) for the out-of-plane mode. This breaks the symmetry and causes the initially degenerate bands to split into two bands. At the O K-edge, we mainly probe the upper gap-edge, which consists of the lower band of the two split bands. In time-resolved measurements, the net ferromagnetic vector oscillates along the $[1\ \bar{1}\ 0]$ direction and the upper gap-edge position is modulated by the magnon frequency. The canting of the Ni^{2+} spins makes the contribution of the oscillatory intensity variation in Fig. 6 of our paper. The oscillation period in our data is consistent with the observation in optical measurements.[1,2] For the second question, it is possible for the oscillations to show opposite phases with another polarisation, as reported in optical measurements.[2] In optical measurements, the transition occurs within the valance band and it might be difficult to explain the complicated inter-valence transitions. In our measurement with X-rays, the transition occurs between the core level and the conduction band, which is easier to analyse.

1 C. Tzschaschel, K. Otani, R. Iida, T. Shimura, H. Ueda, S. Günther, M. Fiebig and T. Satoh, *Phys. Rev. B*, 2017, **95**, 174407 (ref. 20 in the paper).
2 J. Nishitani, T. Nagashima and M. Hangyo, *Phys. Rev. B*, 2012, **85**, 174439 (ref. 44 in the paper).

J. Olof Johansson enquired: In your presentation, you showed oscillations in optical data from another group corresponding to the excitation of a magnon mode. Is that the same magnon mode you observe in your data?

Hermann Dürr replied: Yes, it is the same out-of-plane magnon mode as shown in ref. 1.

1 C. Tzschaschel, K. Otani, R. Iida, T. Shimura, H. Ueda, S. Günther, M. Fiebig and T. Satoh, *Phys. Rev. B*, 2017, **95**, 174407 (ref. 20 in the paper).

J. Olof Johansson asked: Could there be other factors influencing the shift in the oxygen peak position? If you excite sub-gap states, that will most likely influence the electronic environment of the oxygen atoms and therefore their energy levels.

Hermann Dürr answered: The oxygen peak position can be influenced by the alteration of the magnetic exchange. In our experiment, we observed the incoherent shift at the O K-edge, which is an indication that the contribution of the magnetic exchange to the band gap is altered. The oxygen peak position can also be affected by the electric field of the optical pulse. In our previous study,[1] we observed that the line shape at the O K-edge shows a transient transfer of spectral weight mainly during the presence of the laser field.

1 O. Grånäs, I. Vaskivskyi, X. Wang, P. Thunström, S. Ghimire, R. Knut, J. Söderström, L. Kjellsson, D. Turenne, R. Y. Engel, M. Beye, J. Lu, A. H. Reid, W. Schlotter, G. Coslovich, M. Hoffmann, G. Kolesov, C. Schüßler-Langeheine, A. Styervoyedov, N. Tancogne-Dejean, M. A. Sentef, D. A. Reis, A. Rubio, S. S. P. Parkin, O. Karis, J. Nordgren, J.-E. Rubensson, O. Eriksson and H. A. Dürr, *arXiv*, 2020, preprint, arXiv:2008.11115, DOI: 10.48550/arXiv.2008.11115 (ref. 9 in the paper).

Martina Basini questioned: Is the oscillation you measured compatible with the lifetime of the magnon?

Hermann Dürr replied: Due to the limited amount of beamtime, we could only measure the delay up to 2.4 ps with good signal-to-noise ratio and did not extend such measurements to significantly longer time delays. However, in previous optical studies[1–3] it has been observed that the oscillations last for more than 15 ps. We were cautious when extracting the magnon frequency, because we only measured 1–2 oscillation periods. The oscillation frequency in our data at room temperature seems to be slightly lower than the reported value. This is consistent with the softening of the magnon frequency in previous temperature-dependent measurements[1,3] and for the increased spin temperature after excitation.

1 J. Nishitani, K. Kozuki, T. Nagashima and M. Hangyo, *Appl. Phys. Lett.*, 2010, **96**, 221906 (ref. 14 in the paper).
2 C. Tzschaschel, K. Otani, R. Iida, T. Shimura, H. Ueda, S. Günther, M. Fiebig and T. Satoh, *Phys. Rev. B*, 2017, **95**, 174407 (ref. 20 in the paper).

3 T. Kohmoto, T. Moriyasu, S. Wakabayashi, H. Jinn, M. Takahara and K. Kakita, *J. Infrared, Millimeter, Terahertz Waves*, 2018, **39**, 77–92 (ref. 46 in the paper).

Lara Benfatto said: I have a comment related to the possibility of seeing magnon oscillations. Single-magnon excitations can be seen in ordinary Raman scattering in systems with strong spin–orbit interactions. We studied this problem long ago in cuprates.[1] However, when the magnons are very soft it can be difficult to resolve them in spontaneous Raman, since the peak is screened by the elastic response. Nonetheless, they can be seen as coherent excitations in pump–probe, that is sensitive to Raman-active modes.

1 L. Benfatto, M. B. Silva Neto, A. Gozar, B. S. Dennis, G. Blumberg, L. L. Miller, S. Komiya and Y. Ando, *Phys. Rev. B: Condens. Matter Mater. Phys.*, 2006, **74**, 024416.

Hermann Dürr responded: Yes, we agree with the comment. The oscillation period observed in our experiment and many other optical measurements[1–4] corresponds to a frequency of $\sim$35 cm^{-1} ($\sim$1.05 THz), which is indeed close to the elastic line.

1 J. Nishitani, K. Kozuki, T. Nagashima and M. Hangyo, *Appl. Phys. Lett.*, 2010, **96**, 221906 (ref. 14 in the paper).
2 C. Tzschaschel, K. Otani, R. Iida, T. Shimura, H. Ueda, S. Günther, M. Fiebig and T. Satoh, *Phys. Rev. B*, 2017, **95**, 174407 (ref. 20 in the paper).
3 J. Nishitani, T. Nagashima and M. Hangyo, *Phys. Rev. B*, 2012, **85**, 174439 (ref. 44 in the paper).
4 T. Kohmoto, T. Moriyasu, S. Wakabayashi, H. Jinn, M. Takahara and K. Kakita, *J. Infrared, Millimeter, Terahertz Waves*, 2018, **39**, 77–92 (ref. 46 in the paper).

Martina Basini queried: Is the NiO sample a multi-domain sample or single domain?

Hermann Dürr responded: The NiO film used in our experiment was prepared the same way as described in ref. 1. In principle, no magnetic linear dichroism is possible if all the twin domains are present. However, for a thin film the surface and interface have influence on the preference of the spin orientation. For instance, for a (001) surface, the spins have strong preference to form an easy axis perpendicular to the (001) surface. The preferential orientation of the Ni^{2+} spins allowed us to measure the X-ray magnetic linear dichroism line shape at the Ni L_2-edge, and thus investigate the demagnetisation of the antiferromagnetic order which is in agreement with ref. 1.

1 D. Alders, L. H. Tjeng, F. C. Voogt, T. Hibma, G. A. Sawatzky, C. T. Chen, J. Vogel, M. Sacchi and S. Iacobucci, *Phys. Rev. B*, 1998, **57**, 11623 (ref. 7 in the paper).

Aswathy V. Girija enquired: It was mentioned earlier that so far there hasn't been direct observation of fs laser induced order change in antiferromagnetic systems. Is it possible to alter antiferromagnetic order with really intense laser pulses?

Hermann Dürr replied: Previous TDDFT+*U* calculations showed a renormalisation of the Hubbard *U* with an intense laser field.[1] This is the starting point of our previous study,[2] in which we chose 0.6 eV as the excitation photon energy (off-resonant excitation), and did not observe any indication of the change in antiferromagnetic order nor a renormalized *U*. Therefore, in the present study (https://doi.org/10.1039/d2fd00005a), we tuned the optical wavelength to the d–d excitation energy as close as possible and applied the similar field strength as in our previous study. The d–d excitation enhanced the optical absorption in the NiO film and thus induced demagnetisation in the system.

1 N. Tancogne-Dejean, M. A. Sentef and A. Rubio, *Phys. Rev. Lett.*, 2018, **121**, 097402.
2 O. Grånäs, I. Vaskivskyi, X. Wang, P. Thunström, S. Ghimire, R. Knut, J. Söderström, L. Kjellsson, D. Turenne, R. Y. Engel, M. Beye, J. Lu, A. H. Reid, W. Schlotter, G. Coslovich, M. Hoffmann, G. Kolesov, C. Schüßler-Langeheine, A. Styervoyedov, N. Tancogne-Dejean, M. A. Sentef, D. A. Reis, A. Rubio, S. S. P. Parkin, O. Karis, J. Nordgren, J.-E. Rubensson, O. Eriksson and H. A. Dürr, *arXiv*, 2020, preprint, arXiv:2008.11115, DOI: 10.48550/arXiv.2008.11115 (ref. 9 in the paper).

J. Olof Johansson asked: You mention that you rule out coherent 3-photon absorption processes to enable above-gap excitation. However, could it not be possible that the populated gap-states can be excited to reach the conduction band in a 1 + 2 photon excitation step?

Hermann Dürr answered: We considered this possibility while preparing our paper (https://doi.org/10.1039/d2fd00005a). The band gap of NiO is about 4.0 eV, and the tabulated optical absorption coefficient of NiO for sub-gap excitation is 2–3 orders of magnitude lower than the above-gap excitation. In ref. 1 it was argued that two-photon absorption at a fundamental photon energy of 2.16 eV leads to the observed lattice distortions. However, the lower photon energy in our case requires three-photon absorption events. We cannot rule out that this leads to some build-up of a transient electron population in the O 2p conduction bands. However, our O K-edge XAS is sensitive to that resulting in a bleaching of the 1s–2p transition. We do not observe this to happen to a significant degree for the state forming the upper band gap edge.

1 Y. W. Windsor, D. Zahn, R. Kamrla, J. Feldl, H. Seiler, C.-T. Chiang, M. Ramsteiner, W. Widdra, R. Ernstorfer and L. Rettig, *Phys. Rev. Lett.*, 2021, **126**, 147202 (ref. 3 in the paper).

J. Olof Johansson queried: What complementary experiments would you like to do?

Hermann Dürr replied: There are two types of complementary experiments we would like to perform. The first type of experiment is to study the magnon oscillations at different transferred wave-vectors by capturing the X-ray reflection at different incident angles. In the current study (https://doi.org/10.1039/d2fd00005a), we focused on one particular transferred wave-vector. By performing measurements at other transferred wave-vectors, we will be able to measure the dispersion of excited magnon branches. The second type of experiment is to perform time-resolved measurements at the Ni L_3-edge with an improved signal-to-noise level. In our current experiment, we observed the magnon oscillation at the O K-edge. It will be interesting to investigate similar fingerprints for the dynamical evolution of the line shape at the Ni L_3-edge.

requested: Could you comment on the importance of sample preparation and characterisation when comparing results between different groups studying NiO?

Hermann Dürr responded: It is important to perform standard optical spectroscopy measurements to characterise the optical absorption of the NiO samples. In our experiment, we use ultrathin (40 nm) NiO films. For these films epitaxial strain could result in additional defects compared to bulk single crystals. This is known to enhance the dipole forbidden d–d transition intensity detected optically.

Jan Rost opened discussion of the paper by Mohammed Hassan: You obtained an internal delay time by extrapolating the linear dependence of the reflectivity on intensity to zero. How do you interpret this time?

Mohammed Hassan answered: The measured delay response from the reflectivity measurements shows a linear trend as the pump pulse intensity increases, suggesting this delay is more related to Stark shift. We strongly believe that the obtained time by extrapolating the linear dependence of the pump pulse intensity is attributed to the time it takes for the electron to be excited from the valance band to the conduction band in the dielectric system (assuming the linear relation is maintained at low driver field intensity).

Jan Rost asked: What is the relation of your reflection time delays to photo ionization solid-state time delays? In the latter, control of the surface of the target was very important to obtain reproducible results. Does the surface also play a role for the reflectivity time delays?

Mohammed Hassan responded: The study presented in this paper (https://doi.org/10.1039/d2fd00002d) is related to the electronic delay response in the neutral material, not the ionized solid state. The photoionization studies measured the time it takes for an electron to escape to the continuum after absorption of the photon. In the presented study, we measure the time it takes for the electron to be excited from the valence band to the conduction band by absorbing photons in the dielectric material. The quality of the surface is not playing a crucial role in the reflectivity delay measurements.

Martina Basini enquired: You propose a model for which the electrons acquire momentum while moving in the conduction band following the field profile, *i.e.* intraband transition. Do you think that an appreciable change in the interband dipole moment could be associated with the above-mentioned dynamics?

Mohammed Hassan replied: Yes, a light-propagation calculation is performed using a multiscale Maxwell-TDDFT method to obtain the reflected field from a SiO_2 sample as presented in another paper from our group (please see the supplementary information).[1]

1 D. Hui, H. Alqattan and S. Yamada, *et al.*, Attosecond electron motion control in dielectric, *Nat. Photon.*, 2022, **16**, 33–37, DOI: 10.1038/s41566-021-00918-4.

Martina Basini asked: How do you consider the possibility of intraband valence transitions? Can we imagine they are strongly suppressed as the electron mobility is much lower than the conduction band one? Also how do you interpret the phase delay as a function of the laser fluence?

Mohammed Hassan responded: We interpret the phase delay as the time it takes for the electrons to be excited from the valence band to the conduction band due to the increase in the excitation carrier density, the AC Stark effect, and the system's polarizability.

J. Olof Johansson questioned: You say a linear dependence in the time delay means that the Stark shift is involved. Does that not scale quadratically on the electric field?

Mohammed Hassan replied: As mentioned in the paper (https://doi.org/10.1039/d2fd00002d), the delay response may be attributed to the increase in the excitation carrier density, the AC Stark effect, and the system's polarizability. The effect of the Stark shift could be investigated individually in future work.

Aswathy V. Girija said: In the linear plot (relative delay plot) for calcium fluoride (Fig. 4b in the paper; https://doi.org/10.1039/d2fd00002d), we can see that at high and low fluences it follows a linear behaviour, whereas in the centre it is a bit shifted. Please comment on why that's the case.

Mohammed Hassan responded: We believe the general behavior is linear and the two points in the center are in the tolerance of the experimental measurements.

Aswathy V. Girija enquired: One of the major conclusions of this work is that the phase transition dynamics are independent of the type of the material – is looking at just two dielectric systems enough to generalize this?

Mohammed Hassan replied: This work (https://doi.org/10.1039/d2fd00002d) presents a phase transition in two different dielectric systems with two different electronic structures and bandgaps (9 eV, 12 eV), so this would support our conclusion that the phase transition dynamics are field-driven (as reported previously)[1] and independent of the type of the material. Indeed, this could be confirmed with future experiments.

1 D. Hui, H. Alqattan, S. Yamada, *et al.*, Attosecond electron motion control in dielectric, *Nat. Photon.*, 2022, 16, 33–37.

Jan Rost asked: In photoionization attosecond delay experiments the cleanliness of the surface has turned out to influence the delays considerably – before this was known, there was confusion about the measured delays and their interpretation. Is the surface condition also important for reflection measurements or does it not play a role since your measurements are all optical?

Mohammed Hassan answered: Thanks for your question! I believe you are correct, the all-optical reflectivity measurement is less sensitive to the impurities on the surface. Please note that this reflectivity modulation experiment is performed in ambient conditions.

Steven Johnson queried: Is it possible to understand and/or model these observations by assuming complex-valued third or higher-order nonlinear susceptibilities of the samples?

Mohammed Hassan replied: The strong field interaction with dielectric can be expressed and modeled as a combination of multiple Lorentz resonators upon the pump excitation as we explained in another work under review.[1]

1 D. Hui, H. Alqattan, S. Zhang, V. Pervak, E. Chowdhury and M. Th. Hassan, *Attosecond Optical Switching*, 2022, DOI: 10.21203/rs.3.rs-1232650/v1.

J. Olof Johansson enquired: Can the mask used to create the pump and probe beams not cause strange effects in the laser focus (like diffraction rings), which can make interpretation of the results harder?

Mohammed Hassan responded: The beam profile characterization of the pump and probe beams using our mask doesn't show any strange effects in the laser focus.

J. Olof Johansson said: Follow-up on the mask: does this not mean that your weak probe pulse has the same peak intensity as the strong pump?

Mohammed Hassan answered: No, as mentioned in the manuscript (https://doi.org/10.1039/d2fd00002d), the opening of the hole for the probe beam is 1 mm, while it is 3 mm for the pump beam. Accordingly, the beam size at the focus is different, meaning the pump's peak intensity is way higher than the probe beam.

Kota Katsumi enquired: I understand that the phase shift is identified when the pump intensity is increased. How does the "amplitude" of the signal change as a function of the pump fluence? If the amplitude scales linearly to the pump intensity, I guess we can disentangle the origin of the signal from the Kerr effect? Can you comment on the fluence dependence of the signal amplitude?

Mohammed Hassan replied: Thank you so much for this valuable comment. We considered your suggestion and reanalyzed our data to study the amplitude change (for CaF_2) as a function of the pump field intensity. We did that by squaring the reflectivity trace and calculating the integration (area under the curve), then plotted the integration as a function of the field intensity (Fig. 1 here). The curve shows a linear dependence that excludes the measured delay due to a nonlinear process.

Lin Chen asked: You have shown the phase shift as a function of fluence of excitation photons. How did you obtain the waveforms of these fs pulses in the visible region?

Mohammed Hassan responded: The Total Reflectivity Modulation (TRM) trace is obtained by integrating the reflected spectrum at each time instance identical to the vector potential of the measured unknown waveform.[1] The electric field of

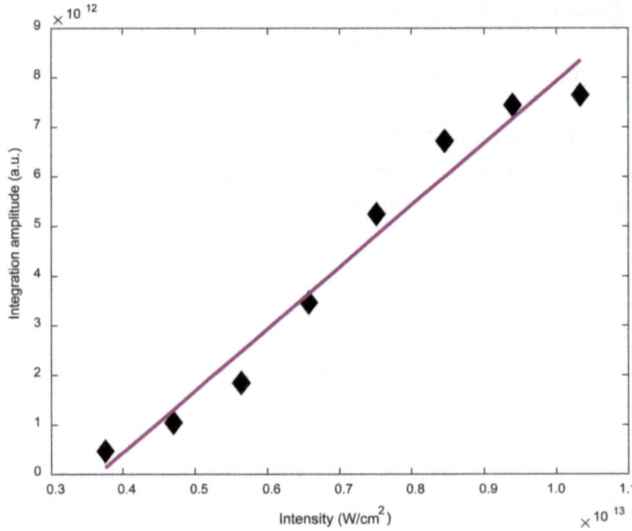

Fig. 1 Integration amplitude as a function of the field intensity for CaF_2.

the strong driver pulses can be retrieved by calculating the time derivative of the vector potential (obtained from the TRM):

$$E(t) = \frac{\partial A(t)}{\partial t} \tag{1}$$

The derivation is carried out in the frequency domain on $A(\omega)$, which is the Fourier transform of $A(t)$. To avoid the high-frequency noise, a bandpass filter is applied, $H(\omega)$, as shown in eqn (2) and (3) below. The spectral window can be specified by changing its width, σ, and center, ω_0, based on the pump pulses' measured spectrum.

$$E(t) = \int_{-\infty}^{\infty} \mathrm{i}\omega \times A(\omega) \times H(\omega) \times \mathrm{e}^{\mathrm{i}\omega t}\, \mathrm{d}\omega \tag{2}$$

where,

$$H(\omega) = \exp\left(-\frac{1}{2}\left(\frac{\omega - \omega_0}{\sigma}\right)^{\text{order}} \right) \tag{3}$$

1 D. Hui, H. Alqattan, S. Yamada, *et al.*, Attosecond electron motion control in dielectric, *Nat. Photon.*, 2022, **16**, 33–37.

Conflicts of interest

There are no conflicts to declare.

Faraday Discussions

ROYAL SOCIETY OF CHEMISTRY

PAPER

Petahertz charge dynamics in a correlated organic superconductor

Shinichiro Iwai, *[a] Yohei Kawakami, [a] Hirotake Itoh [a] and Kenji Yonemitsu [b]

Received 10th January 2022, Accepted 7th February 2022
DOI: 10.1039/d2fd00004k

We report an observation of stimulated emission induced by a nearly single-cycle 6 fs near infrared electric field of 10 MV cm^{-1} in an organic superconductor (κ-(h-ET)$_2$Cu[N(CN)$_2$]Br). The stimulated emission is attributed to a non-linear synchronized coherent charge oscillation. We also report that, in the same organic superconductor, a light-induced current before the scattering time shows up as carrier-envelope phase (CEP)-sensitive second harmonic generation (SHG). This unconventional SHG in the centrosymmetric compound is regarded as a light induced spatial symmetry breaking. These ultrafast optical non-linearities induced by petahertz charge oscillations show anomalous enhancements around the superconducting transition temperature (11.6 K). These results indicate that the microscopic mechanism of superconducting fluctuations is closely related to the Coulomb repulsive interaction in this compound.

Introduction

(A) Ultrafast charge dynamics in organic superconductors within no-scattering time window

Superconductors are characterized by a gap (superconducting energy gap) for the excitation of quasi-particles with an energy of *ca.* meV. However, the electronic properties of the superconductivity are not necessary determined only by such low-energy interactions.[1–3] It is well known that competing energetic factors, such as Coulomb repulsion (U) and inter-site hopping (t), as well as contributions from a spin–orbit interaction and/or an antiferromagnetic interaction, play important roles in strongly correlated electron systems, as shown in Fig. 1. Such a hierarchy on the energy (time) axis is useful for understanding the electronic and magnetic properties of superconductors with strong electron correlations, such as high-temperature superconducting cuprates and organic superconductors.

However, it is not necessarily true that there is no relation between the Coulomb repulsion (>1 eV) and the superconducting gap (~meV), although the former is 3-orders of magnitude larger than the latter. In fact, the contributions of

[a]Department of Physics, Tohoku University, Sendai 980-8578, Japan. E-mail: s-iwai@tohoku.ac.jp
[b]Department of Physics, Chuo University, Tokyo 112-8551, Japan

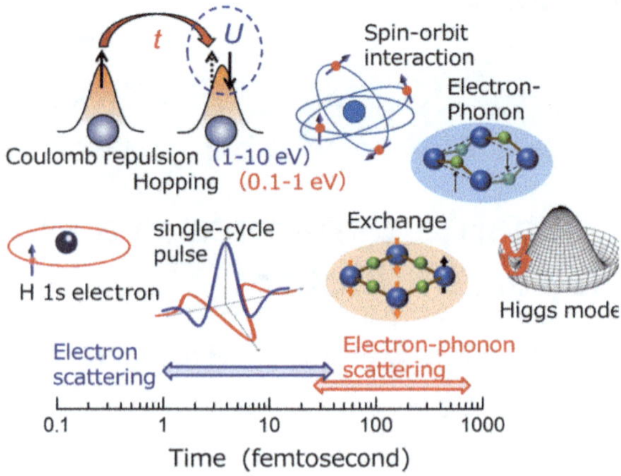

Fig. 1 Time scales of various interactions in strongly correlated systems.

anti-ferromagnetic interactions and Coulomb repulsion are suggested to be important for the high superconducting transition temperature (high-T_{SC}), the isotope effects and the symmetry of the *d*-wave pairing. If we assume that the microscopic origin of the superconductivity is related to the Coulomb repulsion (in the order of 1 eV (organic conductors) and 10 eV (cuprates)), ultrafast responses on the time scales of several hundreds of attoseconds (as) or several femtoseconds are expected to be realized. The photoexcitation of superconductors by visible or near infrared light is known to be a mechanism for superconducting photodetection. This optical response has been understood as the generation of quasi-particles,[4] which necessarily results in an increase in the electron temperature or, equivalently, a disappearance of electronic coherence, losing the controllability of the non-equilibrium state. A low-energy excitation by THz or mid-IR lights is an excellent solution to the problem.[5–8] But, here, we focus on another approach. It is to complete the excitation before the increase in the electron temperature (or before the electron scattering processes become dominant).

In some materials, including organic compounds, the increase in the electron temperature needs *ca.* 40 fs (corresponding to the energy scale of 0.1 eV $= \hbar/40$ fs).[9] If we use a several-femtosecond pulse for those compounds, we can control charge motion before the increase in the electron temperature. In such a no-scattering time window, any degree of freedom hardly reaches equilibrium, and the electronic coherence survives on the time scale from several to several tens of femtoseconds, which corresponds to the energy scale of the inter-site hopping.

This is a new pathway toward the ultrafast control of correlated electrons. In fact, we have realized the coherent modulation of the electronic state before the scattering time by using a 7 or 6 fs, nearly single-cycle pulse in an organic conductor.[10–14] Thus, the excitation within the ultrafast time window is the key issue for realizing coherent charge dynamics which are not disturbed by scattering processes.

Here, we focus on the petahertz charge dynamics in an organic superconductor (κ-(h-ET)$_2$Cu[N(CN)$_2$]Br).[13] We report an observation of stimulated

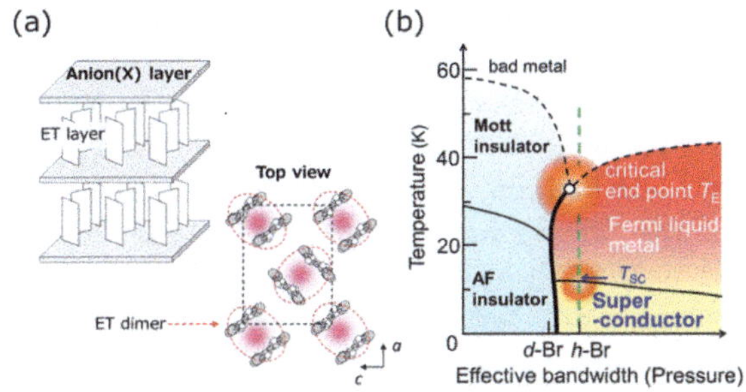

Fig. 2 (a) Schematic illustrations of the crystal structure and triangular lattice of ET dimers in κ-$(ET)_2X$. (b) Temperature–bandwidth phase diagram.

emission induced by a nearly single-cycle 6 fs near infrared electric field. The stimulated emission is attributed to a non-linear synchronized coherent charge oscillation. We also report that, in the same organic superconductor, a light-induced current before the scattering time driven by the 6 fs pulse shows up as second harmonic generation (SHG), which is regarded as a light-induced spatial symmetry breaking in this centrosymmetric compound.[14]

(B) The organic superconductor κ-$(h\text{-}ET)_2Cu[N(CN)_2]Br$ (h-Br)

κ-$(h\text{-}ET)_2Cu[N(CN)_2]Br$ (h-Br) (ET: bis[ethylenedithio]-tetrathiafulvalene), shown in Fig. 2(a), is a well-known layered organic superconductor with a superconducting transition temperature of T_{SC} = 11.6 K.[15–21] The triangular lattice consisting of ET dimers is effectively regarded as a 1/2-filled system, although the averaged charge per ET molecule is +0.5 (3/4 filling). The temperature–bandwidth phase diagram of κ-$(ET)_2X$ (X: anion molecule), where the bandwidth is proportional to the inter-site hopping, is characterized by the first-order Mott transition line and the critical end point (T_{END} = 33 K), as shown in Fig. 2(b). Superconducting fluctuations toward T_{SC} below $2T_{SC}$ have been discussed in terms of a short-range correlation of the Cooper pairs.[22,23]

Experimental

(A) Sample preparation

Single crystals of h-Br (typically 0.7 × 0.5 × 0.8 mm for *a*, *b*, and *c*-axes, respectively) and a thin film of h-Br ($0.5 \times 1.8 \times 10^{-4} \times 0.5$ mm on a CaF_2 substrate with a thickness of 0.5 mm) were prepared using the methods described in previous studies.[20,24]

(B) 6 fs infrared pulse generation and pump–probe measurements

A broadband infrared spectrum covering 1.2–2.3 μm of the 6 fs pulse was obtained by focusing a carrier-envelope phase (CEP) stabilized idler pulse (1.7 μm) from an optical parametric amplifier (Quantronix HE-TOPAS pumped by Spectra-Physics Spitfire-Ace) onto a hollow fibre set within a Kr-filled chamber (Femtolasers). Pulse compression was performed using both active mirror (OKO, 19-ch linear MMDM) and chirped mirror (Femtolasers and Sigma-Koki) techniques.[10–14]

We performed transient reflectivity and absorption measurements for the single crystal and the thin film[24] using the 6 fs pulse. In these measurements, the intensity of the pump pulse was controlled in the wide range from 0.01 to 2 mJ cm^{-2}, with a step of 0.006 mJ cm^{-2}, by a pair of wire-grid CaF_2 polarizers.[12] In the transient reflectivity/absorption measurement, the probe pulse reflected/transmitted from the sample was detected by InGaAs detectors (New-Focus model 2034) after passing through spectrometers (JASCO, M10). We also performed a transient reflectivity measurement with a double-pump pulse, which was generated by a Michelson interferometer.

(C) Measurement and CEP dependence of SHG

We performed SHG and third harmonic generation (THG) measurements for the single crystal using a 6 fs pulse with a reflection geometry (incident angle is smaller than 3°). The fundamental photon energy was 0.75 eV. The intensity and polarization of the fundamental pulse (excitation range: 0.01 to 2 mJ cm^{-2}) were controlled by a pair of wire-grid CaF_2 polarizers.[12] The SHG and the THG were detected by a photomultiplier tube (Hamamatsu R13456) after passing through a spectrometer (JASCO, M10).

The CEP of the fundamental pulse (Fig. 3(a)) was controlled by a pair of glass plates[25,26] (BK-7 with a thickness of 1 mm) with the incident angle of δ (Fig. 3(c))

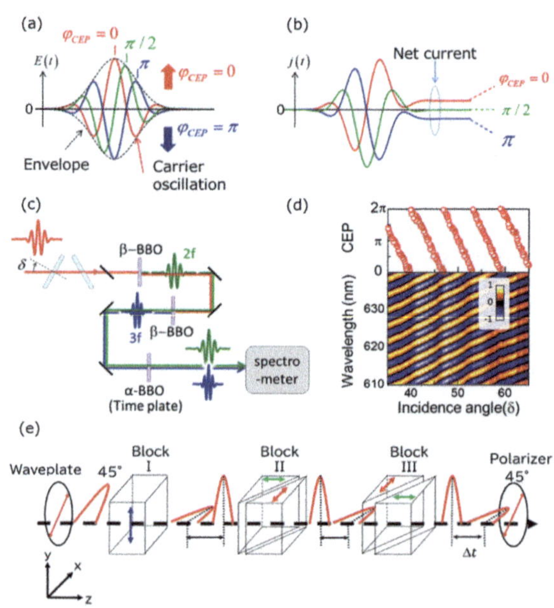

Fig. 3 (a) and (b) Schematic illustrations of single-cycle light field $E(t)$ and current $j(t)$ induced by no-scattering charge acceleration (see the main text), respectively. (c) Set-up of 2f–3f interferometer to detect the relative change in the CEP, which is controlled by a pair of glass plates (δ: incidence angle). (d) Lower panel: interference spectra between 2f and 3f as a function of δ. Upper panel: relative CEP (obtained from the 2f–3f spectra) as a function of δ. (e) Schematic illustration of TWINS[28,29] using α-BBO (negative uniaxial crystal). The blue, red and green arrows indicate the optical axes (fast axis) for the birefringent plate (block I) and wedges (in blocks II and III).

and detected by a 2f–3f interferometer[27] (2f and 3f were generated using β-BBO) (Fig. 3(c) and (d)).

(D) Measurement of 4th order nonlinear polarization

In the measurement of higher (>4th) order harmonic generations (HHGs), absorption losses sometimes prevent the detection because of various optical transitions, such as intramolecular transitions, above the low-energy charge gap in strongly correlated materials. In fact, h-Br has a large absorption loss above 3 eV, although there is a spectral window between 0.6–2.5 eV for SHG and THG. Here, we observe 4th order polarization by measuring SHG and THG that can be modulated by higher order polarizations. This is understood from the concept of 2-D spectroscopy,[28–31] *i.e.*, one axis of the 2 dimensions (2-D) is set up by the conventional dispersion spectrometer, and the other is obtained by the Fourier transform of the interferogram (see also Fig. 11(a) and (b)).

A pair of pulses with a time difference of Δt enters the material and the emitted SHG and THG are detected by changing Δt. Here, Δt is scanned with an accuracy of <100 attoseconds by using the TWINS (translating-wedge-based identical pulses encoding system) technique (Fig. 3(e))[28,29] for a 1–2 micron range. The TWINS consists of birefringent plates and wedges of α-BBO (a negative uniaxial birefringent material). In Fig. 3(e), the y polarization component of incident 45° (in the x–y plane) polarization light advances in comparison with the x polarization component after block I with the fast axis of y (blue arrow). Then, the y polarization component is delayed by propagating in the wedge with the fast axis of x (red arrow) in block II. By moving block II along the x axis, we can change the time difference between the x- and the y-polarized pulses with an accuracy of <100 attoseconds. Block III is used to compensate the changes of dispersions. Finally, the 45° components from both the pulses are extracted to obtain the double-pulse in the same polarization. The interferogram (light intensity as a function of Δt) is detected at the wavelengths of SHG and third harmonic generation (THG). A modulation induced by the 4th order polarization is obtained in the Fourier transformed spectra of the interferograms. That is a part of non-linear 2-D spectroscopy where the interplay between different order harmonics is detected as an off-diagonal component (see Fig. 11(b)).

Results and discussion

(A) Stimulated emission induced by synchronized charge motion

Fig. 4(a) shows the steady-state reflectivity (R) and transient reflectivity ($\Delta R/R$) spectra of a single crystal of h-Br. The polarization of the pump and probe pulses are parallel to the c-axis. The excitation intensity (I_{ex}) and the time delay between the pump and probe pulses (t_d) are 1.0 mJ cm^{-2} (corresponding to 11 MV cm^{-1}) and 10 fs (red line), 200 fs (magenta line), and 500 fs (violet line), respectively. A large increase in R ($\Delta R/R \sim 180\%$ at 0.63 eV) is observed at 10 fs on the higher energy side of the dimer band, which is assigned to the transition from the bonding to antibonding states of an ET dimer in Fig. 2(a). The spectral width of the peak at $t_d = 200$ fs (0.04 eV $= \hbar/(100$ fs)), which is much narrower than that of the dimer band (170 meV), is roughly consistent with the time constant of the fast decay component (70 fs, as described later).

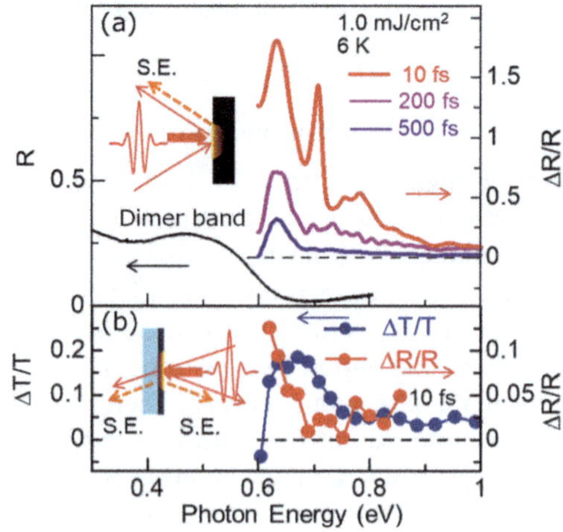

Fig. 4 (a) Transient reflectivity ($\Delta R/R$) and steady-state reflectivity (R) spectra ($\|c$) of a single crystal for $I_{ex} = 1.0$ mJ cm^{-2} at time delays of 10 fs, 200 fs, and 500 fs. (b) Transient reflectivity ($\Delta R/R$) and transmittance ($\Delta T/T$) spectra of 180 nm thin film κ-(BEDT-TTF)$_2$Cu[N(CN)$_2$]Br for $I_{ex} = 1.0$ mJ cm^{-2} at 10 fs. S.E. in the inset shows stimulated emission.

At shorter t_d = 10 fs, the $\Delta R/R$ spectrum has a broad tail with structures reflecting the uncertainty relation between time and frequency. That is a so-called coherent artifact and does not include information of the material. The transient reflectivity and transient transmittance ($\Delta T/T$) of a 180 nm thin film under the same experimental conditions are shown in Fig. 4(b). We notice that $\Delta T/T$ shows a large increase (~20%) at 0.67 eV. We confirmed that $\Delta T/T$ in the CaF_2 substrate is negligible. Note that $\Delta R/R > 0$ and $\Delta T/T > 0$ in the same spectral range in Fig. 4(b) are ascribed to stimulated emission or, equivalently, an optical gain.

Fig. 5(a) shows the time profile of the $\Delta R/R$ peak at $I_{ex} = 1.0$ mJ cm^{-2}, which is reproduced by using the conventional method, as shown by the black line obeying the equation,

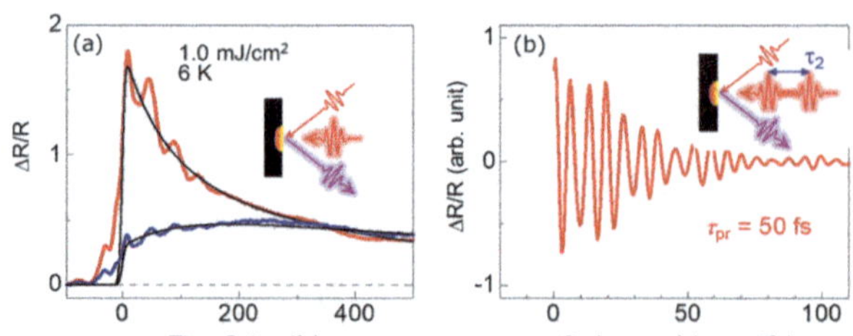

Fig. 5 (a) Time profiles of $\Delta R/R$ peak (0.63 eV) at $I_{ex} = 1.0$ mJ cm^{-2} (red line) and $I_{ex} = 0.1$ mJ cm^{-2} (blue line), which are analyzed by using the conventional method (black curves; see the main text). (b) Oscillating component of $\Delta R/R$ at 0.63 eV as a function of the time difference between the two pump pulses (τ_2) in the transient reflectivity measurements with a double-pump pulse (6 K, 1 mJ cm^{-2}). The delay between the second pump and the probe pulses is set at $\tau_{pr} = 50$ fs.

$$\Delta R(t)/R = \int_{-\infty}^{+\infty} K(t)G(t-t')\mathrm{d}t'$$

$$K(t) = A_{\text{fast}}\left[1-\exp(-t/\tau_{\text{f-rise}})\right]\exp\left(-\frac{t}{\tau_{\text{f}}}\right) + \left[1-\exp(-t/\tau_{\text{s-rise}})\right] \times \sum_{n=1}^{3} A_{\text{slow }n}\exp\left(-\frac{t}{\tau_{sn}}\right)$$

$$G(t) = \exp[-4\ln 2t^2/\{(9\ (\text{fs}))^2\}],$$

where the Gaussian $G(t)$ with a width of time resolution (=9 fs) is an instrumental response function. A_{fast} and $A_{\text{slow }n}$ ($n = 1$–3) represent the coefficients for the fast and slow components, respectively. For the red line in Fig. 5(a) ($I_{\text{ex}} = 1.0$ mJ cm^{-2}), the fast build-up ($\tau_{\text{f-rise}} < 5$ fs) with decay ($\tau_{\text{f}} = 70$ fs) is dominant (65.4%) in comparison with the slow rise ($\tau_{\text{s-rise}} = 90$ fs), with three-component decay $\tau_{\text{s1}} =$ 360 fs (29.7%), $\tau_{\text{s2}} = 4.2$ ps (1.3%), $\tau_{\text{s3}} > 100$ ps (3.6%) and the oscillation (period 43 fs, damping time 70 fs, initial phase, -0.15π). For the blue line in Fig. 5(a) [$I_{\text{ex}} = 0.1$ mJ cm^{-2} ($\times$4)], the fast component (with the same $\tau_{\text{f-rise}}$ and τ_{f}) is smaller (32.2%), while the slow rise is dominant (67.8%; $\tau_{\text{s1}} = 360$ fs (20.8%), $\tau_{\text{s2}} = 2.2$ ps (45.5%), $\tau_{\text{s3}} > 100$ ps (1.5%)). The oscillating component with a period of 43 fs in the time profile is attributed to a coherent intra-molecular vibration [$\nu_{60}(\text{B}_{3\text{g}})$], which strongly interacts with the electronic state.[32,33] Thus, the fast rise ($\tau_{\text{f-rise}} < 5$ fs) and decay ($\tau_{\text{f}} = 70$ fs) component of $\Delta R/R$, reflecting the stimulated emission, becomes dominant for the strong excitation.

Recall that the spectral width of the peak (0.04 eV = $\hbar$/(100 fs)) at $t_{\text{d}} = 200$ fs, which is much narrower than that of the dimer band (0.17 eV) shown in Fig. 4(a), is roughly consistent with the time constant of the fast decay component of 70 fs. This fact suggests that the stimulated emission is induced by only one electronic mode, which is selectively driven in a coherent manner, avoiding dissipation. To confirm such electronic coherence, we perform transient reflectivity measurements with two pump pulses. Such double-pulse experiments have been used for investigating the electronic coherence of the photoinduced phase[34] or coherent phonons.[35] In our case, we can measure the coherence at the origin of the stimulated emission.

Fig. 5(b) shows an oscillating component of $\Delta R(t)/R$ measured at 0.63 eV as a function of the time difference between the two (the first and the second) pump pulses (τ_2) (6 K, 1 mJ cm^{-2} for the respective pump pulses). The delay between the second pump and the probe pulses is set at 50 fs (within the lifetime of the stimulated emission). The observed oscillation shows a dephasing time of *ca.* 40 fs, indicating that the coherence induced by the pump light survives for a much longer time than the pulse duration. It is noteworthy that the coherence time obtained in this measurement is comparable to the above mentioned τ_{f} (70 fs), reflecting the lifetime of the stimulated emission. This fact indicates that the fast rise-and-decay component is governed by the electronic coherence. Considering that the fast component is dominant for the strong excitation, the origin of the stimulated emission is attributed to the coherent charge motion driven by the strong light field. This is very different from the conventional mechanism for the stimulated emission, which is due to population inversion.

To clarify the origin of the non-linear charge motion, *i.e.*, the $\Delta R/R$ peak at 0.63 eV (or equivalently the stimulated emission), we theoretically investigated charge oscillations which are driven by the single-cycle light-field in a two-dimensional (2D) 3/4-filled extended Hubbard model for the 16-site system.[13,36] Fourier transform (FT) spectra of the charge density time profile of a molecule are shown in the lower panel of Fig. 6(a). The weight of the FT spectrum for the field amplitude $F = 0.01$ (V A^{-1}) along the *c*-axis, on-site Coulomb repulsion $U = 0.8$ eV, nearest-neighbor Coulomb repulsions between sites i and j $V_{ij} = 0$ (green dots in the lower panel of Fig. 6(a)) is mainly distributed at $\leq$0.7 eV, which is the main energy range of the steady-state optical conductivity (σ) for polarization along the *c*-axis (upper panel of Fig. 6(a)). The energy scale is slightly larger than the experimental one owing to the smallness of the system treated here. The FT spectrum for $F = 0.01$ corresponds to linear charge density oscillations, including transitions from the bonding to anti-bonding states around a dimer. With an increase in F, the spectral weight at the higher energy ($\sim$0.9 eV) region becomes dominant. This FT spectral weight is attributable to the synchronized charge oscillation[37–39] shown in Fig. 6(b) and (c). Also, for finite V_{ij} (the orange dashed line in the lower panel of Fig. 6(a) for $F = 0.16$), the spectral nature is essentially the same. This oscillation is driven by charge transfers through all the bonds between charge-rich and poor sites, *i.e.*, charges flow into a site denoted by a white ellipse from four neighboring gray sites through the b_1, b_2, and two q bonds (Fig. 6(b)), and then charges flow back to the gray sites (Fig. 6(c)). The observed peak energy of 0.63 eV indicates that the oscillation period is $\sim$6 fs.

The highlight of this section is an enhancement of $\Delta R/R$ near T_{SC} and T_{END}. The temperature dependence of the $\Delta R/R$ spectrum is shown for $t_d = 10$ fs at $I_{ex} =$ 1.0 mJ cm^{-2} (Fig. 7(a) upper panel) and 0.01 mJ cm^{-2} (Fig. 7(a) lower panel). The 0.63 eV peak grows with lowering the temperature below 50 K for both I_{ex}. The ultrafast response of $\Delta R/R$ (0.63 eV) for $t_d = 10$ fs at large I_{ex} (Fig. 7(a) upper panel) indicates an anomalous enhancement near T_{END} ($\sim$30 K) and a further increase at lower temperatures between $\sim 2T_{SC}$ and T_{SC}. It is well known that

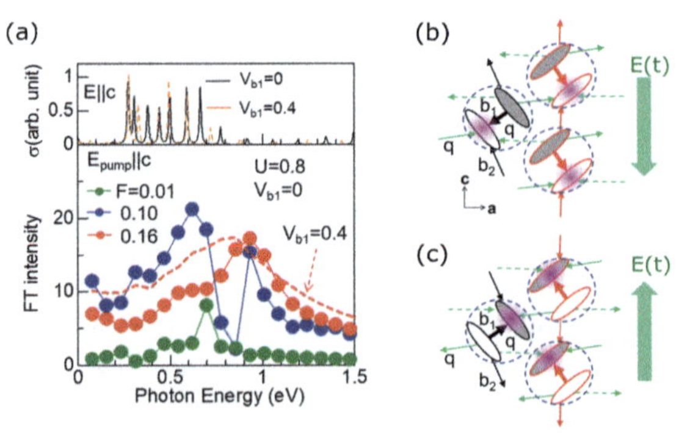

Fig. 6 (a) Upper panel: calculated optical conductivity spectrum (polarization along *c*-axis). Lower panel: calculated FT spectra of the charge density time profile of a molecule for $F = 0.01$ (green dots), 0.10 (blue dots), and 0.16 (red dots). (b) and (c) Schematic illustrations of charge motion through bonds, which are indicated by arrows.

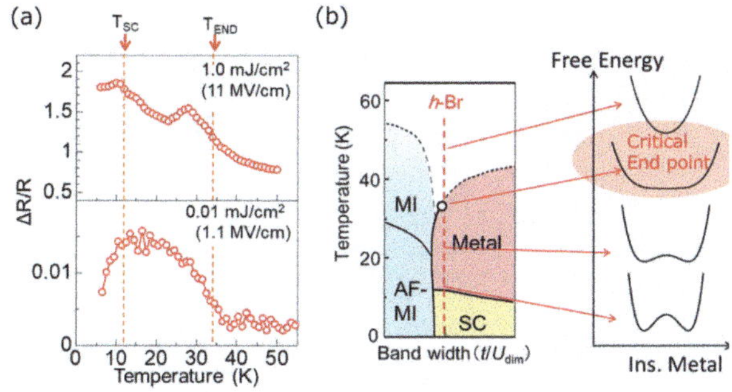

Fig. 7 (a) Temperature dependence of $\Delta R/R$ at 0.62 eV for $t_d = 10$ fs (upper panel) and for $I_{ex} = 1.0$ mJ cm^{-2} (lower panel). (b) Temperature–bandwidth phase diagram and schematic free energy curves at the respective places in the phase diagram.

superconducting fluctuations exist in this temperature range, reflecting the short-range correlation of Cooper pairs.

The application of an $I_{ex} = 1.0$ mJ cm^{-2} (11 MV cm^{-1}) light-field is expected to increase the electron and lattice temperatures. Considering the linear coefficient of the temperature-dependent specific heat $\gamma = 22$ mJ K^{-2} mol^{-1},[40] the electron temperature is finally expected to be 520 K. However, the rise in the electron temperature is negligible at $t_d = 10$ fs (when the anomalies are observed in the temperature dependence (Fig. 7(a))) because electrons are scattered only a few times before in this compound. Note that the anomalies at ~10 K and ~28 K show 2–4 K shifts (at 1.0 mJ cm^{-2}) to the low-temperature side because of a heat accumulation in a 1 kHz operation.[13]

This ~10 fs response is in contrast to the energy scale of the superconducting gap and its fluctuation (~meV from $T_{SC} = 11.6$ K), corresponding to a picosecond timescale. For smaller I_{ex}, on the other hand, the increase in $\Delta R/R$ below $\sim 2T_{SC}$ at $t_d = 10$ fs becomes obscure (0.01 mJ cm^{-2}, lower panel in Fig. 7(a)), while the anomaly at $\sim T_{END}$ remains. No anomalies are observed in the temperature dependence of $\Delta R/R$ in insulating κ-(*d*-BEDT-TTF)$_2$Cu[N(CN)$_2$]Br, which does not have T_{SC} or T_{END} located on the small t/U_{dimer} side of the phase diagram.[13]

The enhancement of the non-linear charge motion at $\sim T_{END}$ can be attributed to the absence of the potential barrier for the first order phase transition (Fig. 7(b)). In such a situation, a strong light field can induce the non-linear charge response shown in Fig. 6(b) and (c) with a large amplitude. Since the Coulomb repulsion is larger than 0.4 eV ($=\hbar/(10$ fs)), the ultrafast ~10 fs response of the anomaly at $\sim T_{END}$ is reasonable, *i.e.*, it confirms that the insulator to metal transition is of electronic origin because the timescale of inter-molecular motion (>200 fs) is much longer.

(B) Unconventional SHG induced by no-scattering current

In this section, we focus on the SHG of h-Br. This compound is centrosymmetric,[16,17] so SHG is generally believed to be forbidden. However, it becomes active for the strong light field because of a light-induced no-scattering current.

Fig. 8(a) shows spectra of the SHG (red ($\boldsymbol{E}_{fund}\|\boldsymbol{c}$, $\boldsymbol{E}_{SH}\|\boldsymbol{c}$), green ($\boldsymbol{E}_{fund}\|\boldsymbol{a}$, $\boldsymbol{E}_{SH}\|\boldsymbol{c}$, ×0.73)) and THG (blue line ($\boldsymbol{E}_{fund}\|\boldsymbol{c}$, $\boldsymbol{E}_{TH}\|\boldsymbol{c}$, ×0.024)) at 6 K. The respective peak energies are 1.5 eV (SHG) and 2.2 eV (THG). $\boldsymbol{E}_{fund}$, $\boldsymbol{E}_{SH}$, and $\boldsymbol{E}_{TH}$ indicate the

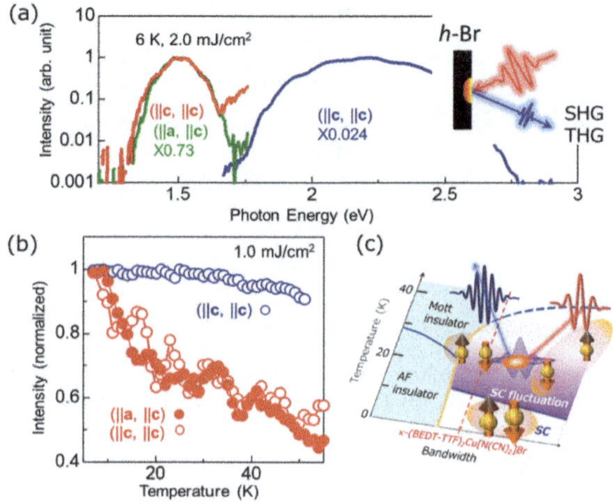

Fig. 8 (a) SHG spectra of h-Br (single crystal) at 6 K (($\boldsymbol{E}_{\text{fund}}\|\boldsymbol{c}$, $\boldsymbol{E}_{\text{SH}}\|\boldsymbol{c}$) (red line) and ($\boldsymbol{E}_{\text{fund}}\|\boldsymbol{a}$, $\boldsymbol{E}_{\text{SH}}\|\boldsymbol{c}$) (green line, ×0.73)). THG spectrum (×0.024) ($\boldsymbol{E}_{\text{fund}}\|\boldsymbol{c}$, $\boldsymbol{E}_{\text{TH}}\|\boldsymbol{c}$) (blue line). (b) Temperature dependence of I_{SH} (closed red circles: $\boldsymbol{E}_{\text{fund}}\|\boldsymbol{a}$, $\boldsymbol{E}_{\text{SH}}\|\boldsymbol{c}$, open red circles: $\boldsymbol{E}_{\text{fund}}\|\boldsymbol{c}$, $\boldsymbol{E}_{\text{SH}}\|\boldsymbol{c}$) and I_{TH} (blue circles, $\boldsymbol{E}_{\text{fund}}\|\boldsymbol{c}$, $\boldsymbol{E}_{\text{TH}}\|\boldsymbol{c}$). Both are normalized by the respective intensities at 6 K. (c) Temperature–bandwidth phase diagram of κ-(ET)$_2$X, which is extracted on the basis of the controlled effective bandwidth.

electric fields of the excitation light, second harmonic and third harmonic lights, respectively. The intensity of the SHG is *ca.* 1/50 times smaller than that of the THG. Note that the SHG is not active for weak fields in this compound because this compound has inversion centers (orthorhombic structure with class P_{nma}), as mentioned above. This SHG is not attributed to a surface effect because of its unconventional dependence on the temperature and the CEP, as shown below.

As shown in Fig. 8(b), the temperature dependence of the SHG (for both $\boldsymbol{E}_{\text{fund}}\|\boldsymbol{c}$ (closed red circles) and $\boldsymbol{E}_{\text{fund}}\|\boldsymbol{a}$ (open red circles)) shows a clear increase toward T_{SC}, although we have no significant temperature dependence for the THG. As mentioned above, superconducting fluctuations exist in the temperature range above T_{SC}, reflecting the short-range correlations of Cooper pairs. The results in Fig. 8(b) clearly indicate that the unconventional SHG is enhanced by the superconducting fluctuations, as schematically shown in Fig. 8(c). Considering that the light field is applied for several femtoseconds, no-scattering charge acceleration induces a current and resultant SHG,[14] as shown below.

According to common sense or if Ohm's law ($j(t) = \sigma E(t)$, j: current, σ: conductivity, E: electric field) holds, an oscillating light field does not give a net current because the time average of the electric field vanishes. However, the light field is applied only for 6 fs. In this case, charge acceleration is realized, resulting in a finite net current before the increase in the electron temperature. If such a finite current is generated during the no-scattering time window of *ca.* 40 fs in this compound, we can detect it as current induced SHG. Note that current-induced SHGs have been observed in graphene[41,42] and BCS superconductors[43] under DC and terahertz fields.

To demonstrate that the SHG is attributable to the no-scattering current, we measure the CEP dependence of the SHG. The time profile of the no-scattering charge current is given by the time integral of the electric field $j(t) \propto \int_0^t E(\tau)\mathrm{d}\tau$.

Fig. 3(b) shows that we have a finite net current depending on the CEP before dephasing processes take place. In this case, the current density should show a one-cycle change when the CEP has a one-period change.

Fig. 9(a) shows the change of the SHG intensity ($\Delta I_{SH}/I_{SH}$, I_{SH}: averaged SHG intensity for varying CEP, where ΔI_{SH} is the difference between the observed SHG and I_{SH}) as a function of the relative change of CEP from 0 to 2π. Fig. 9(b) shows a reference of the conventional SHG in β-BBO (room temperature). We cannot detect any CEP dependence in the reference. In h-Br, we have a two-cycle change in SHG under the one-period change of the CEP. Because SHG cannot distinguish no-scattering currents with opposite directions, the two-cycle change under the one-period change of the CEP is reasonable. Thus, the observed SHG is due to a no-scattering current.

It is noteworthy that the spectral bandwidth (135 meV) is significantly narrower than that of the pump spectrum (*ca.* 500 meV). This fact indicates that the no-scattering current survives for about 30 fs after the application of the light field. This time scale is comparable with that of the scattering time.

To clarify the origin of the SHG in more detail, we theoretically calculate the current density j in a two-dimensional three-quarter-filled Hubbard model for a 98 × 98-site system in the framework of the time-dependent Hartree–Fock approximation. We use the Hubbard model at three-quarter filling, $H=\sum_{\mathrm{i,j}\sigma} t_{\mathrm{ij}}(c^{\dagger}_{\mathrm{i}\sigma}c_{\mathrm{j}\sigma}+c^{\dagger}_{\mathrm{j}\sigma}c_{\mathrm{i}\sigma})+U\sum_{\mathrm{i}} n_{\mathrm{i}\uparrow}n_{\mathrm{i}\downarrow}$, where $c^{\dagger}_{\mathrm{i}\sigma}$ creates an electron in the highest occupied molecular orbital (HOMO) with spin σ at site i, and $n_{\mathrm{i}\sigma}=c^{\dagger}_{\mathrm{i}\sigma}c_{\mathrm{i}\sigma}$. For the on-site Coulomb repulsion, we use $U = 0.8$ eV. The transfer integral t_{ij} depends on the bond ij. The molecular arrangement is taken from the structural data, from which t_{ij} are estimated with the extended Hückel method.[16,17] The initial state is the Hartree–Fock ground state. Photoexcitation is introduced through the substitution $c^{\dagger}_{\mathrm{i}\sigma}c_{\mathrm{j}\sigma} \rightarrow \exp[\mathrm{i}e/\hbar c\boldsymbol{r}_{\mathrm{ij}}A(t)]c^{\dagger}_{\mathrm{i}\sigma}c_{\mathrm{j}\sigma}$ with relative intermolecular position $\boldsymbol{r}_{\mathrm{ij}}=\boldsymbol{r}_{\mathrm{j}}-\boldsymbol{r}_{\mathrm{i}}$. The vector potential we use is $\boldsymbol{A}(t)=\theta(t)F/\omega_{\mathrm{fund}}[\cos(\omega_{\mathrm{fund}}t-\varphi)-\cos\varphi]$ and corresponds to $\boldsymbol{E}(t)=\theta(t)F\sin(\omega_{\mathrm{fund}}t-\varphi)$,with $F=(0,F)\|c$ and the fundamental photon energy $\omega_{\mathrm{fund}}=0.7$ eV. The time-dependent Schrödinger equation

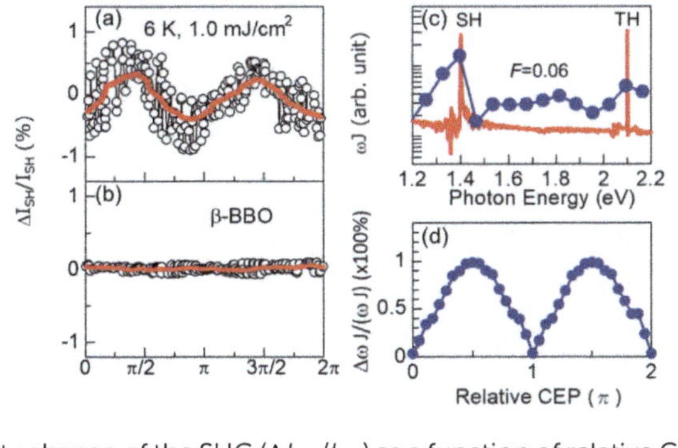

Fig. 9 (a) Intensity change of the SHG ($\Delta I_{SH}/I_{SH}$) as a function of relative CEP. (b) Reference of the conventional SHG in β-BBO (room temperature). In both (a) and (b), the red curve is obtained by averaging the data (guide to the eye). (c) Calculated spectra of ωJ showing the SHG and THG for the light-field (which is polarized parallel to the c-axis) $F = 0.06$ (red line; Hartree–Fock approximation, blue dots; exact diagonalization). (d) CEP dependence of the calculated SHG intensity (peak intensities of ωJ at 1.4 eV (SHG)) for $F = 0.1$.

is numerically solved. We calculate the Fourier transform of the current density $j(t) = -\langle \partial H / N \partial A \rangle$, with N being the number of unit cells for 500 cycles. The absolute value of its Fourier transform is denoted by J. The SHG and THG are evaluated as ωJ (the absolute value of the Fourier transform of dj/dt) at $\omega = 2\omega_{\text{fund}}$ and $\omega = 3\omega_{\text{fund}}$,[44,45] respectively, and shown for $\boldsymbol{E}_{\text{SH}} \| \boldsymbol{c}$ and $\boldsymbol{E}_{\text{TH}} \| \boldsymbol{c}$ with electric field amplitudes (F [V per angstrom]) of 0.06 by the red line in Fig. 9(c). The obtained bandwidth is basically determined by the time window of the Fourier transform in the Hartree–Fock simulation without dephasing. Therefore, we cannot discuss the bandwidth. I_{SH} is sensitive to the CEP (Fig. 9(d)), which is consistent with the result shown in Fig. 9(a). To check the emergence of SHG, we also calculate ωJ by using the exact diagonalization method for a 16-site system[13,36] during 10-cycle irradiation. The Fourier spectrum indicated by the blue dots in Fig. 9(c) clearly shows the SHG peak.

As shown in the upper panel of Fig. 10(a), the SHG is polarized parallel to the c-axis ($\boldsymbol{E}_{\text{SH}} \| \boldsymbol{c}$) for both excitation polarizations ($\boldsymbol{E}_{\text{fund}} \| \boldsymbol{c}$ (red line), $\boldsymbol{E}_{\text{fund}} \| \boldsymbol{a}$ (green line)), although the THG shows the usual polarization, which is the same as that of the fundamental pulse, as shown in the lower panel of Fig. 10(a). Such an unusual polarization dependence of the SHG cannot be reproduced by the theory, *i.e.*, I_{SH} (theory) shows the polarization that is parallel to the fundamental polarization for both $\boldsymbol{E}_{\text{fund}} \| \boldsymbol{c}$ and $\boldsymbol{E}_{\text{fund}} \| \boldsymbol{a}$. The polarization dependence cannot be understood simply by a point group analysis of the $\chi^{(2)}$ tensor for the orthorhombic structure (class $mm2$ (C_{2v}) (after due consideration of the symmetry breaking uniaxially induced by the current j)). To demonstrate a response of $j \| c$ ($\boldsymbol{E}_{\text{pr}} \| \boldsymbol{c}$) under the excitation polarization of $\boldsymbol{E}_{\text{pu}} \| \boldsymbol{a}$, we investigate the polarization dependence of a transient reflectivity (pump–probe) measurement. As shown in Fig. 10(b), the responses of $\boldsymbol{E}_{\text{pr}} \| \boldsymbol{c}$ are larger than $\boldsymbol{E}_{\text{pr}} \| \boldsymbol{a}$ for both $\boldsymbol{E}_{\text{pu}} \| \boldsymbol{a}$ and $\boldsymbol{E}_{\text{pu}} \| \boldsymbol{c}$ ($\boldsymbol{E}_{\text{pu}}$ and $\boldsymbol{E}_{\text{pr}}$ are the electric fields of the pump- and probe-lights, respectively). Therefore, it is clear that light-induced charge motion occurs easily along the c-axis for both $\boldsymbol{E}_{\text{pu}} \| \boldsymbol{a}$ and $\boldsymbol{E}_{\text{pu}} \| \boldsymbol{c}$. The similar polarization dependence has also been observed in κ-(BEDT-TTF)$_2$Cu[N(CN)$_2$]Cl (insulating phase) even under a weaker excitation condition of 3.1 eV (probe energy = 1.55 eV).[46] Thus, the results of the transient reflectivity measurements do not contradict the above results of the SHG.

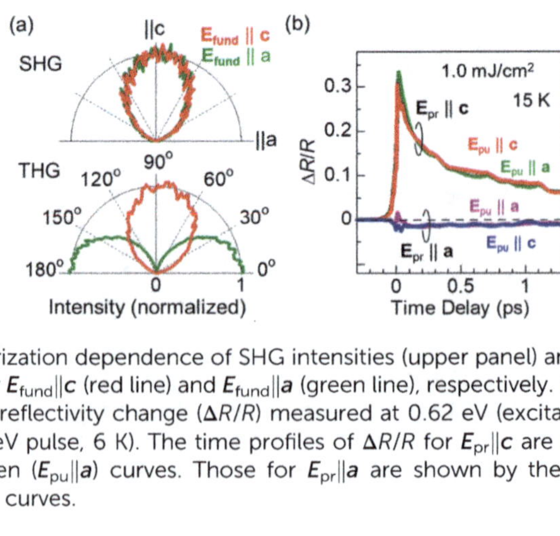

Fig. 10 (a) Polarization dependence of SHG intensities (upper panel) and THG intensities (lower panel) for $\boldsymbol{E}_{\text{fund}} \| \boldsymbol{c}$ (red line) and $\boldsymbol{E}_{\text{fund}} \| \boldsymbol{a}$ (green line), respectively. (b) Time evolution of the transient reflectivity change ($\Delta R/R$) measured at 0.62 eV (excitation by 6 fs, 1 mJ cm^{-2}, 0.6–0.9 eV pulse, 6 K). The time profiles of $\Delta R/R$ for $\boldsymbol{E}_{\text{pr}} \| \boldsymbol{c}$ are shown by the red ($\boldsymbol{E}_{\text{pu}} \| \boldsymbol{c}$) and green ($\boldsymbol{E}_{\text{pu}} \| \boldsymbol{a}$) curves. Those for $\boldsymbol{E}_{\text{pr}} \| \boldsymbol{a}$ are shown by the blue ($\boldsymbol{E}_{\text{pu}} \| \boldsymbol{c}$) and magenta ($\boldsymbol{E}_{\text{pu}} \| \boldsymbol{a}$) curves.

The observation of the SHG induced by the no-scattering current suggests a possibility of even-order HHG. However, absorption losses sometimes prevent the observation of HHG in thick bulk crystals. In fact, intramolecular transitions exist in the spectral region of the HHG in this compound. We attempt to detect a non-linear polarization in the material through the interplay between the polarizations of SHG/THG and HHGs (without a direct detection of HHGs). Here, the essence of 2-D spectroscopy is used. As shown in Fig. 11(a), a pair of pulses with a time difference of Δt enters the material and emitted lights, including the SHG and THG, are detected by changing Δt with an accuracy of 100 attoseconds and by utilizing the TWINS technique,[28–31] as described above. The obtained light intensity as a function of Δt is referred to as an interferogram. In 2-D spectroscopy,[28–31] the Fourier transformed (FT) spectrum of the interferogram is plotted parallel to the ordinate for the respective detection energies in the abscissa, as schematically shown in Fig. 11(b). In the non-linear 2-D spectrum of Fig. 11(b), off-diagonal components, such as (2H, 4H) and (3H, 4H), should show the interplay between 2H and 4H or 3H and 4H, respectively.

Fig. 11(c) shows the Fourier spectra of the interferograms (modulation of the SHG and THG intensities as a function of Δt) detected at 1.5 eV (SHG, red dots) and 2.25 eV (THG, blue dots). We notice the peak at 3 eV corresponding to (2H, 4H) and (3H, 4H) in Fig. 11(b). The results show that a polarization of the 4th harmonic is generated in the material and modulates the lower order polarizations (of SHG and THG). In general, it is difficult to measure HHG in thick bulk crystals because of absorption losses. The method shown here without a direct measurement of HHGs in thin films is expected to be useful for investigating the non-linear polarizations which potentially induce HHGs.

Finally, we consider the relation between the no-scattering current and superconducting fluctuations. The above-mentioned equation, $j(t) \propto \int_0^t E(\tau)d\tau$, is justified only in the ultrafast no-scattering time window. In principle, the no-scattering

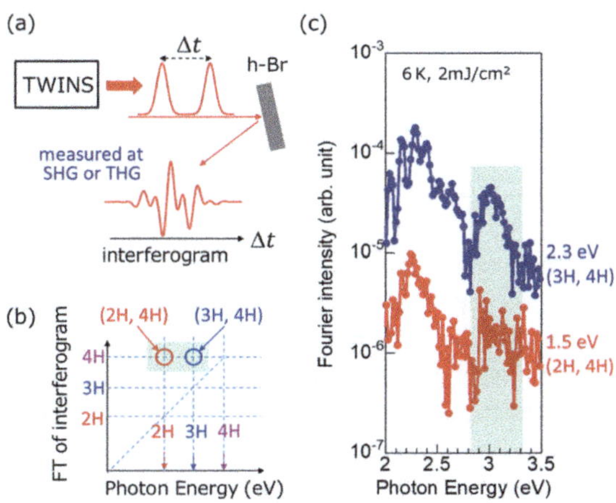

Fig. 11 (a) Schematic illustration of the experimental configuration using TWINS. (b) Schematic illustration of 2-D spectroscopy, where 2H, 3H and 4H represent the energies of the SHG, THG and 4th harmonic generation. (c) Fourier spectra of interferograms (modulation of the SHG and THG intensities as a function of Δt) detected at 1.5 eV (SHG, red dots) and 2.25 eV (THG, blue dots).

mechanism of the SHG discussed above is not necessarily limited to superconducting materials. Indeed, as shown in Fig. 7(a), the SHG increases toward T_{SC}, indicating that the no-scattering current is closely related to superconducting fluctuations. Considering that the time integral of $\boldsymbol{E}(t)$ is the vector potential $\boldsymbol{A}(t)$ in the present choice of a gauge, we notice the similarity between the above equation and the London equation $j(\tau) = -(n_s e^2/m)A$ (m: mass of superconducting electrons, n_s: number density of superconducting electrons, e: elementary charge) phenomenologically describing the Meissner effect. We cannot easily compare the above two equations for j because the no-scattering current and the London equation are based on different conditions (former: local and ultrafast time scale, latter: global and in near-equilibrium). However, the similarity between these two equations suggests that the ultrafast charge dynamics during the no-scattering time window are closely linked to superconducting fluctuations.

Conclusions

The stimulated emission and the SHG in a centrosymmetric organic conductor (κ-(h-ET)$_2$Cu[N(CN)$_2$]Br) under the application of a 6 fs strong light field are induced by a synchronized charge oscillation and a no-scattering current, respectively. These results indicate that correlated charge motion can be driven in a manner which is very different from the conventional photoinduced charge dynamics with scattering processes. Such no-scattering charge dynamics have been hidden behind the increase in the electron temperature. The fact that the stimulated emission and the SHG show anomalous enhancements towards T_{SC} indicates that these ultrafast nonlinear phenomena are related to T_{SC} and/or superconducting fluctuations. The <5 fs rise of the stimulated emission and its temperature dependence indicate that the microscopic mechanism of superconducting fluctuations is related to Coulomb repulsive interactions. Furthermore, we also observe 4th order polarization through the interplay between the polarizations of SHG/THG and HHGs, *i.e.*, without directly detecting the 4th order harmonic generation.

Conflicts of interest

There are no conflicts to declare.

Acknowledgements

The authors thank Professor T. Sasaki (Institute for Materials Research, Tohoku University) and Professor H. M. Yamamoto (Institute for Molecular Science) for providing the single crystals and a thin film of κ-(h-ET)$_2$Cu[N(CN)$_2$]Br. The authors also thank Professor H. Kishida (Nagoya University) for measuring and analyzing the steady state optical spectra. This work was supported by JST CREST (JPMJCR1901), MEXT Q-LEAP (JPMXS0118067426).

Notes and references

1 M. Imada, A. Fujimori and Y. Tokura, *Rev. Mod. Phys.*, 1998, **70**, 1039.
2 D. N. Basov and T. Timusk, *Rev. Mod. Phys.*, 2005, 77, 721.
3 T. Ishiguro, K. Yamaji and G. Saito, *Organic Superconductors*, Springer, 1998.

4 C. Giannetti, *et al.*, *Adv. Phys.*, 2016, **65**, 58.
5 R. Matsunaga, *et al.*, *Science*, 2014, **345**, 1145.
6 A. Cavalleri, *Contemp. Phys.*, 2018, **59**, 31.
7 D. N. Basov, R. D. Averitt and D. Hsieh, *Nat. Mater.*, 2017, **16**, 1077.
8 S. Koshihara, *et al.*, *Phys. Rep.*, 2022, **942**, 1.
9 Y. Kawakami, *et al.*, *Phys. Rev. Lett.*, 2010, **105**, 246402.
10 T. Ishikawa, *et al.*, *Nat. Commun.*, 2014, **5**, 5528.
11 Y. Naitoh, *et al.*, *Phys. Rev. B: Condens. Matter Mater. Phys.*, 2016, **93**, 165126.
12 Y. Kawakami, *et al.*, *Phys. Rev. B: Condens. Matter Mater. Phys.*, 2017, **95**, 201105(R).
13 Y. Kawakami, *et al.*, *Nat. Photonics*, 2018, **12**, 474.
14 Y. Kawakami, *et al.*, *Nat. Commun.*, 2020, **11**, 4138.
15 K. Kanoda, *Hyperfine Interact.*, 1997, **104**, 235.
16 T. Mori, H. Mori and S. Tanaka, *Bull. Chem. Soc. Jpn.*, 1999, **72**, 179.
17 M. Watanabe, *et al.*, *Synth. Met.*, 1999, **103**, 1909.
18 F. Kagawa, K. Miyagawa and K. Kanoda, *Nature*, 2005, **436**, 534.
19 B. J. Powell and R. H. McKenzie, *J. Phys.: Condens. Matter*, 2006, **18**, R827.
20 T. Sasaki, *et al.*, *Phys. Rev. B: Condens. Matter Mater. Phys.*, 2004, **69**, 064508.
21 D. Faltermeier, *et al.*, *Phys. Rev. B: Condens. Matter Mater. Phys.*, 2007, **76**, 165113.
22 M. Lang, F. Steglich, N. Toyota and T. Sasaki, *Phys. Rev. B: Condens. Matter Mater. Phys.*, 1994, **49**, 15227.
23 T. Kobayashi, Y. Ihara, Y. Saito and A. Kawamoto, *Phys. Rev. B: Condens. Matter Mater. Phys.*, 2014, **89**, 165141.
24 H. M. Yamamoto, *et al.*, *Nat. Commun.*, 2013, **4**, 2379.
25 H. Telle, *et al.*, *Appl. Phys. B*, 1999, **69**, 327.
26 F. W. Helbing, *et al.*, *Appl. Phys. B*, 2002, **74**, s35.
27 M. Mehendale, *et al.*, *Opt. Lett.*, 2000, **25**, 1672.
28 D. Brida, C. Manzoni and G. Cerullo, *Opt. Lett.*, 2012, **37**, 3027.
29 J. Rehault, *et al.*, *Rev. Sci. Instrum.*, 2014, **85**, 123107.
30 A. De Sio, *et al.*, *Nat. Commun.*, 2016, **7**, 13742.
31 J. M. Richter, *et al.*, *Nat. Commun.*, 2017, **8**, 376.
32 M. Dressel and N. Drichiko, *Chem. Rev.*, 2004, **104**, 5689.
33 J. E. Eldridge, *et al.*, *Spectrochim. Acta, Part A*, 1996, **52**, 45.
34 Y. Matsubara, *et al.*, *Phys. Rev. B: Condens. Matter Mater. Phys.*, 2014, **89**, 161102(R).
35 H. Sasaki, *et al.*, *Sci. Rep.*, 2018, **8**, 9609.
36 K. Yonemitsu, *J. Phys. Soc. Jpn.*, 2018, **87**, 044708.
37 K. Yonemitsu, *J. Phys. Soc. Jpn.*, 2018, **87**, 124703.
38 T. Shimada and K. Yonemitsu, *J. Phys. Soc. Jpn.*, 2020, **89**, 084701.
39 K. Yonemitsu and P. Werner, *J. Phys. Soc. Jpn.*, 2021, **90**, 044713.
40 Y. Nakazawa, *et al.*, *Phys. Rev. B: Condens. Matter Mater. Phys.*, 2000, **61**, 16295.
41 A. Y. Bykov, *et al.*, *Phys. Rev. B: Condens. Matter Mater. Phys.*, 2012, **85**, 121413(R).
42 M. Tokman, *et al.*, *Phys. Rev. B: Condens. Matter Mater. Phys.*, 2019, **99**, 155411.
43 C. Vaswani, *et al.*, *Phys. Rev. Lett.*, 2020, **124**, 207003.
44 R. E. F. Silva, *et al.*, *Nat. Photonics*, 2018, **12**, 266–270.
45 Y. Murakami, M. Eckstein and P. Werner, *Phys. Rev. Lett.*, 2018, **121**, 057405.
46 S. Tsuchiya, *et al.*, *J. Phys. Soc. Jpn.*, 2019, **88**, 074706.

Faraday Discussions

ROYAL SOCIETY OF CHEMISTRY

PAPER

Local excitation and valley polarization in graphene with multi-harmonic pulses

Ulf Saalmann and Jan Michael Rost

Received 26th January 2022, Accepted 2nd March 2022
DOI: 10.1039/d2fd00017b

We elucidate the mechanism of strong laser pulse excitation in pristine graphene with multi-harmonic pulses, linearly polarized parallel to the line connecting the two different Dirac points in the Brillouin zone and with a maximal vector potential given by the distance of those points. The latter two conditions have emerged from our previous work [Kelardeh *et al.*, *Phys. Rev. Res.*, 2022, **4**, L022014] as favorable for large valley polarization. We introduce a novel compacted representation for excitation, locally resolved in the initial conditions for the crystal momenta. These maps are our main tool to gain insight into the excitation dynamics. They also help with understanding the effect of dephasing. We work out why a long wavelength and a moderate number of overtones in the harmonic pulse generate the largest valley polarizations.

1 Introduction

Illuminating condensed-matter targets with intense short pulses affects non-linear photo-absorption which depends crucially on the band structure of the system and is commonly probed by high-harmonic generation.[1,2] Pioneering work has shown that the basics of the ensuing electron dynamics follows a similar two-step process in reciprocal space,[3,4] as is familiar from atoms and molecules in real space. Firstly, a valence electron tunnels at the gap to the energetically closest conduction band appearing there with wave vector $\boldsymbol{k}$. Subsequently, the wave vector $\tilde{\boldsymbol{k}}$ moves on the conduction band driven by the lasers's vector potential $\boldsymbol{A}$ according to $\tilde{\boldsymbol{k}}(t) = \boldsymbol{k} + \boldsymbol{A}(t)$. Owing to the gap, electronic transitions back to the valence band are exponentially suppressed.

This is not the case for gapless 2D systems such as graphene, rendering the ensuing (coherent) excited electron dynamics potentially more complicated. Moreover, dephasing may alter the final excitation for laser pulses longer than the coherence time. Physical properties of interest are the excitation probability and valley polarization.[5–8] The latter quantifies the difference of excitation originating from initial wave-vectors in the first Brillouin zone (BZ) belonging to the domains of the two Dirac points $\boldsymbol{K}$ and $\boldsymbol{K}'$ of the vanishing gap. These two domains of

Max Planck Institute for the Physics of Complex Systems, Nöthnitzer Str. 38, 01187 Dresden, Germany. E-mail: us@pks.mpg.de; rost@pks.mpg.de

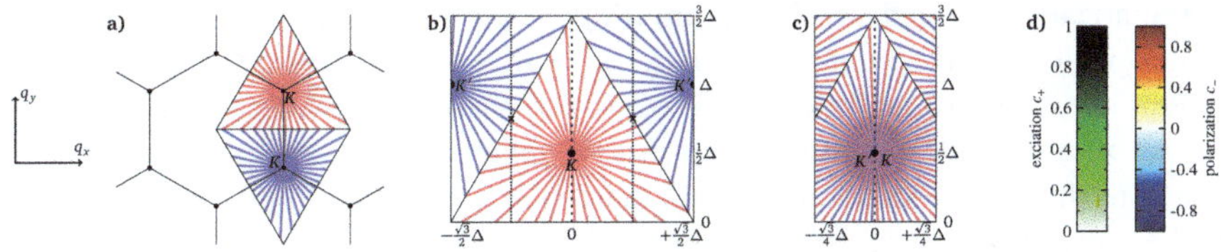

Fig. 1 Sketch for the compact representation of the triangular q-domains around K and K', which we will use below. We fold the Voronoi triangles into a rectangle (a → b). Then the two outer rectangles, confined by the dotted lines, are overlaid with the inner ones (b → c) by a π-rotation around the center between K and K', marked with crosses. Since the dynamics are symmetric with respect to the dashed line, we use the left part for presenting the local excitation $c_+(\boldsymbol{q})$ and the right part for the local polarization $c_-(\boldsymbol{q})$, respectively. The probabilities $c_\pm$ are defined below in eqn (5), their corresponding color scales are given in the right panel of d. The momenta are given in terms of $\Delta = |K - K'|$, the distance between the K-points.

triangular shape form a unit cell of the BZ, well adapted to describe valley polarization, see Fig. 1. Valley polarization (VP), detectable *via* even harmonics,[6] has generated considerable interest, since it promises a means of storing and manipulating quantum information by phase changes only, although this is still quite a distant goal.[8]

In recent work,[9] we have shown that laser fields of sufficient strength to affect excitation, can be chosen such that transitions between bands occur in graphene essentially only close to the Dirac points. Hence, band transitions happen at instants of time when the intense pulse driven electron wave vector $\tilde{\boldsymbol{q}}(t)$ passes by $\boldsymbol{K}$ or $\boldsymbol{K}'$ leading to a sequence of (coherent) electron amplitudes in the conduction band.

As an immediate consequence and in contrast to common belief, linearly polarized pulses can generate substantial valley polarization for a gapless system[9] under two conditions: (i) the linear polarization is parallel to the difference vector $\boldsymbol{\Delta} = \boldsymbol{K} - \boldsymbol{K}'$ of the two Dirac points and the vector potential $\boldsymbol{A}_0$ is of the order of the distance $\Delta = 4\pi/(\sqrt{27}\,a) \approx 0.90$ a.u. between the two Dirac points, where $a = 1.43$ Å is the nearest-neighbor C–C spacing in graphene. (ii) The pulse couples differently to electrons at $\boldsymbol{K}$ and $\boldsymbol{K}'$. This requires an asymmetry† in the driving pulse which is, from a theoretical point of view, most simply achieved with a half cycle pulse.[9] A sufficiently short pulse also ensures that dephasing does not play a role. Experimentally, such pulses can be generated thanks to recent experimental advances,[10,11] requiring, however, special equipment. With one notable exception,[12] to the best of our knowledge pristine (gapless) graphene exposed to light polarized in the graphene plane has only been investigated with linear polarization perpendicular[13,14] to $\boldsymbol{\Delta}$ or with ultrashort circular pulses, where the vector potential's component parallel to $\boldsymbol{\Delta}$ was symmetric,[14,15] with $|\max\{A_\Delta(t)\}| = |\min\{A_\Delta(t)\}|$. These kind of pulses cannot generate VP.

In the following, we investigate the interplay of coherent dynamics and dephasing in graphene as a gapless system with a multi-harmonic pulse whose characteristics in terms of pulse shape, pulse length and wavelength can be

† Asymmetry is here defined through $|\max\{A_\Delta(t)\}| \gg |\min\{A_\Delta(t)\}|$.

systematically changed, while maintaining a basic asymmetry of the pulses to enable valley polarization. The main quantities which will provide detailed insight are local probabilities, resolved w. r. t. the initial conditions in the Brillouin zone. We will analyze excitation and valley polarization. To this end we calculate the time-dependent density matrix in a two-band approximation with a phenomenological dephasing rate γ and we introduce a compact graphical representation for the local probabilities. If the laser pulses are intense as in our case, the two-band approximation is confirmed to provide an excellent description.[14] The model is briefly described in Section 2, where we also explain the multi-harmonic pulse and illustrate with an example, the basic light-driven electronic excitation dynamics.

Section 3 presents the results for different numbers of overtones in the pulses and uncovers how excitation and valley polarization depend on the parameters of the pulse, elucidating how intense pulses steer electrons in graphene. Section 4 uses excitation distributions in connection with the compacted local probabilities to work out the differences between coherent and dephased dynamics and to identify those initial conditions in the unit cell which mainly contribute to excitation and/or VP. The paper ends with a short summary in Section 5. We use atomic units, unless stated otherwise.

2 Prerequisites

2.1 Graphene dynamics including dephasing

The time-dependent Hamilton matrix for the well known two-band approximation for graphene reads in reciprocal space

$$\mathbb{H}(\boldsymbol{q},t) = \frac{E_{\boldsymbol{q}+\boldsymbol{A}(t)}}{2}\sigma_z + \boldsymbol{F}(t)\cdot\left[V_{\boldsymbol{q}}^* i\frac{\mathrm{d}}{\mathrm{d}\boldsymbol{q}}W_{\boldsymbol{q}}\right]_{\boldsymbol{q}+\boldsymbol{A}(t)}\sigma_x \tag{1}$$

with the Pauli matrices σ_x and σ_z. The Hamilton matrix eqn (1), follows from an instantaneous diagonalization of the standard hopping matrix,[16] which defines also the so-called Houston basis.[17] In this basis we obtain the (instantaneous) band gap $E_{\boldsymbol{q}+\boldsymbol{A}(t)}$ and the (instantaneous) eigenvectors $V_{\boldsymbol{q}+\boldsymbol{A}(t)}$ and $W_{\boldsymbol{q}+\boldsymbol{A}(t)}$ for the valence and conduction band, respectively. Find a derivation for $\mathbb{H}$ in appendix A.

Note that both, the vector potential $\boldsymbol{A}(t)$ and the electric field $\boldsymbol{F}(t) = -\frac{\mathrm{d}}{\mathrm{d}t}\boldsymbol{A}(t)$, affect the time-dependent Hamilton matrix $\mathbb{H}(\boldsymbol{q},t)$ in eqn (1) and influence the time evolution of the density matrix through

$$\frac{\partial}{\partial t}\rho_{\boldsymbol{q}}(t) = -i\left[\mathbb{H}(\boldsymbol{q},t)\rho_{\boldsymbol{q}}(t) - \rho_{\boldsymbol{q}}(t)\mathbb{H}(\boldsymbol{q},t)\right] - \frac{\gamma}{2}\left[\rho_{\boldsymbol{q}}(t) - \sigma_z\rho_{\boldsymbol{q}}(t)\sigma_z\right], \tag{2}$$

where γ is the dephasing rate. Products with ρ, $\mathbb{H}$, and $\sigma_{x,z}$ in eqn (2) are matrix–matrix multiplications.

2.2 Compact visualization of differential excitation and valley polarization

Excitation in the triangular $\boldsymbol{q}$-domains ∇K and $\nabla K'$ around $\boldsymbol{K}$ and $\boldsymbol{K}'$ of the initial conditions, representing together a unit cell of the Brillouin zone (see Fig. 1a), are defined as averaged local excitations $c(\boldsymbol{q})$ in the respective domains of area ∇ in $\boldsymbol{q}$-space,

$$P_K \equiv \frac{1}{\nabla}\int_{\nabla K} d^2\boldsymbol{q}c(\boldsymbol{q}) \text{ and } P_{K'} \equiv \frac{1}{\nabla}\int_{\nabla K'} d^2\boldsymbol{q}c(\boldsymbol{q}) \tag{3}$$

with conduction-band occupation c as a function of initial conditions $\boldsymbol{q}$ given by the density matrix element $c(\boldsymbol{q}) = \rho_{\boldsymbol{q},\mathrm{cc}}(t \to \infty)$. In terms of P_K and $P_{K'}$, the total excitation p averaged over the entire unit cell 2∇, and the total valley polarization η are defined by

$$p \equiv \frac{P_K + P_{K'}}{2} \text{ and } \eta \equiv \frac{P_K - P_{K'}}{P_K + P_{K'}} \tag{4}$$

with $0 \leq p \leq 1$ and $-1 \leq \eta \leq +1$.

We obtain $c(\boldsymbol{q})$ by numerically solving eqn (2) for a dense set of initial conditions $\boldsymbol{q}$, sampling each of the two triangular domains with $512 \times 222 = 113\,664$ initial conditions. Therefore we can define excitation p and valley polarization η locally in the initial conditions $\boldsymbol{q}$ with a resolution inversely proportional to the sampling. A natural graphical representation consists of color maps (scales are given in Fig. 1d) filling the area of the two triangles in Fig. 1a. Due to the underlying hexagonal symmetry, we can fold the triangles into a rectangle, *cf.* the transition from Fig. 1a and b.[9]

Since in the end, p and η emerge from sums and differences over local conduction-band occupations $c(\boldsymbol{q})$, one can perform part of this operation graphically by overlaying equivalent areas in the $\boldsymbol{K}$ and $\boldsymbol{K}'$ triangles to arrive at Fig. 1c, which produces an even more compact representation.‡ This second step consists essentially of a π-rotation around the center of the line segment $\boldsymbol{K}$–$\boldsymbol{K}'$, which superimposes each point in $\nabla\boldsymbol{K}$ with one point in $\nabla\boldsymbol{K}'$: $\boldsymbol{q} \to \boldsymbol{q}' = f(\boldsymbol{q})$. By means of this mapping f we can define the local conduction-band occupations for p and η, respectively,

$$c_{\pm}(\boldsymbol{q}) \equiv c(\boldsymbol{q}) \pm c(f(\boldsymbol{q})) \tag{5}$$

They enable a detailed analysis, either in contour plots (Fig. 5, 7, 9) or distribution functions (Fig. 8). Of course, p and η, defined traditionally in eqn (4), follow equivalently from their local contributions eqn (5) as $p = \frac{1}{2\nabla}\int d^2qc_+(\boldsymbol{q})$ and $\eta = \int d^2qc_-(\boldsymbol{q})/\int d^2qc_+(\boldsymbol{q})$. For convenience, below we will use the terms "local excitation" and "local polarization" for c_+ and c_-, respectively.

2.3 Asymmetric pulse shapes from multi-harmonic pulses

Large ionization probability all the way to damaging the sample is easy to achieve by increasing the field strength, in close analogy to atoms or molecules, where damaging corresponds to depletion of the ionized electron orbital. This, however, implies that excitation also occurs further away from the Dirac points at finite band gaps, and eventually, for large enough field strength throughout the unit cell with the consequence of a vanishing VP.

The generation of significant VP hinges upon preserving the vicinity of the Dirac points as they are dominantly responsible for the excitation. This provides the means for selective excitation of the two valley domains upon breaking the equivalence in the excitation *via* the two Dirac points through suitable pulse

‡ Note, that this compactification requires pulses that are mirror-symmetric with respect to Δ.

forms. These requirements select combinations of appropriate field strength, vector potential and frequencies of the light pulse. An obvious choice fulfilling these requirements is a pulse close to a half-cycle, of suitable strength and linearly polarized along $\boldsymbol{\Delta}$, as demonstrated before.[9] To make the interplay of laser parameters for excitation and VP more apparent and also, to assess the effect of dephasing, we will use multi-harmonic pulses. Asymmetric by construction, they look like pulse trains and provide a means to study the interplay of coherent multi-excitation electron dynamics and decoherence. We define these pulse trains as superpositions of harmonic overtones with an envelope determining the overall pulse length T,

$$\boldsymbol{A}(t) = A_0 \boldsymbol{e}_y e^{-2\ln 2t^2/T^2} \frac{1}{H_n} \sum_{k=1}^{n} \frac{\cos(k\omega t - \phi)}{k} \tag{6a}$$

$$\boldsymbol{F}(t) = A_0 \boldsymbol{e}_y e^{-2\ln 2t^2/T^2} \frac{\omega}{H_n} \sum_{k=1}^{n} \sin(k\omega t - \phi), \tag{6b}$$

with the Harmonic number $H_n = \sum_{k=1}^{n} 1/k$ for normalization, *i.e.* $A(0) = A_0$ for $\phi = 0$. The relevant parameters in the pulse eqn (6) are the fundamental frequency ω (defined by the wavelength λ), the amplitude A_0, the (full-width-at-half-maximum) pulse duration T, the number of harmonics n and the harmonic phase ϕ.

For $n > 1$ and $\phi = 0$ the periodic part in eqn (6a), *i.e.* the term given by the sum $S_n(t) = \sum_{k=1}^{n} \cos(kwt)/k$, is highly asymmetric. The contrast between maximum and minimum within one period $T_\omega = 2\pi/\omega$ is given by $C_n = -S_n(t_{\max})/S_n(t_{\min})$ with $t_{\max} = T_\omega$. For odd n, $t_{\min} = T_\omega/2$, for even n there is no explicit analytical expression of $t_{\min}$. Hence, for odd n

$$C_n = \left[\sum_{k=1}^{n} 1/k\right] \Big/ \left[-\sum_{k=1}^{n} [-1]^k/k\right], \tag{7}$$

whereas we have to calculate C_n for even n numerically. Typical values are $C_1 = 1$, $C_2 = 2$, $C_8 \approx 3.69$, and $C_{32} \approx 5.74$. Apparently, the contrast increases with n, see Fig. 2.

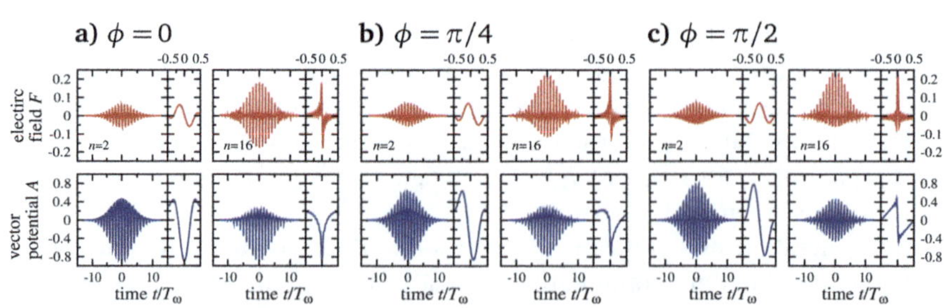

Fig. 2 Field strength (upper row) and vector potential (lower) for multi-harmonic pulses eqn (6) as a function of time scaled by the fundamental period T_ω of the harmonics. We show pulses for three harmonic phases ϕ and two different sizes $n = 2$ and $n = 16$. The right panels show the respective central periods.

2.4 A paradigmatic example of excitation by a multi-harmonic pulse

Before we discuss excitation and VP resulting from different multi-harmonic pulses, we briefly illustrate with Fig. 3 how the strong field excitation mechanism proceeds through the Dirac points with the specific harmonic pulse shown in Fig. 3c. The time-dependent crystal momentum $\tilde{\boldsymbol{q}}(t) = \boldsymbol{q} + \boldsymbol{A}(t)$, with the y-component $e_y \cdot \tilde{\boldsymbol{q}}(t)$ shown in Fig. 3b, is mapped into the unit cell in Fig. 3a with the green line and the initial condition $\boldsymbol{q}$ as a white-filled circle. Maximal and minimal excursion occur close to the maximum of the envelope as marked with arrows. During its evolution, $\tilde{\boldsymbol{q}}(t)$ oscillates back and forth on the green line with the amplitude shown in Fig. 3b. Once this amplitude is large enough, $\tilde{\boldsymbol{q}}(t)$ passes repeatedly by the Dirac point $\boldsymbol{K}'$ (cross in the lower triangle and dashed line in Fig. 3b) as indicated by thin black lines in Fig. 3b–d. At these times, a sequence of transitions occurs between the valence and the conduction band as stated before and is clearly visible in Fig. 3d. Whether they lead to net excitation or de-excitation at a specific encounter of $\boldsymbol{K}'$ depends on the relative phases of the electron excitation amplitudes.

The final excitation $c(\boldsymbol{q})$ for each initial condition is quantitatively indicated by the heat map in Fig. 3a revealing a quite complicated pattern across the unit cell but clearly confined laterally (in the q_x-direction) around the straight line passing through the Dirac points. Overall, the asymmetric pulse affects the $\tilde{\boldsymbol{q}}(t)$ with initial condition in the $\boldsymbol{K}'$ triangle not equivalently to those starting in the $\boldsymbol{K}$ triangle regarding excitation. This holds true despite the fact that the Dirac points, responsible for the transitions, are visited by equal areas of initial conditions from both triangles of the unit cell, irrespective of the pulse shape. Differences in the excitation—the prerequisite for VP—emerge from the intricate coherent dynamics of $\tilde{\boldsymbol{q}}(t)$, sensitively depending on the respective initial conditions (see

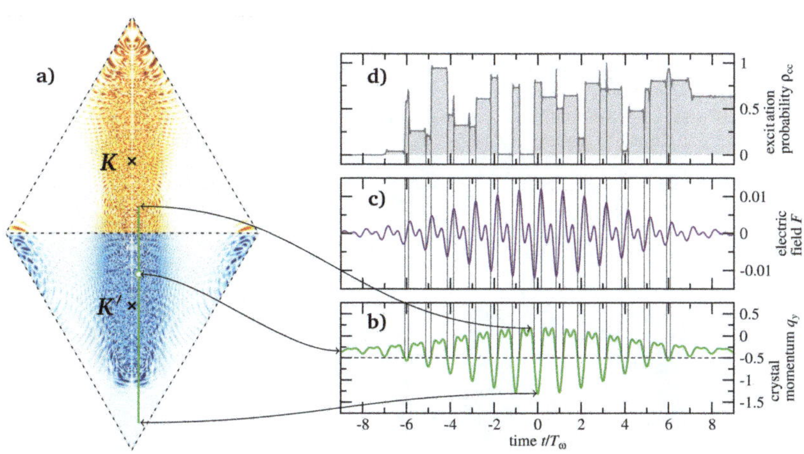

Fig. 3 (a) Local excitation $c(\boldsymbol{q})$ as a function of the initial crystal momentum $\boldsymbol{q}$ for pulses eqn (6), with $\lambda = 4$ µm, $A_0 = \Delta$, $n = 2$, $\phi = 0$. The Dirac points in the center of the triangular unit cells are marked by crosses. (b) Time-dependent crystal momenta $\tilde{q}_y(t) = q_y + A(t)$ in the vertical direction. (c) Electric field $F_y(t)$ in the vertical direction. (d) Corresponding local excitation $\rho_{\boldsymbol{q},cc}(t)$ for $\boldsymbol{q} = \begin{pmatrix} +0.023 \\ -0.271 \end{pmatrix} \Delta$, marked with a white-filled circle in (a).

Fig. 3d). How these VP generate differences and excitation in the first place, depends on global features of the pulse and will be elucidated below.

3 Illumination with harmonic pulses

3.1 Graphene in two-color pulses

For a first impression, we investigate how the averaged excitation p and VP η depend on wavelength and dephasing, we present in Fig. 4 results for harmonic pulses eqn (6) with $n = 2$, containing the fundamental frequency corresponding to the wavelength, and the first overtone. One can see that p decreases while η increases almost linearly and changes sign. Moreover, dephasing reduces p by roughly 1/3, while it hardly affects η. Interestingly, the pulse with the highest η at $\lambda = 8$ µm is quite long with $T = 160$ fs. One can understand these trends better from the local excitation and polarization $c_{\pm}(\boldsymbol{q})$, shown in Fig. 5. Not surprisingly,

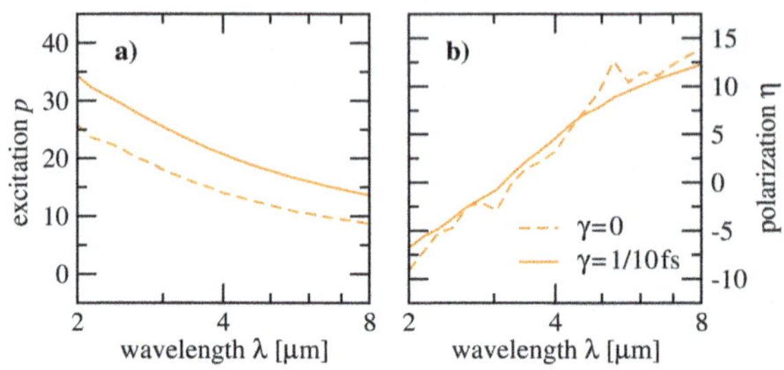

Fig. 4 Excitation p (a) and valley polarization η (b) for a two-color pulse, *i.e.* $n = 2$ in eqn (6), as a function of the wavelength λ. The harmonic phase is $\phi = 0$ and the pulse duration $T = [\lambda/1\ \mu\text{m}]20$ fs. Results for a finite damping time $\gamma = 1/10$ fs are compared to those without any damping.

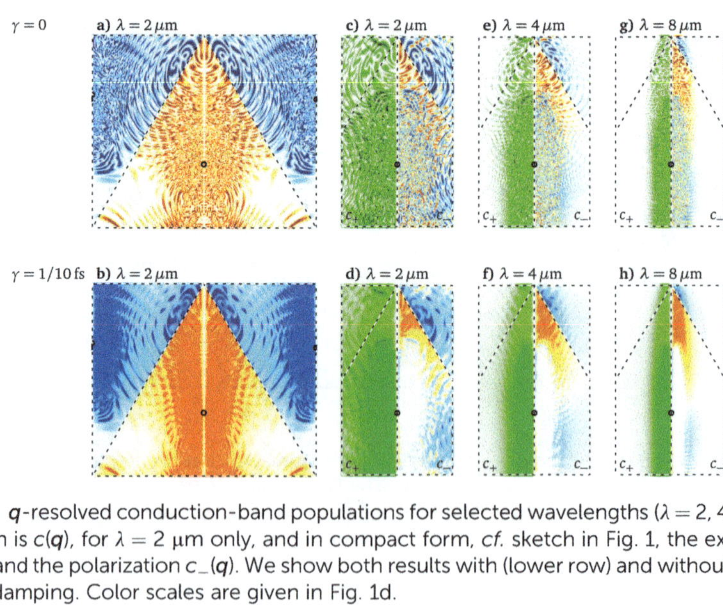

Fig. 5 $\boldsymbol{q}$-resolved conduction-band populations for selected wavelengths ($\lambda = 2, 4, 8$ µm). Shown is $c(\boldsymbol{q})$, for $\lambda = 2$ µm only, and in compact form, *cf.* sketch in Fig. 1, the excitation $c_+(\boldsymbol{q})$ and the polarization $c_-(\boldsymbol{q})$. We show both results with (lower row) and without (upper row) damping. Color scales are given in Fig. 1d.

dephasing blurs the interference pattern leaving less unexcited area (white) in the unit cell, *cf.* upper and lower row in Fig. 5. This is particularly obvious for the green color maps showing c_+ in compact representation, as all the upper panels have more white space than the lower ones.

One notes also, that c_+ is more spread over the unit cell and is overall larger for smaller wavelengths (more green areas). For increasing wavelength, the excitation shrinks laterally towards the $\boldsymbol{K}'$ point (black circle) and offers the selectivity which is necessary for large c_-, and indeed, valley polarization increases towards larger wavelengths with the warm colors dominating the cold colors. Since we keep the vector potential fixed at the distance between the two Dirac points, $A_0 = \boldsymbol{\Delta}$, decreasing the wavelength implies lowering the field strength. Therefore, transitions between the bands happen closer to the Dirac points. Next, we investigate what happens if n is increased systematically beyond 2 in multi-harmonic pulses.

3.2 Multi-harmonic pulses: dependence on number of harmonics

The arguments about the wavelength dependence of p and η given above for $n = 2$ do not explicitly depend on the number of harmonics in the pulse. Therefore, we expect the same qualitative behavior when increasing n. Fig. 6 reveals that this is only the case for the effect of dephasing, which remains generally larger for p than for η and hardly changes for p when n is increased. Surprisingly at first glance, neither p nor η depend monotonically on n. Rather p and η exhibit in most cases a maximum for a specific $n_{\max}(\lambda) \in [2, 32]$, which depends on the wavelength.

Indeed, the VP approaches zero for small, and towards large, n. The limit $n = 1$ corresponds to an ordinary pulse which is symmetric in the vector potential and therefore cannot generate VP. The reason for vanishing VP towards large n is more subtle: large n implies that with large time derivatives of the vector potential's harmonic components $\cos(k\omega t)$ there are large electric fields, *cf.* upper row in Fig. 2. These pulse parts impair the selectivity of the excitation and therefore the ability of the pulse to generate large η in the first place, despite the fact that increasing n produces increasingly asymmetric pulses in the vector potential, *cf.* lower row in Fig. 2. The even more complicated pattern, with two extrema of smaller absolute values of η for $\lambda = 2$ μm in Fig. 6, can be also attributed to lower

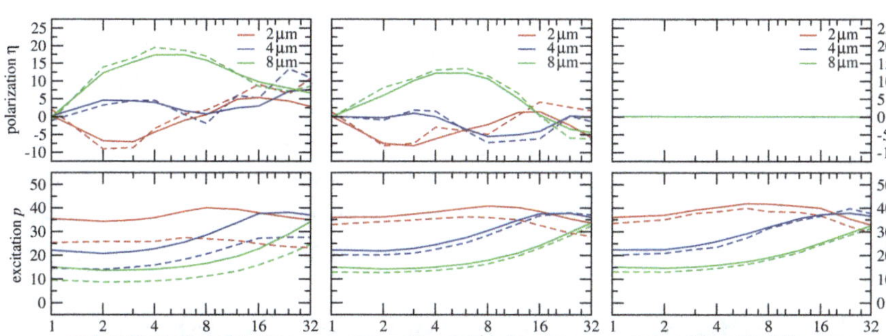

Fig. 6 Excitation p and valley polarization η for multi-harmonic pulses eqn (6), as a function of the number of harmonics n in the pulse for $\phi = 0$ (left), $\phi = \pi/4$ (middle) and $\phi = \pi/2$ (right). The laser wavelengths are $\lambda = 2, 4, 8$ μm and the corresponding pulse durations are $T = 40, 80, 160$ fs, respectively.

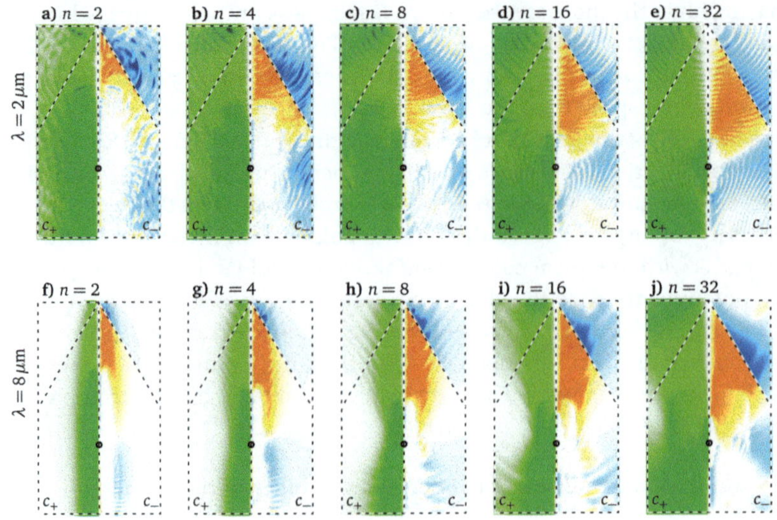

Fig. 7 q-resolved excitation for selected numbers of harmonics ($n = 2, 4, 8, 16, 32$) for $\phi = 0$, which corresponds to Fig. 6a. Again, we show c_+ (left panels) and c_- (right panels). The laser wavelengths are $\lambda = 2, 8$ μm and the pulse duration $T = 40, 160$ fs, respectively. It is $\gamma = 1/10$ fs. Color scales are given in Fig. 1d.

selectivity, since for smaller wavelength, excitation is in general less confined in the unit cell of initial conditions.

The spread of the excitation over the unit cell with increasing n and for small λ can be directly read off the compact color maps for c_+ in Fig. 7, where one sees that the panels for $\lambda = 2$ μm contain more green than those for $\lambda = 8$ μm and that $n = 32$ has in general the most dark green areas. Of course, this dependence on n is only pronounced if the wavelength is long enough to not cover the unit cell already for $n = 2$ in the excitation, which is the case for $\lambda = 2$ μm: the corresponding p is only weakly dependent on n (see Fig. 6).

3.3 Multi-harmonic pulses: dependence on the harmonic phase

For completeness we also show in Fig. 6 different harmonic phases ϕ whose effect is obvious: they hardly influence p, but with increasing ϕ the vector potential within a single period becomes more symmetric (see lower row in Fig. 2) decreasing VP and for $\phi = \pi/2$ with fully symmetric $\boldsymbol{A}(t)$, it is impossible to generate VP. Interestingly, dephasing has less influence for more symmetric vector potentials (the difference between p with and without dephasing is much less for $\phi = \pi/2$ than for $\phi = 0$). Regarding VP, there is in general very little influence of dephasing regardless of the harmonic phase ϕ. Note that the pulse length T varies with the wavelengths between 40 and 160 fs, *i.e.*, ultrashort pulses are not required for VP.

4 The effects of dephasing and excitation distributions

Averaged over local excitation in the valley domains, p and η do not reveal details about dephasing in contrast to the excitation distributions (ED) for $c_\pm$ shown in

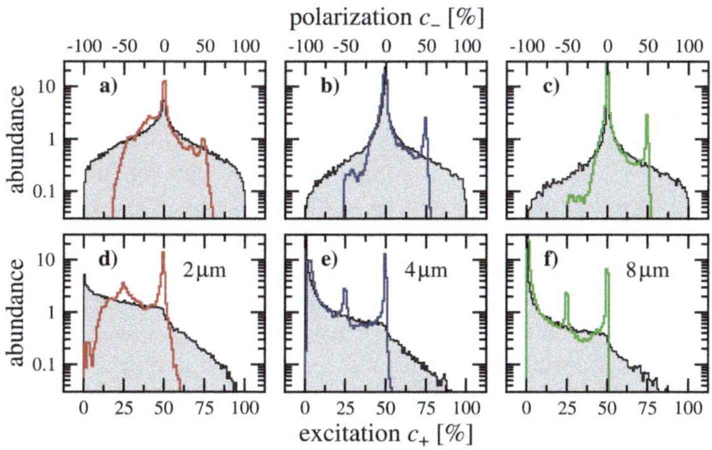

Fig. 8 Distribution of local excitation c_+ and polarization c_- for three wavelengths ($\lambda = 2$, 4, 8 µm) and a two-color pulse ($n = 2$) with maximal contrast ($\phi = 0$). Results for a finite damping time $\gamma = 1/10$ fs (colored lines) are compared to those without damping (gray shaded).

Fig. 8 for $n = 2$. Most striking is the qualitative difference of coherent dynamics with smooth distributions (grey areas) and dephased dynamics exhibiting pronounced peaks (colored lines); note, however, the logarithmic scale. Not surprisingly, $c_- = 0\%$ dominates as VP is difficult to achieve. How is this central peak and the (smaller) peaks at $c_- = \pm 50\%$ related to the dominant peaks in the distributions of c_+? This can be answered by looking at the differential probabilities in Fig. 9. Trivially, all initial conditions which do not lead to excitation (not

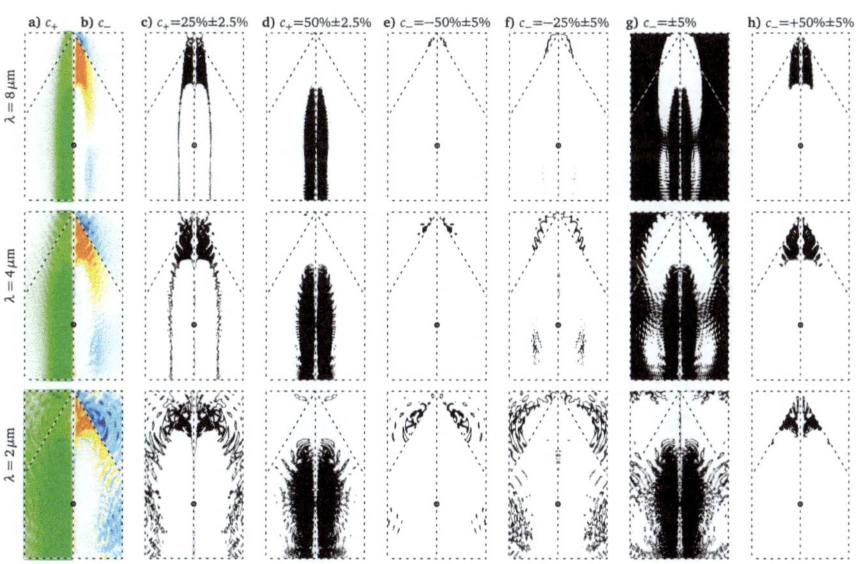

Fig. 9 Local excitation maps corresponding to the peaks in Fig. 8. Initial conditions which contribute to the ranges of c_+ (c and d) and c_- (e–h) given on top are marked in black in the reduced $\boldsymbol{q}$-space. For convenience we also show c_+ (a) and c_- (b).

shown) contribute to the peak at $c_- = 0\%$ in Fig. 8, as can be seen from Fig. 9g and also from the fact that all ED have their maximum at $c_+ = 0$. Note, however, that also the highest abundance of excitation (peak at $c_+ = 50\%$ in Fig. 8d–f) does not contribute to c_-, which becomes clear when looking at Fig. 9a, b and g.

The peak at $c_+ = 25\%$, on the other hand, is the reason for the peak at $c_- = 50\%$. This becomes apparent, when comparing Fig. 9c with Fig. 9h. Although this peak constitutes the main contribution to VP, it does not come from the initial conditions with the highest excitation, which correspond to the green area in Fig. 9a. Rather, the strongest VP comes from excitation at the tip of the triangle in Fig. 9h and the red area in Fig. 9b. The principles that the systematics for $c_- > 0$ follows are confirmed for $c_- < 0$ in Fig. 9, where the corresponding differential probabilities are shown for consistency. Since negative c_- is only dominant for $\lambda = 2$ μm, we do not discuss it here further.

5 Conclusions

We have given a detailed picture of how intense laser pulses affect excitation and valley polarization in graphene, as a representative gapless 2D material. Guided by insight from our previous work,[9] we have focused on pulse trains with asymmetric shapes in each optical period generated by multi-harmonic components, linearly polarized along $\boldsymbol{\Delta} = \boldsymbol{K} - \boldsymbol{K}'$, the line connecting the Dirac points. We have constructed a compact representation of conduction-band occupations in terms of initial conditions in the unit cell, which has given us means to understand the electron dynamics from the local excitation and valley polarization.

The guiding principles to achieve large valley polarization are two-fold: firstly, a pulse, linearly polarized along $\boldsymbol{\Delta}$, and with a maximal vector potential of $\boldsymbol{A}_0 = \boldsymbol{\Delta}$. Secondly, (for fixed $\boldsymbol{A}_0$) a moderate field strength to avoid uncontrolled excitation throughout the unit cell. The latter can be achieved by long wavelengths, and a finite number of overtones n in the harmonic pulse, where n is limited by the requirement of moderate field strength, which—the derivative of the vector potential—becomes large for large n. Although interferences from different excitation amplitudes are omnipresent, they do not influence excitation and valley polarization strongly for the scenarios investigated. This has become clear from the qualitative agreement of results with and without dephasing. It also means, that here dephasing does not pose an obstacle to achieve large valley polarization in gapless 2D materials.

6 Appendix: Hamilton matrix and density matrix equation

For completeness we derive here briefly the Hamilton matrix eqn (1) in the density matrix eqn (2). The starting point is the usual tight-binding two-band Hamilton matrix[16]

$$\overline{\mathrm{H}}_{\boldsymbol{q}} = \begin{pmatrix} 0 & g(\boldsymbol{q}) \\ g^*(\boldsymbol{q}) & 0 \end{pmatrix} \tag{8a}$$

$$g(\boldsymbol{q}) = g_0\left[\text{Å}\mathrm{e}^{\mathrm{i}aq_x} + 2\text{Å}\mathrm{e}^{-\mathrm{i}aq_x/2}\cos\left(\sqrt{3}aq_y/2\right)\right], \tag{8b}$$

where $g_0 = 3.03$ eV and $a = 1.43$ Å. Quantities in the tight-binding basis get an overbar.

In a laser pulse the crystal momentum becomes $q \to \tilde{\boldsymbol{q}}(t) = \boldsymbol{q} + \boldsymbol{A}(t)$. We can write both density matrix and Hamilton matrix by means of the transformation $\mathbb{T}(\tilde{\boldsymbol{q}})$, that makes eqn (8) at instance t diagonal,

$$\bar{\rho}(t) = \mathbb{T}^{\dagger}(\tilde{\boldsymbol{q}})\rho_{\boldsymbol{q}}(t)\mathbb{T}(\tilde{\boldsymbol{q}})\big|_{\tilde{\boldsymbol{q}}=\boldsymbol{q}+\boldsymbol{A}(t)} \tag{9a}$$

$$\overline{\mathrm{H}}(\boldsymbol{q},t) = \mathbb{T}^{\dagger}(\tilde{\boldsymbol{q}})\begin{pmatrix} E_{\mathrm{v}}(\tilde{\boldsymbol{q}}) & 0 \\ 0 & E_{\mathrm{c}}(\tilde{\boldsymbol{q}}) \end{pmatrix}\mathbb{T}(\tilde{\boldsymbol{q}})\Bigg|_{\tilde{\boldsymbol{q}}=\boldsymbol{q}+\boldsymbol{A}(t)}. \tag{9b}$$

The unitary matrix $\mathbb{T}(\tilde{\boldsymbol{q}})$, with $\mathbb{T}^{\dagger}\mathbb{T} = \mathbb{1}$, is often referred to as the Houston basis.[17] The equation of motion for the Houston basis density matrix ρ_q follows from the tight-binding version by means of eqn (10)

$$\frac{\partial}{\partial t}\bar{\rho} = -i[\overline{\mathrm{H}},\bar{\rho}] \to \frac{\partial}{\partial t}\rho = -i[\mathrm{H},\rho] \tag{10}$$

with the Hamilton matrix

$$\mathrm{H} = \begin{pmatrix} E_{\mathrm{v}}(\tilde{\boldsymbol{q}}) & 0 \\ 0 & E_{\mathrm{c}}(\tilde{\boldsymbol{q}}) \end{pmatrix} + i\left(\frac{\partial}{\partial t}\mathbb{T}(\tilde{\boldsymbol{q}})\right)\mathbb{T}^{\dagger}(\tilde{\boldsymbol{q}}). \tag{11}$$

The time the derivative of eqn (11) can be rewritten $i\frac{\partial}{\partial t} = \left(\frac{\partial}{\partial t}\tilde{\boldsymbol{q}}(t)\right)\cdot i\frac{\partial}{\partial \tilde{\boldsymbol{q}}} = \left(\frac{\partial}{\partial t}\boldsymbol{A}(\boldsymbol{t})\right)\cdot i\frac{\partial}{\partial \tilde{\boldsymbol{q}}} = -\boldsymbol{F}(t)\cdot i\frac{\partial}{\partial \tilde{\boldsymbol{q}}}$. In order to arrive at eqn (1) we note that $E_{\tilde{\boldsymbol{q}}} = E_{\mathrm{c}}(\tilde{\boldsymbol{q}}) - E_{\mathrm{v}}(\tilde{\boldsymbol{q}})$ and that the Houston basis $\mathbb{T}(\tilde{\boldsymbol{q}})$ consists of two eigenvectors $V_{\tilde{\boldsymbol{q}}}$ and $W_{\tilde{\boldsymbol{q}}}$, for valence and conduction band, respectively. Whereas these vectors are complex, all matrix elements of the 2nd term in eqn (11) are real. The diagonal elements are identical and can thus be omitted. There we have the expression given in eqn (1).

We introduce a dephasing term with a dephasing rate γ in order to allow for long driving pulses. It will dampen exclusively coherences in the Houston basis density matrix

$$-\frac{\gamma}{2}\left[\rho_{\boldsymbol{q}}(t) - \sigma_z\rho_{\boldsymbol{q}}(t)\sigma_z\right] = -\gamma\begin{pmatrix} 0 & \rho_{\mathrm{vc}} \\ \rho_{\mathrm{cv}} & 0 \end{pmatrix}. \tag{12}$$

Conflicts of interest

There are no conflicts to declare.

Acknowledgements

We thank Hamed Koochaki Kelardeh for discussions in the initial phase of this work We acknowledge financial support by the Deutsche Forschungsgemeinschaft through the Priority Program 1840.

References

1 S. Ghimire and D. A. Reis, *Nat. Phys.*, 2019, **15**, 10.

2 S. Ghimire, A. D. DiChiara, E. Sistrunk, P. Agostini, L. F. DiMauro and D. A. Reis, *Nat. Phys.*, 2011, **7**, 138.

3 G. Vampa, C. R. McDonald, G. Orlando, D. D. Klug, P. B. Corkum and T. Brabec, *Phys. Rev. Lett.*, 2014, **113**, 73901.

4 G. Vampa and T. Brabec, *J. Phys., B*, 2017, **50**, 083001.

5 J. Cayssol, *C. R. Phys.*, 2013, **14**, 760.

6 L. E. Golub and S. A. Tarasenko, *Phys. Rev. B: Condens. Matter Mater. Phys.*, 2014, **90**, 201402.

7 J. R. Schaibley, H. Yu, G. Clark, P. Rivera, J. S. Ross, K. L. Seyler, W. Yao and X. Xu, *Nat. Rev. Mater.*, 2016, **1**, 16055.

8 S. A. Vitale, D. Nezich, J. O. Varghese, P. Kim, N. Gedik, P. Jarillo-Herrero, D. Xiao and M. Rothschild, *Small*, 2018, **14**, 1870172.

9 H. K. Kelardeh, U. Saalmann and J. M. Rost, 2021, arXiv2109.04249.

10 A. Wirth, M. T. Hassan, I. Grguras, J. Gagnon, A. Moulet, T. T. Luu, S. Pabst, R. Santra, Z. A. Alahmed, A. M. Azzeer, V. S. Yakovlev, V. Pervak, F. Krausz and E. Goulielmakis, *Science*, 2011, **334**, 195.

11 P. Krogen, H. Suchowski, H. Liang, N. Flemens, K.-H. Hong, F. X. Kärtner and J. Moses, *Nat. Photonics*, 2017, **11**, 222.

12 M. S. Mrudul, Á. Jiménez-Galán, M. Ivanov and G. Dixit, *Optica*, 2021, **8**, 422.

13 C. Heide, T. Boolakee, T. Higuchi and P. Hommelhoff, *JPhys Photonics*, 2020, **2**, 024004.

14 Q. Z. Li, P. Elliott, J. K. Dewhurst, S. Sharma and S. Shallcross, *Phys. Rev. B*, 2021, **103**, L081102.

15 S. A. Oliaei Motlagh, F. Nematollahi, V. Apalkov and M. I. Stockman, *Phys. Rev. B*, 2019, **100**, 115431.

16 A. J. Leggett, *Graphene: Electronic Band Structure and Dirac Fermions*, University of Waterloo technical report, 2010.

17 W. V. Houston, *Phys. Rev.*, 1940, **57**, 184.

Faraday Discussions

ROYAL SOCIETY OF CHEMISTRY

PAPER

Modelling nonlocal nonlinear spin dynamics in antiferromagnetic orthoferrites†

Yuichi Saito and Rostislav V. Mikhaylovskiy *

Received 7th February 2022, Accepted 6th April 2022
DOI: 10.1039/d2fd00035k

Excitation with an ultrashort light pulse is arguably the only way to control spins in antiferromagnetic materials at both the nanoscale in space and ultrafast time scale. While recent experiments highlighted tantalising opportunities for spin switching and magnonics in antiferromagnets, the theoretical description of antiferromagnetic spin dynamics driven by strongly localised and ultrashort excitation is in its infancy. Here we report a theoretical model describing the nonlocal and nonlinear spin response to the excitation by light. We show that strongly localised ultrafast excitation can drive spin switching, which propagates in space and acts as a source of spin waves. Our theoretical formalism is readily available to describe current and future ultrafast spectroscopy experiments in antiferromagnets.

Introduction

Antiferromagnets, materials supporting magnetic ordering without net magnetization, are the most appealing media for future technologies in ultrafast and energy efficient data storage and computing, based on spintronics and magnonics.[1,2] In particular, antiferromagnets have the magnetic resonance frequency in the sub-terahertz or terahertz (THz) region and therefore produce a much faster coherent response to external perturbation, compared to the conventional ferromagnets, whose resonance frequencies lie in the gigahertz range. Also, strong magneto-optical coupling in many antiferromagnets ensures good susceptibility of their magnetic order to illumination with laser pulses.[3] Therefore, the advent of femtosecond lasers has opened the way to create external stimuli that are shorter than characteristic timescales such as spin-lattice relaxation or spin precession period, thereby enabling ultrafast manipulation of the antiferromagnetic states.[4] Moreover, the development of technology generating intense THz fields, especially the tilted pulse front method,[5] allows researchers to excite spin dynamics in antiferromagnets *via* direct resonant mechanisms.[6,7]

Department of Physics, Lancaster University, Bailrigg, Lancaster LA1 4YW, UK. E-mail: r.mikhaylovskiy@lancaster.ac.uk

† Electronic supplementary information (ESI) available. See https://doi.org/10.1039/d2fd00035k

In most cases optical excitation or THz fields trigger only a linear response of the magnetic order, *e.g.* the spin oscillations at the eigen frequencies of antiferromagnetic resonances.[4,8,9] More interesting from both fundamental and applied points of view is the possibility to trigger spin reorientation,[10] inertial switching[11] and photoinduced phase transition from antiferromagnetic to ferromagnetic states.[12,13] However, for the optical pulses this photoinduced spin-reorientation is thermal in origin and therefore inherently slow.[14] Thermal effects can also be driven by intense THz pulses manifesting as a slight shift of resonance frequency in $TmFeO_3$, thulium orthoferrite.[15] At the same time intense THz pulses can bring spins into a switching regime non-thermally. For instance all coherent spin switching was driven by a THz pulse concentrated and enhanced with a metallic antenna at the transition temperature of spin-reorientation in $TmFeO_3$.[16]

In contrast to this experimental success, the spatial inhomogeneity of the spin dynamics after excitation has been largely unexplored because only an integral magnetic moment along thickness is detected with magneto-optical effects typically used in ultrafast experiments. However, in the experiments with antennas to intensify THz amplitude[15–17] the THz field was distributed non-homogeneously and the spin dynamics is expected to propagate into the sample. Indeed, a very recent experiment revealed that strongly localised optical excitation drives propagating spin waves in an antiferromagnetic $DyFeO_3$,[18] although in a linear regime. It is an open question, however, what kind of propagating spin dynamics one can excite in the nonlinear regime with a strongly localised excitation?

Thus, our research motivation is to reveal temporal and spatial characteristics of the spin dynamics in the nonlinear regime. To this end, we construct the model to solve a partial differential equation which describes spin propagation in antiferromagnets, such as rare-earth orthoferrites. Then, we numerically calculate the spin dynamics triggered by a realistic THz pulse and vary parameters to understand the effect on nonlinearity.

Theoretical formalism

We theoretically explore spin-wave propagation in a nonlinear regime, driven by the intense terahertz pulse. In contrast to previous research, we take into account the spatial inhomogeneity of the exciting pulse and consequent propagation of the spin deflection in space. We use the Lagrangian equation of spin dynamics that yields both time evolution and spatial distribution of the magnetization as the solution.[19] For an antiferromagnet this equation has the form of a sine-Gordon equation. For the indicative case of orthoferrites, the equation of motion is given by:[16]

$$\begin{aligned}&\frac{\partial^2\phi}{\partial t^2}+\alpha\omega_e\frac{\partial\phi}{\partial t}-C_m^2\frac{\partial^2\phi}{\partial y^2}+\omega_e\omega_a\ \sin\phi\cos\phi\left(\eta(T)-\cos^2\phi\right)\\&=\xi E_{THz}^2\ \sin\phi\cos\phi+\gamma\omega_d H_{ext}^x\ \sin\phi+\gamma\omega_d H_{ext}^z\ \cos\phi+\gamma\frac{\partial H_{ext}^y}{\partial t}\end{aligned} \tag{1}$$

where α is the damping factor, C_m is the propagation speed limit of the spin wave, ξ is a coefficient characterising coupling between THz electric field and the spins. The function $\phi(t,y)$ represents the angle of the antiferromagnetic vector **G** in the *ac*-plane, and $\phi = 0$ if **G** is aligned along the *z*-axis; $\omega_e = \gamma H_e$, $\omega_d = \gamma H_d$, and $\omega_a =$

$\gamma K_2/M_{Fe}$ are parameters defined for convenience, where K_2 is the constant for the fourth-order magnetocrystalline anisotropy and M_{Fe} is the magnetization of the iron sublattice. The effective magnetic fields H_d and H_e are attributed to the Dzyaloshinskii–Moriya and the exchange interactions, respectively. The constant γ is the gyromagnetic ratio. The Cartesian *xyz*-system is chosen to coincide with the crystallographic directions *a*, *b*, *c*.

Rare earth orthoferrites exhibit spin-reorientations, so the orientation of the antiferromagnetic vector flips or rotates across a specific temperature point or interval. For instance, part of orthoferrites, such as $TmFeO_3$, $ErFeO_3$ and $SmFeO_3$, have transitions from Γ_2 to Γ_4 phases as temperature increases.[8] The magnetization pointed along the *a*-axis in the Γ_2 phase moves continuously by 90 degrees to the *c*-axis in the Γ_4 phase. We focus on the intermediate Γ_{24} phase with the antiferromagnetic vector in the *ac*-plane, in which considerably large nonlinearity was reported.[20] At that phase, four equilibrium states, given by local minima of the potential energy, are placed in the *ac*-plane in the case of the zero bias field (Fig. 1b). The function $\eta(T)$ in eqn (1) defines the equilibrium orientations and reads

$$\eta(T) = -\frac{K_1}{K_2} = \frac{T_u - T}{T_u - T_l} \tag{2}$$

where the critical temperatures separating Γ_2 and Γ_{24} phases and Γ_4 and Γ_{24} phases are expressed as T_l and T_u, respectively. K_2 is the second-order magnetocrystalline anisotropy constant. The stable angle in Γ_{24} is given by $\cos^2\phi_0 = \eta(T)$.

We consider two processes of excitation *via* the electric field and magnetic field of the THz pulse separately. The first term of the right-hand side of eqn (1) represents the modulation of magnetocrystalline anisotropy by THz electric field. The high-intensity electric field E_{THz} distorts the anisotropy mediated by the orbital states of the rare-earth ions and causes the spin excitation indirectly. Other three terms

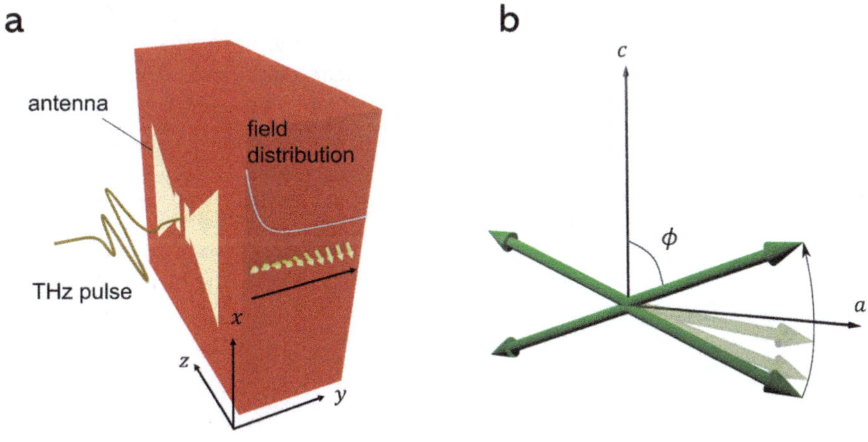

Fig. 1 Model structure for simulation. (a) The THz field is enhanced with metallic antenna fabricated on top of the orthoferrite crystal $TmFeO_3$. The incident pulse travels in the sample from the antenna side. Exponentially localised evanescent field induces spin switching and generates spin waves. (b) There are four stable positions for antiferromagnetic vector **G** in the Γ_{24} phase, especially four-fold symmetric placement in the midpoint between T_l and T_u. Through switching by external stimulation, the **G** jumps to another stable state in the *ac*-plane.

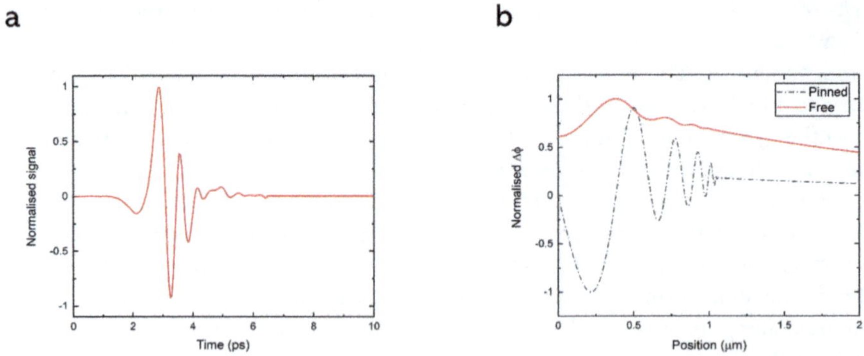

Fig. 2 (a) The THz wave form $s(t)$ used in the calculation was extracted from the real experiment. (b) The calculated snapshot of the spin deflection in the linear regime of excitation (moderate THz field strength). Two cases for different boundary conditions, Dirichlet (red solid) and Neumann (black dash-dot), are shown. The spin wave was generated from the left side where the THz field was localised by antenna enhancement.

describe Zeeman coupling, which is a direct interaction between the external magnetic field $\mathbf{H}_{\text{ext}}$ and the net magnetization. Regarding the magnetic field, we assume two origins for $\mathbf{H}_{\text{ext}}$ in this paper: the static bias field $\mathbf{H}_{\text{bias}}$ is independent of time, which alters the potential energy steadily. As a result, the symmetry in the *ac*-plane is broken; conversely, the dynamic magnetic field $\mathbf{H}_{\text{THz}}$ is caused by the pump terahertz pulse. Since the torques exerted by $\mathbf{H}_{\text{THz}}$ and $\mathbf{E}_{\text{THz}}$ are additive, we compare their effects by solving the equation while only one of these terms is present.

The shape of the terahertz pulse with amplitudes, $\mathbf{H}_{\text{amp}}$ and $\mathbf{E}_{\text{amp}}$, was taken from our measured data for the THz fields generated by tilted-pulse-front excitation and detected by electro-optical sampling.[6] The manual zero filling to the latter data was implemented to generate the longer time scale (normalised data $s(t)$ shown in Fig. 2). Furthermore, we postulated that the electric field and the magnetic field had exponential distribution with homogeneous offset in the substrate to describe the distribution of the near field in the vicinity of the antenna. The antenna fabricated on the surface enhances the terahertz signal and localizes the intensity at the neighbourhood of the antenna.[3] The distribution function is formulated as:

$$\mathbf{H}_{\text{THz}}(t,y) = \left[A\exp\left(-\frac{y}{d}\right)+1\right]\mathbf{H}_{\text{amp}}s(t), \tag{3}$$

where A is the gain of antenna and d is the spatial decay. The electric field was formed in a similar fashion:

$$\mathbf{E}_{\text{THz}}(t,y) = \left[A\exp\left(-\frac{y}{d}\right)+1\right]\mathbf{E}_{\text{amp}}s(t) \tag{4}$$

Results and discussion

Numerical modelling

We adopted a solving method called the Finite Difference Method to analyse the equation numerically. The boundary condition was mainly designated as the

Neumann (free boundary) condition, given by $\partial\phi/\partial y(y = 0) = 0$, to consider the switching phenomena. We applied all parameters for compound $TmFeO_3$, featured in recent experiments on THz spin control.[16,20] The effective fields of exchange interactions were taken from,[21] $H_d = 2 \times 10^5$ Oe and $H_e = 2 \times 10^7$ Oe. The set of transition temperatures of this material is known as approximately $T_u = 90$ K and $T_l = 80$ K. The value $\omega_a = 2.1$ rad ns^{-1}, $\xi = 2.3 \times 10^{18}$ erg^{-1} cm^3 s^{-1} and $\alpha = 1.4 \times 10^{-4}$ was experimentally determined from the data of the quasi-ferromagnetic resonance frequency at $T = 83$ K.[3] Although those two parameters depend on frequency in a strict sense, we regarded them as constants to simplify the calculation. Besides, $C_m = 2 \times 10^6$ cm s^{-1} was used.[5]

In our calculations, the temperature was fixed at the midpoint within the spin reorientation temperature interval, namely $T = 85$ K. One of the equilibrium angles ϕ_0 in the case of the no bias field was accurately set to 45° and the others are located in the fourth-fold symmetric positions. However, the bias field slightly shifts the stationary point of the potential. This means that initial values $\phi(t = 0)$ close to $\phi = 45$ deg should be compensated adequately to remove the excitation by the bias field. After recalculation of the potential to know the new local minima with bias magnetic field, we applied initial conditions, $\phi(t = 0) = \phi_0$ and $\partial\phi/\partial t(t = 0) = 0$ and they are spatially uniform. For parameters for the field distribution, we used $A = 8$ and $d = 4$ μm, which were the values used to model the antenna response in the previous experiment.[15]

Results

At first, we calculated the spin dynamics in a linear regime with a small THz amplitude, to clarify the linear background to the nonlinearity. For the Neumann condition, the THz field $H_{THz} = 45$ kOe and the bias field $H_{bias} = 2$ kOe along the x-axis were applied. In addition, we took the Dirichlet (pinned boundary) condition, expressed as $\phi(y = 0) = 0$, with different amplitude $H_{THz} = 900$ Oe. The pinned condition can be accepted easily for the weak excitation regime, while the pinning on edges should be an unrealistic condition for strong driving and subsequent nonlinear switching phenomena. The result for both cases is illustrated in Fig. 2b. The uniform spin precession, zero-wavenumber spin wave shaped by the antenna was dominant, especially in the Neumann case. Also, the spin wave whose wavenumber was finite, not zero, propagated in spite of the much smaller amplitude than the uniform mode. These results were comparable to a previous simulation on linear propagation excited by an optical pump.[17]

Next, increasing the intensity of THz field, we stepped into the nonlinear regime. The enhanced magnetic field, $H_{THz} = 54$ kOe, induced nonlinear dynamics, involving flipping of the antiferromagnetic vector into another equilibrium orientation (Fig. 3a, see also ESI Movie 1†). This creates a non-stationary domain wall separating the regions with metastable spin orientations. The external bias $H_{bias} = 2$ kOe along the x-axis eliminated the two local minima around $\phi = \pm 135°$ (Fig. 3b). In the first part of the process in about ten picoseconds, the spin around the antenna moves to the other equilibrium position. This phenomenon is caused by inertia motion following the main peak of the THz pulse (at $t < 3$ ps in Fig. 2a). On the other hand, the angle of spins at a distant position from the antenna return to the initial position as a result of damping. This localised reversal forms a domain wall between the different spin

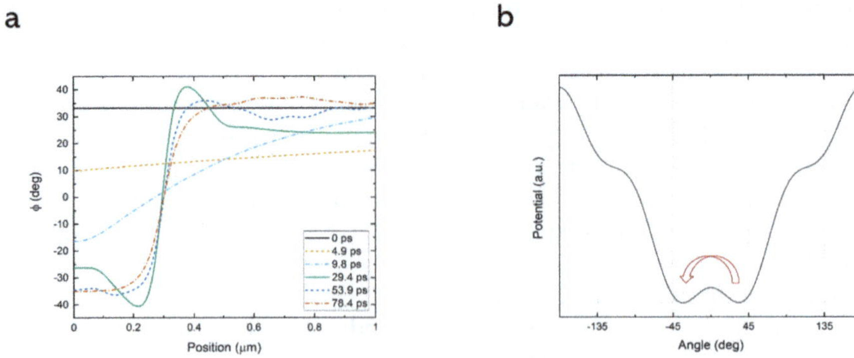

Fig. 3 (a) Snapshots of the spin deflection angle in the strongly nonlinear regime, when the bias magnetic field is applied along the x-axis. After forming the domain wall between the initial and the switched orientations, the various modes of the spin wave are emitted from the wall. (b) The potential landscape set by the bias field with the two stable states. The THz excitation enabled the spin around the antenna to flip to the alternative stable state.

orientations around $y = 0.3$ μm, and the position of the wall after the THz pulse duration is fixed all the time. The position of the domain wall is determined by the parameter d in eqn (3), the penetration depth of the THz field for the chosen wave form and spatial distribution. This position corresponds to the threshold THz amplitude, required to bring the magnetic order parameter over the potential barrier separating two equilibrium orientations (see Fig. 3b). At the same time the sudden appearance of this discontinuity generates spin waves with finite wavenumbers propagating into both positive and negative directions from the domain wall. The spin wave emission mechanisms resemble the linear regime mentioned before (Fig. 2b). The domain wall plays the role of the pinned boundary.

The different orientation of the bias field along the z-axis provide completely different results, accompanied with movement of the domain wall (Fig. 4a, ESI Movie 2†). The comparably weak bias $H_{\text{bias}} = 500$ Oe along the z-axis reduced the barrier to facilitate the switching by the THz pulse with the amplitude, $H_{\text{THz}} = 54$ kOe. The long-distance propagation of the wall follows from the different potential landscape (Fig. 4b). The bias modulated the symmetric potential to non-symmetric one at about $\phi = 0$ and pushed up the initial position ϕ_0 to the metastable state. The spin on this local minimum state fell down to the lower global minimum around $\phi = -45°$. This is the origin of the domain movement. Moreover, the larger amplitude of bias field, for example $H_{\text{bias}} = 1$ kOe, caused the other domain wall between $\phi = -45°$ and $\phi = -135°$, which stayed at the same position (see ESI Movie 3†). The boundary was essentially the same as the result shown in Fig. 3a from the standpoint of the equal potential levels.

To compare dynamics induced by the electric field to one by the magnetic field, we applied $E_{\text{THz}} = 9$ MV cm^{-1} to eqn (1) and neglected terms for the Zeeman effect. We calculated only one case with $H_{\text{bias}} = 2$ kOe along the x-axis. The obtained result was quite similar (ESI Movie 4†) to the one shown in Fig. 3a because the action of E_{THz} can be seen as an effective magnetic field pulse.

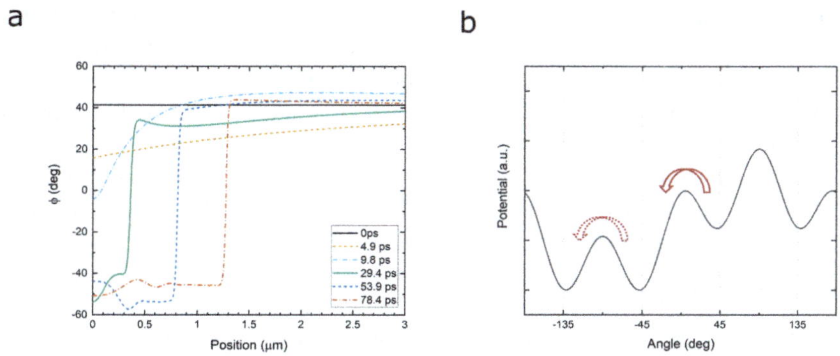

Fig. 4 (a) Snapshots of the spin deflection angle in the strongly nonlinear regime, when the bias magnetic field is applied along the z-axis and the spins are initially aligned in a metastable state. In contrast to the situation shown in Fig. 3, here the domain wall is moving away from the sample boundary. (b) The potential landscape with the bias oriented z-axis show that the main excitation occurred as a result of falling through the slope between −45° and 45°.

Conclusions

In summary, we developed the model for nonlinear and non-local spin dynamics relevant to the recent experiments with THz pump pulses concentrated by the metallic antennas fabricated on top of the antiferromagnetic sample. We predict the formation of the non-stationary and at certain conditions a moving boundary, separating stable orientations of the antiferromagnetic vector. The resulting moving discontinuity can emit spin waves. The propagation of waves through the moving boundary opens a new vistas for the non-stationary effects such as modulation of the spin wave frequencies reminiscent of propagation of the electro-magnetic waves across the relativistic boundary. We believe our theoretical results will inform and inspire future experimental work in the burgeoning field of antiferromagnetic magnonics.

Author contributions

Y. S. has performed the mathematical derivation and numerical calculations and analysis under the guidance of R. V. M., who formulated the project and supervised it.

Conflicts of interest

There are no conflicts to declare.

Acknowledgements

This work was funded by the European Research Council ERC grant agreement no. 852050 (MAGSHAKE).

References

1 T. Jungwirth, X. Marti, P. Wadley and J. Wunderlich, *Nat. Nanotechnol.*, 2016, **11**, 231–241.

2 P. Němec, M. Fiebig, T. Kampfrath and A. V. Kimel, *Nat. Phys.*, 2018, **14**, 229–241.
3 V. V. Eremenko, N. F. Kharchenko, Y. G. Litvinenko and V. M. Naumenko, *Magneto-Optics and Spectroscopy of Antiferromagnets*, Springer, 1992.
4 A. V. Kimel, A. Kirilyuk, P. A. Usachev, R. V. Pisarev, A. M. Balbashov and T. Rasing, *Nature*, 2005, **435**, 655–657.
5 J. Hebling, G. Almasi, I. Kozma and J. Kuhl, *Opt. Express*, 2002, **10**, 1161.
6 T. Kampfrath, A. Sell, G. Klatt, A. Pashkin, S. Mährlein, T. Dekorsy, M. Wolf, M. Fiebig, A. Leitenstorfer and R. Huber, *Nat. Photonics*, 2011, **5**, 31–34.
7 T. Suemoto, K. Nakamura, T. Kurihara and H. Watanabe, *Appl. Phys. Lett.*, 2015, **107**, 042404.
8 A. M. Kalashnikova, A. V. Kimel, R. V. Pisarev, V. N. Gridnev, P. A. Usachev, A. Kirilyuk and T. Rasing, *Phys. Rev. B: Condens. Matter Mater. Phys.*, 2008, **78**, 104301.
9 T. Satoh, S.-J. Cho, R. Iida, T. Shimura, K. Kuroda, H. Ueda, Y. Ueda, B. A. Ivanov, F. Nori and M. Fiebig, *Phys. Rev. Lett.*, 2010, **105**, 077402.
10 A. V. Kimel, A. Kirilyuk, A. Tsvetkov, R. V. Pisarev and T. Rasing, *Nature*, 2004, **429**, 850–853.
11 A. V. Kimel, B. A. Ivanov, R. V. Pisarev, P. A. Usachev, A. Kirilyuk and T. Rasing, *Nat. Phys.*, 2009, **5**, 727–731.
12 T. Li, A. Patz, L. Mouchliadis, *et al.*, Femtosecond switching of magnetism via strongly correlated spin–charge quantum excitations, *Nature*, 2013, **496**, 69–73.
13 D. Afanasiev, B. A. Ivanov, A. Kirilyuk, T. Rasing, R. V. Pisarev and A. V. Kimel, *Phys. Rev. Lett.*, 2016, **116**, 097401.
14 J. A. de Jong, I. Razdolski, A. M. Kalashnikova, R. V. Pisarev, A. M. Balbashov, A. Kirilyuk, T. Rasing and A. V. Kimel, *Phys. Rev. Lett.*, 2012, **108**, 157601.
15 Y. Mukai, H. Hirori, T. Yamamoto, H. Kageyama and K. Tanaka, *New J. Phys.*, 2016, **18**, 013045.
16 S. Schlauderer, C. Lange, S. Baierl, T. Ebnet, C. P. Schmid, D. C. Valovcin, A. K. Zvezdin, A. V. Kimel, R. V. Mikhaylovskiy and R. Huber, *Nature*, 2019, **569**, 383–387.
17 T. Kurihara, H. Watanabe, M. Nakajima, S. Karube, K. Oto, Y. Otani and T. Suemoto, *Phys. Rev. Lett.*, 2018, **120**, 107202.
18 J. R. Hortensius, D. Afanasiev, M. Matthiesen, R. Leenders, R. Citro, A. V. Kimel, R. V. Mikhaylovskiy, B. A. Ivanov and A. D. Caviglia, *Nat. Phys.*, 2021, **17**, 1001–1006.
19 A. K. Zvezdin, *Pis'ma Zh. Eksp. Teor. Fiz.*, 1979, **10**, 605.
20 S. Baierl, M. Hohenleutner, T. Kampfrath, *et al.*, *Nat. Photonics*, 2016, **10**, 715–718.
21 A. M. Balbashov, G. V. Kozlov, A. A. Mukhin and A. S. Prokhorov, in *High Frequency Processes in Magnetic Materials*, World Scientific, 1995, pp. 56–98.

Faraday Discussions

PAPER

Impact of the terahertz and optical pump penetration depths on generated strain waves temporal profiles in a V_2O_3 thin film

Guénolé Huitric, [a] Michael Rodriguez-Fano,[b] Lucas Gournay,[a] Nicolas Godin,[ac] Marius Hervé, [a] Gaël Privault, [a] Julien Tranchant,[b] Zohra Khaldi,[b] Marco Cammarata, [a] Eric Collet, [ac] Etienne Janod[ac] and Christophe Odin [*ac]

Received 20th January 2022, Accepted 11th February 2022
DOI: 10.1039/d2fd00013j

Triggering new stable macroscopic orders in materials by ultrafast optical or terahertz pump pulses is a difficult challenge, complicated by the interplay between multiscale microscopic mechanisms, and macroscopic excitation profiles in samples. In particular, the differences between the two types of excitations are still unclear. In this article, we compare the optical response on acoustic timescale of a V_2O_3 Paramagnetic Metallic (PM) thin film excited by a terahertz (THz) pump or an optical pump, at room temperature. We show that the penetration depth of the deposited energy has a strong influence on the shape of the optical transmission signal, consistent with the modulation of permittivity by the superposition of depth-dependent static strain, and dynamical strain waves travelling back and forth in the sample layer. In particular, the temporal modulation of the optical transmission directly reflects the excitation profile as a function of depth, as well as the sign of the acoustic reflection coefficient between the film and the substrate. The acoustic mismatch between the V_2O_3 layer and the substrate was also measured. The raw data were interpreted with a one-dimensional analytical model, using three fitting parameters only. These results are discussed in the context of triggering phase transitions by ultrafast pump pulses. To the best of our knowledge, this is the first report of the modulation of the optical transmission of V_2O_3 with a THz pump within the acoustic timescale.

1. Introduction

The study of Photo-Induced Phase Transitions (PIPT) is in constant development, owing to new experimental techniques and progress in the physical and

[a]*Univ Rennes, CNRS, IPR (Institut de Physique de Rennes) – UMR 6251, F-35000 Rennes, France. E-mail: christophe.odin@univ-rennes1.fr*

[b]*Institut des Matériaux Jean Rouxel de Nantes (IMN), Univ Nantes, CNRS, F-44322 Nantes, France*

[c]*DYNACOM IRL2015 University of Tokyo - CNRS - UR1, Department of Chemistry, 7-3-1 Hongo, Tokyo 113-0033, Japan*

theoretical interpretation.[1] Moreover, the recent possibility to generate almost single cycle terahertz pulses of high intensity offers appealing perspectives for the excitation and investigation of coherent lattice vibrations.[2–6]

It was soon recognized that, in addition to the microscopic excitation mechanisms, the macroscopic excitation profile is also strikingly important both for the generation of new phases, their propagation in space and time, and for the interpretation of collected data. The finite penetration depth and hence excitation profile of the pump will induce an inhomogeneous distribution of phase nuclei from the surface,[7] which may propagate in the depth of the sample (for instance, photoinduced metallic domains in an insulating phase[8]). The initial spatial pattern is determined by the excitation profile and mechanism, and subsequent evolutions (nucleation, phase front propagation) are driven by corresponding spatial gradients. This spatial and dynamical mixture of sample properties modulates the sample dielectric function, whose effects are probed through time-resolved optical reflectivity or transmission measurements. A model of the full optical detection process was developed in the seminal work by Thomsen *et al.*,[9] pointing out the role and interplay between the pump and probe in the interpretation of the transient signal.

It is admitted that optical pump microscopic excitation processes are mainly mediated by electrons, while terahertz pulses can also interact directly with the lattice through infrared active phonon modes. Among possible transformations that are triggered by such ultrafast strong pulses are coherent phonons or strain-waves.[10,11] In

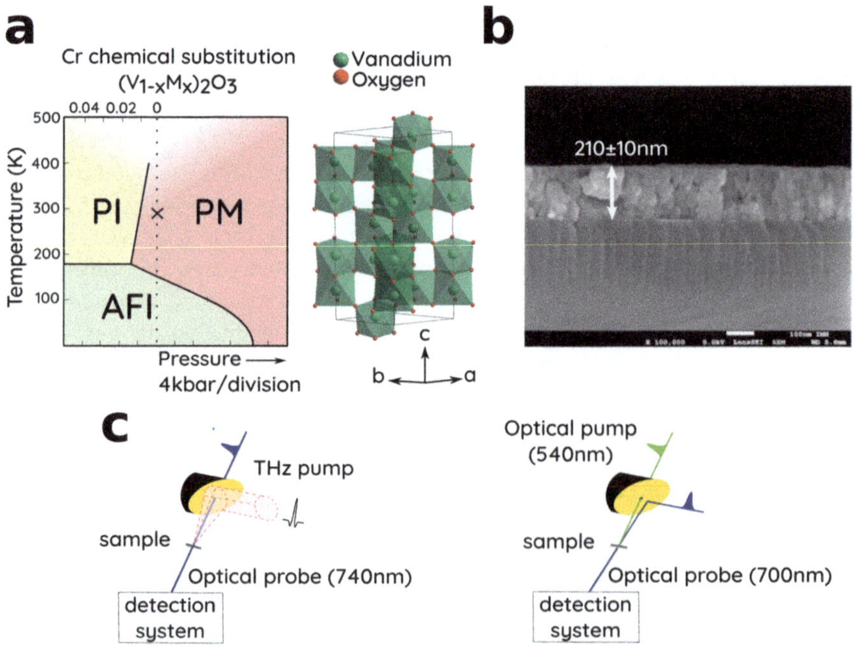

Fig. 1 (a) Phase diagram of V_2O_3 (PM: Paramagnetic Metal; PI: Paramagnetic Insulator; AFI: Antiferromagnetic Insulator) and structure in the PM phase. (b) Scanning Electron Microscope (SEM) image. The film thickness was measured as 210 ± 10 nm. (c) Experimental setups for the terahertz pump or optical pump/optical probe.

addition to possible applications to probe material properties, such strain waves were shown to be able to drive a semi-conductor to metal phase transition.[12]

For these reasons, the comparison of ultrafast pump properties is of primary importance in the understanding of PIPT phenomena, in particular in metal–insulator phase transitions of strongly correlated metals.

Vanadium sesquioxide V_2O_3 is a work horse in the study of Mott–Hubbard phase transitions.[13,14] At ambient pressure, it presents a first order symmetry-breaking Paramagnetic Metal (PM) to Antiferromagnetic Insulator (AFI) phase transition near 160 K, while a third phase (Paramagnetic Insulator, PI) can be obtained by changing pressure or chemical doping (Fig. 1a). At room temperature and pressure, V_2O_3 is in a paramagnetic correlated metallic phase, with a corundum structure with two V_2O_3 formula units in the rhombohedral $R\bar{3}c$ unit cell (Fig. 1a).

The ultrafast responses of such materials in pump–probe experiments are very important both for the understanding of the underlying physics and for potential applications. For instance, optical pump experiments have shown the possibility of inducing metallic nucleation-growth from the low-temperature insulating phase,[15] while very high terahertz-pump fields (>1 MV cm^{-1}) at 4 K achieved insulating to metal transition through electronic tunneling.[16]

In the following, we compare the optical relative transmission transients induced by a terahertz or an optical pump on a V_2O_3 thin film, using the same optical probe, to get information on photoinduced static and dynamical strains within the acoustic timescale.

2. Materials and methods

2.1 Sample preparation

A thin film of V_2O_3 was deposited on a c-cut (0001) sapphire substrate by sputtering of a vanadium target in Ar discharge. Post-deposition annealing at 530 °C in a reducing atmosphere yields a 210 ± 10 nm thick crystallized and single-phased V_2O_3 layer.

The sample thickness of 210 nm was chosen on the one hand because it is higher than the penetration depth of the optical pump (80 nm), and lower than that of THz pump (1300 nm) used in pump–probe experiments presented in this work. On the other hand, a thickness of 210 nm results in an optical density OD $\approx$ 1 (probe penetration depth is 95 nm), which usually provides a good contrast in ultrafast transmission measurements.

2.2 Experimental set-up and simulations

Schematics of the experimental setups are given in Fig. 1c where THz pump/optical probe and optical pump/optical probe techniques are used. The output of a femtosecond regenerative amplifier (Coherent) that delivers 100 fs pulses at 1 kHz repetition rate is split and used for both pump and probe beams. For the THz experiments, the main part (3 mJ) of the amplifier output is used to generate intense THz fields by optical rectification using the tilted-pulse-front technique[2,17] in $LiNbO_3$. We obtain a fluence of 1 μJ mm^{-2} and an electric field reaching 250 kV cm^{-1} (measured by Electro-Optical Sampling, EOS). The remaining output of the amplifier seeds an Optical Parametric Amplifier, OPA (TOPAS Light Conversion) in order to obtain the 740 nm pulse used as probe. The photo-induced signal is recorded with a lock-in detection. For the optical pump experiment, we use the output of an OPA as pump. The wavelength is set to 540 nm and its fluence fixed to 35 μJ mm^{-2}. Due to experimental constraints, in this setup, a 700 nm probe wavelength is preferred. Both experiments are performed in transmission configuration at nearly normal incidence.

In this way, we measured the temporal relative variation of the optical transmission T_r, $\frac{\Delta T_r}{T_r} = [T_r(\text{on}) - T_r(\text{off})]/T_r(\text{off})$, where "on" and "off" mean respectively with and without pump. Eventually, a time-independent constant corresponding to the mean signal at negative times was subtracted. No other data treatments have been performed, nor any part of the signal removed.

Data reduction and all simulations were performed with home-made Python codes that use Numpy[18] and Scipy[19] libraries. Figures were drawn with Matplotlib.[20]

2.3 THz-time domain spectroscopy (TDS)

The V_2O_3 sample was characterized by terahertz time domain spectroscopy. The time traces of the THz electric field, measured by electro-optical sampling with a 1 mm thick ZnTe crystal, are shown in Fig. 2a. From their Fourier transforms, we extracted the complex dielectric function of the film using the Tinkham formula.[21–23] The mean absolute value of the transmission is $T \approx 0.375$, from which a static conductivity of $\sigma_o \approx 1000$ S cm^{-1} was calculated from the Tinkham formula.[21] This estimation is close to those obtained with the same types of samples by THz-TDS.[15,24–26] The frequency dependence of the real and imaginary parts of the dielectric function are shown in Fig. 2b, with the best fit using Drude's model for the conductivity, eqn 1

$$\varepsilon_\omega = \varepsilon_1 + i\varepsilon_2 = \varepsilon_\infty + i\frac{\sigma_\omega}{\omega\varepsilon_o} \quad \text{where } \sigma_\omega = \frac{\sigma_o}{1 + i\omega\tau} \tag{1}$$

where ω is the angular frequency (rad s^{-1}), ε_o is the vacuum dielectric permittivity and ε_∞ the dielectric function at infinite frequency. We obtained an electronic scattering

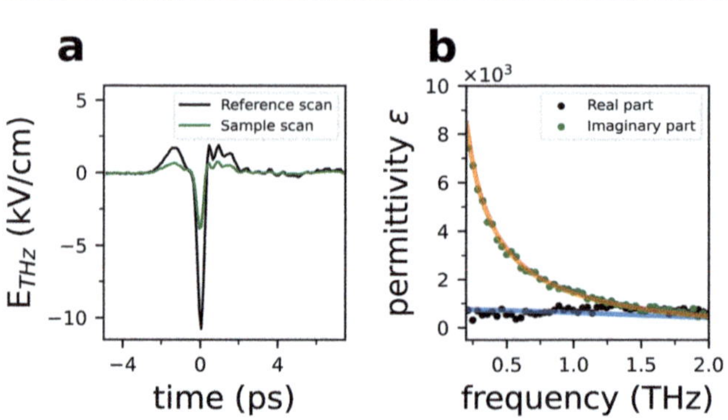

Fig. 2 Terahertz time domain spectroscopy of V_2O_3 at room temperature. (a) Electric field waveform as a function of time: bare substrate (reference scan) and substrate with thin film sample (sample scan). (b) Experimental real ε_1 and imaginary ε_2 parts of the dielectric function as a function of frequency. The continuous lines are the best fits using a Drude model.

rate $\Gamma = 1/\tau \approx 14.3 \pm 0.2$ THz, and a static conductivity of $\sigma_0 \approx 1000$ S cm^{-1}. These values are consistent with those obtained in ref. 26 and 27. These parameters will be used in the following to calculate the sample refractive index at terahertz frequencies.

3. Results and analysis

3.1 Modulation of optical transmission by terahertz or optical pump

In the following we report the comparison of temporal relative transmission variation using either a THz pump or an optical pump. The THz pump EOS temporal trace, shown in Fig. 3c, has a maximum electric field ≈ 250 kV cm^{-1} and typical full duration ≈ 4 ps. The energy power spectrum has its maximum near 0.5 THz (Fig. 3d) with a full width at half maximum FWHM ≈ 0.5 THz. The optical pump at 540 nm has a duration of 100 fs. For experimental reasons, we could not use exactly the same probe wavelengths, 740 and 700 nm respectively for THz and optical pumps. However, it can be shown that the sensitivity function (see below Fig. 4) that determines how the probe interacts with the permittivity modulation induced by the pump are almost equal over the full film thickness, since the ratio

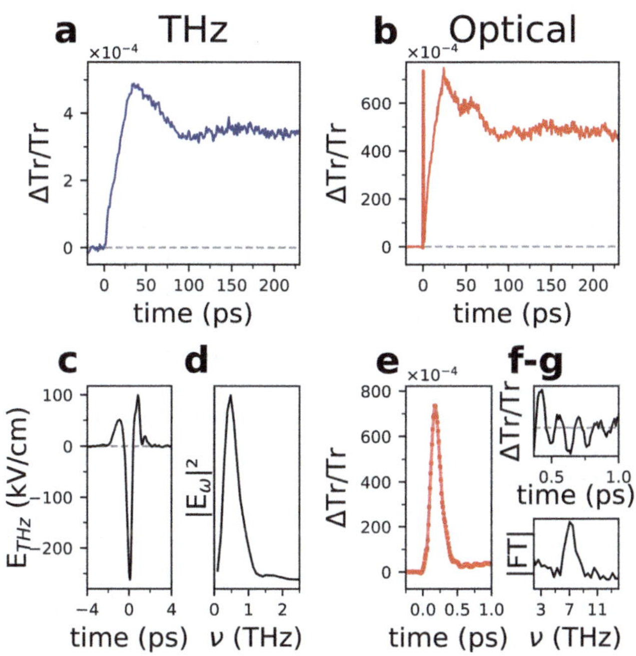

Fig. 3 (a and b) Relative optical transmission as a function of time $\frac{\Delta T_r(t)}{T_r}$. Raw data are presented. (a) THz pump. (b) Optical pump (540 nm). For $t \leq 1.5$ ps the time-step is 0.01 ps, while for $t \geq 2$ ps, the time-step is 0.5 ps. (c and d) Terahertz pump electric field waveform as a function of time, and corresponding squared modulus of the Fourier transform $|E_\omega|^2$. (e–g) Short-time analysis of the optical pump transmission acquisition shown in (b); (e) raw data (points) and best fit (continuous line); (f and g) evidence of coherent phonon excitation: (f) resulting transmission after subtraction of the peak and background as a function of time, and (g) the corresponding modulus of the Fourier transform.

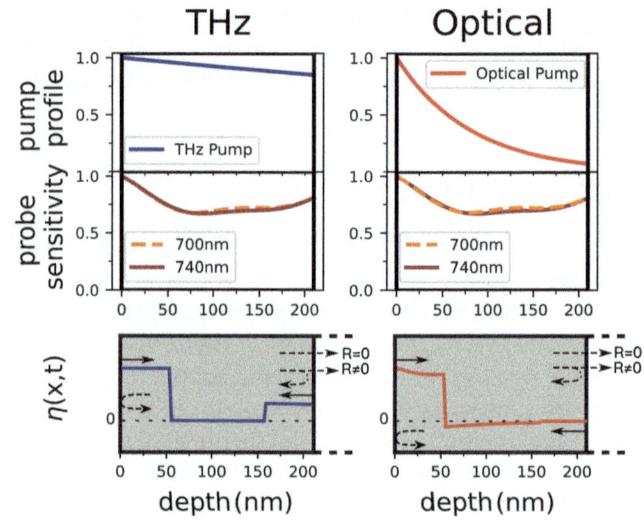

Fig. 4 Left: THz pump and right: optical pump. First line: excitation profiles and modulus of the sensitivity function in the thin film. Parameters correspond to the real parameters of the experiment. All functions were normalized to one at $x = 0$ for clarity. Second line: sketch of the thin-film and substrate interfaces and strain waves.

of the complex index to wavelength $\tilde{n}/\lambda$ are very close (difference of less than 6% for the real part, and 1% for the imaginary part).

The temporal evolution of the raw relative transmission variation $\frac{\Delta T_r(t)}{T_r}$ is presented on Fig. 3 for both pumps, with probes at almost the same wavelength.

Two features are immediately apparent. First, both signals increase within about 30 ps, reaching a maximum before decaying to an almost constant plateau value for delay times larger than 90 ps. In fact, for larger times, small oscillations modulate the plateau. For instance a bump appears near 150 ps in the terahertz pump profile. It is less clear for the optical pump profile. Second, the magnitude of the relative transmission variation with the optical pump is larger than the value with the terahertz pump, as expected from the largest available energy and fluence of the optical pulse.

Most significant is the difference in the profiles within the 35–90 ps time range. While the decay is rather uniform for the terahertz profile, a dip near 44 ps, and second maximum at 56 ps are clearly seen for the profile with the optical pump, making this profile strikingly different from the terahertz profile. Moreover, an additional strong sharp peak signal appears at very short time delays.

By noting that an acoustic timescale is the travel time of a wave from the top surface to the interface of the film with the substrate, that is in our case $\tau_a = d/v \approx$ 29 ps, and an electronic time-scale is typically a few 0.1 ps, we can split our analysis into two parts.

First, we analyse in more detail the optical short-time trace, which was studied with a shorter time-step of 0.01 ps, as presented in Fig. 3e. This photo-induced signal can be divided into three contributions. A first ultrafast response is located just after the pump arrival and lasts for less than 0.5 ps (see Fig. 3e). This feature is related to the photon–electron interaction. At larger time delays, oscillations on a slowly growing background are perceptible. The extraction of the oscillating component is done by removing the two other contributions. The peak is fitted by the convolution of a Gaussian of full width at half height 120 fs, with an exponential decay of 80 fs time constant, while the increasing background is reproduced thanks to a first order polynomial. The Fourier transform of the remaining oscillating component unveils a 7.2 THz mode (see Fig. 3g), in accordance with the frequency of the A_{1g} optical phonon mode previously measured through Raman[28,29] and pump–probe experiments.[29–31] Concerning the THz pump experiment, neither the peak nor any oscillation were detected. The lack of oscillating component is due to the long THz excitation (~1 ps) compared to the period of the phonon mode (~0.14 ps), which washes out the oscillations due to the dephasing.

Second, as discussed before, for times larger than 1 ps, the features of the data fall in the acoustic timescale, the positions of peaks and dips corresponding more or less to multiples of the acoustic travel time $n \times \tau_a$ with n an integer. More precisely, we will show that the measured relative transmission variations as a function of time come from the modulation of the complex dielectric function by photo-induced elastic strain-waves. For metals, the strain waves are usually generated by a thermoelastic mechanism.[9–11,32] In this process, part of the pump pulse energy is absorbed by the electrons, which thermalize in a few hundreds of femtoseconds. During and after the pump pulse, the electronic and lattice degrees of freedom exchange energy through electron–phonon coupling and equilibrate to a common temperature, typically within less than a few picoseconds, a process that is usually described with a two-temperature model.[33] This almost instantaneous heating of the lattice, at the elastic timescale, induces an inhomogeneous thermal stress related to the pump energy deposition profile. The related strain modulates the dielectric function as a function of depth and time, giving rise to a time-dependent optical transmission variation.

In the following, we use the simplest one-dimensional model to interpret our data with the minimum number of fitting parameters.

3.2 Thermoelastic model

The relative transmission variation $\frac{\Delta T_r(t)}{T_r}$ depends both on the pump excitation profile that determines the strain pulse shape $\eta(x, t)$, and on the probe pulse transmission through the perturbed sample. The pioneering work of Thomsen *et al.*[9] with a semi-infinite sample was generalized to a multilayer system by Matsuda *et al.*[34]

In the following, we consider a layer of thickness d on top of a semi-infinite substrate (Fig. 4). Note however that a more complete model, with a finite-width substrate, gives identical results.

Assuming that the permittivity, or refractive index, is modulated by the strain, and using the approximation $\Delta\tilde{n}(x,t) = \frac{\partial\tilde{n}}{\partial\eta}\eta(x,t)$, the relative change of transmission can be written as:

$$\frac{\Delta T_r(t)}{T_r} = \Re e\left(\frac{\partial\tilde{n}}{\partial\eta}\int_0^d \eta(x,t)\tilde{f}(x, \lambda_{pr}, \tilde{n}, \tilde{n}_s, d)\mathrm{d}x\right) \quad (2)$$

Table 1 Parameters used in the simulations. $i^2 = -1$

Parameter	Value	References
Al_2O_3 (sapphire)		
ρ	$3.98\ g\ cm^{-3}$	
v_{001} (sound velocity)	$11.23 \times 10^5\ cm\ s^{-1}$	Winey *et al.*[35]
Z_s (acoustic impedance)	$44.7 \times 10^5\ g\ cm^{-2}\ s^{-1}$	
n (visible)	1.76	Tropf *et al.*[36]
n (THz)	3.07	Grischkowsky *et al.*[37]
V_2O_3 (PM)		
ρ	$4.87\ g\ cm^{-3}$	
v_{100} (sound velocity)	$7.37 \times 10^5\ cm\ s^{-1}$	Yelon and Keem[38]
Z (acoustic impedance)	$35.9 \times 10^5\ g\ cm^{-2}\ s^{-1}$	
$\tilde{n}$ @ 540 nm	1.73 + 0.54 i	Qazilbash *et al.*[39]
$\tilde{n}$ @ 700 nm	1.74 + 0.58 i	Qazilbash *et al.*[39]
$\tilde{n}$ @ 740 nm	1.72 + 0.61 i	Qazilbash *et al.*[39]
ξ_{THz}	1300 nm	From THz-TDS
ν_p (Poisson coefficient)	0.33	
C_p (heat capacity)	$3.25\ J\ K^{-1}\ cm^{-3}$	Keer *et al.*[40]

where $\mathfrak{Re}$ stands for the real part of a complex number, and $\tilde{f}(x, \lambda_{pr}, \tilde{n}, \tilde{n}_s, d)$ is a complex-valued "sensitivity function".[9] This sensitivity function describes the interaction of the probe with the permittivity modulation, and accounts for both absorption and interference effects. The parameters are the probe wavelength λ_{pr}, the film and substrate complex refractive indexes $\tilde{n}$ and $\tilde{n}_s$, and the film thickness d. All these values are known either from the literature or direct measurement (see Table 1).

In this equation, we made the assumption that the derivative of the film photoelastic complex constant $\frac{\partial \tilde{n}}{\partial \eta}$ is not a function of depth. It is an unknown that can only be estimated by fitting the data.

The strain profile $\eta(x,t) = \frac{\partial u(x,t)}{\partial x}$, where $u(x, t)$ is the displacement, can be obtained by solving the two coupled elastic-wave equations:

$$\frac{\partial^2 u}{\partial t^2} = v^2 \frac{\partial^2 u}{\partial x^2} - v^2 \frac{\partial \eta_{th}}{\partial x} \quad \text{for } 0 \le x \le d \tag{3}$$

$$\frac{\partial^2 u_s}{\partial t^2} = {v_s}^2 \frac{\partial^2 u_s}{\partial x^2} \quad \text{for } x > d \tag{4}$$

The subscript s indicates the substrate. The last term in right side of the first equation is the thermal strain, defined as:

$$\eta_{th}(x,t) = \eta_o \exp(-x/\xi_{pump})\Theta(t), \quad \eta_o = \beta_T \Delta T_o \frac{1+\nu_p}{1-\nu_p} \tag{5}$$

$\Theta(t)$ is the Heaviside function, ξ_{pump} the pump penetration depth, β_T the linear expansion coefficient, ΔT_o the initial lattice maximum temperature rise at the surface, and ν_p the film Poisson ratio. The validity domain of such a system of equations is discussed for instance in ref. 9 and 10. The initial temperature jump is related to pump fluence F_p, intensity reflection coefficient R_p and sample heat capacity per unit volume C, by the formula $\Delta T_o = \frac{(1-R_p)F_p}{C\xi_{pump}}$. It is worth mentioning however that the details of the energy exchange mechanisms that produce the initial stress on the lattice, are not important in this model, as long as the initial photoinduced lattice perturbation is very fast compared to the acoustic timescale, leading to an approximate stepwise temporal rise.

Thus, four dimensionless parameters determine the solutions. First, the strain wave reflection coefficient R defined as

$$R = \frac{Z_s - Z}{Z_s + Z} \quad (6)$$

where Z_s and Z are respectively the substrate and thin film acoustic impedances ($Z = \rho v$ with ρ the density and v the sound velocity). Second the ratio $K = \frac{d}{\xi_{pump}}$ of the film thickness d and the pump penetration depth ξ_{pump}. If $K \ll 1$, the excitation profile is almost constant within the film, whereas the exponential decay is strongly marked when $K \gg 1$. Third, $V = \frac{v}{v_s}$ the ratio of sound velocities in the layer and the substrate. And finally the strain at the top-layer surface η_o.

Using Laplace transforms methods, the solutions of eqn (4), $u(x, t)$ and $\eta(x, t)$ in the film and the substrate, were calculated analytically, with the initial conditions of no displacement and strain, as well as zero velocity at $t = 0$, boundary conditions of continuous displacement and stress at the film–substrate interface, and free surface at the air-layer interface (zero-stress condition).

Thus, by combining the solutions of eqn (4), and the formula eqn (2), it is possible to simulate and fit the data, as explained below.

We show in Fig. 4 the normalized excitation profiles for the two pumps and sensitivity functions at the two probe wavelengths, using values from Table 1. As expected, the terahertz excitation is almost constant over the whole film thickness ($\xi_{THz} = 1300$ nm), in contrast to the optical excitation that has a penetration depth of $\xi_{540nm} = 80$ nm. As already said above, the probe sensitivity functions $\tilde{f}(x, \lambda_{pr}, \tilde{n}, \tilde{n}_s, d)$ are very close to each other, and the difference has a negligible effect on the transmission. Thus, we can conclude that if a change of transmission is observed, it is a consequence of the different pump excitations. Interestingly, we also note that the probe sensitivity function is almost constant, despite the $\xi_{probe} = 95$ nm penetration depth of the 700 nm or 740 nm probes. This is a consequence of finite size thickness that leads to constructive interferences within the film.

To get further insight into the interpretation of the data $\frac{\Delta T_r(t)}{T_r}$, we first discuss a rough approximation of eqn (2). Assuming a constant complex-valued sensitivity function $\tilde{f}(x,\ldots) = \tilde{f}(\ldots)$, eqn (2) can be easily simplified to:

$$\frac{\Delta T_r(t)}{T_r} \approx \Re e\left(\frac{\partial \tilde{n}}{\partial \eta}\tilde{f}(\ldots)\int_0^d \eta(x,t)\mathrm{d}x\right) = \Re e\left(\frac{\partial \tilde{n}}{\partial \eta}\tilde{f}(\ldots)\right)\int_0^d \frac{\partial u(x,t)}{\partial x}\mathrm{d}x$$

Thus

$$\frac{\Delta T_r(t)}{T_r} \approx \Re e\left(\frac{\partial \tilde{n}}{\partial \eta}\tilde{f}(\ldots)\right)[u(d,t) - u(0,t)] = \Im \Delta u(t) \tag{7}$$

where $\Delta u(t) = u(d, t) - u(0, t)$ and $\Im = \Re e\left(\frac{\partial \tilde{n}}{\partial \eta}\tilde{f}(\ldots)\right)$. Within this approximation, the relative transmission variation is simply proportional to the variation of the film thickness $\Delta u(t)$.

The behaviour of $\Delta u(t)$ for both terahertz and optical pumps, as well as snapshots of the strain profile $\eta(x, t)$ at typical times, is illustrated in Fig. 5, using values from Table 1, and three values of the strain reflection coefficient $R = 0$ (matched layer), $R = \pm 0.2$. The behaviour of the strain spatial profile is interesting (Fig. 5c and d),

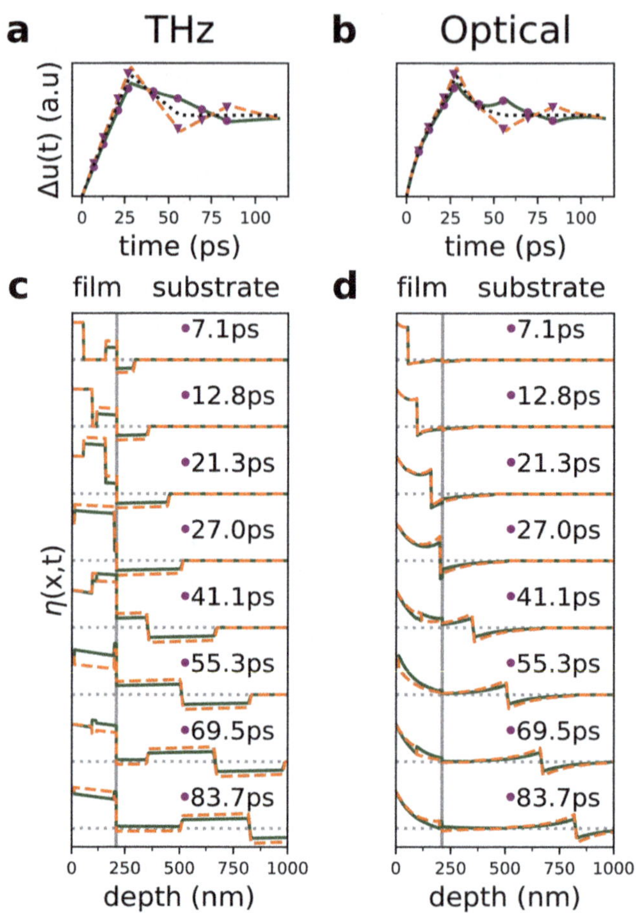

Fig. 5 Temporal evolution of $\Delta u(t)$ (a and b) and snapshots of the strain profile $\eta(x, t)$ (c and d) within the film and the substrate as a function of position at some selected times. Left column (a and c) and right columns (b and d) correspond respectively to THz and optical pump. Panels a and b represent $\Delta u(t)$ with matched layer $R = 0$ (black dotted line) or unmatched $R = \pm 0.2$ ($R = -0.2$ orange dashed line and $R = +0.2$ green continuous line). Markers are placed at times at which corresponding strain wave snapshots are plotted as a function of position. Panels c and d: strain wave profiles $\eta(x, t)$ in the thin film (left of the vertical black line) or in the substrate (right of the vertical black line) at a given time, as a function of position. Horizontal dotted gray lines represent zero strain $\eta = 0$ line for each time. For the simulation, the sound velocity in the substrate was set to $1.5v$.

and can be understood using three contributions. A time-independent contribution corresponding to the initial temperature profile, and propagating and counter-propagating strain waves launched at the free surface and film–substrate surface by the initial stress impulse. The resulting strain is the sum of these three contributions, and evolves depending on the reflections at the free surface or film–substrate interface. At each reflection on the film–substrate interface, the amplitude of the strain-wave is multiplied by R, resulting in a magnitude after n round trips of duration $2\tau_a = 2d/v$ reduced by a factor R^n. For times $\gg 2n\tau_a$ long enough such that $R^n \ll 1$, only the time-independent strain (static strain) survives. Thus, we expect the time-independent strain to determine the magnitude of the plateau. Of course, other phenomena are susceptible to modulate the shape of the transmission signal. For instance, the static strain being related to a temperature profile, and thus to a temperature gradient, heat diffusion, loss and transfer, will smooth and wash away the temperature gradient, and strain, until thermodynamical equilibrium is reached. Such phenomena are negligible on the studied timescale, and are not taken into account in this model.

As remarked by Schick *et al.*[41] who treated the case of a matched layer $R = 0$, the full strain presents positive and negative contributions in the limit $d \gg \xi_{\text{pump}}$. In the opposite limit $d \ll \xi_{\text{pump}}$, the strain has almost the same sign over all the film thickness. It was explained by the fact that the static strain always compensates the strain wave. This conclusion remains valid in our model, at least for the small considered reflection coefficient magnitude $|R| < 0.2$.

Clearly, a comparison of the simulations shown in Fig. 5a and b with the raw data of Fig. 3 shows that the studied V_2O_3 film is not matched to the sapphire substrate. If the layer were matched, $\Delta u(t)$ would reach a plateau after the strain wave launched at the film–substrate interface has travelled back and forth in the film, that is $2\tau_a = 2d/v \approx 58$ ps (black dotted lines of Fig. 5a and b). But experimental data show a variation over more than 90 ps. Considering now the unmatched case, we can determine the sign of the reflection coefficient. For $R < 0$, a dip below the plateau value appears near time $2\tau_a$, which is inconsistent with Fig. 3. Thus we conclude that R is positive, $R > 0$, meaning that the sapphire acoustic impedance is larger than the one of the V_2O_3 film, $Z_s > Z$.

3.3 Analysis

To go further in the analysis of the data, we fitted data obtained from the terahertz and optical pumps with this simplest possible model using the full eqn (2), and the analytical solution of eqn (4).

First, the terahertz penetration depth deserves particular attention, because of the frequency distribution of energy $|E_\omega|^2$ of the terahertz excitation pump pulse, shown in Fig. 6.

From the optical conductivity measured by TDS, and Drude's formula eqn (1), we calculated the complex refractive index (real and imaginary parts in Fig. 6a), and penetration depth $1/\alpha_\omega$ (Fig. 6b). As shown by the figures, these parameters strongly depend on frequency, giving rise to the question of choosing an effective penetration depth in the formula. We model the intensity absorption profile by a weighted average of the absorption profile at each frequency

$$I(x) \sim \int d\omega P_\omega \alpha_\omega e^{-\alpha_\omega x} \sim \frac{1}{\xi_{\text{THz}}} e^{-x/\xi_{\text{THz}}} \tag{8}$$

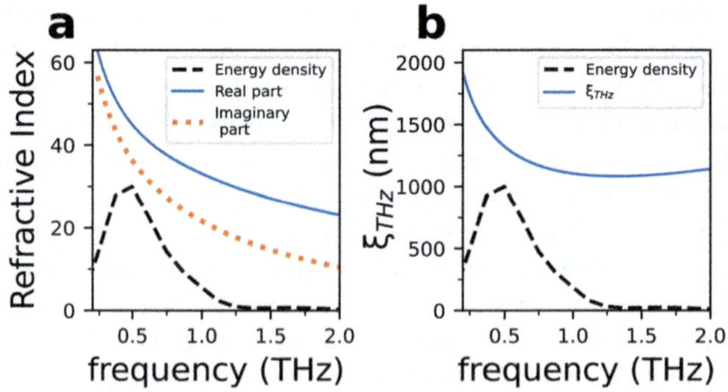

Fig. 6 (a) Real and imaginary parts of the thin film complex index at THz frequencies, as given by the TDS experiment. (b) Corresponding THz penetration depth as a function of frequency. The dashed line represents the terahertz pump pulse spectral density $|E_\omega|^2$ with arbitrary unit.

with probability P_ω of absorbing energy within ω and $\omega + d\omega$ defined as

$$P_\omega = \frac{|E_\omega|^2}{\int d\omega |E_\omega|^2}$$

From the numerical integration of eqn (8), we obtained the effective penetration depth for the terahertz pump $\xi_{THz} \approx 1300$ nm. In fact, the mean absorption profile is almost indistinguishable to the one at single frequency 0.5 THz over all film thickness.

Second, among the fitting parameters, the photoelastic constants $\frac{\partial n}{\partial \eta}$ and $\frac{\partial \kappa}{\partial \eta}$, where n and κ are respectively the real and imaginary parts of the complex refractive index $\tilde{n}$, are usually difficult to estimate. We tried to fit these parameters to our data, without success. But as already mentioned before, the fact that the sensitivity-function is almost depth independent combines the photoelastic constants with the scaling factor, making it difficult or impossible to distinguish both parameters that are strongly correlated. We also verified that a change of the photoelastic constants values could be almost compensated by a change of the scaling factor. Thus, we choose to separate the magnitude from the phase as $\frac{\partial \tilde{n}}{\partial \eta} = \left|\frac{\partial \tilde{n}}{\partial \eta}\right| e^{-i\phi_n}$. The photoelastic magnitude $\left|\frac{\partial \tilde{n}}{\partial \eta}\right|$ is incorporated to the overall scaling factor, while the phase ϕ_n was calculated from the relationship $\frac{\partial n}{\partial \eta} = -1.63 \frac{\partial \kappa}{\partial \eta}$ that was used in ref. 24 and 25.

Therefore, we ended with only three fitting parameters: the acoustic reflection coefficient R, and two scaling factors to adjust the amplitudes to the terahertz or optical-pump results. All the other parameters were obtained from the literature or measured in the laboratory, and fixed to the corresponding values (Table 1).

The best fit is presented in Fig. 7, and shows an overall very good agreement with the data for a three-parameter fit. We would like to stress again that the model fits the overall relative optical transmission variation data without any other corrections, including the rising, decreasing and pseudo-plateau parts. In particular, the result of the exponential excitation profile of the optical pump is well reproduced, as well as the small oscillation coming from the reflection of the

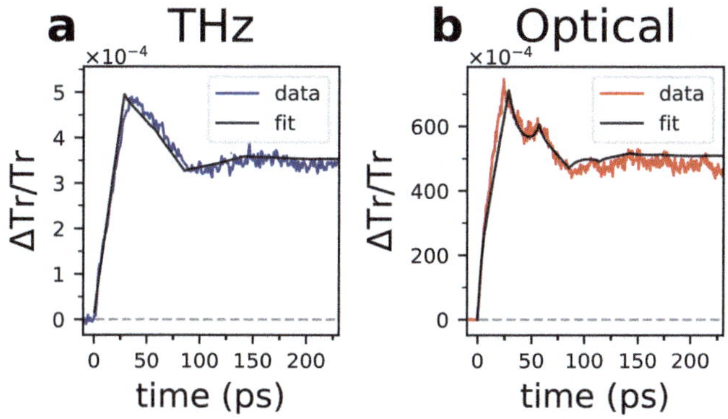

Fig. 7 Relative optical transmission as a function of time $\frac{\Delta T_r(t)}{T_r}$, and best fit results with only three parameters. Experimental raw data are presented. (a) THz pump. (b) Optical pump. Only points every 0.5 ps are shown.

strain pulse at the film–substrate interface. Interestingly, for the optical pump, it shows that the round shape of the rising part when time < 25 ps is also given by the exponential excitation profile, in addition to other possible mechanisms such as non-instantaneous heating, heat diffusion or loss, that are not taken into account in the model.

The value of the fitted acoustic reflection coefficient $R \approx 0.18 \pm 0.02$ is larger than the expected value $R \approx 0.11$ calculated from densities and sound velocities extracted from the literature (Table 1). From R, we obtain the ratio of substrate to film acoustic impedance $Z_s/Z \approx 1.44$. It gives an effective acoustic impedance of V_2O_3 of $Z \approx 31.0 \times 10^5$ g cm^{-2} s^{-1} instead of $Z \approx 35.9 \times 10^5$ g cm^{-2} s^{-1} using density and sound velocity along (110) from the literature, since our sample has a preferential orientation along (110). This small discrepancy of −14% may have different origins.

First, a density change due to internal stress/strain created during the synthesis process might be invoked. However, such effects are usually smaller than 1% in V_2O_3 films (see *e.g.* ref. 42) and cannot account for the observed 14% differences in R values.

Second, values from the literature come usually from single crystal samples that cannot be directly compared to thin films. As a matter of fact, the thin film sample has a heterogeneous structure, with typical grain sizes estimated to be around 25–30 nm, as revealed by the SEM image of Fig. 1b. This dense mixture of nanocrystals introduces orientational disorder, and changes the effective density because of the presence of holes or porosity. Different orientations correspond to different elastic constants and sound velocities. As a result, the measurement over the beamsize would give an effective acoustic impedance by spatial averaging. For instance, in disordered inhomogeneous systems,[43] an effective density and elastic stiffness are obtained by averaging the density, and the inverse of the elastic stiffness. It results in an effective sound velocity that can be smaller than the components' velocities. In our case, the presence of holes or less dense components decreases the effective density, and thus the acoustic impedance.

4. Discussion and conclusion

In this article, the role of the pump profile on the temporal evolution of the ultrafast optical time-resolved transmission response of a thin film of vanadium sesquioxide in the metallic phase was studied for two types of pumps: a terahertz pump and an optical pump. Using raw relative transmission data and only three fitting parameters, the temporal profile of the relative optical transmission for both pumps was analysed within the thermoelastic framework over a 250 ps time range, corresponding to the acoustic timescale of this V_2O_3 film.

From the material point of view, we also measured the acoustic impedance mismatch between the V_2O_3 film and sapphire substrate ($R \approx 0.2$). Such a mismatch was unexpected, and gives another important control parameter in addition to usual ones (X-rays diffraction, four-points resistivity measurements, SEM…), to improve the synthesis methodology and better control the quality of the films in the design of new electronic or opto-electronic devices.

We evidenced that the terahertz absorption profile is almost homogeneous over the film thickness (large penetration depth compared to film thickness). Conversely, the optical pump presents a marked exponential decay within the sample whose shape governs the transmission temporal profile at all times. These differences in excitation profiles are directly observed on the shape of the transmission transients. We showed that the superposition of strain waves to the static strain plays a central role in the interpretation of the data, the former having a contribution that decreases as time increases, the latter determining the plateau value within the acoustic timescale, and for time much smaller than heat transfer timescales. The crucial step in the data interpretation was to use full raw relative optical transmission data with no time-varying "background" removal. Note also that the transmission sensitivity function being almost flat, it can catch the effect of static strain and dynamical contribution with a small distortion, in contrast to reflectivity sensitivity functions.

Among the few studies of the transmission of V_2O_3 in the PM phase near room temperature, we only found one dealing with a terahertz pump and an optical probe, Giorgianni *et al.*[16] However, the transmission data were acquired over a very short time <6 ps, at 175 K, and with a very high field (>1 MV cm^{-1}), making comparison with our data impossible.

Liu *et al.*[24] studied a 400 nm thick V_2O_3 film at 220 K, with an optical pump and terahertz probe within 250 ps, and interpreted the data with a constant sensitivity function. The results are shown in the supplementary material of their article, where two acoustic echoes were evidenced. Similar results, but with a reflectivity setup $\Delta R/R$ and optical pumps, were also presented by Abreu *et al.*[25] and Mansart *et al.*[30] Note that all these authors only presented the oscillatory part of the signal after removal of some time varying background. We would like to point out that the extraction of the "oscillating" contributions may obscure the presence or absence of the contribution of a static strain. Our analysis demonstrated the importance of this static strain in the interpretation of the plateau observed for both excitation processes.

Recently, a comparison of coherent acoustic phonon generation by terahertz and near infrared ultrashort pump pulses (1.55 eV) on chromium metallic thin film ($\approx$14 nm) with the same probe wavelength (400 nm) was carried out.[32] It is demonstrated that terahertz pulse pumping in metals generates coherent phonons by a thermoelastic process, through an ultrafast Joule effect. For the optical pump, this process arises by an intraband hot electron relaxation by heat transfer to the lattice degrees of freedom. Although the mechanisms differ, the oscillatory parts of the transient optical transmissions overlap almost perfectly, within the experimental time range <30 ps, suggesting similar response of the thin metallic film to both excitation processes. In contrast, our observations with a much thicker thin film of V_2O_3, and on much longer times, were clearly able to distinguish between the response to the terahertz and optical pump excitations.

For controlling materials at the macroscopic scale with strong ultrafast pulses, the strain–stress profile is a key feature, and we have shown that terahertz or optical pumps generate different profiles. With this respect, terahertz excitation allows for inducing a more spatially homogeneous excitation over the film depth than optical pump. Thus, we expect thermalization effects that depend on temperature gradients $\frac{\partial T}{\partial x}$ to be minimized. Moreover, in agreement with the matched layer model,[41] our unmatched layer model also predicts that the strain does not change sign when the excitation is almost uniform over the full film thickness, at least in the range $|R| < 0.2$ of our simulations, whereas tensile and compressive strain develop inhomogeneously in the film thickness when the intensity decreases exponentially. This change of sign of the strain, or not, within the sample may be of importance in the study of PIPT where the strain or stress-waves are at the origin of the driving mechanisms that trigger the phase transformation. The comprehension of such phenomena may help in tailoring strain profiles to study ultrafast phase transformations of volume changing materials such as V_2O_3.

Author contributions

GH and CO supervised the project. GH performed pump–probe and THz-TDS experiments. Technical support has been provided by NG, LG, GP, MH and CO prior and during experiments. GH and CO performed data analysis. CO developed the analytical model and carried out the simulations. MR, ZK, JT and EJ synthesized and characterized the sample. MC initiated the THz project and designed the THz setup. LG built the THz source during his PhD thesis. Financial support has been provided by EC and MC. CO and GH wrote the manuscript, with critical reading from EC and EJ. All authors contributed to discussions.

Conflicts of interest

There are no conflicts to declare.

Acknowledgements

The authors gratefully acknowledge Agence Nationale de la Recherche for financial support under grant, ANR-19-CE30-0004 ELECTROPHONE, ANR-19-CE29-0018 MULTICROSS, the University Rennes 1 and CNRS Human Resources Direction, for funding. We acknowledge Maciej Lorenc and Roman Bertoni for experimental support, and fruitful discussions. We would like to thank Pr. H. Cailleau, and Céline Mariette, for useful discussions.

Notes and references

1 S. Koshihara, T. Ishikawa, Y. Okimoto, K. Onda, R. Fukaya, M. Hada, Y. Hayashi, S. Ishihara and T. Luty, *Phys. Rep.*, 2022, **942**, 1–61.

2 H. Hirori, A. Doi, F. Blanchard and K. Tanaka, *Appl. Phys. Lett.*, 2011, **98**, 091106.

3 P. Salén, M. Basini, S. Bonetti, J. Hebling, M. Krasilnikov, A. Y. Nikitin, G. Shamuilov, Z. Tibai, V. Zhaunerchyk and V. Goryashko, *Phys. Rep.*, 2019, **836–837**, 1–74.

4 C. Vicario, M. Jazbinsek, A. V. Ovchinnikov, O. V. Chefonov, S. I. Ashitkov, M. B. Agranat and C. P. Hauri, *Opt. Express*, 2015, **23**, 4573–4580.

5 T. Kampfrath, K. Tanaka and K. A. Nelson, *Nat. Photonics*, 2013, **7**, 680–690.

6 D. Nicoletti and A. Cavalleri, *Adv. Opt. Photonics*, 2016, **8**, 401–464.

7 H. Okamoto, Y. Ishige, S. Tanaka, H. Kishida, S. Iwai and Y. Tokura, *Phys. Rev. B: Condens. Matter Mater. Phys.*, 2004, **70**, 165202.

8 Y. Okimoto, X. Peng, M. Tamura, T. Morita, K. Onda, T. Ishikawa, S. Koshihara, N. Todoroki, T. Kyomen and M. Itoh, *Phys. Rev. Lett.*, 2009, **103**, 027402.

9 C. Thomsen, H. T. Grahn, H. J. Maris and J. Tauc, *Phys. Rev. B: Condens. Matter Mater. Phys.*, 1986, **34**, 4129–4138.

10 O. Matsuda, M. C. Larciprete, R. Li Voti and O. B. Wright, *Ultrasonics*, 2015, **56**, 3–20.

11 P. Ruello and V. E. Gusev, *Ultrasonics*, 2015, **56**, 21–35.

12 C. Mariette, M. Lorenc, H. Cailleau, E. Collet, L. Guérin, A. Volte, E. Trzop, R. Bertoni, X. Dong, B. Lépine, *et al.*, *Nat. Commun.*, 2021, **12**, 1239.

13 N. F. Mott and R. Peierls, *Proc. Phys. Soc.*, 1937, **49**, 72–73.

14 P. Hansmann, A. Toschi, G. Sangiovanni, T. Saha-Dasgupta, S. Lupi, M. Marsi and K. Held, *Phys. Status Solidi B*, 2013, **250**, 1251–1264.

15 E. Abreu, S. Wang, J. G. Ramírez, M. Liu, J. Zhang, K. Geng, I. K. Schuller and R. D. Averitt, *Phys. Rev. B*, 2015, **92**, 085130.

16 F. Giorgianni, J. Sakai and S. Lupi, *Nat. Commun.*, 2019, **10**, 1159.

17 K.-L. Yeh, M. Hoffmann, J. Hebling and K. A. Nelson, *Appl. Phys. Lett.*, 2007, **90**, 171121.

18 C. R. Harris, K. J. Millman, S. J. van der Walt, R. Gommers, P. Virtanen, D. Cournapeau, E. Wieser, J. Taylor, S. Berg, N. J. Smith, R. Kern, M. Picus, S. Hoyer, M. H. van Kerkwijk, M. Brett, A. Haldane, J. F. del Río, M. Wiebe, P. Peterson, P. Gérard-Marchant, K. Sheppard, T. Reddy, W. Weckesser, H. Abbasi, C. Gohlke and T. E. Oliphant, *Nature*, 2020, **585**, 357–362.

19 P. Virtanen, R. Gommers, T. E. Oliphant, M. Haberland, T. Reddy, D. Cournapeau, E. Burovski, P. Peterson, W. Weckesser, J. Bright, S. J. van der Walt, M. Brett, J. Wilson, K. J. Millman, N. Mayorov, A. R. J. Nelson, E. Jones, R. Kern, E. Larson, C. J. Carey, İ. Polat, Y. Feng, E. W. Moore, J. VanderPlas, D. Laxalde, J. Perktold, R. Cimrman, I. Henriksen, E. A. Quintero, C. R. Harris, A. M. Archibald, A. H. Ribeiro, F. Pedregosa, P. van Mulbregt and SciPy 1.0 Contributors, *Nat. Methods*, 2020, **17**, 261–272.

20 J. D. Hunter, *Comput. Sci. Eng.*, 2007, **9**, 90–95.

21 R. Glover III and M. Tinkham, *Phys. Rev.*, 1957, **108**, 243.

22 P. U. Jepsen, D. G. Cooke and M. Koch, *Laser Photonics Rev.*, 2011, **5**, 124–166.
23 J. Neu and C. A. Schmuttenmaer, *J. Appl. Phys.*, 2018, **124**, 231101.
24 M. K. Liu, B. Pardo, J. Zhang, M. M. Qazilbash, S. J. Yun, Z. Fei, J.-H. Shin, H.-T. Kim, D. N. Basov and R. D. Averitt, *Phys. Rev. Lett.*, 2011, **107**, 066403.
25 E. Abreu, S. N. Gilbert Corder, S. J. Yun, S. Wang, J. G. Ramírez, K. West, J. Zhang, S. Kittiwatanakul, I. K. Schuller, J. Lu, S. A. Wolf, H.-T. Kim, M. Liu and R. D. Averitt, *Phys. Rev. B*, 2017, **96**, 094309.
26 Y. Luo, F. Su, C. Zhang, L. Zhong, S. Pan, S. Xu, H. Wang, J. Dai and G. Li, *Opt. Commun.*, 2017, **387**, 385–389.
27 I. Lo Vecchio, L. Baldassarre, F. D'Apuzzo, O. Limaj, D. Nicoletti, A. Perucchi, L. Fan, P. Metcalf, M. Marsi and S. Lupi, *Phys. Rev. B: Condens. Matter Mater. Phys.*, 2015, **91**, 155133.
28 N. Kuroda and H. Fan, *Phys. Rev. B: Solid State*, 1977, **16**, 5003.
29 G. Lantz, B. Mansart, D. Grieger, D. Boschetto, N. Nilforoushan, E. Papalazarou, N. Moisan, L. Perfetti, V. L. Jacques, D. Le Bolloc'h, *et al.*, *Nat. Commun.*, 2017, **8**, 13917.
30 B. Mansart, D. Boschetto, S. Sauvage, A. Rousse and M. Marsi, *Europhys. Lett.*, 2010, **92**, 37007.
31 O. Misochko, M. Tani, K. Sakai, K. Kisoda, S. Nakashima, V. Andreev and F. Chudnovsky, *Phys. Rev. B: Condens. Matter Mater. Phys.*, 1998, **58**, 12789.
32 A. Levchuk, B. Wilk, G. Vaudel, F. Labbé, B. Arnaud, K. Balin, J. Szade, P. Ruello and V. Juvé, *Phys. Rev. B*, 2020, **101**, 180102.
33 S. I. Anisimov, B. L. Kapeliovich and T. L. Perel'Man, *J. Exp. Theor. Phys.*, 1974, **39**, 375–377.
34 O. Matsuda and O. Wright, *J. Opt. Soc. Am. B*, 2002, **19**, 3028–3041.
35 J. Winey, Y. Gupta and D. Hare, *J. Appl. Phys.*, 2001, **90**, 3109–3111.
36 W. J. Tropf, M. E. Thomas and T. J. Harris, *Handbook of optics*, 1995, vol. 2, ch. 33.
37 D. Grischkowsky, S. Keiding, M. Van Exter and C. Fattinger, *J. Opt. Soc. Am. B*, 1990, **7**, 2006–2015.
38 W. Yelon and J. Keem, *Solid State Commun.*, 1979, **29**, 775–777.
39 M. M. Qazilbash, A. A. Schafgans, K. S. Burch, S. J. Yun, B. G. Chae, B. J. Kim, H. T. Kim and D. N. Basov, *Phys. Rev. B: Condens. Matter Mater. Phys.*, 2008, **77**, 115121.
40 H. Keer, D. Dickerson, H. Kuwamoto, H. Barros and J. Honig, *J. Solid State Chem.*, 1976, **19**, 95–102.
41 D. Schick, M. Herzog, A. Bojahr, W. Leitenberger, A. Hertwig, R. Shayduk and M. Bargheer, *Struct. Dyn.*, 2014, **1**, 064501.
42 Y. Kalcheim, C. Adda, P. Salev, M.-H. Lee, N. Ghazikhanian, N. M. Vargas, J. del Valle and I. K. Schuller, *Adv. Funct. Mater.*, 2020, **30**, 2005939.
43 J.-P. Fouque, J. Garnier, G. Papanicolaou and K. Solna, *Wave Propagation and Time Reversal in Randomly Layered Media*, Springer Science & Business Media, 2007, vol. 56.

Faraday Discussions

DISCUSSIONS

THz and laser field excitation processes: general discussion

Martina Basini, Lara Benfatto, Eric Collet, Aswathy V. Girija, Guénolé Huitric, Shinichiro Iwai, J. Olof Johansson, Steven L. Johnson, Shinya Koshihara, Rostislav V. Mikhaylovskiy, Jan Michael Rost, Habib Rostami and Erica Sutcliffe

DOI: 10.1039/d2fd90038f

Steven L. Johnson opened a general discussion of the paper by Shinichiro Iwai: It is an interesting result that you observe bulk second harmonic generation (SHG) in a material with inversion symmetry. Is the reason that this is symmetry allowed related to the carrier-envelope phase of the pulse, *i.e.* the fact that the different frequency components of the pulse have a well-defined phase relationship? Can this phenomenon be explained in terms of nonlinear susceptibilities (*e.g.* χ^2, χ^3)?

Lara Benfatto addressed Shinichiro Iwai and Steven L. Johnson, saying: Concerning Steve's question on the generation of SH. Actually, this is still a χ^3 process, that gives rise to second-harmonic when one field component is a DC current. Indeed, χ^3 processes mediate a current proportional to the third power of the incoming field. However, in the presence of a DC current one field component is at $\omega = 0$, and the non-linear current oscillates at 2ω. So this SH generation does not come from breaking the crystal symmetry, similar effects have been seen in THz spectroscopy of superconductors for a large enough pump field,[1] able to induce such a DC component of the internal electrical field.

I agree with you that if you have a DC component it is easier to understand why you detect SHG.

1 X. Yang, C. Vaswani, C. Sundahl, M. Mootz, L. Luo, J. H. Kang, I. E. Perakis, C. B. Eom and J. Wang, *Nat. Phot.*, 2019, **13**, 707.

Shinichiro Iwai responded: I completely agree with Lara Benfatto's response and have nothing further to add.

Martina Basini asked: Is the described second-harmonic generation due to transient $\chi^{(2)} = \chi^{(3)}E_{\mathrm{THz}}$?

Shinichiro Iwai answered: I agree that the SHG can be described by a $\chi^{(3)}$ process, *i.e.* the $\chi^{(2)}$ process under the DC-like current. It is essential that the DC-like current is driven by the non-scattering current. This is confirmed by the carrier-envelope phase (CEP) sensitivity of the SHG. To the best of our knowledge, a conventional $\chi^{(2)}$ difference frequency generation (DFG) process, cannot explain the CEP sensitivity (the CEP is cancelled in the DFG process). In fact, we have done the reference experiment in BBO crystals. The CEP is independent of the DFG in BBO.

Lara Benfatto added: I would rather say that there is an "effective" χ^2 given by the χ^3 $E(\omega = 0)$, where $E(\omega = 0)$ is a DC component of the electric field generated inside the sample. Then the overall third-order response will give a current χ^3 $E(\omega = 0)E(\omega)E(\omega)$, that oscillates at twice the frequency ω of the THz field.

Jan Michael Rost commented: Your process is non-linear in terms of the number of photons absorbed. What are the laser parameters in terms of intensity, wavelength and pulse duration? How many photons are effectively absorbed during the pulse?

Shinichiro Iwai answered: Non-perturbative light–matter interactions driven by an oscillating strong electric field (such as Floquet and HHG…) have been discussed in terms of $eaF/(\hbar\omega)$, where e: elementary charge, a: intersite distance, F: field amplitude, and ω: field frequency.[1,2] If the parameter is close to 1, a non-perturbative regime is realized. In our case, this condition is satisfied ($F =$ 0.1 V angstrom^{-1}, $a = 7$ angstrom, $\omega = 0.7$ eV). Another important factor is that the excitation is completed before the mean scattering time (*ca.* 40 fs in this material).

1 D. H. Dunlap and V. M. Kenkre, *Phys. Rev. B: Condens. Matter Mater. Phys.*, 1986, **34**, 3625.
2 H. Aoki, N. Tsuji, M. Eckstein, M. Kollar, T. Oka and P. Werner, *Rev. Mod. Phys.*, 2014, **86**, 779.

Eric Collet asked: In the case of the two pumps experiment, which launches coherent oscillations, what happens if you set the second pump out of phase with the previous one? Is coherence destroyed or do both oscillations superpose?

In the double pump experiment (Fig. 5b, https://doi.org/10.1039/d2fd00004k), is it possible to use a second pulse out of phase with respect to initial dynamics to kill coherence? If so, what can we learn from the incoherent or average signal?

Shinichiro Iwai answered: In this experiment, the time difference between double pump pulses is scanned, and that between the second pump and the probe is fixed (50 fs within the lifetime of the stimulated emission). The observation of the interference oscillation indicates that the origin of the stimulated emission is the coherent electronic oscillation. I agree that your suggestion is very interesting: if the time profiles of the stimulated emission for the in-phase double pump and that for the out of phase are different, I can learn from the result.

Jan Michael Rost queried: Is your excitation so non-linear, that it is only sensitive to the absolute pulse maximum? An indicator would be strong sensitivity to the carrier envelope phase. If this is the case, could it be that your physical few-cycle pulse acts like a half-cycle pulse for the excitation?

Shinichiro Iwai replied: I agree with this comment. The field amplitude is also sensitive to the CEP for the nearly single-cycle pulse (not only for the half-cycle pulse). To check this possibility, we have done the same experiment in BBO crystals as a reference. The CEP dependence was absent, indicating that the observed results are not attributed to the change of field amplitude. We would like to do a similar experiment using sub-cycle pulses in the near future.

Lara Benfatto remarked: Can you try to estimate how the SHG scales with the induced DC current? If possible, this would help assigning its origin to a χ^3 process in the presence of a $\omega = 0$ field component.

Shinichiro Iwai answered: I agree that this is very important, but we have never tried. We are planning to do the experiment of SHG under real DC current. We hope that it will help the understanding of our results. (We are afraid that heat accumulation by the DC current disturbs the superconducting fluctuation.)

Jan Michael Rost opened the discussion of the paper by Lara Benfatto: Would a time-dependent perspective in terms of wavepackets, regarding initialization and evolution of the time-dependent electronic dynamics offer an alternative explanation for the phenomena seen?

Lara Benfatto replied: For the kind of phenomena we discuss in our work (https://doi.org/10.1039/d2fd00016d) the perturbative approach we are using seems to work rather well. For example, one predicts the maximum third harmonic generation (THG) to occur when the pump frequency matches the value of $\Delta(T)$, and using the equilibrium value of $\Delta(T)$ allows one to estimate the temperature of maximum THG rather well. On the other hand, experiments with one order of magnitude larger electric field show the emergence of truly metastable states that cannot be captured by this approach. In this case the time-dependent evolution of the wavepackets could probably help understanding the physics. I'm not sure how difficult it would be to develop such an approach in the broken-symmetry state, and it would be interesting to explore such a perspective.

J. Olof Johansson continued the general discussion of the paper by Shinichiro Iwai: Do you have a problem with studying short time delays given that the cryostat windows might stretch your pulse duration?

Shinichiro Iwai responded: Exactly, yes. Our 6 fs NIR pulse has a broad spectrum covering about 1.2–2.2 microns so that the chirping easily occurs within the BaF_2 window (with thickness of 3 mm) of our cryostat. However, the chirped pulse can be pre-compensated using the deformable mirror (DM) in addition to the chirped mirror. The DM enables us to fine-tune the chirp.

Shinya Koshihara asked: Excitation by a light field can interfere with superconductive order parameter vector potential? This would be quite an interesting and important problem for the field of materials science. For this purpose, a CEP locked light pulse pair over a wide spectral region is required (one is near

IR for carrier excitation and the other in the THz region to probe the superconducting gap).

Shinichiro Iwai answered: I completely agree. Correlated materials consist of various interactions with energy scales from meV to 10 eV. The macroscopic nature (or the order parameter) of superconducting states originates from meV interactions, while the microscopic origin may be related to larger energy interactions such as the electron correlation. From such a viewpoint, multi-color experiments such as those involving NIR pumps and THz probes, are powerful for clarifying the interplay between the high and low energy interactions or equivalent hierarchy from microscopic to macroscopic scale. In addition to the suggested CEP stabilized NIR pump and THz probe, double pump (THz pre-pump–NIR pump) experiments are also expected to be an essential approach. As for superconducting materials, this approach enables us to clarify the mechanism of high-T_c superconductors.

J. Olof Johansson commented: You show some interesting new results in your presentation on spin–orbit materials. How sure are you that the pump-helicity dependent measurements are measuring the magnetic response of the material and not just some form of optical Kerr effect?

Shinichiro Iwai replied: The magnetization perpendicular to the *ab*-plane (sample surface) of α-$RuCl_3$ is induced by the circularly polarized light. This is phenomenologically referred to as the inverse Faraday effect (IFE). In the polarization rotation induced by the IFE, the pump light should be strictly circularly polarized inside the material. If the polarization of the pump light becomes elliptic during the propagation of the material, the optical Kerr effect (OKE) also gives a polarization rotation. To confirm the IFE, we need to check that the OKE is really absent. For this purpose, we actually measured the dependence of $\delta\theta$ on the ellipticity of the pump light to confirm that the observed polarization rotation is the IFE: if we have only the IFE, the polarization rotation is maximum for the strictly circular polarization. If not, the OKE is contained. Further evidence is the temperature dependence: there is no reason for a temperature change with the OKE, reflecting the antiferromagnetic phase transition. In fact, we observe OKE under excitation by elliptically polarized light and it is insensitive to temperature. Thus, we can distinguish the IFE and the OKE.

Habib Rostami said: I could not understand what specific symmetry is broken under laser irradiation in your measurement. Did you identify which mirror or rotation is broken and what is the new structural phase after light irradiation?

Shinichiro Iwai replied: The induced current breaks the specific symmetry in the sense that the current $J(t)$ is no longer described by the odd function of $E(t)$ (this can be also described by the asymmetric distribution of electrons in the momentum space[1]). If Ohm's law holds, such DC-like current cannot be induced by the light pulse with multi ($\geq$one)-cycle oscillating fields. But, in our case, the non-scattering mechanism (where the $J(t)$ is given by the time integral of $E(t)$) can cause this. This mechanism is actually confirmed by the CEP sensitivity.

1 J. B. Khurgin, *Appl. Phys. Lett.*, 1995, **67**, 1113.

Aswathy V. Girija opened a general discussion of the paper by Jan Michael Rost: Why is the valley polarisation not affected by dephasing?

Jan Michael Rost replied: Electron dynamics in general is affected by dephasing, and so is VP, although to a lesser degree than excitation. The reason is that dephasing blurs the interference pattern of initial conditions (in the Brillouin zone) increasing ionization simultaneously for the K and K' domains. Since VP is the difference of ionization contributions in the two domains, it is much less affected by this blurring. Moreover, for ultrashort pulses, dephasing is simply too slow to have an effect.

Aswathy V. Girija commented: It is interesting to see that we don't necessarily need circularly polarized light to achieve valley polarization. And you have mentioned in your article that to achieve this experimentally, one needs moderately strong laser pulses (https://doi.org/10.1039/d2fd00017b). Could you quantify this? What would be a moderate pulse, and when would the pulse be too strong?

Jan Michael Rost answered: The peak field strength necessary is set by the maximal vector potential A_0 (in our gauge the integral of the field strength) which should be of the order of the distance of K and K' in k-space, which is 0.9 au for graphene.

For illustration: at a wavelength of 7.2 μm of the laser, the peak intensity should be 10^{12} W cm^{-2}.

J. Olof Johansson asked: How might practical issues affect your results, such as buckling of the graphene sheet when deposited on a surface?

Jan Michael Rost answered: Honestly, we don't know – an experiment will show. However, I am not aware that this would have affected experiments already performed on graphene with strong pulses.

Lara Benfatto remarked: Is this physics only possible in single layer graphene or can it be extended to low gap semiconductors, like *e.g.* transition metal dichalcogenides?

Jan Michael Rost responded: We believe it can be extended to low gap systems and are working on the specific conditions in terms of gap size and laser parameters. A complication is the increasing influence of other bands once the gap increases.

Lara Benfatto said: Do you mean that the symmetry is dictated by the full system, where one accounts for matter and light on the same footing?

Jan Michael Rost answered: Yes.

J. Olof Johansson suggested: Have you thought about twisted bilayers?

Jan Michael Rost replied: Yes, we are thinking of gapped systems and plan to investigate under conditions which the mechanism identified here for gapless systems will prevail. The two-band approximation, excellent for a gapless system will not suffice.

Habib Rostami commented: The question is, because your result directly depends on the choice of the vector potential A, while we know that the vector potential is not a gauged invariant, are we breaking gauge invariance in this example? Can we interpret this effect as a photon-drag phenomenon where the dependence on the vector potential (part of the canonical momentum $\pi = p + eA$) is explained based on a momentum transfer from light to electrons? The final physical result can only depend on either a time or position derivative of A if we have the gauge symmetry.

Lara Benfatto said: Adding to Habib Rostami's question: you said that the effect is controlled by the value of the vector potential A, that should be large enough to connect the K and K' points. But A is not a physical field: how can a physical result depend on its absolute value and not on its derivatives (which define the physical E and B field)? I believe this is the meaning of Habib's question.

Jan Michael Rost answered: Given a specific gauge, the vector potential has physical meaning. We use the length gauge in dipole approximation, and the vector potential is then simply given by the integral of the field strength.

Habib Rostami added: Yes. Thanks Lara. You phrased my question correctly. By the way: can we interpret this effect as a photon-drag phenomenon where the dependence on the vector potential (as part of the canonical momentum $\pi = p + eA$) is explained based on a momentum transfer from light to electrons? The final physical result can only depend on either the time or position derivative of A if we have the gauge symmetry.

Lara Benfatto commented: I agree that including A in the Hamiltonian is the correct way to do the calculation. But the physical results should not depend on how A changes under a gauge transformation – this is the meaning of a gauge-invariant calculation.

Lara Benfatto asked: The point that we don't understand is that the result depends on the absolute value of A, not on its derivatives, so it does not look like a gauge invariant. Perhaps we are misrepresenting something?

Jan Michael Rost answered: A specific gauge choice defines the relation between the vector potential and the field strength, and therefore the former is linked uniquely to the (physical) field strength, see my above answer to Habib Rostami.

Steven L. Johnson said: In some preliminary unpublished studies, we have investigated ballistic field-driven dynamics in low-gap semiconductors with THz radiation which has some results reminiscent of your work. There we see a qualitative difference as a function of field strength due to impact ionization of electron–hole pairs across the bandgap, that activates above a certain THz field. Would you expect something similar to happen for gapped systems in your model?

Jan Michael Rost responded: Typically, one thinks in this case of conduction band electrons executing coherent wavepacket dynamics. Strong sensitivity to laser intensity could point to a regime characterized by the onset of (tunneling) coupling to the valence band at the Gamma point and ensuing decoherence mechanisms which alter the conduction band dynamics.

Steven L. Johnson opened a general discussion of the paper by Rostislav V. Mikahylovskiy: In your model of the switching dynamics, you postulate that the effective double-well potential for a spin order is transiently modified by the THz pulse interaction with low-energy electronic transitions (https://doi.org/10.1039/d2fd00035k). What is the approximate lifetime of that transient modification, and how severely distorted is the potential from its equilibrium value at the time of the interaction (or just after it)?

Rostislav V. Mikhaylovskiy responded: The low energy electronic transitions are short-lived, *i.e.* they are excited at the timescale of the pump pulse. It can be deduced from the spectroscopy data, which shows that at the temperature interval of interest (80–90 K) the corresponding absorption bands are very broad. That means that the transient modification exists for $\sim$1 ps and the distortion of the potential is small. The resonant frequency is not shifted significantly meaning that after excitation the potential is practically unchanged.

Steven L. Johnson asked: Does the transient effective spin potential still have two stable minima, or is it really a single-well potential?

Rostislav Mikhaylovskiy answered: Yes, it does have two minima as the action of the THz field is more like a "kick" (instantaneous change of a curvature/slight shift in the equilibrium position), *i.e.* it is impulsive.

Steven L. Johnson prompted: Can you comment on the effective "force" that is exerted on the system that gives it enough energy to overcome the barrier for switching? It seems like this force must be very large.

Rostislav Mikhaylovskiy replied: The switching works only in the narrow temperature interval, in the vicinity of the two phase transition temperatures, at which the anisotropy barrier is very small (in the simplest model it is even zero at the points of the phase transitions). Hence the force is not large here.

Martina Basini queried: Is it possible, by means of the combined approach you studied, to create a grating of switched/unswitched spin within the sample?

Rostislav Mikhaylovskiy responded: This is a very good point, and potentially it should be possible. For example, one may use pulse-shaping techniques. It is also possible to use the fact that for a very high driving field back-switching should be possible. It could well happen in the actual experiment,[1] but we could not detect it with a relatively large probe pulse.

1 S. Schlauderer, C. Lange, S. Baierl, T. Ebnet, C. P. Schmid, D. C. Valovcin, A. K. Zvezdin, A. V. Kimel, R. V. Mikhaylovskiy and R. Huber, *Nature*, 2019, **569**, 383.

J. Olof Johansson said: In general, why are the resonance frequencies of antiferromagnets so much higher than those of ferromagnets?

Rostislav Mikhaylovskiy answered: Due to the "exchange enhancement". While in the ferromagnets the resonance frequency ("gap frequency") is proportional to the magnetic anisotropy field (typically kG or smaller), in the antiferromagnets it is proportional to the square root of the product of the anisotropy field and the exchange field (up to 1000 tesla). Hence several orders of magnitude difference.

J. Olof Johansson further commented: What are the challenges in using antiferromagnets for data storage applications?

Rostislav Mikhaylovskiy answered: I think the main challenge is that antiferromagnetic materials are mostly dielectrics and it is not easy to integrate them into conventional electronics. They are also not well studied and understood yet. The reading of the antiferromagnetic domains is also not straightforward (but can be made easier if using canted antiferromagnets such as orthoferrites).

J. Olof Johansson remarked: Would it be possible to image spin waves in your materials using magneto-optical microscopy?

Rostislav Mikhaylovskiy answered: This will be very challenging if not impossible (large wavelength mismatch). Detection is possible in the reciprocal space though, due to the scattering of the reflected light by magnons.[1]

1 S. Baierl, M. Hohenleutner, T. Kampfrath, A. K. Zvezdin, A. V. Kimel, R. Huber, and R. V. Mikhaylovskiy, *Nat. Photonics*, 2016, **10**, 715.

Habib Rostami asked: Electric couples to spin in the second order while magnetic field couples linearly. I guess under low intensity laser light coupling to the *B* field of light is more important, while for an intense laser the coupling to the *E* field matters more. Is there a universal characteristic intensity from *E*-dominated to *B*-dominated coupling regimes?

Rostislav Mikhaylovskiy replied: Yes, you are right, however the relative strength of the nonlinear and linear contributions depends on the nonlinear coefficient and strength of the electric/magnetic field. The coefficient is material specific and hence there is no universal characteristics.

Steven L. Johnson remarked: The dynamics you simulate here remind me of similar calculations done in charge density wave systems where the density wave was destroyed and reformed but with phase inverted domains that vary with depth.[1] If you extend your simulations to higher excitation levels would you see similar dynamics: *i.e.* would you see at some point at higher intensities the formation of domains that flip back to their original magnetization orientation?

1 R. Yusupov, T. Mertelj, V. V. Kabanov, S. Brazovskii, P. Kusar, J.-H. Chu, I. R. Fisher and D. Mihailovic, *Nat. Phys.*, 2010, **6**, 681.

Rostislav Mikhaylovskiy replied: Thank you for this suggestion, we will look at the paper and see whether a similar scenario can be realised in an antiferromagnet.

Steven L. Johnson added: Is it possible to perform an experiment similar to the paper by Guénolé Huitric and colleagues (https://doi.org/10.1039/d2fd00013j) presented in this session to study the motion of magnetic domain walls using optical interference (essentially time-resolved Brillouin scattering)?

Rostislav Mikhaylovskiy responded: This is a very interesting suggestion and I think it is feasible. The time-resolved Brillouin scattering has been already used to detect propagating spin waves in an antiferromagnet.

Erica Sutcliffe remarked: Do you observe any domain wall motion or generation of spin waves in the x–y plane? Or only along the z-axis?

Rostislav Mikhaylovskiy answered: We have not modeled the dynamics in the xy-plane as the characteristic sizes of the excitation spot ($\sim$10 μm) are much larger than the magnon wavelengths ($\sim$100 nm).

Shinya Koshihara asked: Is it possible to design the form of electrodes located on the surface of crystals to enhance the irradiation light field (to expand the obtained results into the excitation by light with various polarizations)? Lately, the design of sub-micron scale patterns of thin metal have been developed from the viewpoint of meta-material based optics. In the THz region, many design ideas have been proposed and tried (because the fabrication size is sub-mm or micrometer). Even in the visible region, this kind of design for meta-material based optics is in progress (one example is a meta-len(s)z in the iPhone). Of course, for this purpose, the improvement in wavelength accuracy will be essential. Collaboration with the semiconductor industry may be fruitful as with the meta-len(s)z case.

Rostislav Mikhaylovskiy replied: Thank you for this suggestion. We actually tried several different antenna designs in the experiments and the bow-tie was empirically found to be the best for spin switching.

Lara Benfatto opened a general discussion of the paper by Guénolé Huitric: What you measure are the variations of the transmittance as a function of the pump–probe delay. How do you go from $\Delta T/T$ to strain? How can you have one to one relationship between these?

Guénolé Huitric responded: The strain modifies the dielectric constant $\Delta\varepsilon(z,t) \sim (\partial\Delta\varepsilon/\partial\eta)\eta(z,t)$, that, in turn, modulates the transmission as a function of time, as explained in our article (https://doi.org/10.1039/d2fd00013j). The relationship is not one-to-one, but the clear agreement between the simulations and the experimental data for both types of excitation (optical or terahertz), with only a very few parameters, give confidence that most of the observed transmission variations comes from the induced strain. The strain $\eta(z,t)$ is obtained from the simulations.

Lara Benfatto asked: Concerning the previous question: how do you know that there is no electronic contribution? Are you always pumping below the optical gap? In general any pump–probe experiment detects a variation on reflectivity, that appears as a peak followed by some decay. This is often attributed to electronic features. Are you saying that the strain modifies the band structure and this in turn affects the electronic contribution? The connection between the two is still not clear to me.

Guénolé Huitric answered: There is no optical gap as the sample is in the paramagnetic metallic phase.

An electronic contribution is clearly observed in the case of optical excitation. The peak occurring at very short delays (see Fig. 3e, https://doi.org/10.1039/d2fd00013j) is attributed to a fast non-thermal electronic contribution and decays to zero within less than 500 fs, in agreement with the literature[1].

Afterwards, the excited electronic gas transfers its energy to the lattice degrees of freedom within a few ps, and launches a slower process – an elastic wave propagation in the sample.

However, in the case of excitation by THz pulses (4 meV), no fast signal has been observed. According to us, this absence is caused by the weakness of the perturbation in terms of photon energy and in excitation density.

1 G. Lantz, B. Mansart, D. Grieger, D. Boschetto, N. Nilforoushan, E. Papalazarou , N. Moisan, L. Perfetti, V. L. Jacques and D. Le Bolloc'h, *et al.*, *Nat. Commun.*, 2017, **8**, 13917.

J. Olof Johansson addressed Guénolé Huitric and Lara Benfatto: a further comment on Lara Benfatto's question; as the electron system heats up, it may well affect the measured transmittance.

Shinichiro Iwai remarked: May I confirm the experimental conditions – in particular, the relation between penetration depth and film thickness? The figure showing this looks strange because THz penetration depth is longer than that for optical excitation. The absorption coefficient for THz is much smaller than the optical region even in metal?

Guénolé Huitric answered: Indeed THz-TDS experiments performed at the lab enabled us to extract a static conductivity of about 1000 S cm^{-1} (in accordance with the literature[1,2]), from which a 1300 nm THz penetration depth (ξ_{THz}) was estimated (see Fig. 6b, https://doi.org/10.1039/d2fd00013j). This value confirms the lower absorption of the THz compared to the optical beam ($\xi_{opt} = 80$ nm), in the PM phase of V_2O_3.

Using a 210 nm thick sample, we can therefore compare two cases: a quasi-flat pump profile (ξ_{THz} > thickness in the THz case) and an exponentially decaying pump profile (ξ_{opt} < thickness in the optical case).

1 E. Abreu, S. Wang, J. G. Ramírez, M. Liu, J. Zhang, K. Geng, I. K. Schuller and R. D. Averitt, *Phys. Rev. B*, 2015, **92**, 085130.
2 M. K. Liu, B. Pardo, J. Zhang, M. M. Qazilbash, S. J. Yun, Z. Fei, J.-H. Shin, H.-T. Kim, D. N. Basov and R. D. Averitt, *Phys. Rev. Lett.*, 2011, **107**, 066403.

Shinichiro Iwai commented: What is the photon energy of the optical excitation and the probe light?

Guénolé Huitric answered: Concerning the optical pump optical probe experiment, the photon energy of the pump is 2.3 eV (540 nm), and 1.8 eV (700 nm) for the probe beam.

Shinichiro Iwai remarked: Are the time profiles for the THz excitation and the optical excitation the same or different? In the result for the optical excitation, you may have additional structures?

For both THz excitation and optical excitation, the small difference in the results is just attributed to the penetration depth of the respective cases?

Guénolé Huitric replied: Looking at acoustic timescales, the time profiles can be considered as similar. Even if they follow the same trend, the "structures" occurring close to a integer number of round trips are much more visible in the optical case due to the exponential decay of the pump profile. Indeed, in the THz case, the flat profile leads to smooth line breaks that may be difficult to see. For instance in the case $R < 0$, the line breaks are much clearer (see Fig. 5a and b, https://doi.org/10.1039/d2fd00013j).

Thus, we would not say that there are additional features, but that the underlying structure (in particular wave reflections) coming from the strain are better seen with the optical pump due to the sharpness of the excitation profile.

Aswathy V. Girija commented: From a materials point of view, we can see that there is an impedance mismatch between the film and the substrate. There is also a slight disagreement in the value of reflection coefficient calculated, compared to the value from the literature, though most of the previously reported studies are done in single crystals. How challenging is it to tune material properties such as the crystallinity of film here?

Guénolé Huitric replied: This kind of sample deposition is challenging and will impact not only the crystallinity, but also the stoichiometry and the residual strain in the film, parameters which might affect the electronic properties. The oxidation process at high temperature controls both the homogeneity and the stoichiometry through the film thickness. As an example, low-stoichiometry of a few % in the V : O ratio completely modifies the low temperature properties of V_2O_3, with the total disappearance of the antiferromagnetic insulating phase occurring around 150 K in a stoichiometric sample. Also, a relevant maximum temperature for the baking process has to be selected to prevent any internal stress: a too high annealing temperature might indeed lead to tensile strain which drives V_2O_3 from a metal to an insulator.

Aswathy V. Girija asked: How are the films prepared? Is it a temperature dependent method such as thermal evaporation?

The reason for my previous questions on the film preparation is that in the case of most organic systems, by changing parameters such as temperature and deposition rate, one can tune the crystallinity of films. Is a similar approach applicable/relevant here so that one can change the material properties and do a systematic structure–function study?

Guénolé Huitric replied: The V_2O_3 films were prepared in two steps. A film is first deposited by magnetron sputtering of a vanadium target in Ar plasma, without intentional heating of the substrate. Then the films are annealed at 530 °C over 10 hours in a controlled atmosphere in order to crystallize and achieve the proper V : O stoichiometry. In this case, the temperature is indeed a control parameter in the oxidation process. Choosing the correct baking temperature will prevent an oxygen concentration gradient being established through the film.

Steven L. Johnson asked: How do the transient strains/pressures you generate in this study compare to the strains or pressures that would be needed to trigger an insulator-to-metal transition? Is the transient temperature jump induced by the laser heating a problem? If so, is it possible to use a separate optically absorbing layer as a transducer to avoid heating the sample itself?

Guénolé Huitric responded: At this stage of the study, we cannot give an estimation of strain. We estimated that the temperature jump induced by laser heating is quite small for an optical pump (less than 10 K) and even smaller with a terahertz pump. Since the experiments were carried out at atmospheric pressure and room temperature, we are far from both transition lines. Thus, we do not expect the transient temperature jump caused by the laser heating to have a significant impact on the material.

The question of using an external, compared to an internal, generated strain from an absorbing layer was also considered in some studies. However, the external generated strain is bipolar, a feature that we try to avoid in our study.

Habib Rostami asked: The question is if we can measure the light-induced strain directly to ensure the mechanism of optical response modulation, what is the primary mechanism of optical transmission modulation: light-induced stress, photothermal, or hot electron effects? Can you directly measure strain using other approaches such as SHG, Raman spectroscopy, or X-ray diffraction? How much is the strain?

Guénolé Huitric responded: Several mechanisms are responsible for optical transmission modulations. These contributions usually occur on different time-scales. First the fast non-thermal electron contribution occur with very short delays (500 fs), followed by coherent excitation of an optical phonon mode (A_{1g}) and the launch of a dynamic strain propagation (elastic wave). The direct quantitative measurement of the light-induced strain would be obtained using time resolved X-ray diffraction.

Nevertheless, we expect to be able to obtain an order of magnitude of the strain in later studies.

Martina Basini remarked: What were the experimental constraints for the use of 700 nm and 740 nm probes?

Guénolé Huitric responded: Those two experiments were performed with two different setups special requirements. Nevertheless, their very similar sensitivity functions (see Fig. 4, https://doi.org/10.1039/d2fd00013j) allow us to conclude that this 40 nm spectral shift does not play any role in the probed effects.

Conflicts of interest

There are no conflicts to declare.

Faraday Discussions

ROYAL SOCIETY OF CHEMISTRY

PAPER

Concluding remarks: Directing and controlling materials and molecules

Graham R. Fleming [abc]

Received 30th June 2022, Accepted 13th July 2022
DOI: 10.1039/d2fd00133k

Controlling the macroscopic properties of materials, particularly quantum materials, *via* external inputs such as optical fields is a key goal of modern physical science. The *Faraday Discussion* presented a cross section of current experimental and theoretical progress with mostly ultrashort pulse excitations with frequencies ranging from the X-ray to the THz regions of the spectrum. This paper offers a perspective on the meaning of control in different scientific and engineering contexts. Despite the enormous challenge of implementing full feedback control on the types of material of interest in this discussion, I sketch such a system taken from a photosynthetic context to provide inspiration for future development in control of materials.

1. Introduction

The word control or controlling has appeared only three times in the titles of *Faraday Discussions* over the past 25 years, in 1999, 2011 and 2022. For example, the 2011 discussion on Coherence and Control in Chemistry used the term to describe the use of ultrashort light pulses and/or interference effects to modify photochemical reaction yields in "Coherent Control." This discussion appears to be the first confronting control of materials, no doubt indicating the difficulty of such control. Nonetheless the control of material and molecular properties, and responses, by external means such as by using ultrashort optical or THz pulses, is a major goal of condensed phase physical science.[1,2] One stimulus for this focus is in a 15 year old report for the Basic Energy Sciences Division of the Office of Science of the U.S. Department of Energy.[1] The report in 2007 put forth the view with respect to materials and molecular systems that we were at the threshold of a transformation to "control science" and noted that new tools would be required to achieve this transition, in particular tools that provide multimodal information on the timescales of quasi particles, electrons, and nuclear motions, on the length scale of bond lengths, defects, and lattice spacing, and with an energy resolution on a scale appropriate to the specific system under study. The papers in this

[a]Department of Chemistry, University of California, Berkeley, USA. E-mail: grfleming@lbl.gov
[b]Molecular Biophysics and Integrated Bioimaging Division, Lawrence Berkeley National Laboratory, USA
[c]Kavli Energy NanoScience Institute, Berkeley, CA 94720, USA

discussion build on advances in creating attosecond and femtosecond light sources from X-ray to THz frequencies, and on advances in microscopies such as electron microscopy. However, rather less progress has been made in methods that combine both temporal and spatial dimensions. Significant progress has also been made in theoretical methods, though many problems remain in electron-correlated systems.[3]

The report cited above shied away from using controlling in its title, settling instead for "Directing Matter and Energy: Five Challenges for Science and the Imagination." The five challenges are:

- How do we control materials and processes at the level of electrons?
- How do we design and perfect atom- and energy-efficient synthesis of new forms of matter with tailored properties?
- How do remarkable properties of matter emerge from complex correlations of atomic and electronic constituents and how can we control these properties?
- Can we master energy and information on the nanoscale to create new technologies with capabilities rivaling those of living systems?
- How do we characterize and control matter away—especially very far away—from equilibrium?

A follow-on report in 2015 described progress on the five "Grand Challenges" and the challenges and opportunities for continued progress towards true control of matter and energy.[2] The challenges listed above are daunting but, aside from the fourth challenge, they map directly onto the topics described in four sessions of this discussion.

In the context of this discussion ultrashort light pulses play a key role in most of the directing and controlling topics discussed. Perhaps the most apparent control method is to use the ultrahigh time resolution to outpace interfering phenomena, such as decoherence when coherence is required for the desired output. The development of optical pulses across the optical spectrum enables nonlinear, generally third-order spectroscopies, which probe the system response to intense applied electric and magnetic fields. Such studies, in particular multi-dimensional coherent spectroscopies, have dramatically changed our ability to study complex, often partly disordered, material and molecular systems. However, except in rare cases these nonlinear spectroscopies do not provide complementary spatial information. As Burghardt *et al.* (https://doi.org/10.1039/D2FD00014H) described in their contribution higher order spectroscopies such as fifth order responses, *via* interaction between elementary excitations (*e.g.* excitons), have the potential to provide complementary spatial input into the dynamics of extended systems. Such a program will require the fifth-order signal to be cleanly separated from both lower and higher order responses. The intense electric and magnetic fields of ultrashort pulses also enable the application of non-thermal effects such as modification of the system's potential energy surface to allow, for example, the creation of phases that are not accessible thermally. Non-thermal pathways created by ultrashort light pulses to control the macroscopic properties of quantum materials have recently been discussed in detail by de la Torre *et al.*[4] from which Fig. 1 is taken. Optical switching into metastable states, for example, can enable new applications to low temperature memories or to microelectronics.[4]

Turning to molecular systems two papers – Kim, Castellano, Li, Chen *et al.* (https://doi.org/10.1039/D2FD00009A), and McCusker *et al.* (https://doi.org/

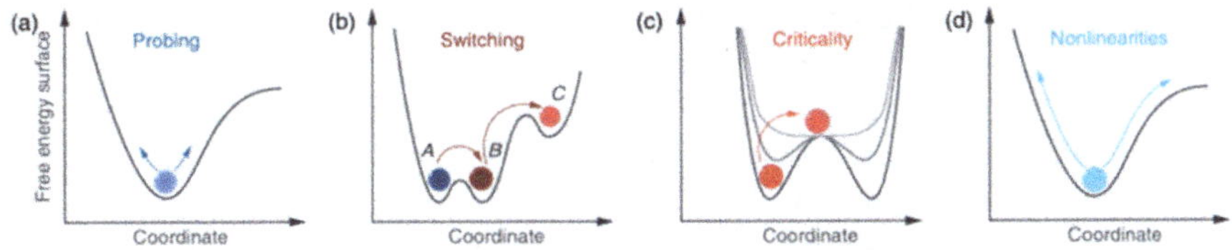

Fig. 1 Illustration of the range of applications of increasingly intense short light pulses to material systems. (a) Probing quasi particle interactions. (b) A more intense pulse populates the degenerate g.s. or moves the system to a metastable state. (c) Modifying the PES to induce nonequilibrium critical behavior. (d) Moving out of the harmonic region to create nonlinearities. (Figure reprinted with permission from A. De La Torre, D. M. Kennes, M. Claassen, S. Gerber, J. W. McIver and M. A. Sentef, *Rev. Mod. Phys.*, 2021, **93**, 041002. Copyright 2021 by the American Physical Society).[4]

10.1039/D2FD00106C) – take up the question of whether vibrational or vibronic wavepackets generated by ultrashort pulses can provide insights to guide synthetic control of photochemical reaction pathways – a long-standing goal of chemistry. Distinguishing spectator modes from modes directly involved in the dynamics remains highly challenging and system specific. As yet no general rules have emerged, though in one case modification of the structure with the goal of attenuating and perhaps decoupling a specific large amplitude motion from the reaction coordinate had the predicted effect of slowing the dynamics (McCusker *et al.*, https://doi.org/10.1039/D2FD00106C). The analysis of vibronic interactions in two-dimensional electronic spectra (2DES) of excitonic systems by Burghardt *et al.* (https://doi.org/10.1039/D2FD00014H) suggests that specific excitation of low frequency ("soft") modes could be used to optimize transport properties *via* the control of dynamical disorder to guide quasi-particle motion.

2. Directing and controlling

The word control or controlling carries differing meanings in different communities. To an engineer, control aimed at improving stability, robustness, and performance involves closed loop feedback systems. A familiar example is cruise control to keep the speed of a vehicle steady at a set value. The Shorter Oxford English Dictionary gives two main meanings for control: (1) the act or power of directing or regulating and (2) the action of holding in check or prevention of something unwanted. The papers in this discussion use the first definition, in other words to make something happen that would not happen spontaneously, for example, thermally. Examples are a quantum phase change or a translocation of an atom or group from one location to another. "Photodoping", transient n-type doping of a semiconductor by optical excitation of deep electronic defects, as described by Gierster *et al.* (https://doi.org/10.1039/D2FD00036A), raises the possibility of transient control of material properties, such as conductivity in practical devices. The second meaning of control, implying that a material or system can limit the extent of a change in response to rapid changes in external parameters to optimize (smooth) output or minimize damage is obviously far more challenging. Such control, however, can be met in some cases by passive design rather than active feedback. For example, Arp *et al.*[5] use a network model to show that the absorption spectra of natural photosynthetic light harvesting systems from a variety of organisms adapted to widely differing light conditions (*e.g.* full sunlight, filtered by seawater, or other organisms), and with different sets of absorber molecules, minimize the sensitivity of the output to fluctuations in light level. The model is based on spectral separation of two (groups of) absorbers with appropriate wavelength separation and reproduces the actual spectra of plant and bacterial antenna systems. As will be discussed briefly below, to minimize oxidative damage from excess light (a daily occurrence for plants and algae) feedback systems more directly resembling the engineering control loops described above are utilized for photoprotection by land plants and algae. Control theory and application in the engineering sense has been extensively discussed in a domain intermediate between materials and natural systems – that of synthetic biology.[6,7]

What phenomena are of most interest for control or prevention in materials in which delicate quantum phenomena such as conventional and unconventional superconductivity occur, or in quantum information/computing systems? Coherence destroying processes such as electron scattering, or spin-lattice relaxation should be outrun or minimized. The very high electric fields and few-cycle duration pulses now available with THz sources bring new opportunities, several of which were harnessed in this discussion. For example, Iwai *et al.* (https://doi.org/10.1039/D2FD00004K) avoid the generation of quasi particles which lead to an increase in electron temperature and destroy electronic coherence by using 6 fs THz pulses to create synchronized coherent charge oscillation and a scattering-free current, on a timescale faster than that of electron scattering, in an organic superconductor. The scattering-free current had previously been hidden by the increase in electron temperature and the enhancements of the nonlinear optical response towards T_{SC} suggest a relationship to T_{SC} and/or superconducting fluctuations. Udina *et al.* (https://doi.org/10.1039/D2FD00016D) note that the nonlinear Raman responses to THz pulses and to visible pulses are not equal so that excitation of collective modes of *e.g.* the superconducting order parameter, requires sophisticated methods to interpret experimental data.

Overall, the discussion provided a view of a rapidly developing field where new light sources such as free electron lasers, lab-based XUV and THz sources are being applied to highly complex materials, where even the nature of the measured quantity is not fully clear in some cases. What is clear, however, is that enormous technological advances could be realized with the ability to control quantum materials *via* external, non-thermal means. Similarly, the ability to design or exploit vibronic coherence for control of photochemical dynamics could also have far reaching applications. The development of experimental methods capable of removing obscuring signals such as those from heating, is rapidly emerging in both optical and THz spectroscopies.

3. Looking forward

New light sources, particularly in the X-ray and XUV regions, will create new opportunities.[8] High repetition-rate sources will enable mapping of charge distributions and reaction dynamics on the molecular length scale. They should enable element specific mapping of quantum coherences, and multidimensional spectroscopies in the soft and hard X-ray regions and perhaps two-color (hard-soft) two-dimensional spectroscopy. Time resolved X-ray scattering will go some way to filling the needs for combined spatial and temporal resolution.

New light sources hold great promise for the study of quantum materials, *via* resonant inelastic X-ray scattering, for example. The spontaneous fluctuations and nanoscale heterogeneity typical of strongly correlated materials with multiple phases can be addressed with X-ray photon correlation spectroscopy with chemical specificity, for example.

The methods alluded to above likely require facility-level free electron laser systems to produce highly intense sources. There is a second area utilizing much weaker sources – quantum light spectroscopy based on single photon pairs. Single photon pairs, which may or may not be quantum mechanically entangled in *e.g.* polarization, or frequency, enable a great number of new control parameters. Examples are entanglement time, joint spectra, different frequencies between the pairs, *etc.*[9] Frequency–time entangled pairs can break the Fourier transform limit of simultaneous temporal and spectral resolution. In other words, resolution of highly congested spectra with very rapid relaxation timescales should become possible. For imaging, quantum light provides the potential for:

- Sub shot noise measurement microscopy
- Low intensity non-linear imaging of fragile systems
- Imaging with one wavelength and detecting at another wavelength (*e.g.* to reduce scattering or detect at a wavelength with more sensitive detectors)
- Imaging without illuminating the sample
- Counting biomolecules for improved localization
- Enhanced resolution in super-resolution microscopy by photon statistics
- Entangled electron pair microscopy

The final bullet point deals not with photons but with entangled electrons which could alleviate the sample damage problems of conventional electron microscopy.

For dynamical studies single photon pairs offer the following tantalizing possibilities:

- Photon statistics to characterize quantum *vs.* classical dynamics
- High time and spectral resolution
- High time and spatial resolution
- Linear two-photon absorption
- New molecular control strategies
- Exploring polariton bio-chemistry (molecular cavities)

In addition many possibilities can be imagined by combining a classical pulse with an entangled photon pair.[9]

Much initial interest centered on the possibility of using single photon pairs to carry out two photon absorption that is linear in photon number for microscopy. Currently this remains a controversial topic[10] but new sources such as bright squeezed vacuums, may change the picture in the future. The possibility of using the quantum state of light as a parameter for the coherent control of material properties was described theoretically by Ishida *et al.* (https://doi.org/10.1039/D2FD00007E). The application of such selected states, for example, length squeezed vacuums, provides a tantalizing glance of future possibilities.

From a practical application perspective, femtosecond lasers, let alone free electron lasers, are unlikely to produce a scalable technology. Real world devices will likely use sunlight directly or indirectly *via* electricity as their energy input. If the device uses sunlight directly this brings up a question which has been raised in studies of photosynthesis: are the coherences observed with femtosecond light sources relevant to actual photosynthesis in sunlight (an incoherent thermal source)? We have recently demonstrated using quantum light that a photosynthetic cycle of absorption, energy transfer to different absorbers, multiple more energy transfer steps and finally emission of fluorescence starts and proceeds at the single quantum level. Although questions remain about the spatio-temporal nature of the initial excitation, this result strongly implies that yes, the ultrafast spectroscopic results are directly applicable to systems in sunlight. Such a picture will also apply to material systems.

4. Towards full control

The fourth "Grand Challenge" listed in the introduction suggests looking to the biosphere for inspiration for what is possible in full control of materials. With the goal of provoking consideration of application of feedback control to optimize output, minimize damage *etc.* I provide a brief description of the type of control system used by land plants and algae.

To operate the only TW energy conversion system on earth, which is intrinsically and necessarily a nanoscale system,[11] plants and algae must balance two contradictory requirements. For optimal efficiency through the diurnal cycle hundreds of antenna molecules (mostly chlorophylls) surround a reaction center that initiates charge separation, the first chemical event of photosynthesis.[12] To provide excitation energy to the reaction center, the exciton must move hundreds of times with each step very much faster than any dissipation timescale in the

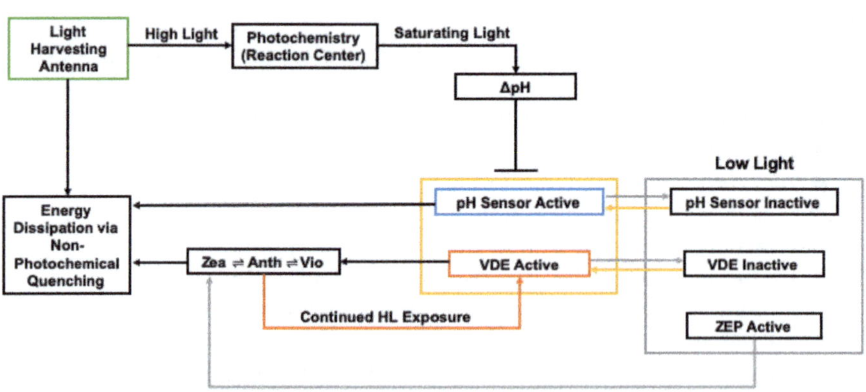

Fig. 2 A schematic of the dual roles of light harvesting antennae, which switch between energy harvesting (photochemistry) and energy dissipation (non-photochemical quenching) based on the light levels. When the reaction centers (RC) are saturated, energy must be dissipated quickly before reactive oxygen species can form. Normal action of photosystem II leads to a pH-gradient forming across the thylakoid membrane; but when excess light is present and the internal pH drops below a fixed value, pH sensitive proteins are triggered. Violaxanthin de-epoxidase (VDE) is one of these pH-activated proteins that acts as an actuator (orange box), regulating the amount of Zea to efficiently dissipate excess energy. When low light conditions return (grey box), VDE is inactivated, and zeaxanthin epoxidase (ZEP) reverses the conversion of Zea back to Vio.

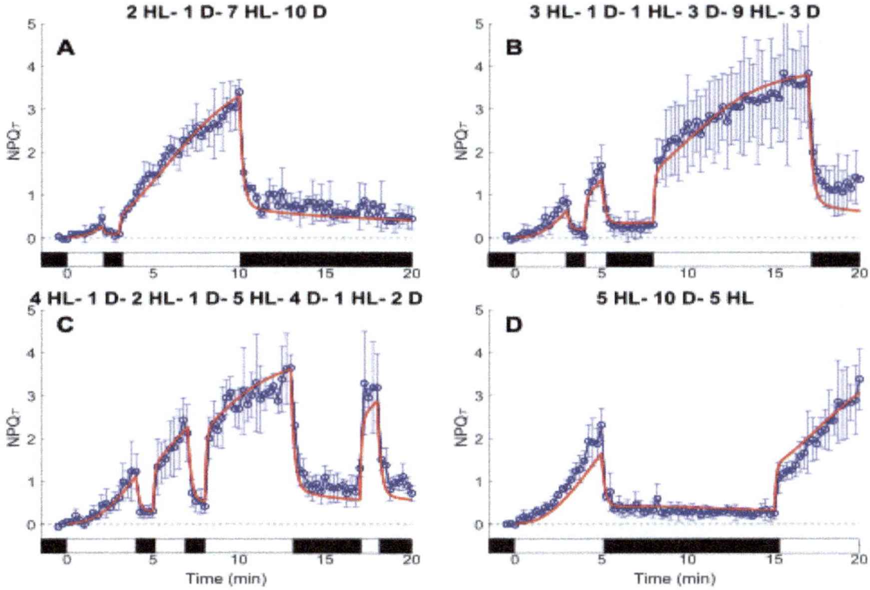

Fig. 3 Fluorescence lifetime quenching (NPQτ) traces for four irregular fluctuating light sequences in *N. oceanica* algal cells (A–D).[13] The experimental NPQτ (blue, open circles) and model predicted NPQτ curves (red line) are shown. The black boxes at the bottom of each plot represent the dark periods while the white boxes represent HL exposure. Overall, the model is able to reproduce the rapid switch on/off with the transitions between light/dark and *vice versa*. 95% confidence intervals are represented by error bars ($n = 3$). (Data from A. H. Short, *et al.*, *J. Chem. Phys.*, 2022, **156**, 205102, Copyright 2022 Author(s) under a Creative Commons Attribution 4.0 license).

absorbers. On the other hand, overexcitation of the reaction center leads to the formation of reactive oxygen species including singlet oxygen which destroy the proteins holding the antenna molecules and reaction center components. All green photosynthetic organisms have developed a feedback scheme to respond to fluctuations in light level (from clouds, wind moving leaves above, waves in water producing transient lenses, *etc.*). A dissipation mechanism based on a sensor actuator feedback cycle is reversibly turned on in response to excess light and off in low light. The simplest feedback scheme we have found is in a micro alga called *Nannochloropsis oceanica* illustrated in Fig. 2.

Photosystem II of plants and algae generates protons (from water splitting) in the lumen (inner) compartment of the chloroplast and therefore a pH gradient across the photosynthetic membrane. Low lumen pH also activates an enzyme (the actuator) which converts one carotenoid (violaxanthin) to another (zeaxanthin). Zeaxanthin, likely in consort with the pH sensor protein, initiates dissipation of antenna excitations before they reach the reaction center. When the light level drops, the pH gradient dissipates and a second enzyme converts the zeaxanthin back to violaxanthin which does not produce quenching of excitation.

Fig. 3 shows the response of *N. oceanica* to varying periods of high light and dark along with a kinetic model of the feedback scheme sketched above.[13] Being a "bottom up" model, the model makes predictions for the carotenoid concentrations during such illumination sequences which are in reasonable accord with separate experimental measures of the predicted concentrations.

Although plants have more complex control systems,[14] the elements described above are responsible for the rapid response of plants to excess light. As an important aside, optimizing this control system for crop plants will be important for food production by 2050 as population outpaces production, echoing the 1898 prediction of physicist William Crookes[15,16] which led to the Haber–Bosch process and the development of ammonia-based fertilizers.

Can such sophisticated controls be applied to material systems? A simple pH reactive antenna system has been demonstrated.[17] With soft materials passive control of temperature has been achieved.[18] To continue this development, as ref. 2 notes, creating complexity and functionality will require coupling of chemical, optical, electronic, magnetic, mechanical and thermal degrees of freedom. The extraordinary possibilities imaginable for, *e.g.*, energy materials or even quantum materials provide an exciting vista for the future.

Conflicts of interest

There are no conflicts of interest to declare.

Acknowledgements

I am very grateful to Ms Audrey Short for providing Fig. 2 and 3 and Dr Collin Steen for helping me assemble the manuscript. I gratefully acknowledge support from the U.S. Department of Energy, Office of Science, Division of Basic Energy Sciences, Chemical Sciences, Geosciences and Biosciences Department *via* FWPs 449A and 449B.

References

1 *Directing Matter and Energy: Five Challenges for Science and the Imagination*, https://www.osti.gov/servlets/purl/935427, 2007, DOI: 10.2172/935427.

2 *Challenges at the Frontiers of Matter and Energy: Transformative Opportunities for Discovery Science*, https://www.osti.gov/servlets/purl/1283188, 2015, DOI: 10.2172/1283188.

3 A. Alexandradinata, N. P. Armitage, A. Baydin, W. Bi, Y. Cao, H. J. Changlani, E. Chertkov, E. H. da S. Neto, L. Delacretaz, I. El Baggari, G. M. Ferguson, W. J. Gannon, S. A. A. Ghorashi, B. H. Goodge, O. Goulko, G. Grissonnanche, A. Hallas, I. M. Hayes, Y. He, E. W. Huang, A. Kogar, D. Kumah, J. Y. Lee, A. Legros, F. Mahmood, Y. Maximenko, N. Pellatz, H. Polshyn, T. Sarkar, A. Scheie, K. L. Seyler, Z. Shi, B. Skinner, L. Steinke, K. Thirunavukkuarasu, T. V. Trevisan, M. Vogl, P. A. Volkov, Y. Wang, Y. Wang, D. Wei, K. Wei, S. Yang, X. Zhang, Y.-H. Zhang, L. Zhao and A. Zong, 2020, arXiv:10.48550/arxiv.2010.00584.

4 A. De La Torre, D. M. Kennes, M. Claassen, S. Gerber, J. W. McIver and M. A. Sentef, *Rev. Mod. Phys.*, 2021, **93**, 041002.

5 T. B. Arp, J. Kistner-Morris, V. Aji, R. J. Cogdell, R. Van Grondelle and N. M. Gabor, *Science*, 2020, **368**, 1490–1495.

6 D. Del Vecchio, A. J. Dy and Y. Qian, *J. R. Soc. Interface*, 2016, **13**, 20160380.

7 V. Hsiao, A. Swaminathan and R. M. Murray, *IEEE Control Syst*, 2018, **38**, 32–62.
8 *New Science Opportunities Enabled by LCLS-II X-Ray Lasers*, https://www.osti.gov/servlets/purl/1630267, 2015, DOI: 10.2172/1630267.
9 K. E. Dorfman, F. Schlawin and S. Mukamel, *Rev. Mod. Phys.*, 2016, **88**, 045008.
10 A. Mikhaylov, R. N. Wilson, K. M. Parzuchowski, M. D. Mazurek, C. H. Camp, M. J. Stevens and R. Jimenez, *J. Phys. Chem. Lett.*, 2022, **13**, 1489–1493.
11 E. Edri, S. Aloni and H. Frei, *ACS Nano*, 2018, **12**, 533–541.
12 R. E. Blankenship, *Molecular Mechanisms of Photosynthesis*, Wiley, Chichester, 3rd edn, 2021.
13 A. H. Short, T. P. Fay, T. Crisanto, J. Hall, C. J. Steen, K. K. Niyogi, D. T. Limmer and G. R. Fleming, *J. Chem. Phys.*, 2022, **156**, 205102.
14 J. Zaks, K. Amarnath, D. M. Kramer, K. K. Niyogi and G. R. Fleming, *Proc. Natl. Acad. Sci. U. S. A.*, 2012, **109**, 15757–15762.
15 W. Crookes, *Science*, 1898, **8**, 561–575.
16 P. Horton, S. P. Long, P. Smith, S. A. Banwart and D. J. Beerling, *Nat. Plants*, 2021, **7**, 250–255.
17 Y. Terazono, G. Kodis, K. Bhushan, J. Zaks, C. Madden, A. L. Moore, T. A. Moore, G. R. Fleming and D. Gust, *J. Am. Chem. Soc.*, 2011, **133**, 2916–2922.
18 X. He, M. Aizenberg, O. Kuksenok, L. D. Zarzar, A. Shastri, A. C. Balazs and J. Aizenberg, *Nature*, 2012, **487**, 214–218.

Poster titles

Coherent control of chiral spin states in spin-phonon triangles, **N. Aoyagi, H. Matsueda and K. Ishida,** *Utsunomiya University, Japan*

The photo-induced dynamics of $(NPr_4)_2[Fe_2(Cl_2An)_3]$ suggesting the appearance of hidden state, **S. Banu, M. Kato, K. Takubo, T. Ishikawa, Y. Okimoto, S. Koshihara, N. Eguchi, W. Kosaka and H. Miyasaka,** *Tokyo Institute of Technology, Japan*

Ultrafast photoinduced Jahn-Teller switch in Mn(III) terpyridine complexes, **K. Barlow, J. Eng, I. Ivalo, M. Coletta, E. K. Brechin, T. J. Penfold, and J. O. Johansson,** *University of Edinburgh, UK*

Terahertz electric-field driven magnetism in paraelectric $SrTiO_3$, **M. Basini, M. Pancaldi, M. Udina, B. Wehinger, T. Terumasa, M. C. Hoffmann, A. V. Balatsky and S. Bonetti,** *Stockholm University, Sweden*

Intra-chain exciton dynamics in conjugated polymers: Reduced-dimensional approaches to the computation of 2DES signals, **D. Brey, R. Binder, R. Martinazzo and I. Burghardt,** *Goethe University Frankfurt, Germany*

Terahertz-driven nonlinear dynamics: New opportunities arising from accelerator-based THz sources, **J. -C. Deinert, I. Ilyakov, A. Ponomaryov, M. Chen, T. de Oliveira, and S. Kovalev,** *Helmholtz-Zentrum Dresden-Rossendorf, Germany*

Contribution of collective excitations to third harmonic generation in superconducting MgB_2, **J. Fiore, G. Seibold and L. Benfatto,** *Sapienza University of Rome, Italy*

Time resolved THz dynamics in thermotropic liquid crystals, **P. Friebel, M. Lutz and L. Cattaneo,** *Max-Planck-Institut für Kernphysik, Germany*

The structure of the photoinduced and X-ray induced state in spin-crossover hybrid system, **T. Ishikawa, S. Nakano, R. Takahashi, K. Takubo, Y. Okimoto, S. Koshihara, A. Miyawaki, K. Matsumoto, K. Takahashi, T. Yokoyama, K. Ichiyanagi, R. Fukaya, S. Nozawa and A. Sato-Tomita,** *Tokyo Institute of Technology, Japan*

Photoexcited nonequilibrium dynamics in an excitonic insulator candidate Ta_2NiSe_5, **K. Katsumi, A. Alekhin, S. M. Souliou, A. -A. Haguiguirad, M. Le Tacon, S. Houver, M. Cazayous, A. Sacuto and Y. Gallais,** *Université Paris Cité, France*

Optical generation and detection of propagating spin waves in antiferromagnets, **R. A. Leenders, R. V. Mikhaylovskiy, A. V. Kimel and D. Afanasiev,** *Lancaster University, UK*

Applying DFTB approach to study X-ray-induced phase transitions in solids, **V. Lipp, V. Tkachenko, M. Stransky, B. Aradi, T. Frauenheim and B. Ziaja,** *Deutsches Elektronen-Synchrotron DESY, Hamburg, Germany*

Collective-mode emission from photo-driven superconducting-interface in non-uniform charge density wave system, **Y. Matsubayashi, Y. Masaki and H. Matsueda,** *Tohoku University, Japan*

Ultrafast kinetics of magnetic phase transition in FeRh/Pt bilayer, **R. Medapalli, G. Li, R. V. Mikhaylovskiy, S. K. K. Patel, T. Rasing, O. Kolosov, A. V. Kimel and E. E. Fullerton,** *Lancaster University, UK*

Exciton dynamics in regioregular P3HT: diffusion and comparison of H and HJ aggregate models, **M. Mondelo-Martell, D. Brey and I. Burghardt,** *Goethe University Frankfurt, Germany*

Topological transitions and anomalous Hall effect in periodically driven Dirac semimetals, **S. Okumura, A. Mitra, and T. Oka,** *The University of Tokyo, Japan*

Large itinerant electron exchange coupling in the magnetic topological insulator $MnBi_2Te_4$, **H. Padmanabhan, V. A. Stoica, P. Kim, M. Poore, T. Yang, X. Shen, A. H. Reid, M. -F. Lin, S. Park, J. Yang, H. Wang, N. Z. Koocher, D. Puggioni, L. Min, S. -H. Lee, Z. Mao, J. M. Rondinelli, A. M. Lindenberg, L. -Q. Chen, X. Wang, R. D. Averitt, J. W. Freeland and V. Gopalan,** *Harvard University, USA*

Can photonic CMOS replace the silicon photonics for system-on-chip integration?, **J. Pan,** *Northrop Grumman Corporation, USA*

Studying wavepacket motion in Mn(III) 1,4,8,11-tetraazacyclotetradecane complexes using ultrafast transient absorption spectroscopy, **R. Phelps, A. Etcheverry-Berrios, E. K. Brechin and J. O. Johansson,** *University of Edinburgh, UK*

Theory for coherent (shear) phonon dynamic in 2D layered materials, **H. Rostami,** *Nordita/Stockholm University, Sweden*

THz emission spectroscopy of ferro/antiferromagnet heterostructures with exchange bias, **Y. Saito, F. N. Kholid, E. Karashtin, A. Fraerman, and R. V. Mikhaylovskiy,** *Lancaster University, UK*

Broadband, ultrafast magneto-optics in layered ferromagnet $Cr_2Ge_2Te_6$, **E. Sutcliffe, X. Sun, I. Verzhbitskiy, G. Eda and J. O. Johansson,** *University of Edinburgh, UK*

Theory of THz-driven rare-earth dynamics in $RFeO_3$, **N. R. Vovk, E. V. Ezerskaya and R. V. Mikhaylovskiy,** *Lancaster University, UK*

Ultrafast nonlinear spectroscopy in electronic ferroelectric $LuFe_2O_4$, **H. Yu, K. Takubo, T. Ishikawa, S. Koshihara, M. Fukada, G. Nishida, N. Ikeda, K. Fujihara and Y. Okimoto,** *Tokyo Institute of Technology, Japan*

The affiliation of the presenting author is listed above.

The PCCP Poster Prize for the best poster was awarded to Ms Erica Sutcliffe of The University of Edinburgh, UK, for their poster on broadband, ultrafast magneto-optics in a layered $Cr_2Ge_2Te_6$ ferromagnet.

List of participants

Mr Marc Alías-Rodríguez, *Univeristat Rovira i Virgili/ Univeristy Aix-Marseille, France*
Mr Noriyuki Aoyagi, *Utsunomiya University, Japan*
Mr Martin Appleby, *University of Sheffield, United Kingdom*
Mr Samiran Banu, *Tokyo Institute of Technology, Japan*
Mr Kyle Barlow, *University of Edinburgh, United Kingdom*
Dr Martina Basini, *Stockholm University, Sweden*
Professor Dr Lara Benfatto, *Sapienza University of Rome, Italy*
Mr Yorrick Boeije, *University Of Cambridge, United Kingdom*
Mr Dominik Brey, *Goethe University Frankfurt, Institute for Physical and Theoretical Chemistry, Germany*
Dr Wibke Bronsch, *Elettra - Sincrotrone Trieste, Italy*
Dr Irene Burghardt, *Goethe University Frankfurt, Germany*
Dr Andrew Burnett, *University of Leeds, United Kingdom*
Professor Lin Chen, *Department of Chemistry, Northwestern University/Argonne National Laboratory, USA*
Professor Eric Collet, *Institut de Physique de Rennes CNRS University Rennes 1, France*
Miss Carri Cotton, *Royal Society of Chemistry, United Kingdom*
Mr Rory Cowin, *University of Sheffield, United Kingdom*
Mr Ashley Crane, *Amplitude, United Kingdom*
Dr Jan-Christoph Deinert, *Helmholtz-Zentrum Dresden-Rossendorf, Germany*
Professor Hermann Durr, *Department of Physics and Astronomy, Uppsala University, Sweden*
Professor Ilya Eremin, *Ruhr-University Bochum, Germany*
Mr Jacopo Fiore, *Sapienza University of Rome, Italy*
Professor Graham Fleming, *University of California, Berkeley, USA*
Mr Patrick Friebel, *Max-Planck-Institut für Kernphysik Heidelberg, Germany*
Dr Omair Ghafur, *Daresbury Lab, Science Technology and Facilities Council, United Kingdom*
Mr Pratyush Ghosh, *University of Cambridge, United Kingdom*
Ms Aswathy V. Girija, *University of Cambridge, United Kingdom*
Mr Stuart Govan, *Royal Society of Chemistry, United Kingdom*
Professor Mohammed Hassan, *University of Arizona, USA*
Professor Dwayne Heard, *University of Leeds, United Kingdom*
Dr Hamoon Hedayat, *University of Cologne, Germany*
Mr Sam Howell, *Royal Society of Chemistry, United Kingdom*
Mr Guénolé Huitric, *University of Rennes, CNRS, IPR - UMR 6251, France*
Professor Dr Kunio Ishida, *Utsunomiya University, Japan*
Dr Tadahiko Ishikawa, *Tokyo Institute of Technology, Japan*
Professor Kyoko Ishizaka, *University of Tokyo, Japan*
Professor Dr Shinichiro Iwai, *Department of Physics, Tohoku University, Japan*
Dr J. Olof Johansson, *University of Edinburgh, United Kingdom*

Professor Dr Steven Johnson, *ETH Zurich / Paul Scherrer Institute*, *Switzerland*
Dr Kota Katsumi, *Université Paris Cité*, *France*
Dr Pyosang Kim, *Northwestern University and Argonne National Lab*, *USA*
Dr J. Michael Klopf, *Helmholtz-Zentrum Dresden-Rossendorf (HZDR)*, *Germany*
Professor Shinya Koshihara, *Tokyo Institute of Technology*, *Japan*
Mr Ruben Leenders, *Lancaster University*, *United Kingdom*
Dr Vladimir Lipp, *Institute of Nuclear Physics, Polish Academy of Sciences*, *Poland*
Mr Yukihiro Matsubayashi, *Tohoku University*, *Japan*
Professor James McCusker, *Michigan State University*, *USA*
Dr Rajasekhar Medapalli, *Lancaster University*, *United Kingdom*
Dr Rostislav Mikhaylovskiy, *Lancaster University*, *United Kingdom*
Dr Manel Mondelo Martell, *Goethe University Frankfurt*, *Germany*
Dr Christophe Odin, *CNRS University Rennes 1*, *France*
Dr Shun Okumura, *Institute for Solid State Physics, The University of Tokyo*, *Japan*
Dr Hari Padmanabhan, *Harvard University*, *USA*
Dr James Pan, *Northrop Grumman Corporation*, *USA*
Dr Daniel Payne, *Springer Nature*, *United Kingdom*
Dr Ryan Phelps, *University of Edinburgh*, *United Kingdom*
Dr Matteo Puviani, *Max Planck Institute for Solid State Research*, *Germany*
Professor Dr Jan M Rost, *Max Planck Institute for the Physics of Complex Systems*, *Germany*
Dr Habib Rostami, *Nordita/Stockholm University*, *Sweden*
Mr Yuichi Saito, *Lancaster University*, *United Kingdom*
Professor Elton Santos, *The University of Edinburgh*, *United Kingdom*
Mr Ralph Sauer, *The Sauer Consultancy Limited*, *United Kingdom*
Ms Erica Sutcliffe, *University of Edinburgh*, *United Kingdom*
Mrs Sarah Thirkell, *Royal Society of Chemistry*, *United Kingdom*
Ms Izzy Tibbetts, *Royal Society of Chemistry*, *United Kingdom*
Mr Mattia Udina, *Sapienza University of Rome*, *Italy*
Mr Mykola Vovk, *Lancaster University*, *United Kingdom*
Mr Julian Wagner, *Universität zu Köln, II. Physikalisches Institut*, *Germany*
Professor Julia Weinstein, *University of Sheffield*, *United Kingdom*
Dr Tatchamapan Yoskamtorn, *The University of Oxford*, *United Kingdom*
Dr Hongwu Yu, *Tokyo Institute of Technology*, *Japan*
Dr Serhane Zerdane, *Paul Scherrer Institut*, *Switzerland*